The Mechanics of Solder Alloy Interconnects

Edited by

D. R. Frear
S. N. Burchett
H. S. Morgan
J. H. Lau

VNR VAN NOSTRAND REINHOLD
New York

Library of Congress Catalog Card Number 93-30138
ISBN 0-442-01505-4

Printed in the United States of America

Van Nostrand Reinhold
115 Fifth Avenue
New York, NY 10003

International Thomson Publishing
Berkshire House, 168-173
High Holborn, London WC1V 7AA
England

Thomas Nelson Australia
102 Dodds Street
South Melbourne 3205
Victoria, Australia

Nelson Canada
1120 Birchmount Road
Scarborough, Ontario
M1K 5G4, Canada

International Thomson Publishing Germany
Königswinterer Str. 418
53227 Bonn
Germany

International Thomson Publishing Asia
221 Henderson Building #05-01
Singapore 0315

International Thomson Publishing Japan
Kyowa Building, 3F
2-2-1 Hirakawacho
Chiyoda-Ku, Tokyo 102
Japan

BBR 16 15 14 13 12 11 10 9 8 7 6 5 4 3 2 1

Library of Congress Cataloging-in-Publication Data

The Mechanics of solder alloy interconnects / [edited by] Darrel
Frear, Harold Morgan, and Steven Burchett
p. cm.
Includes index.
ISBN 0-442-01505-4
1. Solder and soldering. I. Frear, D. R. II. Morgan, Harold
(Harold S.) III. Burchett, Steven IV.. Lau, John.
TS610.M43 1993.
671.5'6—dc20

93-30138
CIP

Contents

Foreword

In 1992, Congress passed the Defense Manufacturing Engineering Education Act with the intent of encouraging academic institutions to increase their emphasis on manufacturing curricula. The need for this incentive to integrate the academic and industrial communities was clear: gaps in manufacturing science were inhibiting the evolution of new manufacturing technologies that are required for the U.S. to maintain a competitive posture in the world marketplace.

The Army Research Laboratory and Sandia National Laboratories sought to contribute to the congressional intent by initiating a new series of graduate level college textbooks. The goal was to focus next generation scientists onto issues that were common to the needs of the commercial market, the affordability of DoD weapons systems, and the mobilization readiness of the U.S. Armed Forces.

The textbook, *The Mechanics of Solder Alloy Interconnects*, was written in this spirit by nationally renowned scientists for academe and industry. Researchers using the book are encouraged to formulate programs that will establish scientific correlations between manufacturing process controls and product reliability. Such correlations are essential to the building of a new electronics industry which is based upon the futuristic concepts of Virtual Factories, Prototyping, and Testing.

George K. Lucey
U.S. Army MANTECH Manager
for Soldering Technology

Acknowledgments

The editors want to especially acknowledge the professional effort put forth by the team of authors that made this book. We asked a great deal from each of them to write a seamless critical appraisal of the subject of the mechanics of solder alloy interconnects. Each of the authors went above and beyond what is normally expected and responded in a timely and enthusiastic manner. We would also like to thank the many reviewers whose excellent comments improved the quality of the book.

Much of the technical insight resulted from both the formal and informal discussions at the Mechanics of Solder Alloy Interconnections Workshop that was held in Santa Fe, New Mexico, in September of 1992. We acknowledge the attendees whose participation helped stimulate the ideas and concepts that are summarized in this book.

The enthusiastic support of George Lucey of the U.S. Army Research Laboratories brought this project to fruition. George strongly believes in the concept of bringing fundamental knowledge to improve the quality and reliability of solder interconnections, and it is our hope that with this book we have supported his efforts.

This book was reviewed, and the quality was greatly improved, by the professional editors of Technically Write. We would like to acknowledge the great efforts put in by Jennifer Scofield, who proofread and edited the manuscripts. Jeff Eckert, at J. K. Eckert and Company, did an excellent job in compositing the manuscript and putting the text into the correct format. We would also like to thank the people at Van Nostrand Reinhold for believing in this project and helping with the many details in putting a volume of this size together. Specifically, we would like to mention the efforts of Marjorie Spencer, who coordinated the effort at VNR, with the assistance of Dawn Wechsler, and Chris Grisonich, who worked out all the details to get the manuscript ready for the printer.

Finally, the editors want to acknowledge the direct and indirect support of the U.S. Army Research Laboratories and Sandia National Laboratories, without which this book could never have been written.

Albuquerque, New Mexico

Chapter Authors & Affiliations

S. N. Burchett
Sandia National Laboratories
Department 1561
P.O. Box 5800
Albuquerque, NM 87185

L. A. Foster
University of North Texas
Center for Materials Characterization
P.O. Box 5303
Denton, TX 76203-5308

D. R. Frear
Sandia National Laboratories
Center for Solder Science & Technology
Department 1832
P.O. Box 5800
Albuquerque, NM 87185

J. L. Freer Goldstein
Lawrence Berkeley Laboratory
1 Cyclotron Road
Mail Stop 66-325
Berkeley, CA 94720

S. M. Heinrich
Department of Civil Engineering
Marquette University
1515 W. Wisconsin Avenue
Milwaukee, WI 53233

J. H. Lau
Hewlett-Packard Laboratories
1501 Page Mill Road
Palo Alto, CA 94304

J. L. Marshall
University of North Texas
Center for Materials Characterization
P.O. Box 5303
Denton, TX 76203-5308

Z. Mei
Lawrence Berkeley Laboratory
1 Cyclotron Road
Mail Stop 66-325
Berkeley, CA 94720

H. S. Morgan
Sandia National Laboratories
Department 1561
P.O. Box 5800
Albuquerque, NM 87185

J. W. Morris, Jr.
Lawrence Berkeley Laboratory
1 Cyclotron Road
Mail Stop 66-325
Berkeley, CA 94720

M. M. Rashid
Sandia National Laboratories
Department 1561
P.O. Box 5800
Albuquerque, NM 87185

J. A. Sees
University of North Texas
Center for Materials Characterization
P.O. Box 5303
Denton, TX 76203-5308

H. D. Solomon
General Electric R&D
Ki-3251
P. O. Box #8
Schenectady, NY 12301

D. S. Stone
University of Wisconsin–Madison
Dept. of Materials Science and Engg.
1509 University Avenue
Madison, WI 53706

1

Introduction: The Mechanics of Solder Alloy Interconnects

D. R. Frear, S. N. Burchett, and H. S. Morgan

1.1 INTRODUCTION

Solder joints were initially intended to be simple electrical interconnections between mechanically interlocked components in electronic packages. As technology advanced, electrical component size decreased, and the number of input/output terminations increased. To accommodate these changes, the numbers of solder joints per package have increased, while the joint dimensions have decreased. Mechanically interlocked components are being replaced by surface mount packaging where the solder joint is both the electrical and mechanical interconnection. Additionally, solder joints with smaller and smaller feature size are being incorporated in the electronics packages. The benefits of shrinking solder joint dimensions are numerous (e.g., increased speed, greater packaging density, etc.) but reliability concerns are increasing exponentially. Reliability is critical because electronic packages are being used to control operational and safety functions in aerospace and automotive applications. With the critical nature of these small electrical-mechanical interconnections, the question arises: Just how reliable are the solder joints in a modern electronic package?

A key issue in the long term reliability of solder joints is joint failure during thermal cycling. The individual components that are soldered together in an electronic package (including the solder itself) typically have differing thermal expansion coefficients. When the soldered assembly undergoes a temperature change, strain is imparted to the solder joint. These thermally induced strains are often cyclic in nature and result in thermomechanical fatigue and often failure of the solder joints. The failure (electrical open) of a single solder joint could render an entire electronic assembly inoperable. The reliability of solder joints under conditions of thermomechanical fatigue becomes more critical as new electronic packaging technologies emerge. Larger silicon devices and the rapidly evolving multi-chip modules have substantially increased the number of joints with potentially larger strains induced. The service environment of solder joints is also becoming more severe. Some of the thermal environments expected for electronic packages are shown in Table 1-1. New power devices dissipate more heat and are mounted closer together, thereby increasing the possibility of fatigue failures in personal and mainframe computer applications. In the automotive industry, electronic packages are positioned under the hood and closer to the engine for faster response and greater cost effectiveness. For increased fuel efficiency, styling is more streamlined and the number of air inlets for cooling purposes is decreasing. This results in higher operating temperatures under the hood and greater strains in the solder joints. In aerospace applications, electronic packages used for sensing, control, and telemetry are exposed to severe ambient environmental temperature swings. The severity and frequency of those environmental cycles create large cyclical strains in the solder joints. It is essential that the device, and in particular the solder joints, survive the thermal cycles for the device's projected lifetime. Thus, the reliability of a device is a function of its lifetime and an accurate lifetime predictive capability is essential. The most accurate methodology to determining solder joint reliability is to build a statistically significant number of devices and subject them to actual real time environments and determine when the joints fail. Obviously, this methodology is unacceptable in today's marketplace because it is geometry-dependent and device-specific, cannot be extrapolated from one system to another, and is also extremely time consuming and expensive. Alternatively, a number of lifetime predictive methodologies have been employed for solder joints. While these methods have found some usefulness for materials such as steel, they are all

TABLE 1-1 Thermal Environments for Solder Joints in a Variety of Electronic Packages

Use Conditions	*Thermal Excursion (°C)*
Consumer Electronics	0 to 60
Computers	15 to 60
Telecommunications	–40 to 85
Commercial Aircraft	–55 to 95
Military Aircraft	–55 to 125
Space	–40 to 85
Automotive—Passenger Compartment	–55 to 65
Automotive—Under the Hood	–55 to 150

based on the behavior of a homogeneous continuum and do not provide reliable life prediction for solder. The majority of the current methodologies are based on empirical curve fitting techniques. Normally the solder joints on a test vehicle (which may be actual product) are exposed to conditions of accelerated thermomechanical fatigue. Data are taken in the form of the number of cycles to failure as a function of imposed strain amplitude. A curve through those data serves as the "model." The "strain" in these curves is actually the computed displacement difference between the two (rigid) parts being soldered divided by a characteristic length and is, at best, an average strain. Based upon this "model," lifetime predictions are extrapolated. The problems with this methodology are:

1. The tests are geometry sensitive—a new model/curve must be developed for minor changes in design. Therefore, the models are design checks that cannot be incorporated into design for reliability.
2. The tests are material sensitive—a new model/curve must be developed for each change in material (solder, device, printed wiring board, etc.).
3. The tests must be performed under accelerated conditions and acceleration factors are not well understood.
4. Statistically significant numbers of hardware must be built and tested which can be very expensive especially if several designs are being evaluated.
5. Essentially no information on the thermomechanical response of the device and solder joints other than the number of cycles to failure is obtained from the test.
6. The methodology assumes that the failure location is always in the solder.

While this empirical methodology to determine product reliability is the mainstay of the electronics industry, it is unsatisfactory.

Predictive mechanical modeling of solder interconnections offers the potential to move from empirically based design checks to design for reliability. The technology to computationally evaluate solder interconnections in a thorough, realistic and verifiable manner without significant simplifying assumptions, however, is just coming into place primarily through the use of the finite element method. The difficulty in computationally predicting solder thermomechanical response is the same difficulty experienced by those employing empirically based techniques; the time-temperature dependent (creep) response of solder. The strain history (and fatigue life) in a solder interconnection depends upon mean temperature, cyclic temperature range, dwell times, dwell temperatures, loading and unloading rates, evolving solder microstructure, presence of intermetallics, presence of alloying and trace elements, joint geometry, soldering defects, and residual strains due to processing (Dasgupda et al., 1992). In essence, the thermomechanical response of solder is highly nonlinear and is difficult to characterize either experimentally or computationally. In order to move to computationally based design for reliability, the challenge is to develop a verified computational capability.

The ultimate goal in this field is to computationally design a reliable solder interconnect system with sufficient confidence to minimize or eliminate prototype development and testing. This lofty goal will not, in our opinion, occur in the near future (three years). Alternatively, a combined computational/experimental approach to reliability design offers the most promise in the near term. In this approach, computational parameter studies are performed on proposed interconnect designs and the most promising designs are selected for prototype testing, thus minimizing expensive testing. Even for this interactive approach, the need to improve confidence in computational capability is required.

It must be kept in mind that a multidisciplinary approach is required for a successful solder interconnect design (Figure 1-1). Computational modeling is only one of the required activities which could influence the design. Having a thermomechanically sound interconnect design that cannot be processed is untenable.

It is essential that there be a strong link between the understanding of the materials behavior of solder with the mechanics modeling aspects. This involves joining two groups that traditionally have not worked closely together: materials scientists and mechanics modelers. The modeling group has always used material behavior input for their computational simulations, but the two-way street of interaction has not been present until recently. To be able to develop a meaningful reliability predictive capability, the materials scientists and mechanics modelers

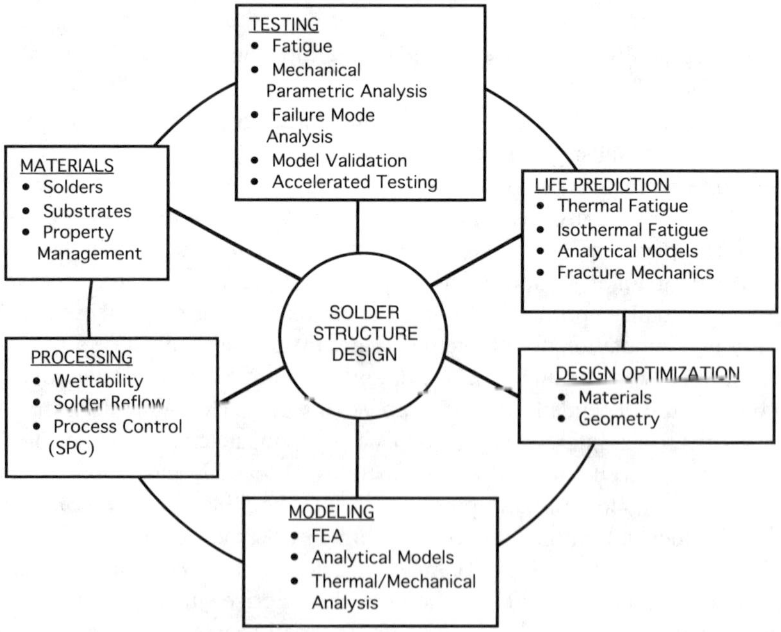

FIGURE 1-1 The multidisciplinary approach to solder joint design.

must be able to communicate their needs so that both groups can work toward the same goal. This book represents a coherent effort by the modeling and materials communities to delineate what is known and what is needed to develop a reliability assessment model for solder joints in electronic packages.

This volume represents a summary assessment of the state of the art in current understanding of the mechanical modeling of solder interconnects. The book is one in a series that are derived from workshops on the mechanics of solder. The first workshop, the First International Workshop on Materials and Mechanics Issues of Solder Alloy Applications, was held in June, 1990 and covered all areas of soldering science and technology. That workshop was summarized in the volume, *Solder Mechanics: A State of the Art Assessment* (TMS publications, 1991). The Second International Workshop on Solder Wetting and Spreading Assessment was held in June, 1992. The second workshop was summarized in the volume, *The Mechanics of Solder Alloy Wetting and Spreading* (Van Nostrand Reinhold, 1993). The third workshop was held in September, 1992 and dealt specifically with mechanics and modeling issues in solder alloy interconnects. Present at the third workshop were scientists and engineers recognized for their expertise in materials science, testing, and mechanics. The goal of the workshop was to review the state of the art in solder mechanics. This book represents the general consensus of attendees at the Third International Workshop on Materials and Mechanics Issues of Solder Alloy Applications.

1.2 PREVIEW OF CHAPTERS

To implement a computational model, four input parameters must be known or assumed: the response of each material in the assembly, the geometry of the assembly, the loading conditions, and the boundary conditions (Figure 1-2). In this volume we try to cover each of these aspects in detail. In Chapter 2, the basic material properties of solder alloys are discussed. Included are both the metallurgical aspects of solder as well as the mechanical response of solder specimens in test environments. In Chapter 3, the formation, the importance, and the mechanical properties of intermetallics and interface materials are discussed. In this volume we purposely limit the materials discussion to solders and intermetallics. It must kept in mind that all materials in the assembly (device, printed wiring boards, etc.) need to be characterized and in fact can also fail when subjected to mechanical or thermomechanical environments. In Chapter 4, the constitutive models of solder alloys are presented. This chapter will discuss how the observed behavior of solder alloys can be mathematically described.

One of the difficulties in computationally modeling solder interconnects is that the joint geometry is not known beforehand, and in fact forms during the soldering process. Computational simulation of interconnects requires that the geometry be specified. In order to develop a completely predictive computational capability,

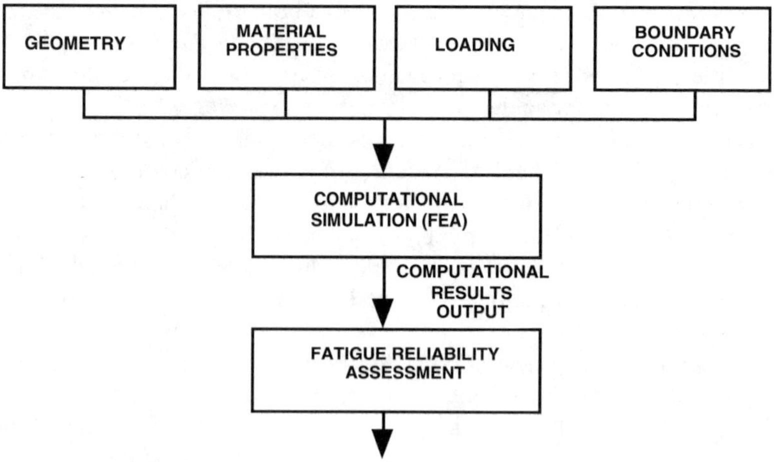

FIGURE 1-2 Input parameters for computational modeling of solder joints.

the joint geometry must be *a priori* predicted. Chapter 5 discusses the state of the art in predicting the shape of the joint from the melt.

In Chapter 6, the aspects of thermomechanical modeling of solder interconnects are discussed; these include geometry approximations, boundary conditions, and loading considerations as well as modeling techniques.

In Chapter 7, lifetime prediction through accelerated testing techniques is discussed in detail. Currently, the output from computational models is the computed displacement, stresses, and strains in the assembly, i.e., progressive failure is not generally included in the analyses. It is still necessary to evaluate reliability based upon computed response and available accelerated test results. Research is ongoing to develop progressive failure techniques; however, this technology is in its infancy.

Finally, in Chapters 8 and 9, specific applications in electronic packages are discussed. These include through-hole solder joints, surface mount solder joints, and multichip-modules.

This book is intended to serve as a resource, and a state of the art assessment, to be used in developing a solder joint reliability assessment. Each chapter is written to be used as a "stand-alone" resource for a particular aspect of materials and modeling issues. With this gained understanding, the reader in search of a solution to a solder joint reliability problem knows where to go in the materials and mechanics modeling communities for the appropriate answer.

REFERENCE

Dasgupda, A., Oyan, C., Barkar, D., and Pecht, M. 1992. *Trans. ASME, J. of Electron. Packag.*, vol. 114.

2

Microstructural Influences on the Mechanical Properties of Solder

J. W. Morris, Jr., J. L. Freer Goldstein, and Z. Mei

2.1 INTRODUCTION

The main intent of this book is to review analytic methods for predicting the behavior of solder joints. These are based on continuum mechanics. The solder is treated as a continuous, homogeneous body, or composite of such bodies, whose mechanical behavior is uniform and governed by constitutive equations that are simple enough to produce mathematically tractable models.

However, it is well known that real solids, and particularly solder joints, are not continuous media. They are microcomposites with complex microstructures. The constitutive equations that are used to describe their behavior are macroscopic relations that predict the cumulative effect of complex deformation processes that occur on the microstructural level. At the present state of the science it is not possible to construct tractable theories that incorporate the microstructure directly. However, it is important to recognize that the microstructure exists, and that it is ultimately responsible for the macroscopic behavior that is modeled.

The microstructure of a solder joint influences its mechanical properties in three generically different ways.

The microstructure governs deformation and failure.

The normal mechanical properties of a solder joint depend on the motion of dislocations and the growth and reconfiguration of grains. The processes are more sensitive to the microstructure of the solder than to its chemical composition. Since a joint of given composition may have any one of several qualitatively different microstructures, depending on how it is solidified, aged or deformed, the microstructure must be given, at least implicitly, as part of the specification of the material that underlies the choice of a particular constitutive relation to describe its mechanical properties. There is no such thing as the mechanical behavior of a particular solder; it depends on the microstructure, which depends, in turn, on how the solder is processed.

Common solders deform inhomogeneously.
Solder joints ordinarily have complex microstructures, are used at high homologous temperatures, and deform under relatively low loads. These three factors have the consequence that the plastic deformation of a solder joint is rarely uniform. It is concentrated in well defined shear bands that are accentuated as deformation proceeds. From the macroscopic perspective, solder often exhibits strain-softening. This is not necessarily because the solder inherently softens (though it may) but because the inhomogeneous pattern of the deformation has the consequence that very little of the solder volume actually participates in plastic deformation. From the microscopic perspective, it is insufficient to know the local mechanical properties of the solder. One must also have a sense of how the inhomogeneous deformation pattern develops.

Common solders are microstructurally unstable.
Good solders are low-melting alloys and, hence, are ordinarily eutectics or other two-phase mixtures. They bond to the substrate by reacting chemically to form intermetallic compounds. They are ordinarily used in the as-solidified or slightly aged condition. These factors ordinarily lead to a microstructure that is thermodynamically unstable. As the joint is aged, thermally cycled or deformed, its microstructure evolves so that its mechanical behavior changes with time.

In the following sections, we shall develop these principles by briefly reviewing the variety of microstructures that are often found in solder joints and discussing some of the ways in which the microstructure influences the common types of high-temperature mechanical behavior. We shall not attempt to be comprehensive. A great many different solder compositions are used in modern electronic devices, only a few of which have been studied in metallurgical detail.

2.2 SOLDER MICROSTRUCTURES

The most common solder joints contain lead-bearing solders (Sn-Pb, Sn-In, Sn-Bi, Sn-Ag, or ternaries of these) that join surfaces that are predominantly copper or nickel, and may be coated with gold. We first consider the microstructure of the solder itself, and then consider the changes that are caused by its reaction with the metallization at the interface.

Pb-Sn solder joints provide a good general model, since they exhibit most of the important types of behavior that are found in general in solder joints.

The Pb-Sn binary phase diagram is drawn in Figure 2-1. It is of a type that is called a simple eutectic phase diagram, since the lowest value of the melting point of the system falls at a eutectic point where the liquid solidifies into a mixture of two solids, α, a lead-rich solid solution, and β, a tin-rich solid solution. The other common tin-bearing solders have phase diagrams that resemble this one in their most pertinent details.

The microstructures that result from solidification in a system that has a phase diagram like that shown in Figure 2-1 depend on the composition of the liquid and the rate at which the liquid is cooled. The influence of composition can be described in terms of the three typical cases that are indicated in Figure 2-1: eutectic (labeled (1) in the figure), off-eutectic (labeled (2)), and solvent-rich (labeled (3)).

2.2.1 *Eutectic Microstructures*

If a material with a eutectic composition (labeled (1) in Figure 2-1) is solidified by cooling relatively slowly, it forms a microstructure like that shown in Figure 2-2. The two solid solutions that make up the eutectic grow simultaneously and parallel to one another in grain-like colonies like those shown in the figure.

If the two phases are present in roughly equal volume fraction, they form parallel plates within each colony. If one phase predominates, the second ordinarily forms rods in the matrix of the other. The former case is illustrated by the Sn-Bi eutectic that is shown in Figure 2-3a [Freer-Goldstein and Morris, a]. The latter case is illustrated by the Sn-In eutectic shown in Figure 2-3b [Mei et al. 1991]. When one phase is very strongly predominant the second may appear in the form

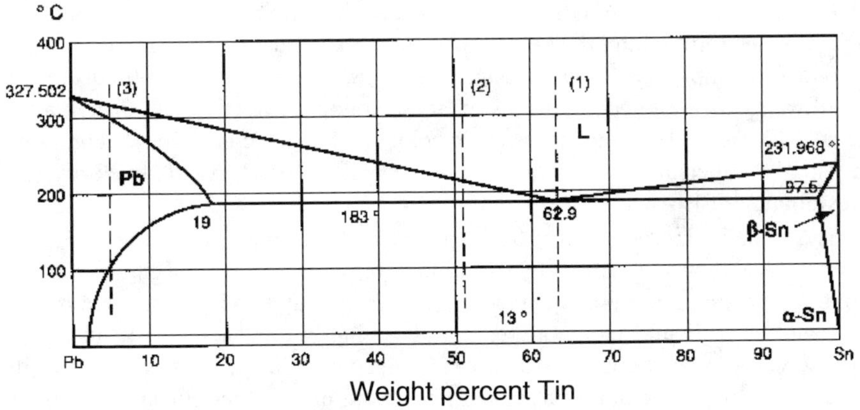

FIGURE 2-1 Sn-Pb phase diagram.

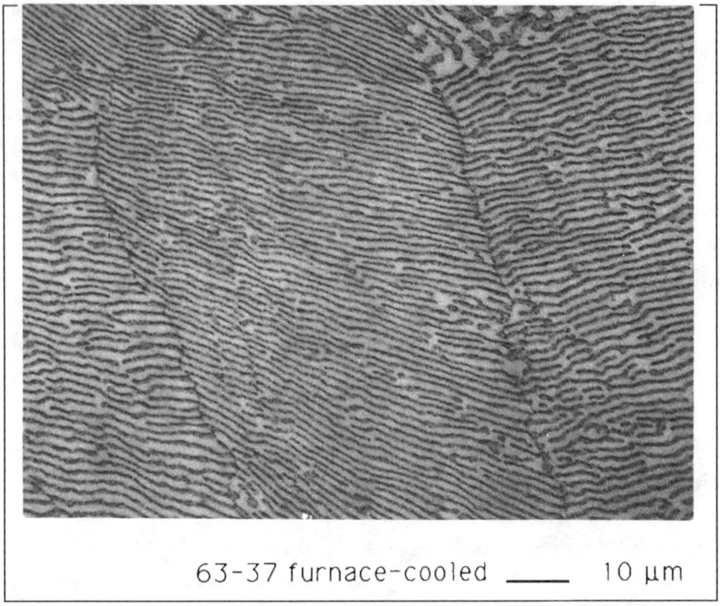

FIGURE 2-2 Optical micrograph of a eutectic Sn-Pb solder, showing colonies and colony boundaries.

of acicular needles, as seen in the Sn-Ag eutectic in Figure 2-3c [Freer-Goldstein and Morris, b].

The eutectic microstructure is often complicated by precipitation within one or both of the eutectic constituents when the sample is cooled below the eutectic solidification temperature. This phenomenon is observed, for example, in the Sn-Pb [Frear et al., 1989a] and Sn-Bi eutectics. As illustrated in Figure 2-4, small Bi-rich particles form within the Sn-rich phase [Freer-Goldstein and Morris, b]. This secondary precipitation is due to the fact that the solubility of Bi in tin decreases significantly as the temperature is lowered below the eutectic point. When a eutectic solder is cooled below its eutectic point, the Sn-rich constituent becomes increasingly supersaturated with Bi. The supersaturation is relieved by the precipitation of Bi-rich phases within the tin-rich phase.

The eutectic microstructure has a very high surface area per unit volume, and is, therefore, thermodynamically unstable with respect to reconfiguration into a mixture of coarser or more equiaxed grains. If a freshly solidified eutectic is simply aged at moderate temperature, the phases grow, and eventually rearrange themselves into an equiaxed structure. In most cases this process is dramatically accelerated by plastic deformation. In Pb-Sn and many other eutectics, accumulated plastic deformation couples with the large surface area to promote recrystal-

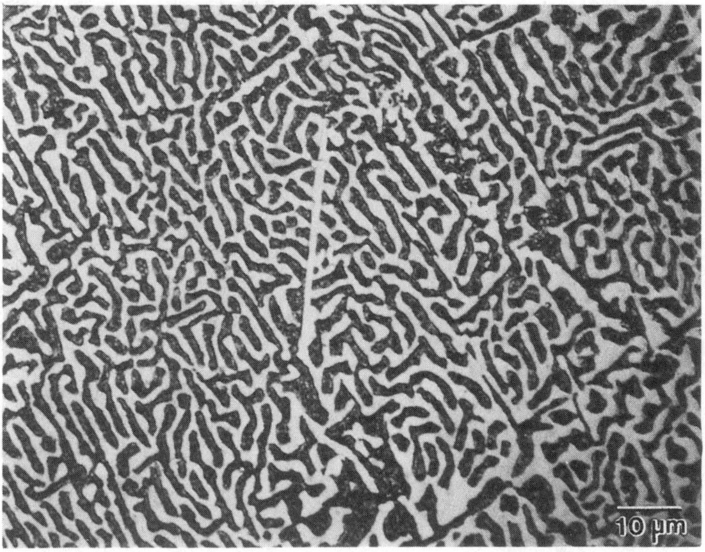

(a)

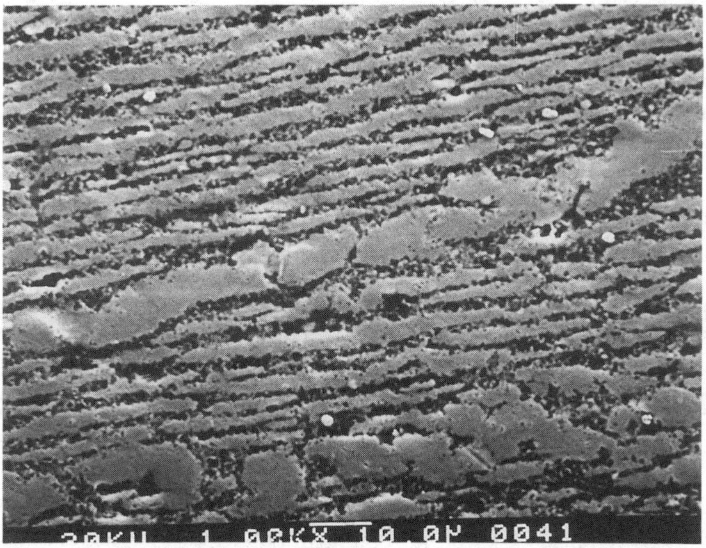

(b)

FIGURE 2-3 Optical micrographs of (a) 58Bi-42Sn, a eutectic with similar volume fractions of both phases; (b) 52In-48Sn, a eutectic in which one phase predominates; (c) 96.5Sn-3.5Ag, a eutectic with the major phase comprising over 90 percent by volume (continued next page).

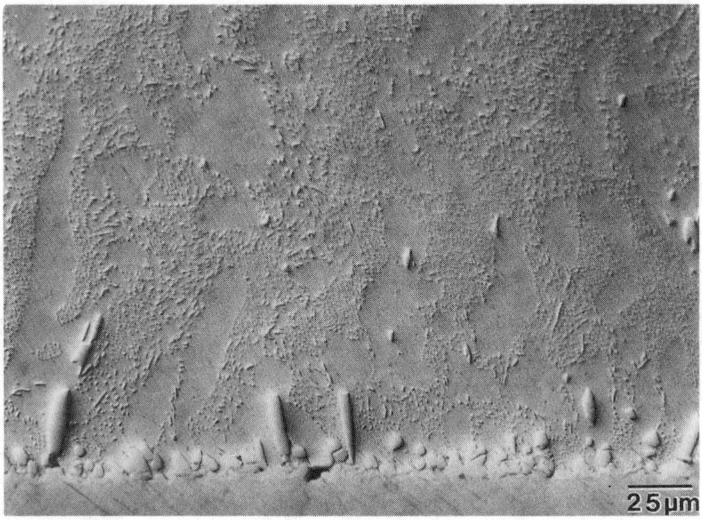

(c)

FIGURE 2-3 continued.

lization into a microstructure of fine, equiaxed grains. The recrystallized structure of the Pb-Sn eutectic is illustrated in Figure 2-5 [Mei and Morris, 1991]. Once the structure has recrystallized, the grains gradually increase in size to further decrease the surface area per unit volume.

A eutectic structure that strongly resembles the recrystallized microstructure can be obtained directly by cooling quickly from the melt. When solidification happens quickly under substantial undercooling, the classic eutectic structure is suppressed and the material solidifies directly into an equiaxed two-phase mixture. The structure of rapidly solidified Pb-Sn is illustrated in Figure 2-6 [Mei and Morris, 1991].

2.2.2 Off-Eutectic Microstructures

The vertical line labeled (2) in Figure 2-1 is an example of an off-eutectic composition. When a solder of this composition is solidified by cooling at a slow to moderate rate, the α-phase forms as discrete islands at temperatures within the α+L field. The liquid that remains when the eutectic temperature is reached solidifies into the eutectic microstructure. The final microstructure resembles that shown in Figure 2-7, with islands of pro-eutectic phase embedded in a matrix of eutectic [Summers, 1991].

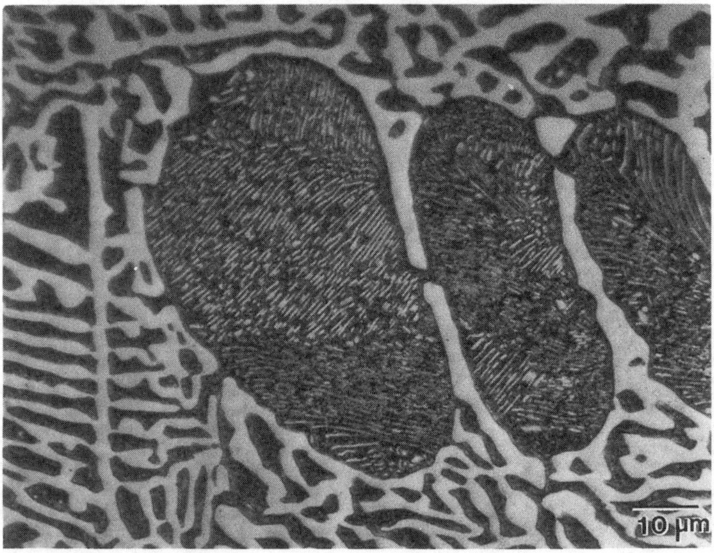

FIGURE 2-4 Optical micrograph of a region of a Bi-Sn eutectic, showing Bi (light) precipitating within the tin-rich phase (dark).

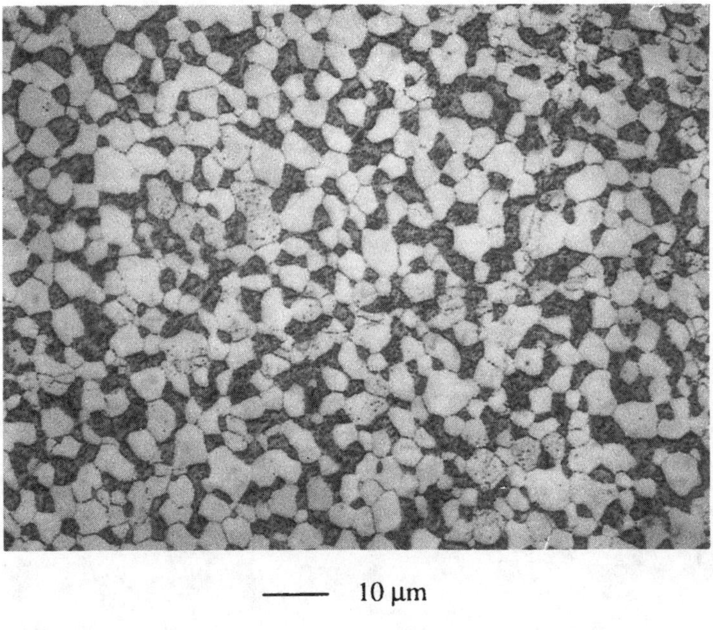

10 μm

FIGURE 2-5 Optical micrograph of a eutectic Sn-Pb solder after cold work and recrystallization showing equiaxed grains.

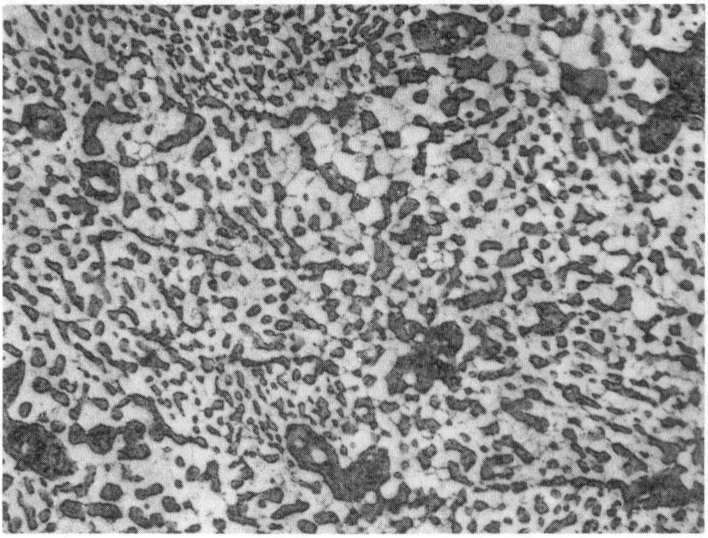

FIGURE 2-6 Optical micrograph of rapidly solidified 60Sn-40Pb, with a primarily equiaxed microstructure.

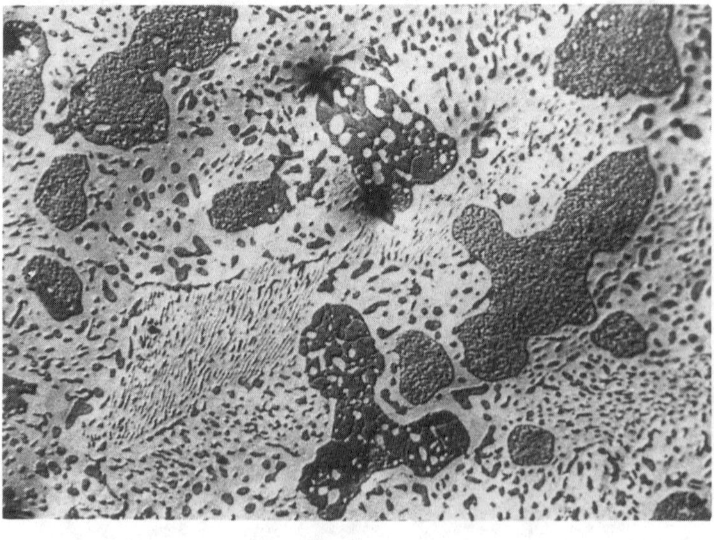

FIGURE 2-7 Optical micrograph of off-eutectic 50Sn-50Pb, with Pb-rich islands in a eutectic matrix.

2.2.3 Precipitated Microstructures

A composition like that labeled (3) in Figure 2-1 leads to a qualitatively different microstructure. A solder of this composition solidifies into a polygranular microstructure of a single phase (α in the example given in the figure). On further cooling, the solder penetrates the two-phase ($\alpha + \beta$) region, which causes the precipitation of particles of β-phase within the matrix of α. 95Pb-5Sn is a widely used solder that has this solidification behavior.

When precipitated microstructures are aged at relatively high homologous temperatures, both the primary α grains and the embedded β precipitates gradually coarsen. If the microstructure is thermally cycled, however, it may exhibit more dramatic instabilities. If the solubility of the minor species is a strong function of temperature, as it is in lead-rich Pb-Sn solders, for example, then the volume fraction of precipitate decreases rapidly on heating, and increases on cooling. A periodic thermal cycle causes the periodic dissolution and re-precipitation of solute, and may lead to the reconfiguration of the precipitate phase into grain boundaries or regions of high internal stress.

Similar instabilities are observed if the material is stressed at high temperature. If the solute-rich precipitate causes a volume or shape change when it forms, then a reconfiguration of the precipitate distribution causes a mechanical strain. At high temperature, solute diffusion is relatively rapid and the precipitate distribution may change in a way that serves to relax the imposed stress.

2.2.4 Intermetallics at the Interface

The composition of a solder is chosen so that it forms a strong bond to the metal surfaces it is intended to join. Bonding is promoted by chemical reactions between the solder and the substrate that form intermetallic compounds. The most common interfacial intermetallics are Cu-Sn and Ni-Sn compounds that are formed by reaction between tin in the solder and copper or nickel in the substrate.

The most thoroughly studied intermetallics are the Cu-Sn compounds that form when Sn-Pb solders wet copper. There are two common compounds: Cu_3Sn and Cu_6Sn_5. Cu_3Sn forms preferentially when there is an excess of copper; for example, it is often found at interfaces between copper and lead-rich solders such as 95Pb-5Sn. Cu_6Sn_5 forms preferentially in excess tin, and is, therefore, the usual interfacial intermetallic in eutectic Pb-Sn solder joints (63Sn-37Pb) and at Cu-Sn interfaces. Since Cu_3Sn is more stable at higher temperatures than Cu_6Sn_5, it is also found at interfaces that were exposed to tin-bearing solders at high temperature. If a relatively thick intermetallic layer is formed by reacting eutectic Pb-Sn with copper, as happens when a joint is reflowed for significant time or when it is aged after solidification at a temperature near the melting point, the intermetallic layer is often a composite, with Cu_3Sn at the copper interface and Cu_6Sn_5 at the solder interface (Figure 2-8) [Sunwoo et al., 1992]. The relative amounts of the

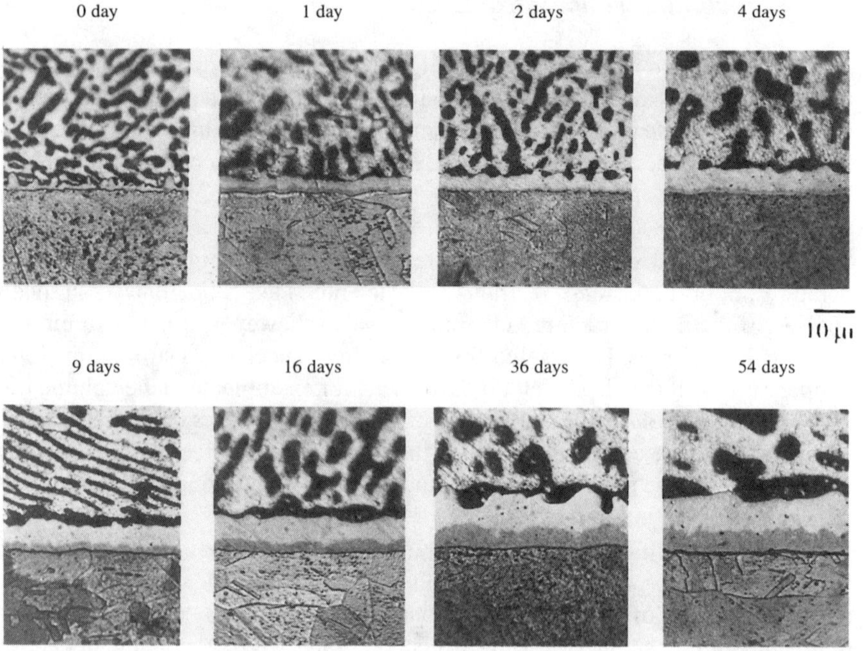

FIGURE 2-8 Series of optical micrographs showing growth of intermetallic layers at the interface between Sn-Pb solder and copper. There is a layer of Cu_3Sn (darker) on the copper side and Cu_6Sn_5 on the solder side.

two intermetallics increase with the growth temperature in a predictable way [Mei et al., 1992].

The morphology of the intermetallic layer reflects the crystal structure of the intermetallic. Cu_3Sn is orthorhombic in structure, and tends to form a tightly coherent layer with nearly equiaxed grains, as illustrated in Figure 2-9 [Grivas et al., 1986]. Cu_6Sn_5 is hexagonal, and tends to form a rough layer with knobs or hexagonal prisms that extend into the solder, as illustrated in Figure 2-10 [Tribula, 1990].

2.2.5 Intermetallics within the Bulk

In addition to the intermetallic layer that forms at the wetted interface, intermetallic phases are also often found within the bulk of the solder joint. These bulk intermetallics have two common sources.

First, intermetallics of the type that form at the interface may also be found in the bulk. The most common examples are the Cu-Sn intermetallics. Long, hexag-

FIGURE 2-9 Cu_3Sn layer at the interface between 95Pb-5Sn solder and copper.

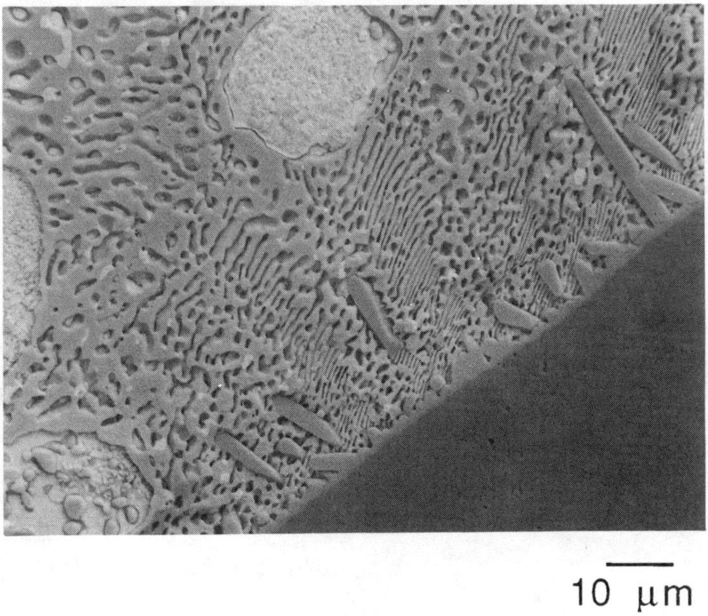

FIGURE 2-10 SEM micrograph showing Cu_6Sn_5 intermetallics growing into Sn-Pb solder.

onal rods of Cu_6Sn_5 are sometimes found in solders immediately after solidification [Tribula, 1990]. An example is shown in Figure 2-11. These apparently grow from the interface into the molten solder and break away into the bulk while the solder is still molten. Intermetallics also form ahead of the interface during the aging of copper joints with thin, pre-tinned layers [Mei et al., 1992]. An example is shown in Figure 2-12. In this case, the intermetallic is due to copper diffusion into the solder, which gradually creates a supersaturated condition.

Second, when gold films are used to protect the substrate prior to wetting, gold-based intermetallics appear within the bulk of the solder. These typically form while the solder is still molten; gold dissolves into the solder and reacts with tin (or nickel, in the case of In-Sn solders) to form intermetallics. An example is the dense distribution of $AuSn_4$ intermetallics shown in Figure 2-13, which formed when a copper substrate protected by gold was wet with a eutectic Pb-Sn solder [Kramer, 1992]. If the gold layer is not completely dissolved by molten solder during wetting, gold-rich intermetallics continue to grow in the solid state until the layer is consumed. An example in which Sn-In partially wets gold is shown in Figure 2-14.

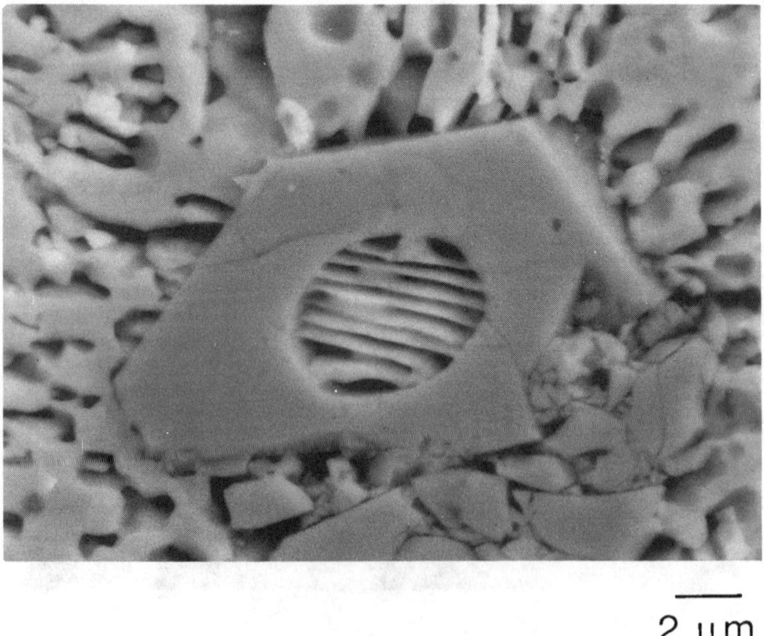

FIGURE 2-11 SEM micrograph showing the cross-section of a rod of Cu_6Sn_5 within the Sn-Pb solder joint. It has presumably broken off from the interface during solidification.

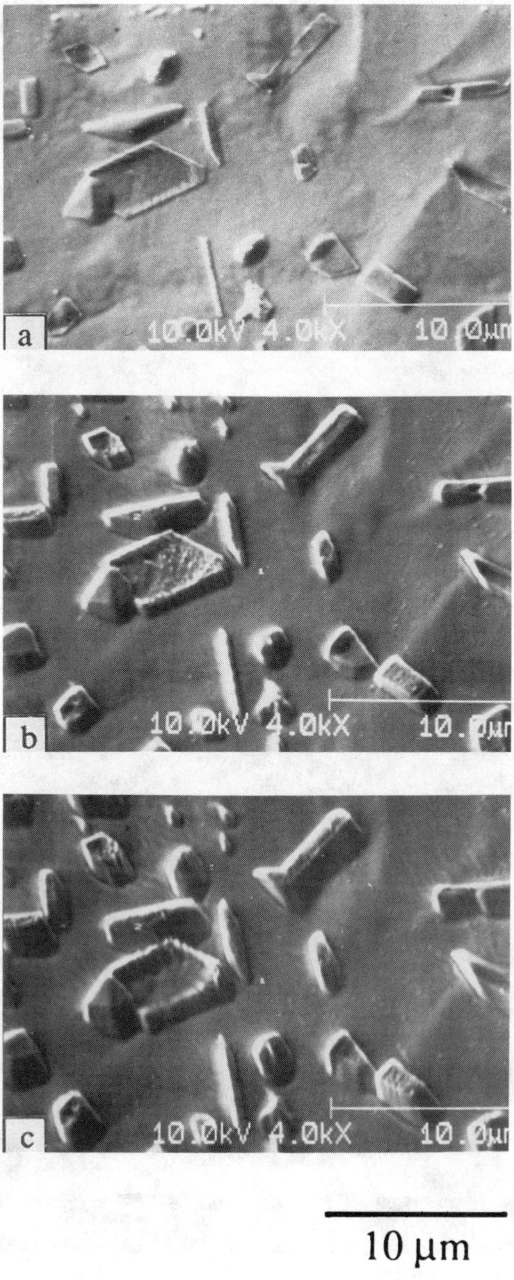

FIGURE 2-12 SEM micrographs of the surface of a pre-tinned sample after successively longer sputter times. Intermetallic particles can be seen protruding from the surface of the solder.

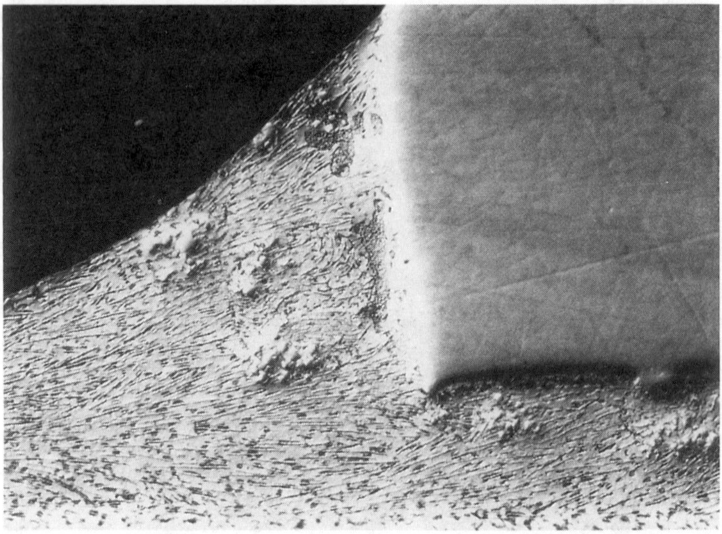

25μm

FIGURE 2-13 Optical micrograph of a eutectic Sn-Pb solder joint formed on gold plating, showing the $AuSn_4$ intermetallics which form within the solder.

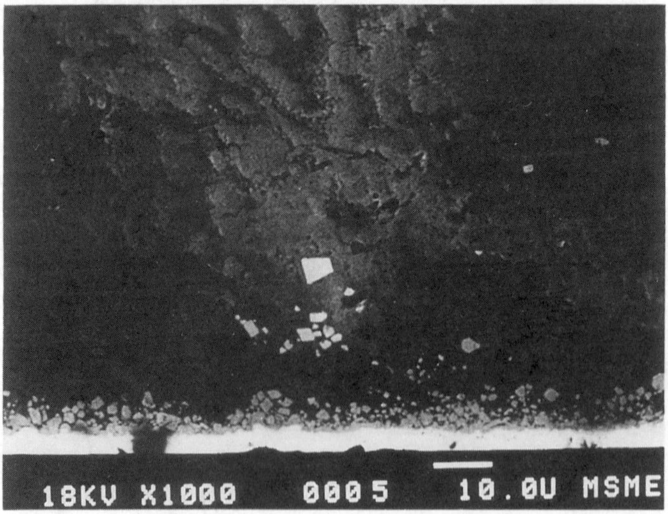

10 μm

FIGURE 2-14 Backscattering SEM image of 52In-48Sn solder on gold-coated copper plate, showing both In_2Au intermetallic particles and undissolved Au underneath (bottom of micrograph).

2.2.6 Changes in the Solder through Intermetallic Formation

The formation and growth of intermetallic precipitates changes the composition of the solder by depleting it of the constituent that joins the intermetallic. If the intermetallic forms while the solder is still molten, the composition change may change the solidified microstructure of the solder. For example, if eutectic Sn-Pb wets a gold-coated interface the composition of the molten solder shifts to the Pb-rich side of the eutectic value, and the solidified solder contains islands of pro-eutectic lead as well as Au-Sn intermetallic precipitates. When the intermetallic (Cu-Sn for example) grows in the solid state, the depletion of tin creates a lead-rich layer ahead of the growing intermetallic, and produces a chemically inhomogeneous solder.

2.3 THE INFLUENCE OF MICROSTRUCTURE ON MECHANICAL PROPERTIES

The mechanical properties that are important to the service behavior of solder joints include: 1) their resistance to fatigue under the thermal excursions encountered during normal operation, and 2) their resistance to deformation and fracture during installation and transport. Since the melting points of common solders are only slightly above room temperature, the deformation mechanisms that govern these properties are high-temperature mechanisms such as creep, recrystallization, and recovery. In the following sections we illustrate how the microstructure affects these properties by reviewing examples of microstructural influences on high-temperature creep, fatigue, and overload fracture.

2.3.1 Creep

Because solders are used at high homologous temperatures, creep deformation plays a role in almost all aspects of their mechanical behavior. A typical high-temperature creep curve is illustrated in Figure 2-15 [Mei and Morris, 1992a], which shows how the plastic strain increases with time under constant applied stress. The creep response is conveniently divided into three regimes, called primary, steady-state, and tertiary creep. During primary creep the strain rate decreases with strain, reaching a steady-state value that is preserved for some period of time, but eventually begins a monotonic increase that leads to fracture. The parameter that is most commonly used to characterize creep behavior is the steady-state creep rate, which ordinarily obeys an equation of the form [Bird et al. 1969]

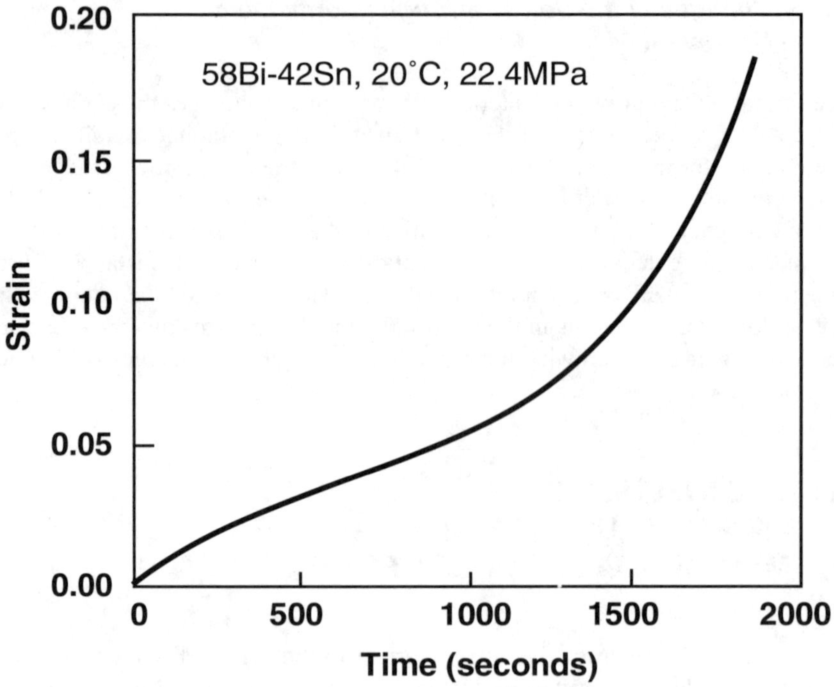

FIGURE 2-15 A typical creep curve for a eutectic Bi-Sn solder joint.

$$\dot{\gamma} = A\tau^{n}\exp\left(-\frac{Q}{RT}\right) \qquad (9.1)$$

where $\dot{\gamma}$ is the shear strain rate, τ is the shear stress, n is called the stress exponent, and Q is an activation energy. The stress exponent and the activation energy change with the dominant creep mechanism, and may have different values in different regimes of the applied stress.

The microstructure affects the steady-state creep rate in two ways. It influences the pre-factor, A, and it governs the range of stress over which particular creep mechanisms dominate at given temperature.

High-Stress Creep Mechanisms

At sufficiently high stresses, all metals creep by a mechanism whose rate is determined by the motion of dislocations through the bulk of the crystal grains. While the micromechanical details of this universal creep mechanism are still not well understood, it leads to a high stress exponent (n in the range 3-7) and has an activation energy that is close to that for self-diffusion in the bulk, which suggests that dislocation climb is the critical step in the process. The values of n and Q that pertain to this process are sensitive to composition, but do not ordinarily have a strong microstructural dependence.

The absence of a strong microstructure effect in this creep regime is consistent with the deformation mechanism. If the deformation rate is controlled by the resistance to dislocation motion in the interiors of grains, then this step must be slow compared to the processes that transmit strain from grain to grain. As a consequence, the strain rate is relatively insensitive to microstructural details.

The creep properties that remain sensitive to microstructure at high stresses are the transition to tertiary creep and the total elongation to failure. At least two different mechanisms govern the transition to tertiary creep: the onset of cavitation damage at grain boundaries, and plastic instability leading to inhomogeneous deformation.

Microstructural Influence on Elongation at High Stress

Cavitation is largely responsible for tertiary creep in bulk solder samples that are tested in tension. Cavities nucleate, primarily at three- or four-grain junctions; grow with strain, and join to cause failure. This process is promoted by increased grain size, which increases the stress concentration at grain junctions, by irregular grain shapes, which provide sites of unusual stress concentration; and by intergranular precipitates, which constrain deformation at grain boundaries. Hence, the elongation is sensitive to the microstructure.

Plastic instability is often responsible for tertiary creep of solder samples that are tested in shear. The strain in these samples becomes progressively inhomogeneous through the development of well-defined shear bands. Particularly at low strain rates, these tend to follow planes of microstructural weakness in the material, such as phase boundaries and colony boundaries in eutectic materials (Figure 2-16). The development of inhomogeneous shear bands is especially pronounced in solders that have unstable, eutectic microstructures that are easily recrystallized, such as the Sn-Pb [Tribula, 1990] and Sn-Bi eutectics [Freer-Goldstein and Morris]. In these materials, incipient shear bands cause recrystallization, leading to the development of the well defined recrystallized bands that are commonly observed in eutectic Sn-Pb solders that are crept or fatigued in shear (Figure 2-17) [Tribula, 1990]. However, plastic instability also develops in soft solders, such as the Sn-In eutectic, that do not undergo obvious recrystallization. In this case shear bands develop, but are not decorated with recrystallized material.

Unlike tertiary creep induced by cavitation, tertiary creep due to plastic instability does not necessarily lead to rapid failure. Soft solders tested in shear often have very large tertiary strains. As illustrated in Figure 2-15, even the Sn-Bi eutectic, which is one of the hardest of the low-melting-point solders, undergoes over 50% of its creep strain to failure at 20°C in the tertiary regime [Mei and Morris, 1992].

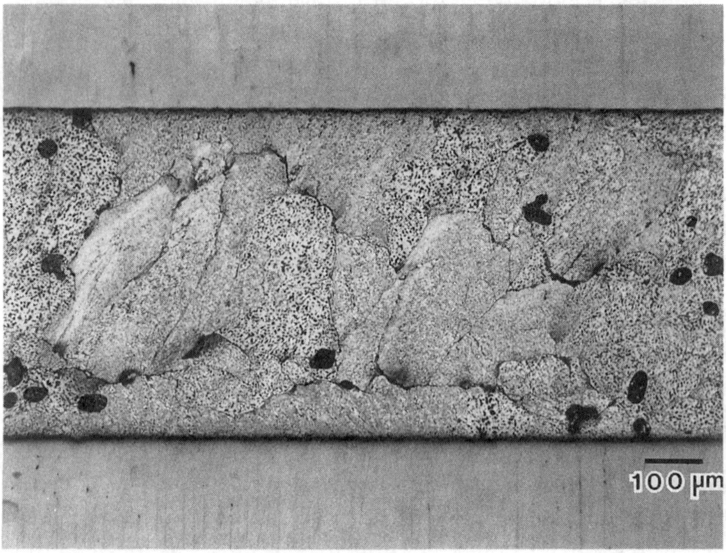

FIGURE 2-16 Optical micrograph of deformed 60Sn-40Pb solder joint, showing that the deformation pattern follows microstructural features such as colony boundaries.

The onset of plastic instability can be delayed by modifying the microstructure to inhibit the development of shear bands. At least three mechanisms are known to be effective: refining the grain size, which tends to homogenize the deformation [Mei and Morris, 1991] adding chemical species that disrupt the development of eutectic colonies [Tribula, 1990], and introducing pro-eutectic grains or precipitate particles that disrupt the growth of shear bands [Summers, 1991; Kramer et al.].

Intergranular Creep

The most striking effect of solder microstructure on creep is the intrusion of intergranular creep when the microstructure is fine-grained, the stress is low, and the temperature is high. The intergranular creep mechanism changes the creep behavior in two ways. First, the intergranular creep mechanism has a stress exponent (n, Equation (2.1)) near 2, and an activation energy that is approximately 50% lower than that observed at higher stresses, and is close to the activation energy for grain boundary diffusion. Second, the intergranular creep mechanism leads to a very stable and non-damaging plastic deformation. This creep mechanism is responsible for the most dramatic type of superplasticity, in which a material undergoes creep strains of several hundred percent prior to failure.

The principal microstructural requirement for intergranular creep is that the grain size be small, nearly equiaxed, and sufficiently stable that it is preserved dur-

FIGURE 2-17 Optical micrographs of a deformed Sn-Pb solder joint. The top micrograph shows the surface relief, which is concentrated in a narrow band along the bottom of the joint. The bottom micrograph, the same region after light polish, shows the band of coarsened and recrystallized material which developed in the highly deformed region.

ing deformation. When this is the case, the material deforms primarily by grain boundary sliding, in which grains are displaced with respect to one another by slipping along the boundary between them. The importance of a high grain boundary density is reflected in the constitutive equation for intergranular creep. The pre-factor, A, in this equation decreases with the grain size approximately as d^{-2}. Intergranular creep is negligibly slow unless the mean grain size is less than a few microns.

Since grains are three-dimensional objects, it is impossible to deform an intact material by grain boundary sliding alone. The grain boundary sliding mechanism must act in series with a bulk deformation mechanism that maintains contact along the boundaries and, particularly, along three-grain junction lines. This bulk mechanism is not well understood, but, at least in the case of eutectic Sn-Pb, it obeys a constitutive law of the form shown by Equation (2.1) with n « 3 and Q equal to the activation energy for bulk diffusion. The rate of creep is governed by the slowest of mechanisms that act in series. Since the bulk mechanism has a higher stress exponent, it dominates at low stress and limits the range of stresses over which intergranular creep is observed.

The combination of the three creep mechanisms leads to a steady-state creep relation like that shown in Figures 2-18a and b [Grivas, 1974], which are taken from research on recrystallized eutectic Sn-Pb, with the dominant creep mechanism changing from the high-stress creep mechanism to intergranular creep to the compatible bulk creep mechanism as the stress is lowered. However, since the low-stress bulk creep mechanism has a higher activation energy than the intergranular creep mechanism, the intergranular creep regime is gradually pinched off as the temperature decreases. Intergranular creep only appears at intermediate stresses, only when the temperature is high, and only when the grain size is small.

Microstructural Effects in Intergranular Creep

The classic method for obtaining a material that exhibits intergranular creep, and, hence, superplastic behavior, is to recrystallize a eutectic to obtain a fine, equiaxed grain structure that is stabilized against grain growth because the grains are a fine mixture of two different phases. Eutectic Sn-Pb and eutectic Sn-Bi solders are two of the systems in which this is most easily accomplished. As a consequence, they have been extensively studied as prototypical examples of superplastic materials. More recently, it has been realized that similar microstructures can be achieved by solidifying the eutectic at a sufficiently rapid rate, and that the necessary solidification rates are naturally achieved in eutectic Sn-Pb during the normal processing of small solder joints [Mei et al., 1990]. Further work has shown that small joints of Sn-In eutectic solidify in a microstructure that exhibits intergranular creep at temperatures above about 60°C, and that non-eutectic solders such as 40In-40Sn-20Pb also form and maintain fine-grained microstructures that exhibit intergranular creep [Mei and Morris, 1992b]. On the other hand, fine-grained microstructures are difficult to achieve in Sn-Bi eutectic solders by solidification; these develop a more regular eutectic microstructure when they are solidified at rates that are typical for solder processing [Mei and Morris, 1992a].

The fine-grained microstructures that lead to intergranular creep have the advantage of providing exceptional creep ductility. As we shall discuss below, they also have excellent fatigue resistance. The disadvantage is that they are unstable with respect to grain growth. Grain growth will cause a gradual deterioration in creep properties and a gradual change in the constitutive equation that

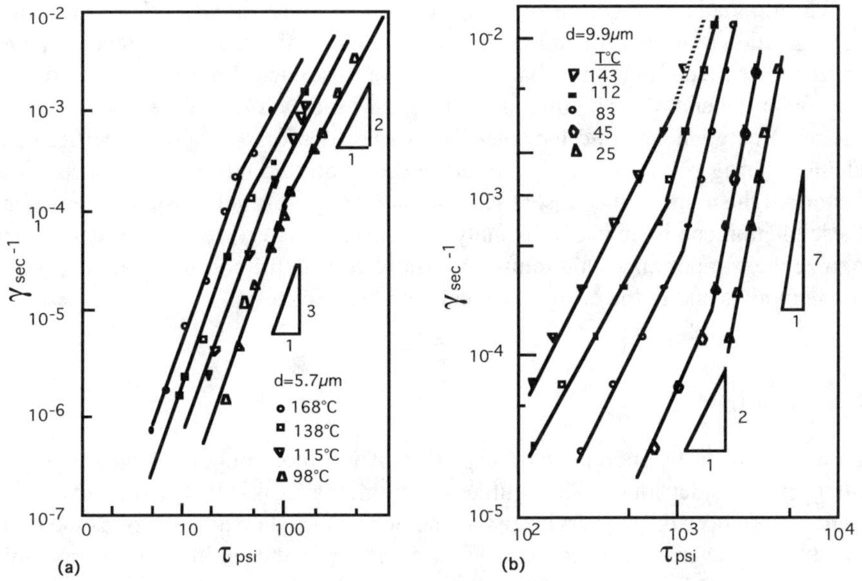

FIGURE 2-18 Creep data for eutectic Sn-Pb, showing (a) transition from dislocation climb-controlled creep to grain boundary-controlled creep, and (b) transition from grain boundary-controlled creep to dislocation glide-controlled creep.

governs creep behavior [Mei et al., 1991]. The rate of grain growth increases with the temperature at which the solder is used, and is also affected by the microstructure, since grain growth is inhibited in a fine, two-phase mixture. To further complicate the issue, deformation of the solder, for example, by thermal excursions during service, can induce a recrystallization that refines the grain size and re-establishes intergranular creep [Pan, 1992].

Intergranular creep can also be suppressed by the introduction of precipitates or dispersants into the solder matrix, including the formation of pro-eutectic grains. An example that is important in practice concerns the use of eutectic solders on gold-coated substrates. As discussed earlier, the gold dissolves into the solder when it is molten and forms intermetallics. These intermetallics decorate the matrix (as illustrated in Figure 2-13) and, if the net gold concentration is appreciable, change the composition of the solder by depleting the constituent that participates in the intermetallic. The formation of a dense distribution of intermetallics, like that shown in Figure 2-13, suppresses intergranular creep in eutectic Sn-Pb [Kramer et al.], even when the tin depletion is not great enough to make a dramatic change in the underlying microstructure. A higher gold concentration would deplete the solder of tin, promoting the formation of large islands of lead-rich phase that would also suppress intergranular creep.

The dramatic changes in creep properties that are introduced by the rapid solidification or grain refinement of many common solders are a serious barrier to the accurate prediction of mechanical behavior in service. The initial properties of the solder depend on its solidification rate; thus they vary with the size, location, thermal treatment, and interface metallurgy of solder joints. These properties then change during service in ways that depend on both the thermal and mechanical history of the joint. At the present state of knowledge, it is not possible to represent these phenomena in an accurate analytic theory. However, it is important to recognize their importance, and to use that knowledge in the construction of approximate models and in the design of accelerated service tests.

2.3.2 Fatigue

Failure induced by thermal fatigue is one of the most important threats to the integrity of solder joints. The source of the fatigue load is the thermal cycle the joint experiences during service, as the device is turned on and off, or as the temperature of the environment changes. In many cases the strain cycle is predominantly in shear. In almost all cases the joint is cycled at a high homologous temperature, so the cyclic deformation that causes fatigue is largely creep.

Most of the mechanistic studies that have been done on the creep fatigue of solder materials have used fatigue tests that have involved cyclic shear loads on model solder joints [Solomon, 1989; Shine and Fox, 1988], cyclic tensile loads on model joints [Frear et al.,1989] or cyclic tensile loads on bulk solder specimens [Vaynman et al., 1988]. Most of the work has been done under isothermal conditions, though thermal fatigue studies are also included. It is the authors' opinion that the cyclic shear of model joints provides the best representation of the service behavior of typical solder joints; at least, it produces the types of metallurgical damage that are revealed by post-mortem analysis of simulated service failures.

Fatigue Mechanisms in Shear

The high-temperature fatigue of solder joints in cyclic shear is strongly affected by the heterogeneity of creep deformation. As discussed above, the creep deformation of typical solders in shear tends to be concentrated in well-defined shear bands. Cyclic shear deformation in fatigue is concentrated in such bands, and eventually leads to crack nucleation and failure. In almost all cases studied to date, creep fatigue damage is located within the bulk solder. While the shear bands that lead to failure may lie close to the interface, the brittle intermetallic layer at the interface does not ordinarily participate in failure unless it is so thick that it essentially bridges the solder joint, or unless it is internally flawed.

The shear fatigue mechanism in 63Sn-37Pb with the eutectic microstructure is particularly well documented [Tribula, 1990; Frear, 1987; Schmitt-Thomas and Wege, 1986; Wolverton, 1987]. In this case the concentration of shear within planar shear bands causes recrystallization of the Sn-Pb eutectic, which replaces the eutectic microstructure with a fine-grained, equiaxed material that gradually coarsens as fatigue progresses. The cyclic strain is concentrated within this recrystallized material, and the joint eventually fails along shear bands as illustrated in Figure 2-19. Bands of coarsened, equiaxed grains that resemble these are observed in eutectic Sn-Pb joints that failed in service or under simulated service conditions [Keller, 1986; Lynch et al., 1983; Wild, 1972].

Reasonably well defined shear bands form during the fatigue of most coarse-grained solders, but are not always accompanied by recrystallization, even when the initial microstructure is eutectic. For example, the 58Bi-42Sn eutectic recrystallizes within shear bands, while the 52In-48Sn eutectic apparently does not [Frer-Goldstein and Morris].

The development of damage in inhomogeneous shear bands has important implications for both the analysis of solder fatigue and the metallurgical search for fatigue-resistant solders. From the analytic perspective it has the consequence that only a small fraction of the solder cross-section actually participates in cyclic deformation, and undergoes strain cycles that are much larger than the nominal strain on the joint. At least a few analytic studies have attempted to account for this phenomenon and have shown its importance for fatigue life [Winterbottom]. Of course, the analytic problem is complicated further if the sheared material recrystallizes, since this changes the mechanical properties of the material that experiences strain.

Metallurgical Routes to Improved Fatigue Resistance in Shear

The observation that strain concentration in shear bands leads to failure suggests how fatigue resistance may be improved. If the concentration of shear can be minimized or suppressed, the fatigue resistance should increase substantially. Three microstructural approaches have been used to accomplish this: the production of fine-grained microstructures that exhibit intergranular creep, the use of chemical additives to disrupt the microstructure, and the addition of dispersants to interrupt shear bands.

The most successful of these is the fine-grained microstructure, which can be produced by rapid solidification and is naturally achieved in many small solder joints. While there has been very little research on the creep fatigue of superplastic materials, it was hypothesized some years ago [Morris and Mei, 1991] that cyclic deformation by intergranular creep should lead to little or no microstructural damage. This hypothesis is largely confirmed by data like that shown in Figure 2-20 [Mei and Morris, 1992c] which compares the shear fatigue behavior of 63Sn-37Pb solder joints in two microstructures: a normal eutectic microstructure, and a refined microstructure made by rapid solidification. When the joints are tested at strain rates that lie primarily within the intergranular creep regime, the fatigue life of the joint with a fine-grained microstructure is greater by at least three times.

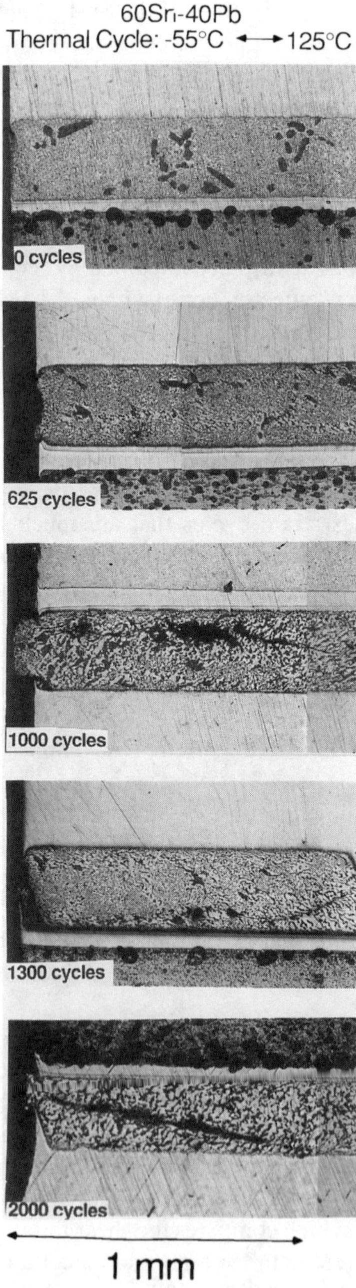

FIGURE 2-19 Series of optical micrographs of 60Sn-40Pb solder joint microstructure as a function of number of cycles in thermal fatigue.

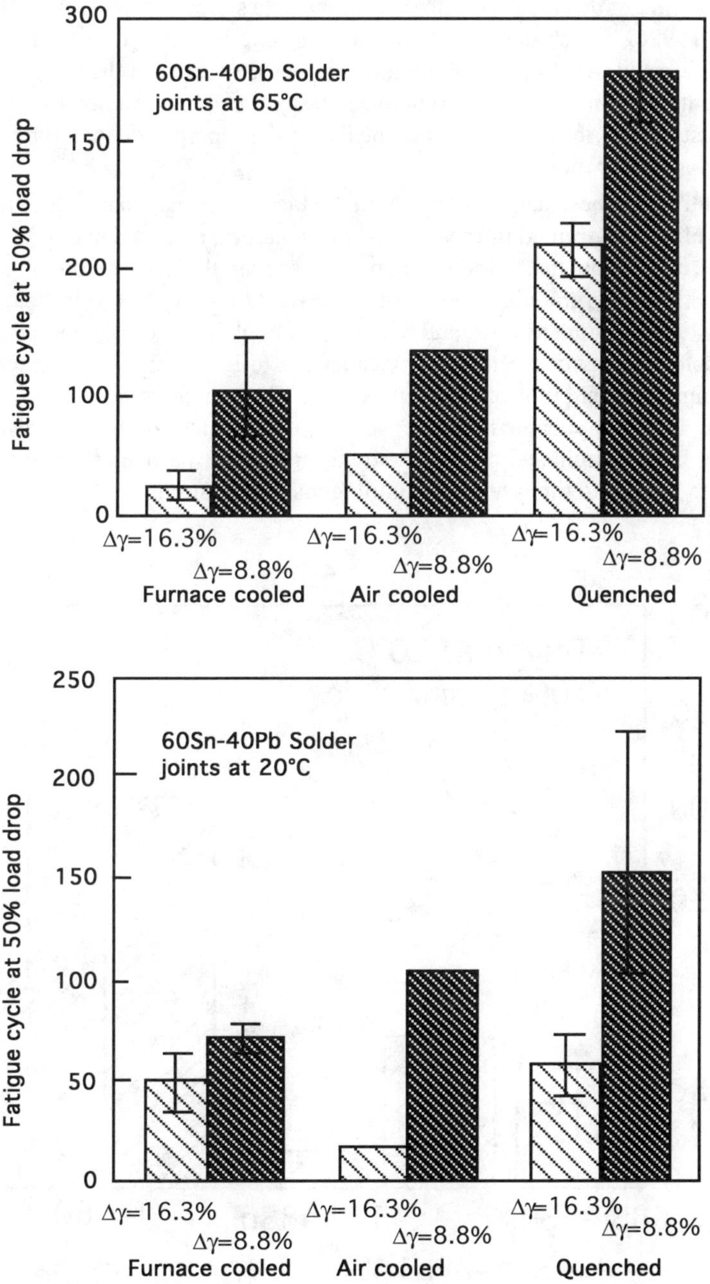

FIGURE 2-20 Comparison of isothermal fatigue life for solder joints cooled at various rates, tested at two different temperatures.

The generality of this result is indicated by the data shown in Figure 2-21 [Mei and Morris, 1992c], which compares the shear fatigue behavior of a Sn-Bi eutectic joint, which solidified in a normal eutectic microstructure, to that of a geometrically identical 40In-40Sn-20Pb joint that solidified in a fine-grained structure with superplastic behavior. Again, the fatigue life of the superplastic material is greater by a factor of 3 or more.

The utility of these data lies largely in the fact that many small solder joints either develop fine-grained microstructures as a natural consequence of their manufacture or can be made to do so by increasing the rate of cooling during solidification. Intergranular creep almost certainly contributes to the fatigue life of many existing solder joints, and may help to explain the observation that small Sn-Pb solder joints in electronic devices seem to have better fatigue resistance than design tests on bulk solder would suggest. On the other hand, this technique is not obviously applicable to thicker solder joints, which solidify at slower rates, and may be compromised by grain growth in service, particularly if the joint is subject to long hold times at high homologous temperatures.

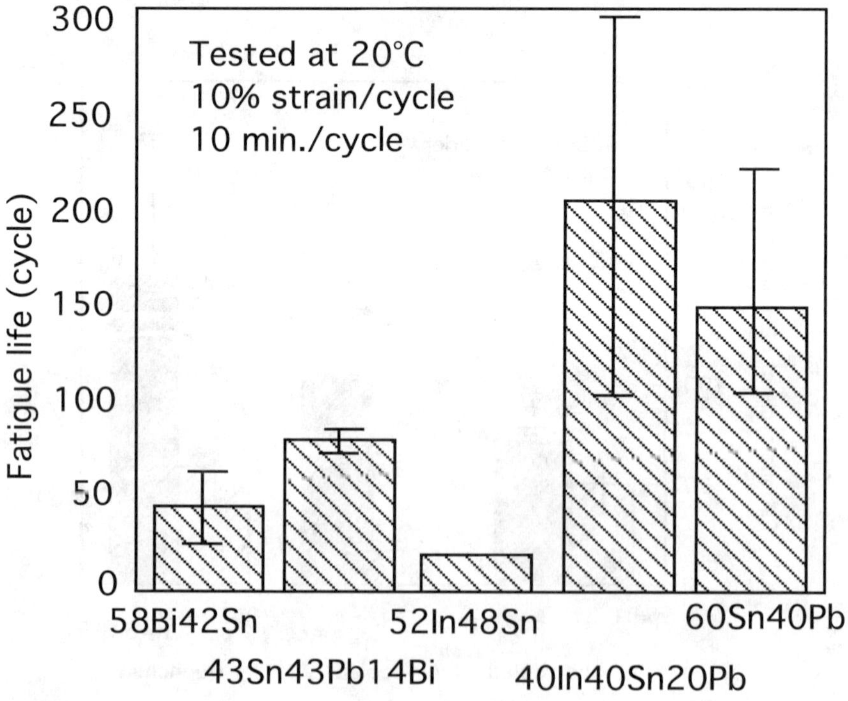

FIGURE 2-21 Comparison of isothermal fatigue life for five solder alloys.

The second approach to improving the fatigue life is the use of compositional changes to create an irregular microstructure that resists the formation of planar shear bands. Two methods have been successfully used. First, one may simply adjust the solder composition to an off-eutectic value that produces a mixed microstructure that contains relatively large pro-eutectic grains (for example, Figure 2-7). In a bimodal microstructure of this sort, each constituent disrupts the normal deformation pattern of the other. It has been known for some time that off-eutectic Sn-Pb solders have better fatigue resistance than eutectic material [Bangs and Beal, 1978; Inoue et al., 1986; Jarboe, 1980; Wild, 1975]. Recent research has documented the mechanistic connection between the improvement in fatigue life and the change in the pattern of deformation [Summers and Morris, 1990]. Second, one may use chemical additives to disrupt the formation of the eutectic microstructure, which accomplishes essentially the same effect. In prior research at Lawrence Berkeley Laboratory, small additions of indium and cadmium to 63Sn-37Pb were shown to produce a significant improvement in shear fatigue life. Metallographic analysis shows that they inhibit the completion of the eutectic microstructure, leading to a structure that has broad, featureless bands along the eutectic colony boundaries [Tribula, 1990].

The third approach is to introduce dispersants into the solder to disrupt the deformation pattern. Research along these lines is underway in several laboratories. While the research is certainly worthwhile, the approach is problematical. Assuming that a suitably dense distribution of dispersants can be successfully introduced, the fatigue properties of the solder will reflect the constraints on deformation that are imposed by the hard dispersants as well as the benefit from dispersing the matrix slip. In one study recently completed in this laboratory we used the fact that the addition of gold to 63Sn-37Pb solder naturally produce a dense distribution of dispersants (Figure 2-22) [Kramer et al.] to explore its consequences for the fatigue life. While small gold additions do seem to improve the fatigue life, these do not produce dense dispersant fields. When the gold addition is sufficient to produce dispersants, as in Figure 2-22, the fatigue resistance is lowered rather than raised [Kramer et al.]. At least in this case, the deleterious influence of the distribution of hard precipitates apparently outweighs any benefit from the dispersion of slip in the solder.

Thermal Fatigue in Shear

The basic mechanisms of shear fatigue are the same whether the temperature is held constant or cycled to generate thermal fatigue like that encountered in service. However, the temperature cycle does produce noticeable differences in the deformation pattern. The shear bands are more coarse in samples subjected to thermal fatigue, and the material within them appears to coarsen more rapidly. A likely source of this difference is the asymmetry of the deformation that is produced by the thermal cycle. When the temperature is increased, the creep rate increases exponentially, with the consequence that deformation is much easier in

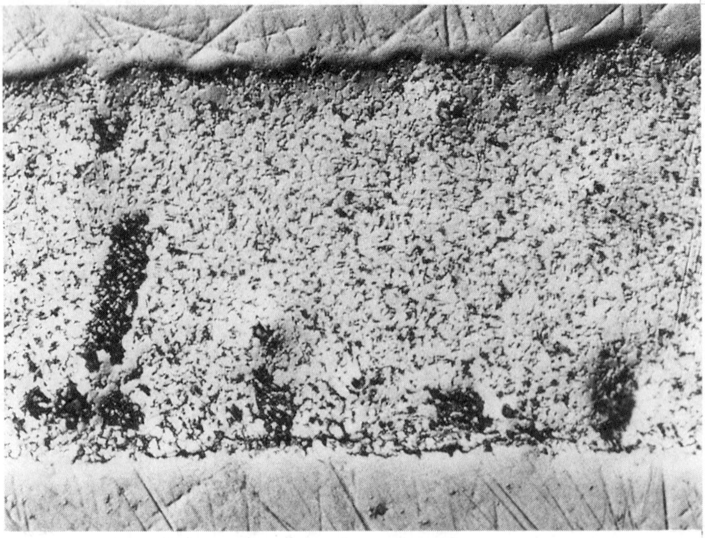

25μm

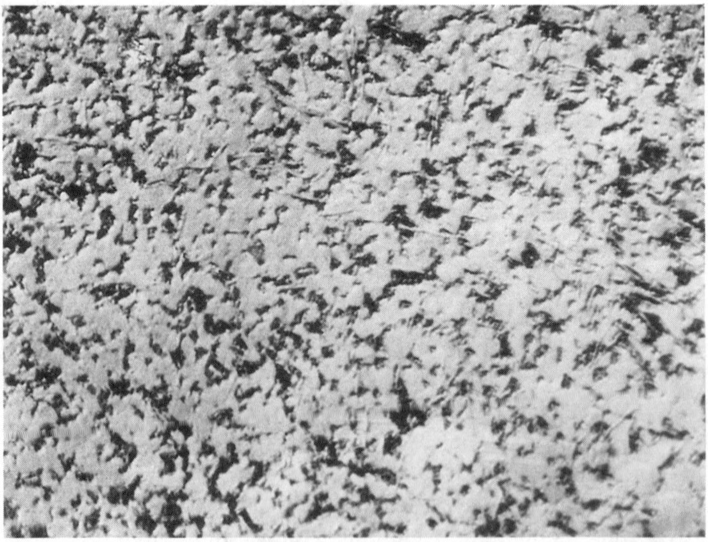

25μm

FIGURE 2-22 Optical micrographs of solder joint made from eutectic Sn-Pb plus 2.5 weight percent Au.

the high-temperature phase of the thermal cycle than it is in the low-temperature phase. The sample effectively softens during deformation in the direction of increasing temperature, and hardens during its reverse deformation as the sample cools. As a consequence, the forward and reverse deformation paths will not ordinarily be exactly the same, and the deformation pattern should be more diffuse. Finite element analysis [Chen, 1989] has shown that the thermal strains can lead to irreversibilities that produce a progressive macroscopic deformation of the body of the solder. The increased coarsening observed in thermal fatigue has a similar explanation. Microstructural defects that are preferentially stored in the material during the low-temperature step provide an additional driving force for microstructural change during the high-temperature step where grain growth is relatively easy.

Fatigue in Tension

Research on the tensile fatigue of solders has also revealed interesting microstructural effects. For example, tensile fatigue studies on model solder joints of eutectic Sn-Pb solder uncovered a tendency for these joints to fail within the intermetallic layer [Frear et al., 1989b]. This result is consistent with the results of tensile tests that will be reviewed below. It is presumably due to a combination of continuum and microstructural effects. Since the solder joint is thin in the direction of the applied load, the solder has a relatively high effective yield strength, which transfers the load to the intermetallic. The intermetallic is brittle, and hence provides relatively easy propagation paths for fatigue cracks that nucleate at or near the interface. A second interesting result was obtained from cyclic tension tests on bulk specimens of lead-rich solders [Frear, 1987]. These have microstructures that are made up of relatively equiaxed grains of lead-rich phase. When they are tested in air under small cyclic loads, they tend to fail by intergranular fracture. The intergranular fracture mode is believed to be promoted by oxygen adsorption on grain boundaries. It should, hence, be inhibited by grain refinement, which creates a more tortuous intergranular path, or by the addition of chemical species that act to getter oxygen.

2.3.3 Deformation and Fracture

Since a solder joint must resist loads that are applied during manufacture and transport, its mechanical response to relatively large mechanical loads that are applied at strain rates like those used in ordinary tensile tests are also of interest. However, since solders are ordinarily used and tested at high homologous temperatures, their response is affected by the same microstructural mechanisms that govern creep at high stresses. The deformation of solder joints is particularly sensitive to the geometry of the load, and they behave differently in tension and in shear.

When solder joints are strained to failure in shear, the deformation is primarily within the solder. Assuming a high homologous temperature, the joint yields under very low load, hardens to an ultimate tensile strength, and then softens dramatically with further strain. As an example, Figure 2-23 shows shear stress-strain curves for a model joint of eutectic Sn-In on copper at several different temperatures [Frear-Goldstein and Morris, b]. In this case, strain softening appears to be due to progressive heterogeneity in the deformation pattern. The solder becomes weaker because less and less of its volume participates in the strain. Eutectic Sn-Pb shows a generally similar behavior, but in this case the strain softening is enhanced by dynamic recrystallization, which not only provides an effective mechanism of recovery, but leads to a microstructure that is considerably softer than the eutectic one.

Soft solders have similar behavior in tensile tests, but strain softening is less pronounced, which reflects the more homogeneous deformation of the specimens. The most striking microstructural influence on the tensile behavior is the aging of the solder microstructure. Tensile properties vary with time. Eutectic Sn-Pb softens noticeably for several months after casting, and continues to change, though at a slower rate, for the life of the specimen. These changes are associated with

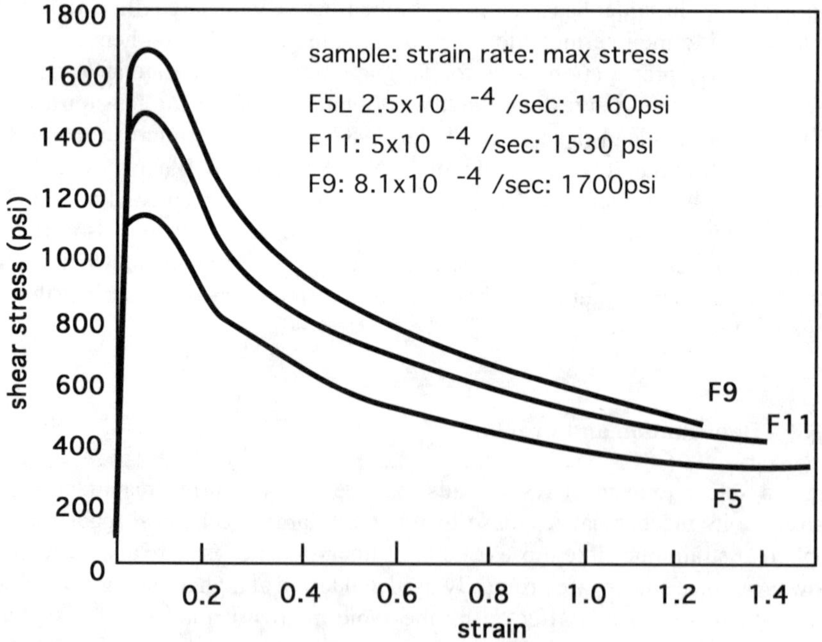

FIGURE 2-23 Stress-strain curves for eutectic In-Sn on Cu tested at 40° C and three different strain rates.

normal reconfigurations of the microstructure. These occur at a measurable rate at room temperature, which is a relatively high homologous temperature for the alloy. The changes involve precipitation of tin within the lead-rich phase, grain growth, and reconfiguration from the eutectic structure toward a more equiaxed arrangement of grains.

When solder joints are tested to failure in tension, the ultimate fracture may occur either within the body of the solder or in the intermetallic layer at the interface. The site of the failure depends on the strength of the solder, the nature and thickness of the intermetallic, and the microstructure of the solder. Since the intermetallic is normally brittle, the overall tensile strength of the joint is higher and more nearly reproducible if the microstructure is changed to promote fracture through the solder bulk.

These principles are illustrated by tensile tests on 60Sn-40Pb and 95Pb-5Sn solder joints on copper [Quan et al., 1987]. The 95Pb-5Sn solder is the weaker of the two. In the as-solidified condition, it fractures within the solder through void nucleation at tin-rich precipitates in the lead-rich grains. When the joint is reflowed to thicken the Cu_3Sn intermetallic layer at the interface, tin is depleted from the matrix. The fracture site shifts to the solder-copper interface, and the fracture strength decreases by more than 50%.

The behavior of eutectic 60Sn-40Pb on copper is quite different. When it is solidified at slow to moderate rates, this solder has a basically eutectic microstructure with a double-layer of intermetallic at the interface (Cu_6Sn_5 on Cu_3Sn, as in Figure 2-8). In this case the solder fails through the Cu_6Sn_5 intermetallic layer. The apparent reason is the greater strength of the solder, which increased the load on the intermetallic. Reflow treatments that increase the thickness of the intermetallic cause a significant decrease in strength, with the same fracture mode. Small changes in the composition of the solder that do not change the fracture mode have almost no effect on the joint strength. However, if the solidification pattern of the joint is such that long rods of Cu_6Sn_5 intermetallic are released into the body of the solder, as in Figure 2-24 [Frear et al., 1987], these may serve as void nucleation sites that shift the fracture from the interface into the solder. When this happens, the ductility of the solder has the consequence that the tensile strength increases slightly.

These results suggest the importance of maintaining a thin intermetallic layer to preserve the tensile strength of the solder joint. They also show that inclusions (or, by extension) small voids in the solder are not necessarily sources of weakness. They may act to preserve strength by ensuring that the crack path runs through the solder.

A final interesting fracture behavior is observed when eutectic Sn-Pb solder joints on copper are fractured in liquid nitrogen. The fracture path runs along the solder-Cu_6Sn_5 interface, as shown in Figure 2-25 [Quan et al., 1987], even though this interface is irregular and leads to a tortuous fracture path. The appearance of the fracture surface suggests that the lead-rich phase remains well bonded to the

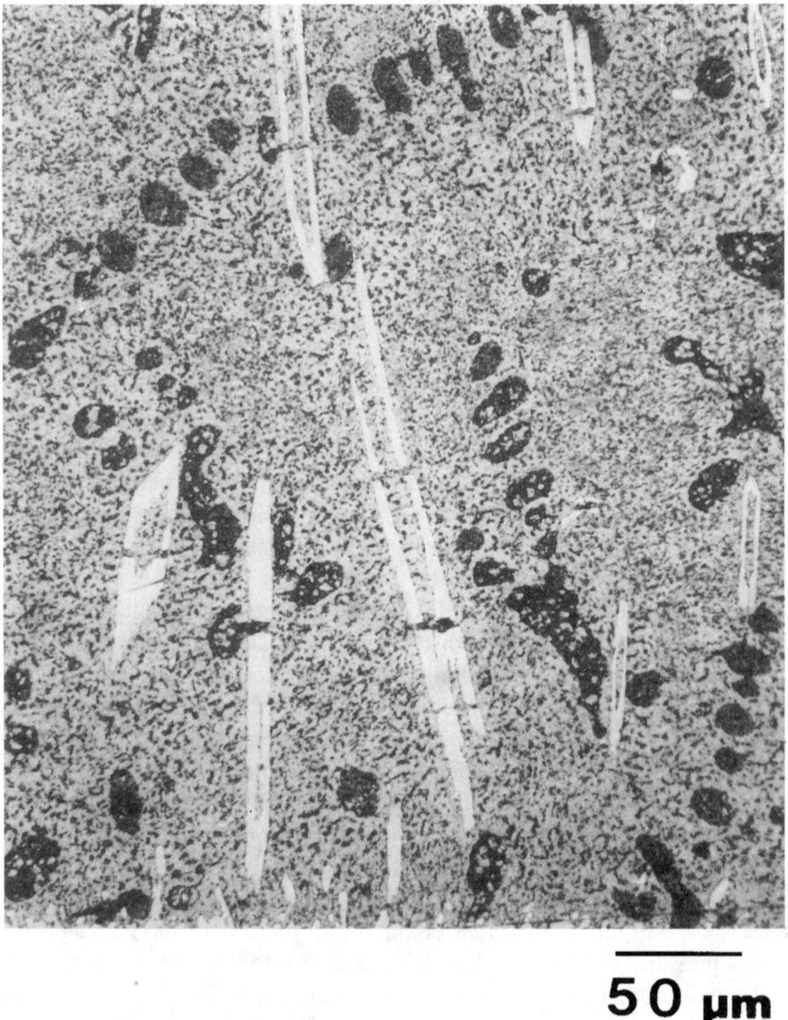

FIGURE 2-24 Optical micrograph showing Cu_6Sn_5 intermetallics within Sn-Pb joint.

intermetallic even at 77K; islands of lead-rich material adhere to the intermetallic. The source of this curious fracture mode seems to be the de-bonding of tin. The probable reason for this behavior is the anisotropic thermal contraction of metallic tin, which has a complex, tetragonal crystal structure. This thermal contraction produces strains at the interface with the rigid intermetallic, causing the two to separate cleanly.

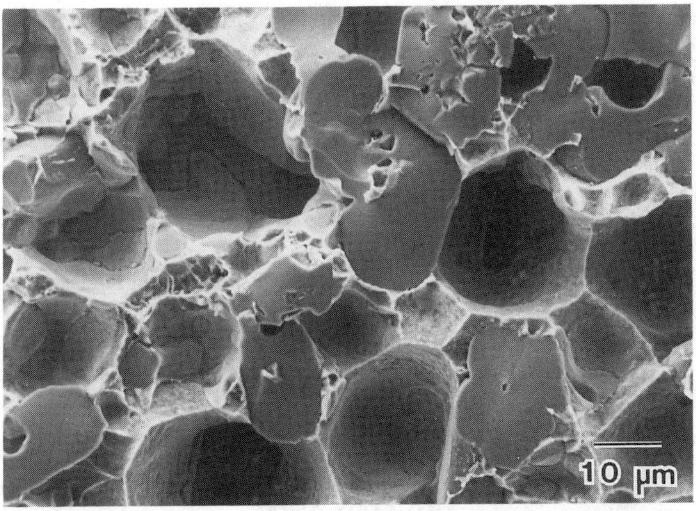

FIGURE 2-25 SEM micrograph of 60Sn-40Pb sample on copper failed in tension.

2.4 CONCLUSION

The examples given above should suffice to show that the microstructure of a solder joint cannot be ignored if its mechanical properties are to be understood. Changes in the microstructure can and do cause dramatic changes in mechanical properties and failure modes.

On the other hand, the current state of materials science does not offer the level of theoretical guidance that is needed to construct microstructure-based mechanical models. It appears that for the foreseeable future these will have to be based on fairly straightforward constitutive relations.

While the metallurgy does not provide an accurate theory, it does provide scientific guidance that ought to be useful to those who must construct predictive models. Given the microstructural complexity of solder behavior, the constitutive equations used in these models should be taken from tests on solder joint configurations that bear the closest possible microstructural resemblance to those that are addressed by the model and are subject to load geometries that resemble the loads of interest. It is incorrect, and may be badly misleading, to adopt a constitutive relation simply because it was measured on a material that has a particular chemical composition.

Similar considerations pertain to the design of accelerated tests that are used to confirm or refine predictive models. An accelerated test is probative to the extent that it accelerates the mechanism that governs the behavior of interest. If the method of acceleration changes the microstructure or the microstructural mechanisms that operate, the test may mislead rather than inform. For example, accelerating creep by increasing the applied stress may change the basic mechanism of creep unless one is careful; accelerating creep by increasing the temperature may trigger microstructural changes that have the same undesirable effect.

2.5 ACKNOWLEDGMENTS

The authors acknowledge helpful contributions from Pamela Kramer, LBL Anne Sunwoo, LLNL and Darrel Frear, Sandia National Labs. The work reported here draws on research funded by Sandia National Laboratories, Digital Equipment Corporation, Western Digital Corporation, and by the Director, Office of Energy Research, Office of Basic Energy Sciences, U.S. Department of Energy, under Contract NO. DE-AC03-76SF00098.

REFERENCES

Bangs, E. R. and Beal, R.E., 1978, *Welding Journal (Res. Suppl.)*, 54:377s.

Bird, J.E., Mukherjee, A.J. and Dorn, J. F., 1969, *Quantitative Relation Between Properties and Microstructure*, 255, Israel University Press.

Chen, F. K., Ph.D. Thesis, University of California at Berkeley, 1989.

Frear, D. R., Posthill, J. B., and Morris, J.W., Jr., 1989a, *Metallurgical Transactions A*, 20A 325.

Frear, D. R., Grivas, D. and Morris, J.W., Jr., 1989b, *J. Electronic Mater.*, 18:671.

Frear, D. R., Grivas, D. and Morris, J. W. Jr., 1987, *J. Electronic Mater.*, 16:181.

Frear, D. R., Ph.D Thesis, University of California at Berkeley, 1987.

Freer Goldstein, J. L. and Morris, J. W., Jr., (a) submitted to *Metall. Trans. A*.

Freer Goldstein, J. L. and Morris, J.W., Jr., (b) unpublished work.

Grivas, D., Frear, D. R., Quan, L. and Morris, J.W. Jr., 1986, *J. Electronic Mater.*, 15:355.

Grivas, D. M.S. Thesis, University of California at Berkeley, Jan. 1974.

Inoue, H., Kurihara, Y. and Hachino, H., 1986, *IEEE Transactions on Components, Hybrids, and Manufacturing Technology*, 9:190.

Keller, J. N., 1986, *IEEE Transactions on Components, Hybrids, and Manufacturing Technology*, 4:132.

Kramer, P. A., M.S. Thesis, University of California at Berkeley, May 1992.

Kramer, P. A., Glazer, J. and Morris, J.W. Jr., submitted to *Metall. Trans. A*.

Kramer, P. A., Mei, Z., and Morris, J. W., Jr., unpublished work.

Lynch, J. T., Ford, M.R. and Boetti, A., 1983, *IEEE Transactions on Components, Hybrids, and Manufacturing Technology*, 6: 237.

Mei, Z. and Morris, J.W. Jr., 1992a, *J. Electronic Mater.*, 21:599.

Mei, Z. and Morris, J.W. Jr., 1992b, *J. Electronic Mater.*, 21:401.

Mei, Z. and Morris, J. W., Jr., 1992c, *Trans. ASME, J. Electronic Packaging*, 114:104.

Mei, Z. and Morris, J.W. Jr., 1991, *J. Electronic Mater.*, 20:599.

Mei, Z., Shine, M.C. and Morris, J.W. Jr., 1991, *ASME Journal of Electronic Packaging*, 113:109.

Mei, Z., Grivas, D., Shine, M.C. and Morris, J.W. Jr., 1990, *J. Electronic Mater.*, 19:1273.

Mei, Z., Sunwoo, A.J. and Morris, J.W. Jr., 1992, *Metall. Trans. A*, 23A:857.

Morris, J. W., Jr. and Mei, Z., 1991, in Solder Mechanics, A State of the Art Assessment, D.R. Frear, W.B. Jones, K.R. Kinsman eds., TMS.

Pan, T-Y., Ford Motor Company, Third International Workshop on Materials and Mechanics Issues of Solder Alloy Applications, Santa Fe, September, 1993.

Quan, L., Frear, D. R., Grivas, D., Morris, J. W. Jr., 1987, *J. Electronic Mater.*, 16:203.

Schmitt-Thomas, K. G. and Wege, S., 1986, *Brazing and Soldering*, 11:27.

Shine, M.C. and Fox, L. R., 1988, ASTM STP 942, H.D. Solomon, et al., ed., p 588.

Solomon, H. D., 1989, *J. of Electronic Packaging*, 111:75.

Summers, T. S. E., Ph.D. Thesis, University of California at Berkeley, May 1991.

Summers, T. S. E. and Morris, J. W. Jr., 1990, *Trans. ASME, J. Electronic Packaging*, 112:94.

Sunwoo, A. J., Morris, J.W. Jr., and Lucey, G.K. Jr., 1992. *Metall. Trans. A*, 23A:1323.

Tribula, D., Ph.D. Thesis, University of California at Berkeley, June 1990; and Tribula, D., and Morris, J. W. Jr., 1990, *ASME Journal of Electronic Packaging*, 112:87.

Vaynman, S., Fine, M.E., Jeannotte, D.A., 1988, *Metall. Trans. A*, 19A:1051.

Wild, R.N., 1972, *Welding Journal (Res. Suppl.)*, 51:521.

Winterbottom, W. L., private communication.

Wolverton, W. M., 1987, *Brazing and Soldering*, 12:33.

Jarboe, D. M., 1980, Technical Communications Bendix, Kansas City: BDX-613-2341.

Wild, R. N., 1975, IBM Technical Paper No. 74Z000481.

3

Interfaces and Intermetallics

James L. Marshall, L. Ann Foster, and Jennifer A. Sees

3.1 INTRODUCTION

The American Welding Society defines soldering as a joining process wherein a coalescence between metal parts is produced by heating to temperatures below 800°F using filler metals (solders) having melting temperatures below those of the base metal [Fenton, 1959]. After its deposition, the solder is then evenly distributed between the properly fitted surfaces of the joint by capillary action. Solders act by (a) wetting the basis metal surfaces forming the joint; (b) flowing between these surfaces so as to fill the space between them; and (c) metallurgically bonding to the surfaces when solidified [Thwaites, 1982]. This metallurgical bonding occurs, in a solder joint, between the solder and the substrate at a boundary area termed the interface [Morris et al., 1991]. This interface consists of a thin layer of an alloy formed during the soldering process. The alloy—termed the intermetallic—is generated by metals from both the solder and the substrate [Morris et al., 1991]. Figure 3-1 shows a typical cross section of a solder joint displaying the intermetallic.

In this chapter we will characterize intermetallics and discuss their potential impact upon solder reliability. When intermetallics occur at the interface of a soldered surface, they can grow and increase in thickness. In large solder joints, which are properly soldered and which function in benign environments, intermetallics have no deleterious effect upon long-term reliability of solder joints. How-

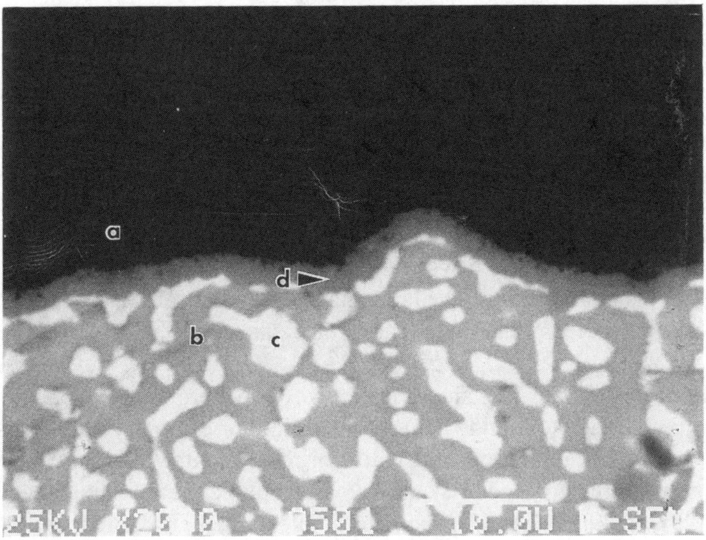

FIGURE 3-1 SEM (scanning electron micrograph) of cross section of 60% Sn-40% Pb solder joint, manifesting the copper-tin Cu_6Sn_5 intermetallic, about 1 μm thick. Top half, dark region: copper substrate (a). Bottom half: darker domains, tin (b); lighter domains, lead (c). Intermediate region, Cu_6Sn_5 intermetallic (d). X2000, backscattered mode.

ever, the intermetallic may play a role as the solder joint continues to miniaturize and as the metallic system becomes more complex.

3.2 INTERMETALLICS IN THE SOLDER JOINT

3.2.1 Characterization of the Intermetallic

When two metal parts are joined by solder, a metallic continuity is established as the result of two interfaces that form where the solder is bonded to both metal parts [Manko, 1979]. This metallic continuity, or joining interface, is referred to as the intermetallic layer or layers. Manko defines an intermetallic compound as a distinguishable homogeneous phase having a relatively narrow range of compositions with simple stoichiometric proportions (i.e. atom ratios) [Manko, 1979]. Intermetallic compounds grow at the interface of the solder and the substrate during long term storage at ambient temperatures and more rapidly at high temperatures [Davis, Warwick, and Kay, 1982; Woodgate,1983].

Examples of typical intermetallic layers include: tin/gold, tin/silver, tin/nickel, and tin/copper. The growth of the intermetallic compounds between tin and gold is very rapid (a total layer thickness of 50 μm will form after only 300 hours at 150°C) and include AuSn, $AuSn_2$, and $AuSn_4$ [Wassink, 1989]. Ag_3Sn is the most common intermetallic layer that forms between silver and tin [Wassink, 1989]. Because nickel is often used as a barrier layer between tin and copper, $NiSn_3$ and Ni_3Sn_4 are also often observed in electronic systems [Wassink, 1989]. Cu/Sn intermetallics are probably the most commonly studied systems because of the frequency with which the Sn/Pb system is soldered onto copper substrates. In Cu/Sn systems, two layers can form: the layer that forms adjacent to the copper substrate is Cu_3Sn (ε-phase), and the layer that forms adjacent to the solder is Cu_6Sn_5 (η-phase).

There are a few unusual situations, however, where intermetallic compound growth may not be required to achieve bonding. For example, Belser reports that alloys with a high indium content will coat many nonmetallic materials such as glass, quartz, porcelain, and brick, as well as the oxides of aluminum, copper, iron, and nickel [Manko, 1979; Belser]. In this application, the indium does not flow but must be smeared over the surface to be wetted, while maintaining a temperature only slightly above the melting point of indium [Belser; Chalco, 1983; Keeler, 1987].

Indium cold-welds easily because it is soft and has a low tensile strength. The term "cold-weld" describes the joining of two metallic surfaces mechanically by pressure so that the surfaces at the interface of the bond have deformed and fused into one another. Hence, a metallic continuity can develop [Manko, 1979].

3.2.2 Copper-Tin Intermetallics

The existence of the ε- and η-phases of Cu/Sn intermetallic are evident from the Cu/Sn phase diagram (Figure 3-2) [Hansen, 1958]. Strictly speaking, the η-phase that exists at room temperature is η', while the high temperature form (above 186° C) is η [Metals Handbook, 1973]. Both forms exhibit $B8_2$ symmetry; the low temperature form exhibits an ordered distribution of 20% copper atoms and 80% vacancies, and the high temperature form shows a disordered distribution of 20% copper atoms and 80% vacancies [Gangulee, 1973]. Figure 3-3 illustrates the crystal structure of the Cu_6Sn_5 η- or η'-phase.

The ε-phase is orthorhombic (the crystal structure is presented in Figure 3-4). There exists some evidence for both a disordered structure (called ε) and a long-term superlattice (called ε'), the proportions of which may vary at room temperature depending upon mechanical and thermal treatment [Brooks, 1970]. The physical properties of these two forms are almost identical. It is possible that in the future careful transmission electron microscopy (TEM) characterization might further distinguish between these two phases.

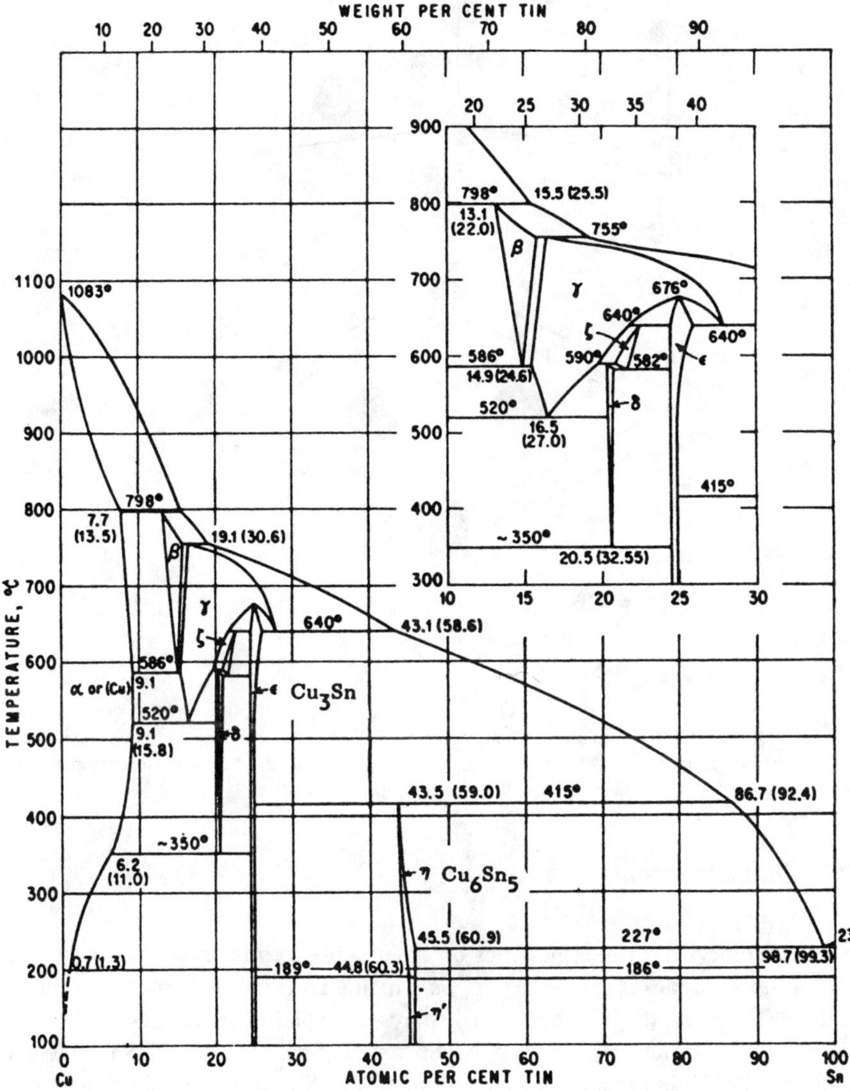

FIGURE 3-2 Phase diagram of copper/tin (Cu/Sn). This complex system exhibits four stable phases at room temperature: copper, ε (Cu_3Sn), η (Cu_6Sn_5), and tin. These four phases are observed in a copper/tin solder system.

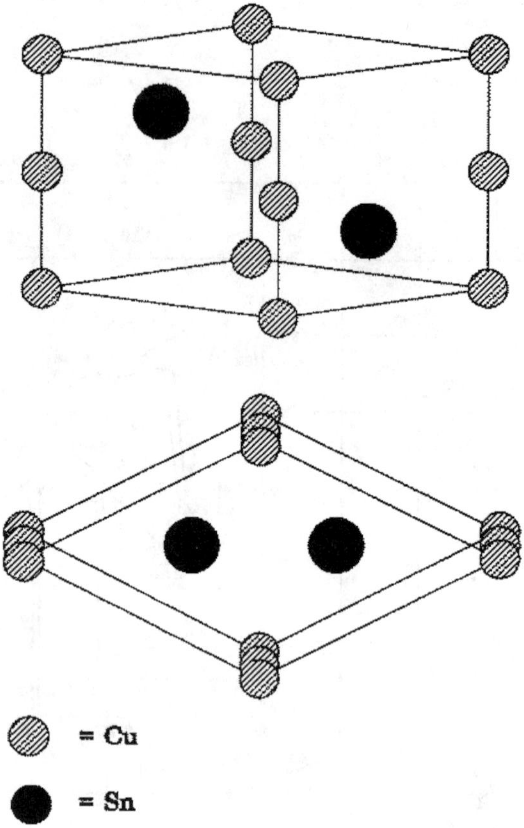

FIGURE 3-3 Crystal structure of the η-phase (Cu_6Sn_5). This lattice has $B8_2$ symmetry. Inside the unit cell occurs a copper atom with 20% frequency—either in an ordered fashion (low temperature phase, η') or in a disordered fashion (high temperature phase, η).

3.2.3 Other Intermetallics

Tin-based solders form intermetallics on other substrate materials. For example, solder joints on a nickel substrate (e.g. on a nickel barrier) can form Sn/Ni intermetallics. The phase diagram of Ni/Sn displays three phases (Figure 3-5) [Hansen, 1958], with the stoichiometries $\beta = Ni_3Sn, \gamma = Ni_3Sn_2$, and $\partial = Ni_3Sn_4$ [Metals Handbook, 1973]. In practice, the stoichiometry Ni_3Sn_4 and also a tin-rich acicular phase with a stoichiometry corresponding to $NiSn_3$ have been observed [Romig et al., 1991]. On silver (e.g., on a silver-metallized leadless component) tin-based solders can form Sn/Ag intermetallics with the stoichiometry Ag_3Sn [Romig et al., 1991], and this intermetallic is reflected in the phase diagram as the

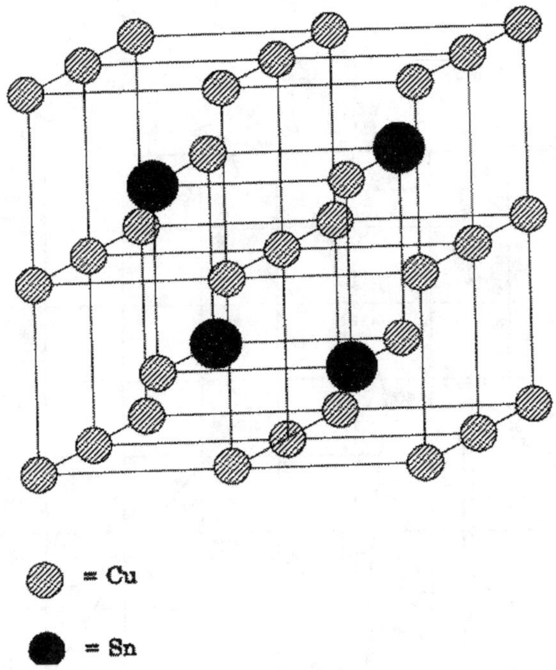

FIGURE 3-4 Crystal structure of the ε-phase (Cu_3Sn). The crystal type is orthorhombic.

ε-phase (Figure 3-6) [Hansen, 1958]. Several studies have been conducted which have consistently observed these phases of Sn/Ni and Ag/Sn [Kay and MacKay, 1973]. These studies have shown that Ag/Sn intermetallic formation is faster, and Sn/Ni intermetallic formation is slower than that of Cu/Sn [Kay and MacKay, 1973; Haimovich, 1989].

Less common examples include indium and bismuth solders, commonly used in low-temperature applications. Lead-indium solders on copper form a copper-indium intermetallic. The phase diagram (Figure 3-7) suggests the β-phase (In_3Sn) may predominate [Marshall and Walter, 1987; Marshall, 1988]. For Sn/In solders, a Cu/Sn/In intermetallic develops [Marshall and Walter, 1987; Marshall, 1988]. Unfortunately a ternary phase diagram of Cu/Sn/In is not available, and the stoichiometry of the intermetallic is not known. Sn/Pb/Bi solders form a Cu/Sn intermetallic (the Bi/Pb forms a separate phase, along with the tin, in the bulk of the solder) [Marshall, 1991). Other studies have been conducted on other substrates, involving brass, bronze, nickel, palladium, cobalt, and iron [Kay and MacKay, 1979; Keller, 1986; Warwick and Muckett, 1983].

All of these intermetallics, like Cu_6Sn_5, cause no reliability problems if properly soldered. However, tin on gold can form excessively brittle Sn/Au interme-

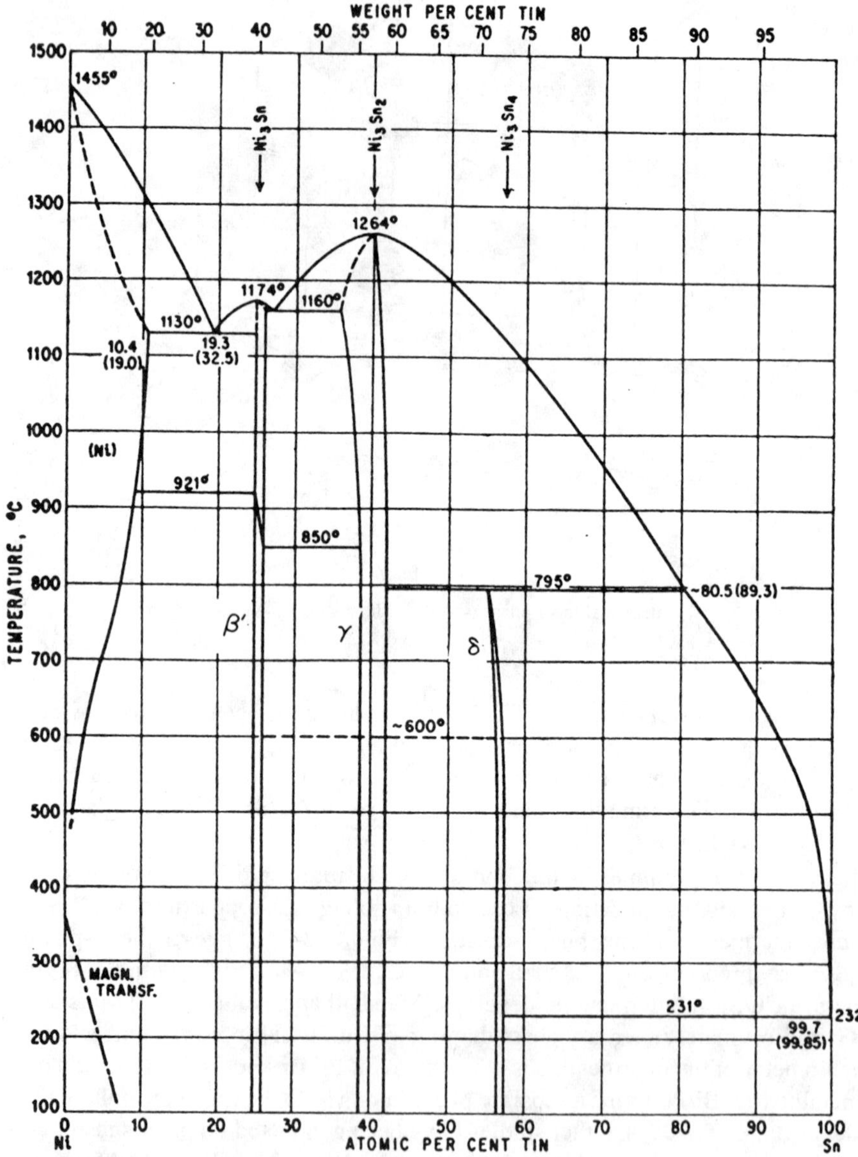

FIGURE 3-5 Phase diagram of nickel/tin (Ni/Sn). This intricate system shows five possible stable phases at room temperature: Ni, β' (Ni_3Sn), γ (Ni_3Sn_2), ∂ (Ni_3Sn_4), and tin. In nickel/tin solder systems, all these phases are observed except the γ (Ni_3Sn_2) phase.

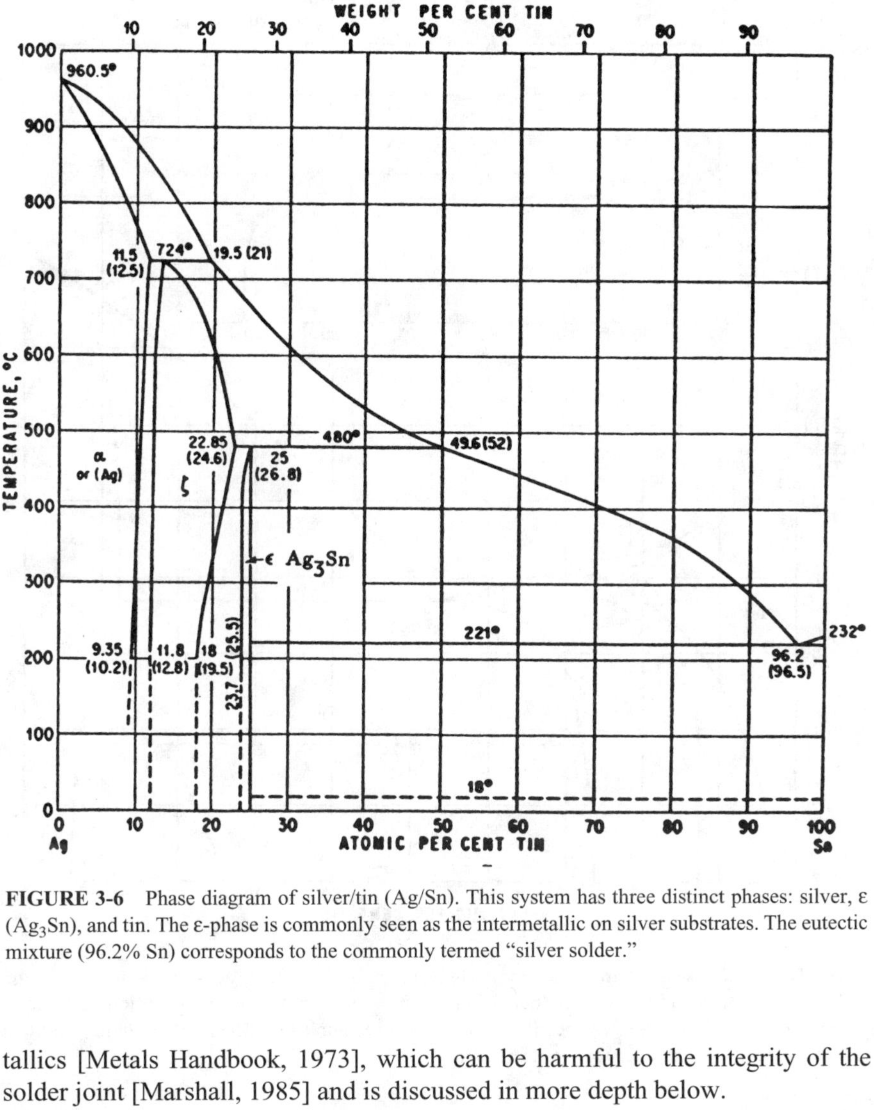

FIGURE 3-6 Phase diagram of silver/tin (Ag/Sn). This system has three distinct phases: silver, ε (Ag_3Sn), and tin. The ε-phase is commonly seen as the intermetallic on silver substrates. The eutectic mixture (96.2% Sn) corresponds to the commonly termed "silver solder."

tallics [Metals Handbook, 1973], which can be harmful to the integrity of the solder joint [Marshall, 1985] and is discussed in more depth below.

3.2.4 Gold Intermetallics

The phase diagram of Au/Sn displays several intermetallics [Hansen, 1958] (Figure 3-8): β = $AuSn_4$, γ = $AuSn_2$, ∂ = AuSn, and ζ = 90% Au. The ζ-phase corresponds roughly to commercial gold solder which performs well when used to solder to gold or gold-plated substrates in the absence of tin. However, when tin is

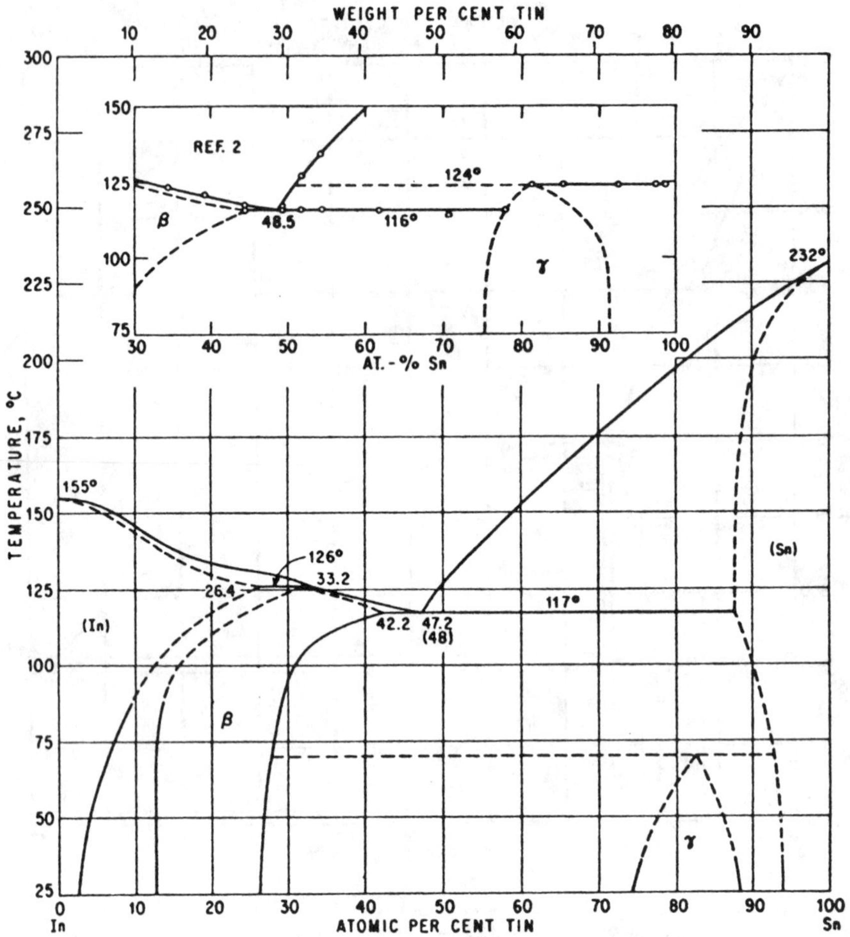

FIGURE 3-7 Phase diagram of indium/tin (In/Sn). Four stable phases occur at room temperature: indium, β (In_3Sn), γ ($InSn_4$), and tin. Indium-tin intermetallics are observed in solder joints, but careful characterization of these intermetallics has not been accomplished. The In/Sn eutectic (close to a 50:50 composition) has a very low melting point of 117°.

present, the other phases β, γ, and ∂ can form, with very damaging effects. The major culprit to solder reliability is the β-phase, $AuSn_4$, which forms in abundance whenever Sn is used in soldering to gold or gold-protected substrates. The $AuSn_4$-phase is extremely brittle, and soldering to gold can be very risky, because the Au/Sn intermetallics can fail (Figure 3-9) [Marshall, 1985]. This failure occurs if the Au/Sn intermetallics are in direct contact with the substrate (Figure 3-10). Upon thermal cycling or other mechanical stresses, failure can occur directly at the inter-

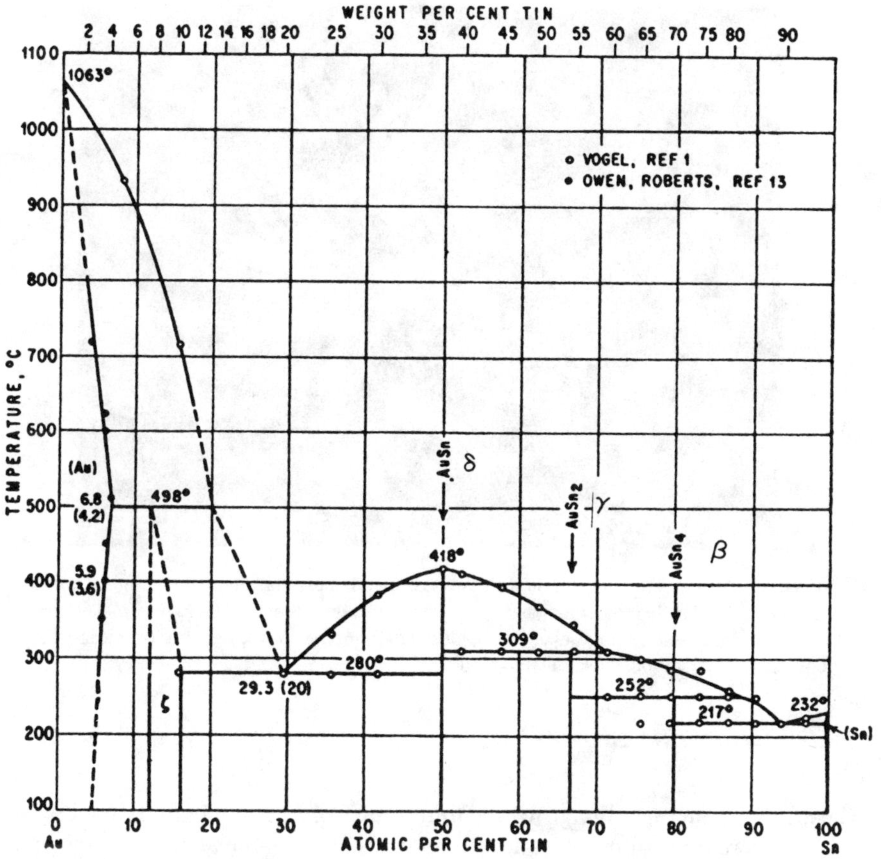

FIGURE 3-8 Phase diagram of gold/tin (Au/Sn). Although not as complex as other phase diagrams, nevertheless a large number of phases (six) can occur at room temperature: gold, ζ (ca. 90 wt.-% Au), β (AuSn), γ ($AuSn_2$), ∂ ($AuSn_4$), and tin. The ζ-phase corresponds to "gold solder," used on gold substrates in the absence of tin solders. The β, γ, and ∂ phases are all observed in solder systems, but the bulk typically consists of ∂ ($AuSn_4$), since an excess of tin is usually present.

metallic, which separates from the substrate [Marshall, 1985]. This separation occurs between the gold and the $AuSn_4$ intermetallic; whether this separation is due to the brittleness of the $AuSn_4$, an inherently weak bond, porosity formation, or Kirkendall porosity, etc., is not clear. In order to produce an acceptable solder joint to a gold surface, the gold intermetallics must be absorbed into the bulk solder (Figure 3-11) such that the interface is Sn/Ni or Sn/Cu intermetallic (Figure 3-12). This can be guaranteed if the original gold layer is no more than 1 μm in thickness [Marshall, 1985].

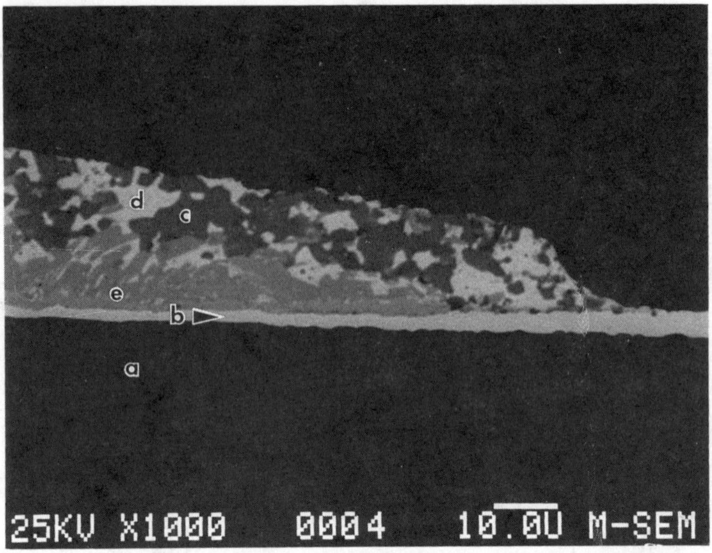

FIGURE 3-9 SEM of a cross section with excess original thickness of gold. The original gold layer was 5 μm thick, and in this solder joint has been reduced to 3 μm. Bottom, dark region: nickel substrate (a). White horizontal strip: gold layer (b). Solder: dark regions, tin (c); light regions, lead (d); gray regions, $AuSn_4$ (e). X1000, backscattered mode.

3.2.5 Intermetallic Growth

The effect of intermetallic growth within solder joints is not entirely understood. While the presence of intermetallic compounds are an indication that a good metallurgical bond has formed, the fact that these compounds are brittle may also make them deleterious to a joint's mechanical integrity. When these compounds form as continuous layers at the solder-substrate interface, the intermetallics can interrupt electrical currents with their high resistivity, effectively isolating the metals that were to be electrically joined [Brothers,1981]. In addition, if these intermetallic compounds become too thick, the reliability of the joint can be jeopardized due to cracking. This can be a problem especially if the joint is exposed to any mechanical forces, such as expansion or contraction of the printed wiring board (PWB) laminate caused by variations in temperature [Woodgate, 1983]. Exposure of these intermetallic compounds to the atmosphere via microtunnels formed through the solder during the tinning or aging process has also been recently postulated as allowing for the environmental degradation of the intermetallics located on the surface of the base metal [Marshall, 1991].

In extreme cases, the intermetallic compound growth reaches the surface of the solderable coating over a significant area of the sample and renders the joint unsolderable with mild fluxes [Brothers, 1981]. Finally, the interdiffusion processes

FIGURE 3-10 SEM of cross section of solder joint with excess original thickness of gold, and reflowed with excess solder in an attempt to scavenge all metallic gold from the substrate. However, large amounts of $AuSn_4$ still adhere to the substrate. Bottom, dark region: nickel substrate (a). White horizontal tongue, leading from right: gold layer (b). Solder: dark regions, tin (c); light regions, lead (d); gray regions, $AuSn_4$ (e). X1000, backscattered mode.

which produce the intermetallic layers can also produce Kirkendall porosity (voids) which can also degrade the mechanical properties of the connection [Romig et al., 1991].

It appears that there is some optimal thickness of these intermetallics, below which a good metallurgical bond has not been achieved, and above which they become a liability to the system. A study by Davis et al. [1982] indicates that solderability can be reduced when intermetallic compound growth between Cu and Sn on Sn/Pb protective coatings reaches a thickness of 2 to 4 μm, especially when there is little remaining coating after aging. At the same time, it is evident that the parameters by which solderability is judged are also improved if a thicker remaining coating is present after high-temperature exposure. The application of a sufficient thickness of solder as a protective coating tends to enhance solderability, even when exposure to elevated temperatures is involved.

An understanding of their phase formation is important because of the important contribution that the various intermetallics make to the stability of the solder connection. The solder/substrate reaction falls into one of two categories: molten solder/substrate reactions and solid state growth of intermetallic compounds [Romig et al., 1991]. During wetting, molten solder comes into contact and reacts with the solid substrate. At this point, two processes occur simultaneously: the base

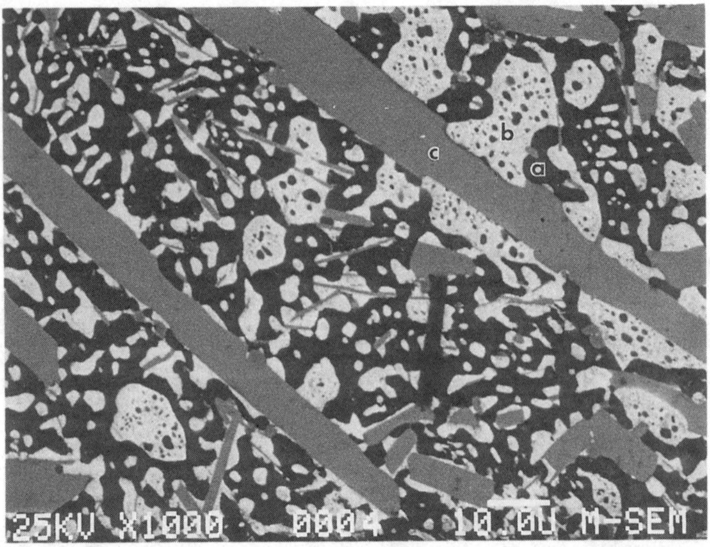

FIGURE 3-11 SEM of a cross section of the bulk of a well-soldered joint to gold-plated substrate. The $AuSn_4$ lamellae have drifted well into the bulk of the solder. Dark regions: tin (a). Light regions: lead (b). Gray regions: $AuSn_4$ (c). X1000, backscattered mode.

metal dissolves into the molten metal, and the active constituent in the solder reacts with the base metal. Both processes will form intermetallic compounds in the interfacial region between the molten solder and the base metal.

The amount of base metal that dissolves into the molten solder is dependent on its solubility in that material, while the intermetallic compound formation is dependent on the solubility of the active element in the base metal [Romig et al., 1991]. Many studies have been published that examine the reaction between the molten solder and substrate [Romig et al., 1991; Okamoto, 1986; Saperstein and Howes, 1968; Howes and Saperstein, 1969; Bader, 1969].

The solid state growth of intermetallic compounds is generally more complex and includes the problem of long term growth of intermetallic layers at the solder/substrate interface during the service life of the connection. The solid state growth of intermetallics can be inhibited either by the use of diffusion barriers or by changing one or more constituents in the solder alloy or the base metal in order to slow growth [Romig et al., 1991].

3.2.6 *Kinetics of Intermetallic Growth*

As described above, tin-based solders on a copper substrate form two Cu/Sn intermetallic compounds with two distinct stoichiometries at room temperature:

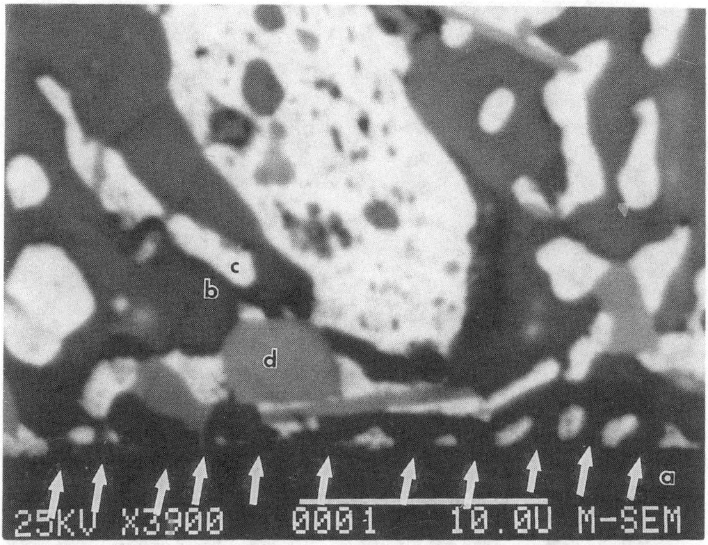

FIGURE 3-12 SEM of a cross section of well-soldered joint to gold-plated substrate, including the interface. The gold-tin intermetallics have drifted far from the interface (see Figure 3-9), leaving the tin to form the normal tin-nickel intermetallic at the interface (arrows). Bottom, horizontal dark area: nickel substrate with nickel-tin intermetallic (a). Solder: dark areas, tin (b); light areas, lead (c); gray areas, $AuSn_4$ (d). X3900, backscattered mode.

Cu_6Sn_5 (η-phase) and Cu_3Sn (ε-phase) [Hansen, 1958]. In a typical manufacturing solder operation, if the solder joint is cross-sectioned by the analytical laboratory and studied by optical microscopy or scanning electron microscopy, only a Cu_6Sn_5 layer is observed, because Cu_6Sn_5 grows much faster than Cu_3Sn in the molten state. In optical microscopy, Cu_6Sn_5 appears as a creamy-white layer, about 1 μm (micrometer) thick, scarcely contrasting visually with the metallic white solder. The Cu_3Sn layer is visible only upon prolonged heating as a pastel blue layer, and can be used as a sleuth for a high-temperature history of a solder joint. Even though only the Cu_6Sn_5 intermetallic is observed in freshly-soldered joints by optical and electron microscopy, at higher magnification some Cu_3Sn can be observed. Recent transmission electron micrographs with greatly increased resolution show that there in fact exists a very thin layer of Cu_3Sn in a newly formed solder joint, a few hundred angstroms thick, Figure 3-13.

Two types of solid state growth are distinguished: linear and parabolic growth kinetics. Linear growth implies that the growth rate is limited by the reaction rate at the growth site (reaction controlled). Parabolic growth implies that growth is limited by bulk diffusion of elements to the reaction interface (diffusion controlled). Systems consisting of Cu substrates in contact with Sn/Pb solder alloys

FIGURE 3-13 TEM of cross section of interface of typical (unaged) solder joint. Four distinct phases can be observed: the tin (top), Cu_6Sn_5 (columns of varying height, about 3 µm in diameter), Cu_3Sn (an irregular band less than 1.0 µm thick), and copper (bottom). All phases were identified using SAD (selected area diffraction). X20,000. (For an SEM of the same solder joint, at much lower magnification, still displaying the same Cu_6Sn5 columns, see Figure 3-30).

are usually approximated by parabolic or subparabolic growth kinetics [Wassink, 1989; Romig et al., 1991; Kay and MacKay, 1976].

Review of Intermetallic Studies

The kinetics of copper substrate/Sn-Pb solder systems have been more widely studied than those of any other system. Zakraysek [1971] examined the morphology and growth of intermetallic phases that formed at elevated temperature between a 95Sn-5Ag solder alloy and tough-pitch copper straps, both in the presence and absence of a statically and dynamically induced stress. Results indicated that the interfacial intermetallics formed upon solidification of the solder indicating a high nucleation and growth rate. Continued exposure to the elevated temperature resulted in thickening of the interface envelope and in the appearance and lengthening of side plates. After additional aging, two distinct phases of the intermetallic were observed (whose compositions were unverified), along with spheroidization of the side plates, resulting in massive intermetallics within the matrix. Total intermetallic thicknesses were measured with values ranging from 0.67 μm for the as-soldered sample to 11.1 μm for the sample aged 42 days at 150°C. The thicknesses and general morphology of the intermetallics in the prestressed samples were virtually identical to those of the unstressed samples, although fractures were observed to occur both in the solder fillet as well as along intermetallic interfaces.

A subsequent study [Zakraysek, 1972] examined the morphology and growth of intermetallic compounds that formed at elevated temperatures between copper and four types of solder: 97.5Sn-2.5Sb, 95Sn-5Ag, and 60Sn-40Pb. The dual-composition intermetallic thickness ranged from 1 μm or less for samples aged at 100°C to 68.8 μm for the 97.5Sn-2.5Sb alloy aged at 200°C for 32 days. The author noted little dependence of growth rates on composition and that the dominating factor was the maximum exposure temperature. He further attributed the growth phenomena encountered to established diffusion theory [Zakraysek, 1972; Purdy, 1972].

Tu [1973] examined interdiffusion and reaction in bimetallic Cu/Sn thin films. Layered polycrystalline Cu/Sn thin films of different layer thicknesses were prepared in vacuum at room temperature. Annealing was then conducted at –2, 20, 60, and 100°C. While Cu_6Sn_5 was detected in all specimens, Cu_3Sn was found only in samples annealed at 60 and 100°C. The author also reported a transformation of the η-phase to η' at 170°C with the η'-phase existent at the lower temperatures. Tin whiskers were observed to grow spontaneously on the free surface of tin in the bimetallic Cu/Sn thin films but not on tin films without the copper underlayer. The driving force for this growth was attributed to density inequalities resulting from interdiffusion and reaction occurring in these films; that is, density of tin is less than Cu_6Sn_5 is less than copper.

Kay and MacKay [1976] studied the growth of intermetallic compounds on common basis materials coated with Sn and Sn/Pb alloys. Investigated were 60Sn-40Pb, 30Sn-70Pb, 40Sn-60Pb, 10Sn-90Pb on copper, brass, and silver substrates heat treated at 20, 70, 100, 135, and 170°C for up to 1700 days. Ag_3Sn was the one layer identified between all coatings and silver. Cu_3Sn and Cu_6Sn_5 were observed for all coatings on copper and at room temperature; the only growing layer was identified as Cu_6Sn_5. Two phases were identified at the solder/brass interface: a ternary Cu/Sn/Zn alloy grew in contact with the brass base metal while a compound corresponding closely to Cu_6Sn_5 (containing some zinc) was observed at the solder interface. The major result was that at any temperature, comparing coatings of the same composition on the various substrates, the growth rate for the total compound thickness was the fastest on silver, followed by copper, and then brass. The 60Sn-40Pb coating showed the highest growth rate on all base materials.

Kay and MacKay [1979] also undertook a study of the effect of barrier layers on diffusion of active species and growth of intermetallic compounds. Copper and brass coupons were plated first with eight separate metals or alloys (Sn/Ni, Cu_6Sn_5, lead, silver, nickel, cobalt, iron, or Ni/Fe) that were to act as diffusion barriers and then with an excess thickness of Sn. Samples were then annealed at 170°C. The lead and Cu_6Sn_5 barriers also allowed for the formation of Cu/Sn intermetallics. After formation of the Ag_3Sn intermetallic between the silver barriers and tin coating, the two Cu/Sn intermetallic layers formed at the copper and brass interfaces. The nickel and Sn/Ni barriers formed $NiSn_3$ then Ni_3Sn_4 compounds, and then again the tin diffused through these barriers to form the Cu/Sn intermetallics. The Ni/Fe barrier formed a series of complex compounds before the Cu/Sn intermetallics formed adjacent to the base metal. The iron barrier layer gave the best result in the trials; $FeSn_2$ formed extremely slowly, not using all of the 5 μm coat even after 400 days at 170°C.

A series of experiments involving Sn/Pb and Sn/Pb/Sb solders on copper and gold substrates was undertaken to determine the relationships of these compounds to the soldering process [Wassink, 1989]. Gold-plated substrates for soldering applications presented special problems involving the phenomena of dewetting and porosity. Au/Sn intermetallics were found to be particularly susceptible to separation from their substrates because of their brittle character. There was also evidence that separation of solder from gold-plated substrates occurred spontaneously upon cooling from reflow temperatures. This was attributed to large differences in the thermal expansion properties of the compound and substrate. The interaction of tin with impurities such as iron and arsenic was also investigated and found to result in intermetallic compound formation.

Another study [Davis, Warwick, and Kay, 1983] demonstrated that solderability was reduced when compound growth between copper and tin or Sn/Pb protective coatings reached a thickness of only 2-4 μm as determined by area-of-spread and surface tension balance tests. It was further discovered that solderability could be improved if a thicker remaining coating was present after high temperature exposure.

Another study [Tu et al., 1982] investigated the kinetics of interfacial reactions in bimetallic Cu/Sn thin films. Thin films of the metals were again deposited onto quartz disks in high vacuum at room temperature. Cu_6Sn_5 was found to form immediately. No Cu_3Sn was detected at room temperature. In room-temperature growth of Cu_6Sn_5, copper atoms were found to be the dominant diffusing species based on displacement of tungsten diffusion markers. Finally, in the growth of

Cu_3Sn between copper and Cu_6Sn_5 films in the temperature range 115 to 150°C, a parabolic reduction rate of Cu_6Sn_5 was observed with an activation energy of 0.99 eV.

Electroplated tin and Sn/Pb coatings frequently undergo a reflowing process in electronics applications before soldering operations are performed. Davis, Warwick, and Muckett [1983] investigated what effect the reflowing process has on intermetallic growth and solderability. Coatings of tin and Sn/Pb were electroplated on sheets of copper, followed by a hot oil reflow. Solderability was assessed by area-of-spread and wetting balance methods. Cu_6Sn_5, but not Cu_3Sn, was found in unaged samples. Cu_3Sn appeared after thermal aging at 135°C. After one day of aging, thicknesses of the two phases were found to be comparable to those of nonreflowed samples. The growth of Cu/Sn intermetallic compounds caused a deterioration in the solderability of tin and 60Sn-40Pb on copper for reflowed coatings only when a significant portion of the coating had been consumed. This was in contrast to a previous report [Davis, Warwick, and Kay, 1982] that the presence of 2-4 μm of intermetallic reduced solderability, in nonreflowed samples, regardless of the overlying residual coating thickness.

Attempts to control intermetallic compound growth have been explored. Substrate conditions (hardness, rolled surfaces) were found [Warwick and Muckett, 1983] to involve only second order effects on growth and therefore could not be used to reduce it. At a storage temperature of 170°C, Fe was the only barrier layer that displayed an acceptably low reaction rate with Sn, in agreement with previous results [Kay and MacKay, 1979], and it was found that an Fe barrier of 2 μm would greatly increase the lifetime of a solderable coating.

An experimental procedure was developed [Walker et al., 1984] whereby a chip soldered to a substrate was cross-sectioned in order to investigate intermetallic compound formation *in situ.* It was found that some metals, like gold and palladium, disperse immediately into the solder while others, like nickel and copper tend to form continuous layers of intermetallics at interfaces for normal joining cycles. Continuous layers of intermetallics tended to give higher stress regions in the joint. This stress arose from coefficient of thermal expansion mismatch and/or volume changes during intermetallic formation.

In another novel experimental procedure [DeHaven, 1985], a method was detailed to study the reaction occurring at the molten solder/metal interface during the soldering operation. The method used high temperature x-ray diffraction to study the reaction of copper and nickel with 60Sn-40Pb solders. The most obvious result was the much slower reaction rate of nickel as compared to that of copper. In addition, a sample quenched after one minute was found to contain only Cu_6Sn_5, but in a sample quenched after 5 minutes, both Cu_6Sn_5 and Cu_3Sn were present.

The effects of single crystal copper orientation on the growth rates of Cu_3Sn and Cu_6Sn_5 intermetallic compounds and the influence of the growth of these compounds on the tensile strength of butt joints was examined [Dunn et al., 1985].

The activation energy for growth of Cu_3Sn on polycrystalline copper was found to be 1.27 eV and 0.47 eV for Cu_6Sn_5. Diffusion couples showed that Cu_3Sn and Cu_6Sn_5 had lower activation energies for growth on the copper (110) face than on either the copper (100) or the copper (111) faces. This effect of copper orientation on growth was not explained. Again, intermetallic growth appeared to be diffusion controlled.

Finally, a study of Sn/Pb solder on a copper substrate [Dirnfeld and Ramon, 1990] revealed typical Cu/Sn intermetallic conclusions: two distinct phases were identified, time-at-temperature dependence of layer thickness was found to obey $t^{1/2}$ kinetics, and very thin and relatively thick intermetallic compound layers were associated with mechanically weaker joints. In addition, crystallographic morphologies of the intermetallic compounds were identified by electron diffraction analysis using TEM.

3.2.7 Physical Properties of Intermetallics

The physical properties of intermetallics contrast significantly with those of the solder or the base metal—intermetallics are less ductile, and less thermally and electrically conductive. Physical properties of Cu/Sn and Cu/Ni intermetallics are included in Table 3-1 [Frederiske et al., 1991; Fields et al., 1991]. The intermetallic material used to generate this data was produced by rapid solidification by supersonic, inert gas, metal atomization, and resulted in single-phase, fine-grained material. Other literature values for intermetallics are in general agreement—elastic properties [Cabaret et al., 1949; Subrahmanyan, 1972; Ostrovskaya et al., 1985], microhardness [Olsen et al., 1975; Wassink, 1989; Revay, 1977; Dunn, 1989], toughness [Hoyt, 1987], and coefficient of expansion [Reichenecker,

TABLE 3.1 Room Temperature Physical Properties of Intermetallics

Property	Cu_6Sn_5	Cu_3Sn	Ni_3Sn_4
Vickers Hardness (kg/mm,^{2})	378±55	343±47	365±7
Toughness MPa•m$^{-1/2}$	1.4±0.3	1.7±0.3	1.2±0.1
Youngs Modulus GPa	85.56±1/65	108.3±4.4	133.3±5.6
Shear Modulus GPa	50.21	42.41	45.0
Thermal Expansion x10^{-6}/°C	16.3±.03	19.0±0.3	13.7±0.3
Thermal Diffusivity cm^2/sec	0.145±0.015	0.24±0.024	0.083±0.008
Heat Capacity J/gm/deg	0.286±0.012	0.326±0.012	0.272±0.012
Resistivity μohm-cm	17.5±0.1	8.93±0.02	28.5±0.1
Density gm/cc	8.28±0.02	8.9±0.02	8.65±0.02
Thermal Conductivity watt/cm-deg	0.341±0.051	0.704±0.098	0.196±0.019

1981]. However, in these previous examples, generally bulk samples were obtained by casting and annealing, probably producing a coarse, multiphase material, sometimes with defects. Hence, the data of Table 3-1 [Frederiske et al., 1991; Fields et al., 1991] is considered more reliable.

To gain an appreciation of the significance of the values of Table 3-1, selected properties are presented in Table 3-2 for copper, tin, and nickel [Lynch, 1989]. The hardness is greater for the intermetallics, indicating their greater brittleness. With these high values of hardness for the intermetallics, no plastic deformation should be expected under normal levels of stress in solder joints. The elastic modulus is likewise greater for the intermetallics, giving rise to a lower ductility.

The thermal coefficient of expansion for the intermetallics is similar to that of metals, and therefore it would seem that the intermetallics would expand much like their associated substrate material.

Thermal and electrical conductivities are generally lower for the intermetallics, but for a thin intermetallic layer should pose minimal problems.

All these properties of intermetallics—mechanical, thermal, and electrical—should give rise to no significant difficulties in a solder joint wherein the thickness of intermetallic is low compared with that of the solder. It is possible, however, as solder joints miniaturize, that these physical properties could impact the performance of circuitry.

3.3 EFFECT OF INTERMETALLICS ON THE INTEGRITY OF SOLDER JOINTS

3.3.1 *Solderability Issues*

Normally, the intermetallic does not play a role in the failure of a solder joint [Morris et al., 1991]. However, if a solder joint is improperly soldered because of poor wetting, then failure can occur directly at the interface. That this failure

TABLE 3.2 Room Temperature Physical Properties of Elements

Property	*Cu*	*Sn*	*Ni*
Vickers Hardness Kg/mm^2	30	100	15
Youngs Modulus GPa	117	41	213
Thermal Expansion x10^{-6}/°C	17.1	23 (solder 20-29)	12.9
Heat Capacity J/gm/deg	0.385	0.227	0.439
Resistivity μohm-cm	1.7	11.5	7.8
Density gm/cc	8.9	7.3	8.9
Thermal Conductivity watt/cm-deg	3.98	0.67	0.905

occurs at the intermetallic, and not in the bulk solder, can readily be proven by chemical analysis of the surface of the fractured joint by energy dispersive x-ray analysis (EDX) [Heinrich and Newbury, 1986]. Poor solderability is usually due to contamination or corrosion of the substrate before soldering. Sometimes this defect is caused by faulty soldering material, e.g. aged solder paste. Particularly harmful are corroded intermetallics [Marshall, 1991]—principally oxides—which are originally present during the soldering process [Boettinger, Handwerker, and Smith, 1991]; oxidized Cu_6Sn_5 has been shown to experience serious deterioration of its wettability much more rapidly than copper itself [Boettinger, Handwerker, and Smith, 1991; Marshall, 1991]. This problem frequently occurs when an insufficient layer of protective tin has been coated over the base metal, through which intermetallics can grow and eventually be exposed on the surface of the tin [Marshall, 1991].

3.3.2 Fatigue Failures of Solders—Bulk

The intermetallic is more brittle than the bulk solder, and one might predict that failure therefore would occur at this intermetallic. However, low-cycle thermomechanical fatigue and ultimate failure of a properly soldered soft solder joint occurs through the bulk of the solder [Morris et al., 1991]. Even when visually it may appear that failure occurs at the interface in a cross-sectional analysis, closer inspection shows that the separation in fact occurs within the bulk of the solder. Frequently a high magnification scanning electron micrograph [Verhoeven, 1986] is necessary to see that in fact the failure occurs in the solder near the interface (Figure 3-14). This failure in proximity to the interface may be due to stress concentrations at or near the interface between solder and the intermetallic, arising from plastic inhomogeneity between solder and the substrate metal [Fields et al., 1991].

Failure occurs in the bulk solder because the deformation during the fatiguing process occurs within the bulk solder, via the "heterogeneous coarsening" mechanism [Morris et al., 1991]. Crack propagation occurs close to the interface for two reasons: (1) the stress concentrates there, and (2) tin-depletion during the soldering process weakens this area and subjects it to greater strain during the fatiguing process.

Examples of cases where failure occurs through the interfacial intermetallic of solder joints are discussed below.

3.3.3 Fracture of Aged Solder Joints—Through Intermetallic

Failure can occur at the intermetallic layer in the fracture of solder joints subjected to extended periods of high temperature. For solder joints held at 170° C for thirty

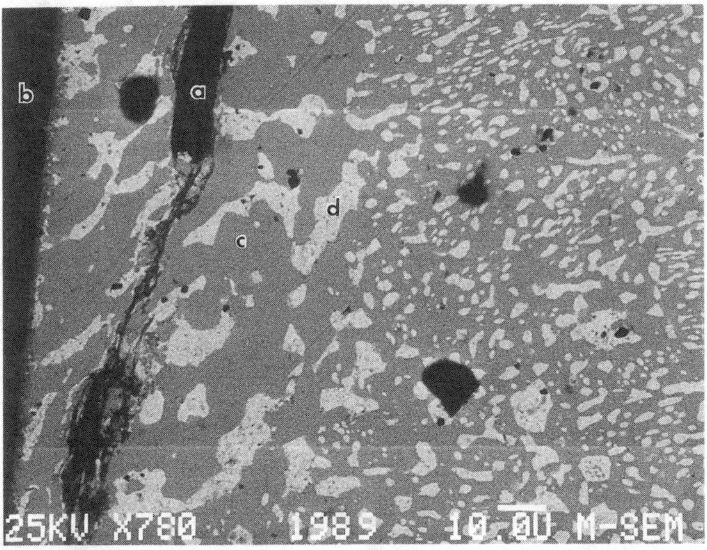

FIGURE 3-14 SEM of fatigued 60Sn-40Pb solder joint. The area of deformation, and ultimate failure, is parallel to the solder joint and is observed here as the vertical crack (left) (a) to the right of the copper (dark; extreme left) (b). Solder: dark areas, tin (c); light areas, lead (d). Sometimes it is impossible to see by optical microscopy what is apparent here: the crack is *close*, but not *directly on*, the interface; hence the failure is through the bulk solder. X780, backscattered mode.

days, the intermetallic layer grows to a thickness of almost 20 μm, and the fracture toughness decreases almost fourfold [Stromswold, 1992]. Concurrently, the failure mechanism evolves from bulk failure (Figure 3-15) to intermetallic failure (Figure 3-16).

3.3.4 Fracture of Hard Solders—Through Intermetallic

Silver solder (95Sn-5Ag) can experience failure at the interface (Figure 3-17), because this harder solder resists deformation and thus avoids the usual low-cycle thermomechanical fatigue mechanisms [Marshall, 1985]. Furthermore, the failure mechanism is different—tin depletion is not a factor, and heterogenous coarsening/strain mechanisms are not operative as for Sn/Pb solders [Stromswold].

3.3.5 Bridging of Intermetallic Across Solder Joint

If intermetallic continues to grow during the high temperature treatment of a solder joint, then it may eventually bridge across a solder joint. In this case, deforma-

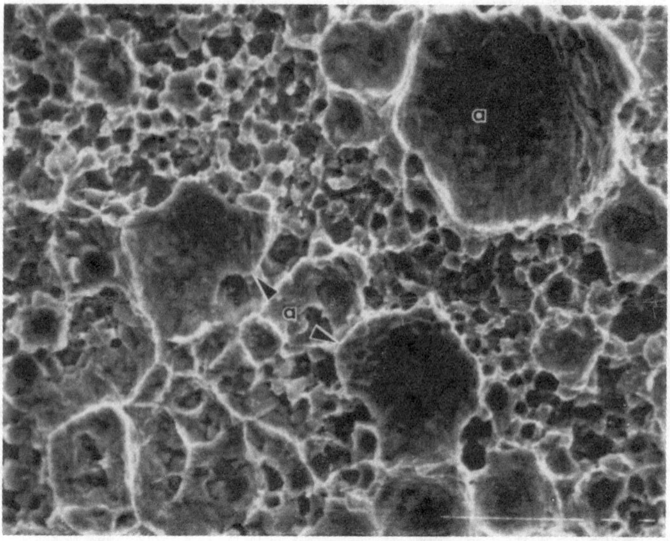

FIGURE 3-15 SEM of fracture surface of unaged solder joint. The failure is through the bulk of the solder: only the solder can be seen, which is adhering to the intermetallic layer below. X350, secondary mode. The "craters" (a) are voids.

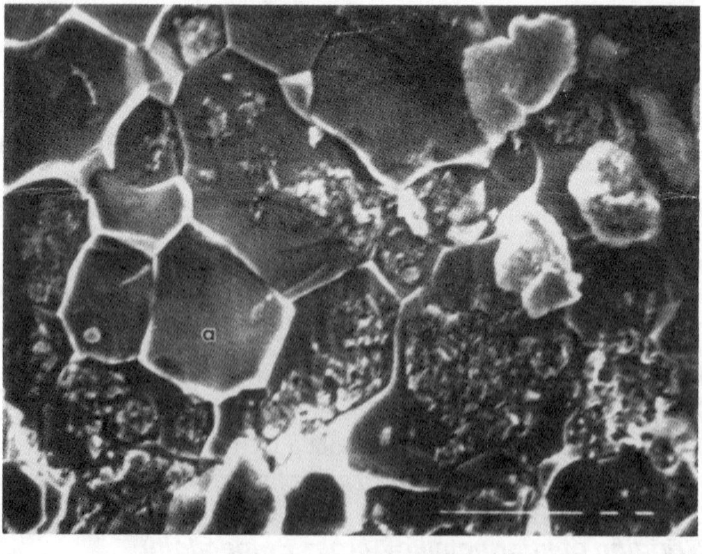

FIGURE 3-16 SEM of fracture surface of solder joint which has been aged at 170° C for 30 days. The failure is now at the intermetallic: the hexagonal cross sections of the Cu_6Sn_5 rods (a) are clearly discernible. X350, secondary mode.

tion must be accommodated by the intermetallic. Such situations are unusual, because the Cu_6Sn_5 hexagonal rods that protrude into the molten solder tend to break off and appear as intermetallic particles in the bulk of the solder joint (Figure 3-18) [Morris et al., 1991]. However, as the thickness of solder joints is reduced, it is possible that this can become more of a consideration as solder joints are continually miniaturized because any failure must be through the intermetallic.

3.3.6 Conclusions: Failure Through the Intermetallic

Modeling of the failure of solder joints must include the intermetallic if:

- The joint is not properly soldered (i.e., good wettability is not ensured and failure can occur directly at the weakened interface)
- The solder is a hard solder or does not creep (e.g., 95% Sn-5% Ag)
- The solder joint has been subjected to heating for extended periods of time (e.g., 100° C for several months or 150° C for several days) or
- The cross section of the solder is thin (less than 10 μm), so that the intermetallic constitutes a significant portion of the cross section of the solder joint.

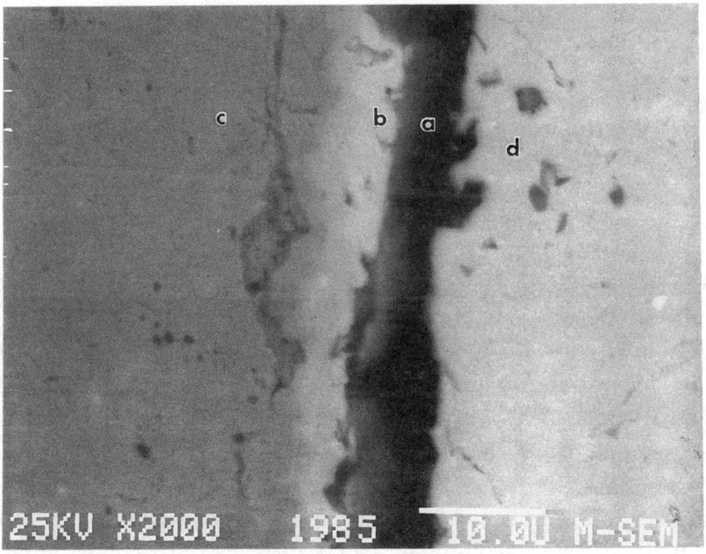

FIGURE 3-17 SEM of failed 95Sn-5Ag solder joint. The failure (crack = a) for hard solders such as this occurs at the intermetallic itself (b). Left: copper pin (c). Right: solder (d). X2000, backscattered mode.

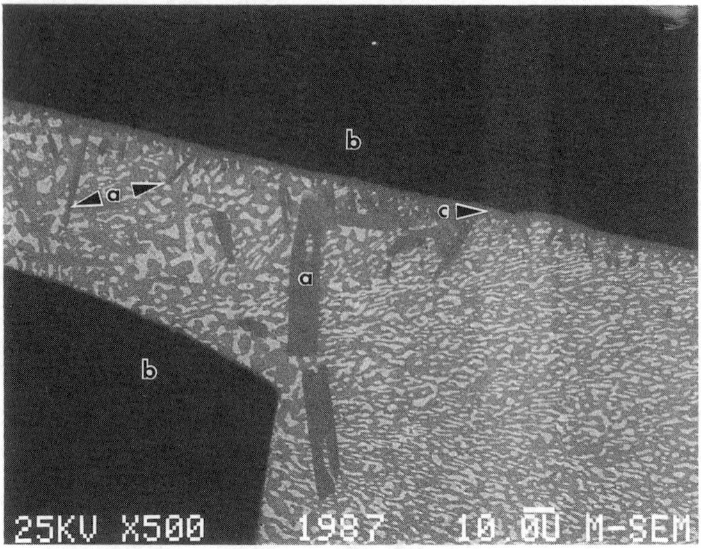

FIGURE 3-18 SEM of cross section of solder joint, excessively heated; protruding rods of Cu_6Sn_5 (a) have broken off and are drifting in the bulk solder. Copper substrate = b, top and bottom. Cu_6Sn_5 intermetallic (c) can also be seen next to substrate. X500, backscattered mode.

3.4 RECENT ADVANCES IN MICROSTRUCTURAL ANALYSIS OF INTERMETALLICS

3.4.1 Transmission Electron Microscopy Foil Preparation

To observe thin intermetallic layers, transmission electron microscopy offers higher resolution than SEM [Eddington, 1976]. Unfortunately, the preparation of solder specimens is "perhaps the most difficult in the entire realm of TEM in materials science" [Romig et al., 1991] because an electron beam must pass through a thin (less than 1000Å thick) sample to form an image on a fluorescent screen or a photographic negative [Romig et al., 1991]. Nevertheless, a recent methodology has been developed which is described here [Jacobs et al., 1993].

Ultramicrotomy or diamond-knife sectioning has been used for a variety of materials [Malis and Steele, 1992] and can be used successfully to produce ultrathin (30 nm or less) foils of copper-solder joints for TEM analysis of the metallurgical phases in the sample [Jacobs et al., 1993]. Copper-solder-copper joints are made by applying 0.1 g of solder paste to a 10.0 x 35.0 x 0.5 mm copper strap. A second strap is placed on top of the solder paste and the sandwich is positioned on top of a ceramic hot plate, heated to 250°C, and left on the hot plate until the solder remelts. The resulting copper-solder-copper joint (Figure 3-19) is removed from the hot plate, cooled to room temperature and rinsed in isopropyl alcohol.

The copper-solder joint requires support during sectioning with the diamond knife to prevent scrolling. This support, portrayed in Figure 3-20, is provided by first cleaning the sample in acetone to remove all traces of oil which would prevent the resin from adhering to the metal surface. The joints are then coated with a thin layer of carbon in a vacuum evaporator, which improves the adherence of the resin to the metal surface. At this point, the carbon-coated sample is dipped in liquid cyanoacrylate ester adhesive and this polymer coat is allowed to dry. The sample is set in a plastic mold and liquid epoxy resin is poured into the mold to cover the sample. The resin is polymerized by placing the mold in an oven at 30°C overnight.

The sample, embedded in the polymerized epoxy resin block, is pretrimmed with a razor blade to reduce the face of the sample for sectioning to less than 0.5 x 0.5 mm square (Figure 3-21). The first trimming cuts are made at the tip of the

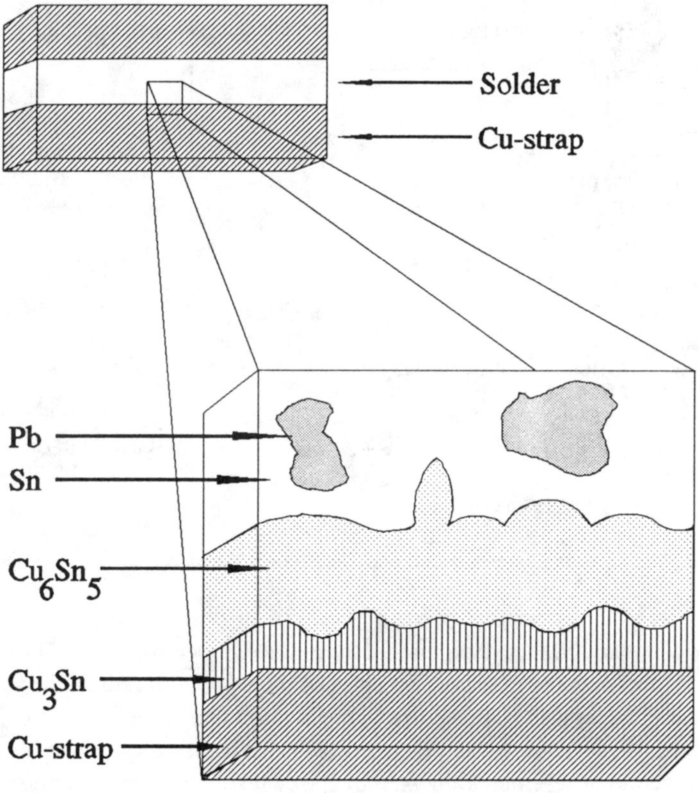

FIGURE 3-19 Copper-solder-copper joint provided for ultramicrotomy, to be used in TEM and SEM sample preparation.

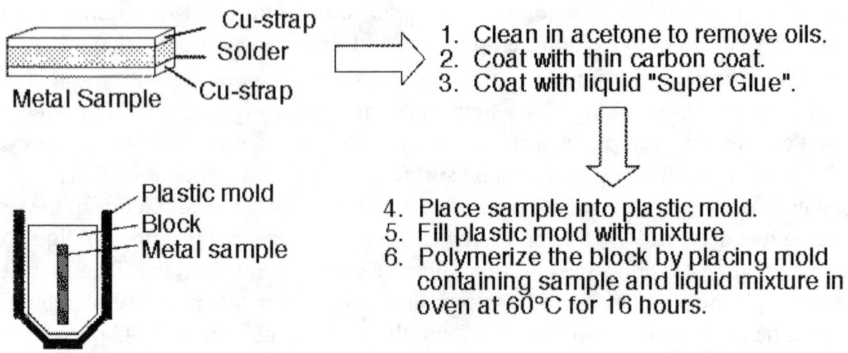

FIGURE 3-20 Sample preparation for ultramicrotomy.

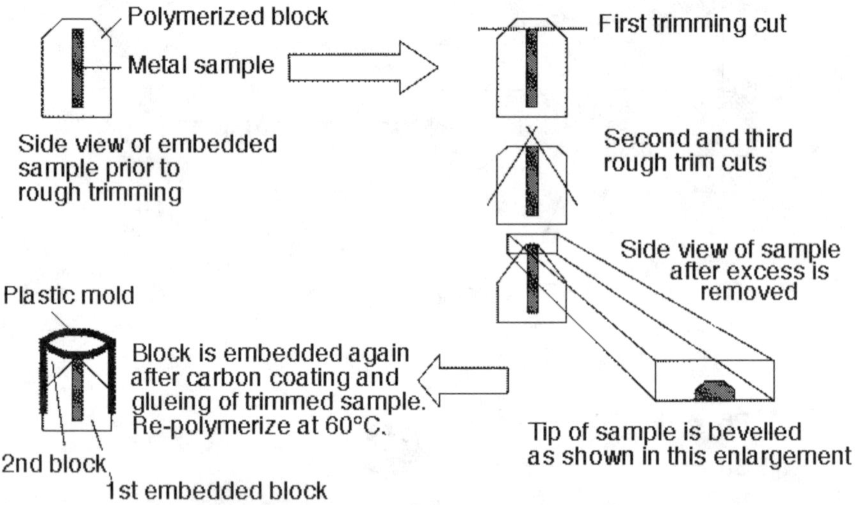

FIGURE 3-21 Rough trimming and second embedding of metal sample.

sample to make a flat surface that includes one copper-solder interface. The second and third cuts are made on the sides of the sample to reduce the surface area of the sample that is to be presented to the diamond knife during sectioning. These pretrimming cuts remove polymerized resin from the sides of the block and require a second or re-embedding step in epoxy resin and polymerizing the resin in an oven at 30°C overnight. The re-embedded joint is trimmed carefully with a razor blade to remove the excess resin at the tip and to expose the sample face for ultramicrotomy.

Figure 3-22 summarizes the sectioning process. The sample is mounted in the ultramicrotome with the sample surface containing one copper-solder interface perpendicular to the diamond knife edge and is oriented with the interface at a 90° angle to the knife edge. The knife has a diamond blade mounted in the knife holder at a 35° angle and is fastened in the ultramicrotome knife stage at a 6° tilt angle and the knife "boat" is filled with distilled water. Thin sections are removed from the sample face by the diamond knife as the sample passes the knife edge. These sections float on the surface of the water in the knife boat. Floating sections are removed from the water surface with a thin metal loop which collects a drop of water and holds the sections on the surface of the drop. The drop of water is placed on the surface of a copper grid and the excess water wicked away with a piece of filter paper. The sections are left on the surface of the grid and allowed to air dry.

All the metallurgical phases of the sample (Figure 3-23) can be visualized in one section and can be identified by selected area diffraction (SAD) [Eddington, 1976]. The Cu_3Sn intermetallic is identified as the intermetallic layer next to the copper surface; it is seen as slightly less transparent to the electron beam and has a coarser grain structure than the Cu_6Sn_5 intermetallic above the Cu_3Sn intermetallic layer. The Cu_6Sn_5 intermetallic exists between the solder matrix and the Cu_3Sn intermetallic and has a columnar grain structure. The elongation of the grain structure in the Cu_6Sn_5 intermetallic is an example of rebound compression and is a result of mechanical damage introduced in the section by the diamond

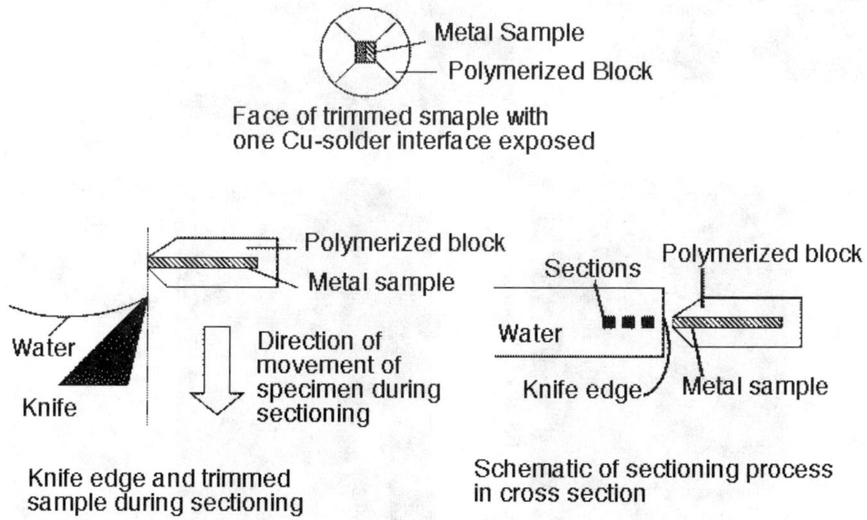

FIGURE 3-22 Ultramicrotomy of embedded metal sample.

(1) Cu, (2) Pb-rich phase, (3) Sn-rich phase, (4) Cu_6Sn_5 and (5) Cu_3Sn.

FIGURE 3-23 Identification of metallurgical phases in 63/37 Sn/Pb solder with SAD.

FIGURE 3-24 TEM (transmission electron micrograph) of copper-solder interface after aging at 110° C for 64 days. X19,000. a = Cu; b = Cu_3Sn; c = Cu_6Sn_5; d = Pb; e = Sn.

knife during ultramicrotomy. Tin-rich phases in the solder matrix surround lighter lead-rich phases. After aging at 110°C for 64 days, the grain structure of the Cu_3Sn intermetallic layer (Figure 3-24) appears to coarsen when compared to the unaged sample (Figure 3-25) and the distinction in grain structure between the two intermetallic layers is more apparent.

The increase in size of the lead-rich regions is seen as the most dramatic effect of thermal aging of the solder matrix. The lead-rich regions are smaller in unaged samples (Figure 3-26) and appear as holes in the Sn-rich region. After aging at

FIGURE 3-25 TEM of copper solder matrix from unaged sample (enlarged from Fig. 3-13) X29,000. a = Cu, b = Cu_3Sn, c = Cu_6Sn_5, d = Sn.

110°C for 64 days, they appear much larger and include small islands of Sn in the Pb-phase (Figure 3-27).

Using this technique the metallurgical phases can be identified, and changes in the microstructure of copper-solder joints can be characterized as a function of thermal aging.

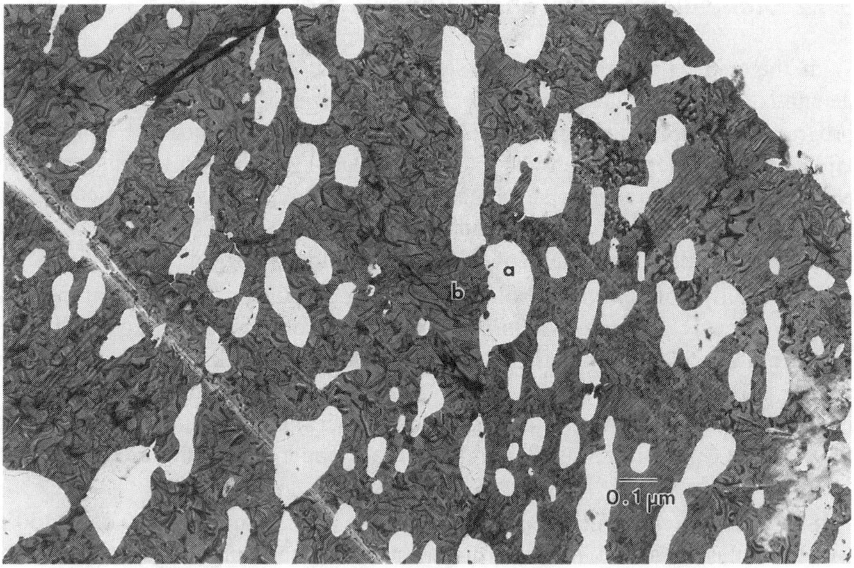

FIGURE 3-26 TEM of Sn-Pb solder matrix from an unaged sample. X29,000. a = Pb; b = Sn.

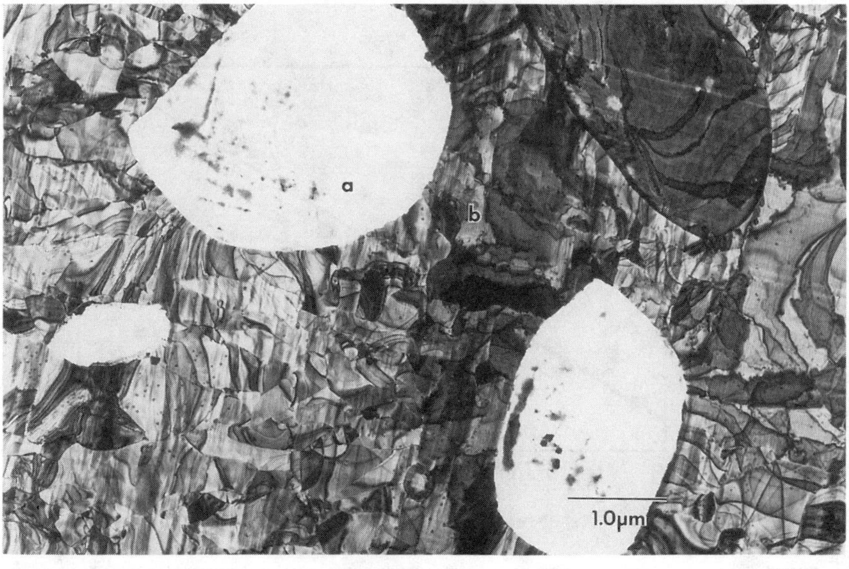

FIGURE 3-27 TEM of Sn-Pb matrix after aging at 110° C for 64 days. X10,000. a = Pb; b = Sn.

3.4.2 Activation Energy and Composite Solders

With the increasing trend toward surface mounted devices, greater structural demands are being placed on solder as a mechanical bond. As miniaturization progresses, the role of the intermetallic in a solder joint may become more important as the thickness of the bulk solder is reduced. In an effort to understand the role of various extraneous substances upon intermetallic growth, metallic and intermetallic particles have been added to eutectic 63Sn-37Pb solder to create "composite solders" that may have the effect of improving joint reliability.

The creation of composite solder by mixing particles of foreign materials has been previously attempted, using copper-coated carbon fibers [Ho and Chung, 1990] and Cu_6Sn_5 intermetallic [Frear, 1987]. For this work, Cu_6Sn_5, Cu_3Sn, and copper particles were added to eutectic Sn/Pb solder [Marshall et al., 1991] and exposed to elevated temperatures for extended periods of time in order to monitor Cu_6Sn_5 and Cu_3Sn intermetallic growth at the copper/solder interface [Pinizzotto et al., 1992].

Samples were annealed for 0 to 64 days at temperatures of 110°C to 160°C. SEM samples were prepared according to standard metallurgical techniques [Marshall, 1991]. Cu_6Sn_5 and Cu_3Sn intermetallic thicknesses were measured at each time and temperature and then plots of each layer thickness vs. the square root of anneal time were obtained. The good linear correlation of the results indicates that the formation of these intermetallics is a diffusion controlled process. Figure 3-28 presents a typical plot [Pinizzotto, 1993].

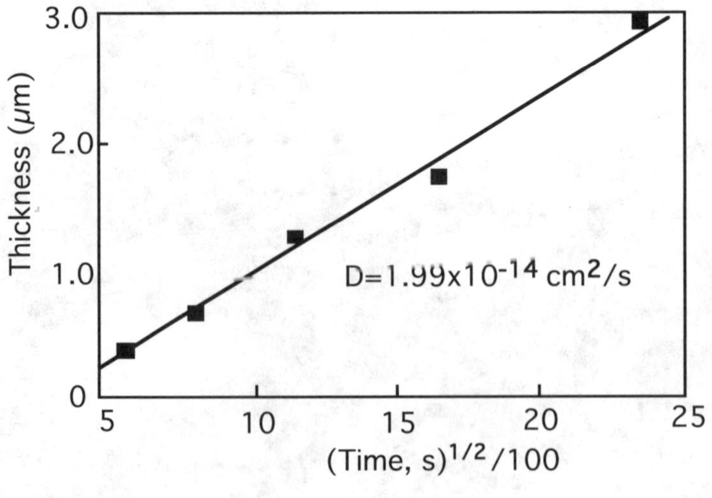

FIGURE 3-28 Diffusion coefficient value for Cu_3Sn intermetallic layer in 63Sn-37Pb solder aged at 150° C for a maximum of 64 days, derived from equation (3-1).

Diffusion coefficients were then calculated using the simple growth model [Wassink, 1989]:

$$d = \sqrt{Dt} \tag{3.1}$$

where: d = layer thickness, D = diffusion coefficient, and t = time

Plots of ln D vs. 1/T were obtained, and activation energies for the formation of the Cu_6Sn_5 and Cu_3Sn intermetallic layers were then calculated from least squared fits of the slopes using the Arrhenius relationship [Wassink, 1989]:

$$D = D_0 e^{-\frac{Q}{kT}} \tag{3.2}$$

where: D = diffusion coefficient, D_o = diffusion constant, Q = activation energy, k = Boltzmann constant, and T = absolute temperature.

Figure 3-29 shows an example of derivation from equation 3-2 of activation energies. Table 3-3 is a summary of the results.

It is apparent that the particle additions have a dramatic effect on the growth patterns of the intermetallic layers. The activation energy for formation of Cu_6Sn_5 (η-phase) in solder is 0.80eV and 1.69 for Cu_3Sn (ε-phase) growth. The addition of 20 wt.% Cu_6Sn_5 increases the energy requirement for the η-phase to 1.18eV

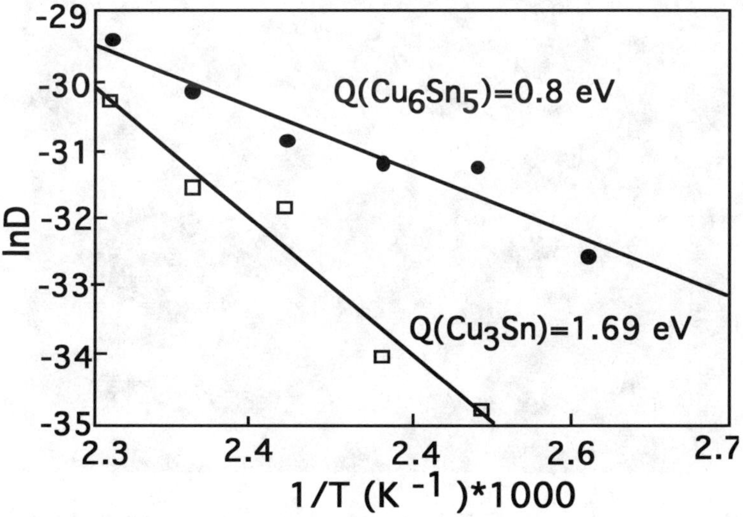

FIGURE 3-29 Example of derivation from equation (3-2) of activation energies—in this case, for Cu_6Sn_5 and Cu_3Sn in 63Sn-37Pb solder.

TABLE 3.3 Activation Energies Q of Composite Solders in 63/37 Sn/Pb [Pinizzotto, 1993]

Composite Solder	*Q, Cu_6Sn_5, eV*	*Q, Cu_3Sn, eV*
No Addition	0.80	1.69
20 wt. % Cu_6Sn_5	1.18	0.95
20 wt. % Cu_3Sn	1.42	0.80
7.6 wt. % Cu	0.96	1.15
4 wt. %Ni	2.17	very large

while reducing that for ε-phase formation to 0.95. The addition of 20 wt.% Cu_3Sn particles has an even more dramatic effect: activation energy for η-phase formation is further increased to 1.42eV while reducing the energy requirement for ε-phase formation to 0.80. Cu additions have a similar but less pronounced effect: 0.96 eV for η-phase formation and 1.15 for ε-phase formation.

Figures 3-30 and 3-31 are backscattered SEM images of the η and ε-phases in unaged and aged (64 days at 150°C) eutectic Sn/Pb solder. Several observations can be made. The first is the coarsening of the lamellar substrate with time. Also noticeable is the increased thickness and smoothness of the η-phase in the 64 day

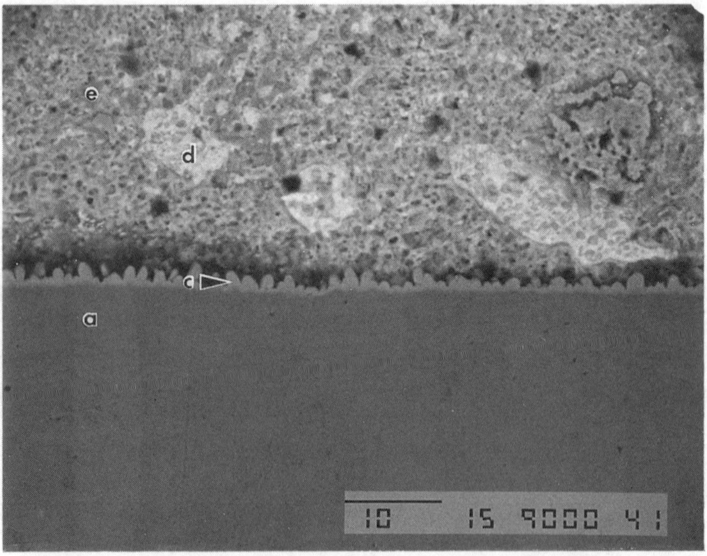

FIGURE 3-30 SEM of interface of unaged 63Sn-37Pb solder. Phase identification: a = Cu; b = Cu_3Sn; c = Cu_6Sn_5; d = Pb; e = Sn. X1500, backscattered mode. (For a TEM of the same solder joint at X20,000, see Figure 3-13).

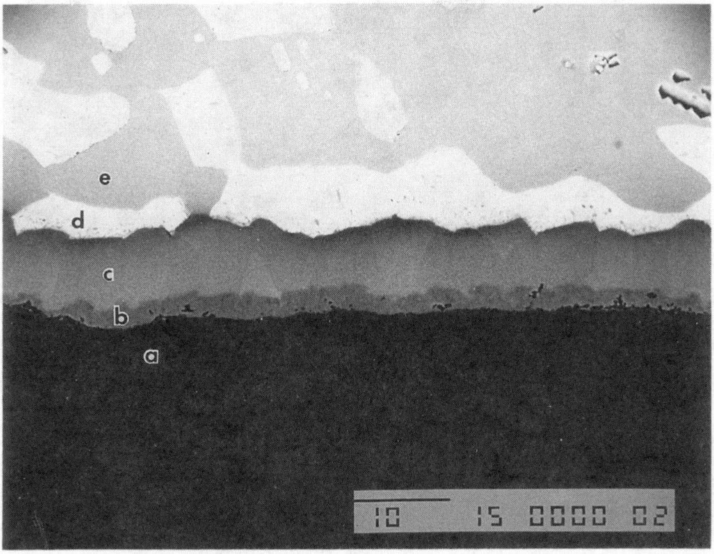

FIGURE 3-31 SEM of interface of 63Sn-37Pb solder, aged at 150° C for 64 days. Phase identification: a = Cu; b = Cu_3Sn; c = Cu_6Sn_5; d = Pb; e = Sn. X1500, backscattered mode.

sample. The ε-phase is also noticeable in the aged sample but not in the unaged sample.

Figures 3-32 and 3-33 are images of 20 wt.% Cu_6Sn_5 and 20 wt.% Cu_3Sn aged for 64 days at 130°C, respectively. Noticeable is greater ε-phase thickness in the micrographs as compared to the eutectic solder sample. Figure 3-34 is that of the interfaces of 7.6 wt.% Cu composite solder aged, again, for 64 days at 130°C. The ε-phase thickness is also enhanced at the expense of the η-phase. It is thought that these particle additions are acting as tin-sinks, thereby depleting the interfacial regions of the tin necessary for reaction with the Cu_3Sn in order to produce the tin-rich Cu_6Sn_5 layer. The tin-poor ε-phase then grows at the expense of the Cu_6Sn_5.

Figure 3-35 depicts Cu3Sn particles in the unaged Sn/Pb matrix. The Cu_3Sn particles are large and spherical surrounded by a thin, irregular layer of Cu_6Sn_5. Figure 3-36 shows the interface and matrix of a 64 day, 110°C aged 20 wt.% Cu_3Sn sample. The Cu_3Sn particle adjacent to the interface is now surrounded by a thick layer of Cu_6Sn_5. The particle has acted as a tin-sink and has depleted the sample of tin. Figures 3-37 and 3-38 further represent this phenomenon. Figure 3-37 is of an unaged copper particle in the Sn/Pb matrix. The majority of the particle exists as copper. After being aged for 64 days at 130°C, the particle is mostly converted to the Cu_3Sn and terminal Cu_6Sn_5 phases (Figure 3-38). Lamellar coarsening is also evident in the composite samples, as it was in the eutectic solder.

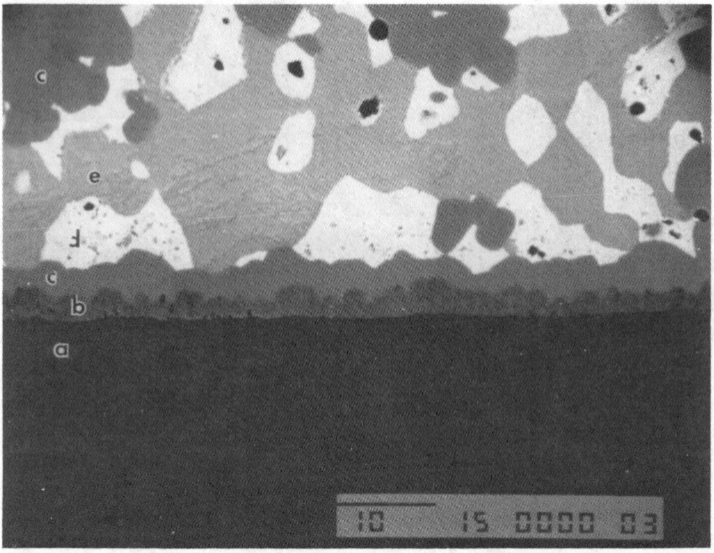

FIGURE 3-32 SEM of interface of composite solder, 20 wt% Cu_6Sn_5 in 63Sn-37Pb, aged at 130° C for 64 days. Phase identification: a = Cu; b = Cu_3Sn; c = Cu_6Sn_5; d = Pb; e = Sn. X1500, backscattered mode.

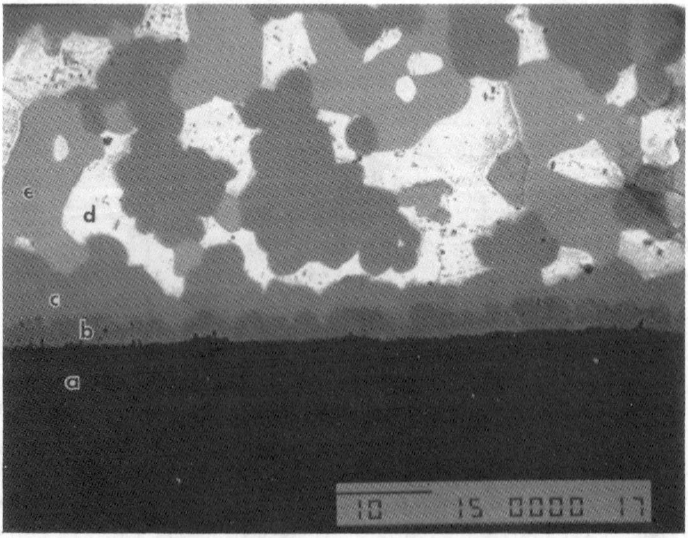

FIGURE 3-33 SEM of interface of composite solder, 20 wt% Cu_3Sn in 63Sn-37Pb, aged at 130° C for 64 days. Phase identification: a = Cu; b = Cu_3Sn; c = Cu_6Sn_5; d = Pb; e = Sn. X1500, backscattered mode.

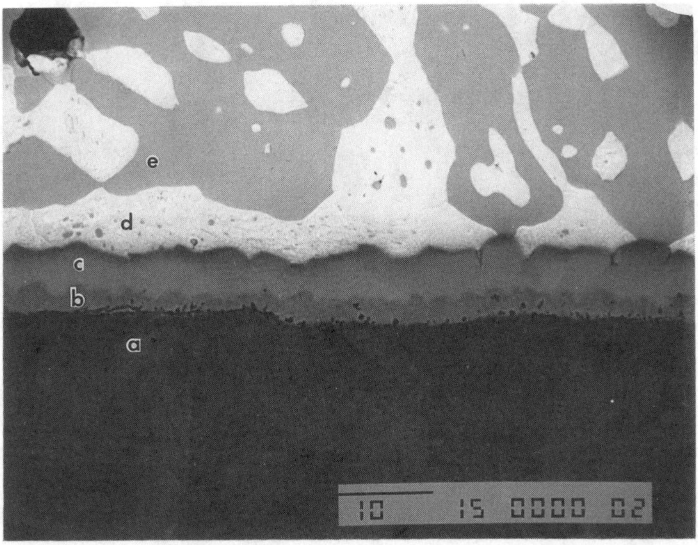

FIGURE 3-34 SEM of interface of composite solder, 7.6 wt.% Cu in 63Sn-37Pb, aged at 130° C for 64 days. Phase identification: a = Cu; b = Cu_3Sn; c = Cu_6Sn_5; d = Pb; e = Sn. X1500, backscattered mode.

These studies demonstrate that metal or intermetallic particle addition to bulk solder can render a dramatic change in the interface-intermetallic growth and development. The copper-containing composites all increase the activation energy of Cu_6Sn_5 formation and therefore reduce its thickness. These particles also decrease the activation energy for Cu_3Sn formation and enhance its thickness. This is consistent with the particles reducing the diffusion of Sn through the solder.

There are two mechanisms that must be considered in analyzing the intermetallic thickness data. First, the copper and Cu_3Sn particles act as tin-sinks. They act like a sponge that removes tin from the solder and reduces the amount of tin available for reaction at the substrate interface. The particles continue to react with the tin until the terminal phase Cu_6Sn_5 is formed. Therefore, the amount of tin that reaches the solder/substrate interface in a given period of time is reduced. The kinetic balance is pushed towards the copper portion of the Sn/Cu phase diagram, favoring Cu_3Sn (with a larger Cu/Sn ratio) over Cu_6Sn_5.

If tin-sinks were the only mechanism responsible for reduced intermetallic formation, pure copper particles should have a larger effect than the Cu_3Sn particles, since they require more tin to form the Cu_6Sn_5 terminal phase. Unfortunately, it is not possible to compare the two composite solders because the amount of Cu_6Sn_5 and copper fillers was not the same.

The second mechanism was believed to arise since Cu_6Sn_5 particles also reduce the amount of tin available for reaction at the interface, even though Cu_6Sn_5

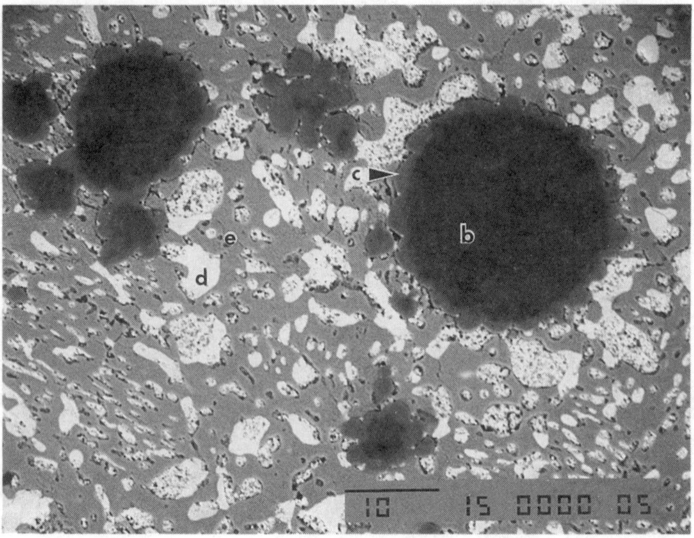

FIGURE 3-35 SEM of matrix of composite solder, 20 wt% Cu_3Sn in 63Sn-37Pb, unaged. Phase identification: a = Cu; b = Cu_3Sn; c = Cu_6Sn_5; d = Pb; e = Sn. X1500, backscattered mode.

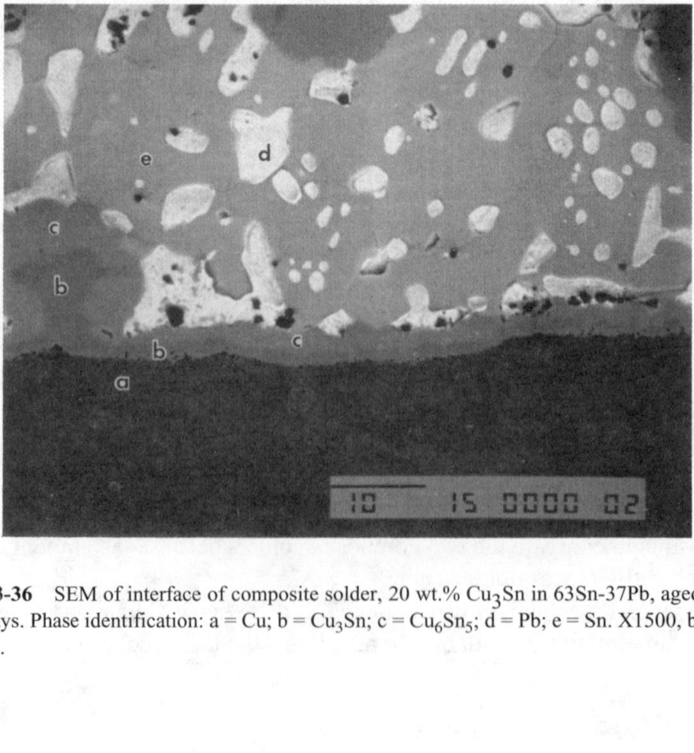

FIGURE 3-36 SEM of interface of composite solder, 20 wt.% Cu_3Sn in 63Sn-37Pb, aged at 110° C for 64 days. Phase identification: a = Cu; b = Cu_3Sn; c = Cu_6Sn_5; d = Pb; e = Sn. X1500, backscattered mode.

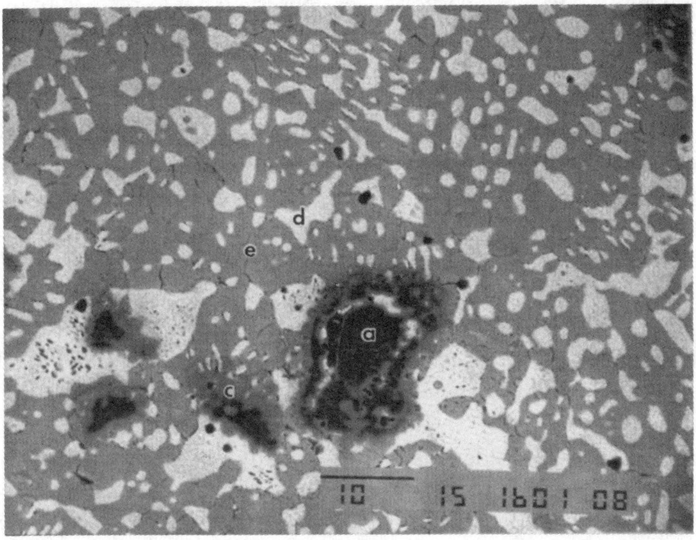

FIGURE 3-37 SEM of matrix of composite solder, 7.6 wt% Cu in 63Sn-37Pb, unaged. Phase identification: a = Cu; b = Cu_3Sn; c = Cu_6Sn_5; d = Pb; e = Sn. X1500, backscattered mode.

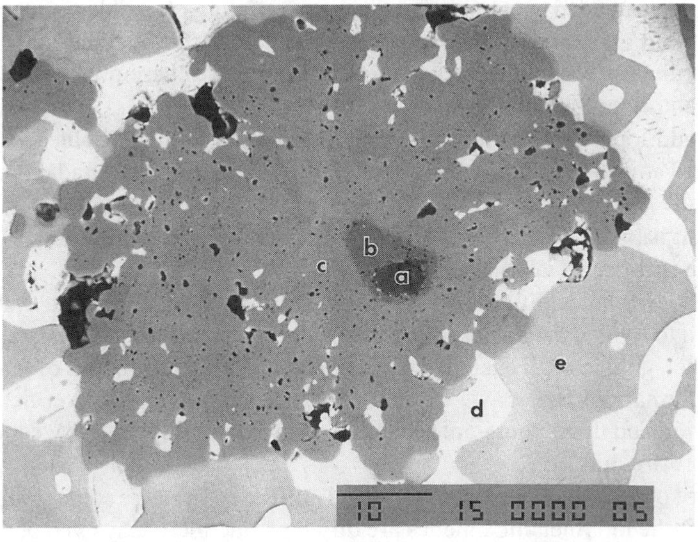

FIGURE 3-38 SEM of matrix of composite solder, 7.6 wt% Cu in 63Sn-37Pb, aged at 130° C for 64 days. Phase identification: a = Cu; b = Cu_3Sn; c = Cu_6Sn_5; d = Pb; e = Sn. X1500, backscattered mode.

is the terminal phase. These particles were observed not to react with the solder, and it was believed that the Cu_6Sn_5 formation was inhibited by reduction of the cross-sectional area of the solder.

Intermetallic formation for the nickel-composite solder was quite different than for the copper-containing solders. The activation values for both Cu/Sn intermetallics increased significantly: the Cu_3Sn formation was almost completely suppressed; the thickness of the Cu_6Sn_5 layer was increased, but the layer was very porous and a substantial fraction of the volume was void space. It appeared that the nickel was acting as a diffusion barrier, but the exact mechanism could not be ascertained.

During each of these experiments, the lead-rich phase of the solder at the solder/substrate interface increases with anneal time from approximately 30% to an equilibrium value of approximately 90%; hence, a lead-rich region forms just as in normal, unfilled solder.

Recent research with solders has shown that the solder system is an extremely complex system. The following considerations and areas for future research are suggested:

- As the thickness of a solder is reduced, and the integrity of the intermetallic may play a more important role, what concern should we attach to the effect of extraneous metals and intermetallics?
- Is the solder joint to nickel and other base metals a much different phenomenon than to copper? Do reliability questions arise?
- During extended heat treatment (such as annealing of a solder joint, or during the field operation in harsh environments), does the changing nature of the intermetallic, including development of porosity, cause potential long-term reliability problems?
- Should long-term reliability testing be performed of thin solder joints (less than 1 mil), ascertaining the effect of different base metals and environmental stress?
- What happens in the field? Do high temperature annealing studies calibrate with real-life situations?
- We know that additional metals affect the behavior of the entire system. We need to look at the solder joints as one unified metallurgical system; copper, nickel, gold, etc. all change the microstructure and the kinetics of intermetallic formation. Hence, we cannot make assumptions about intermetallic behavior, and thus about joint reliability and solderability, unless we look at the microstructure.
- The solder joint to nickel is completely different from the solder joint to copper—the intermetallic kinetics are different, and there may even be a change in the intermetallic phase distributions. Thus, any data on conventional Pb/Sn solder joints, if extrapolated to nickel and gold-plated joints, are done so at the peril of the investigator.

- Reliability studies need to be done. The porosity in the nickel-containing solders have an unknown effect.
- Long-term reliability and aging studies should be performed on thin solder joints. The system is dynamic and interactive. We know, for example, that the relative concentrations of Pb and Sn change at the interfaces. Probably the same is true of extraneous metals in the solder, or of composite additions. Thus, thin joints may actually be material-limited and may not act as infinite sources and sinks of metals. This could lead to a change in the kinetics.

REFERENCES

Bader, W. G., *Welding Journal, Research Suppl.*, 1969, vol. 48, (2), P. 551.

Belser, R. B., "Intermetallic Indalloy Solders," The Indium Corporation of America.

Boettinger, W. J., Handwerker, C. A., and Smith, L. C.,"On the Wetting of the Intermetallics Cu_6Sn_5 and Cu_3Sn by Ph-Sn Alloys," in *The Metal Science of Joining*, Cieslak, M. J., Perepezko, J. H., Kang, S., and Glicksman, M. E., eds., 1991, The Minerals, Metals, and Materials Society, Warrendale, Pennsylvania 15086, October, pp. 165-174.

Brooks, P. L., and Gillam, E., *Acta Met.*, 1970, vol. 18, p. 1181.

Brothers, E. W., *The Western Electric Engineer*, 1981, Spring/Summer, p. 49.

Cabarat, R., Guillet, L., and LeRoux, R., "The Elastic Properties of Metallic Alloys," *J. Inst. Metals*, 1949, vol. 75, p 391.

Chalco, P. A., *Int. J. Hybrid Microelectr.*, 1983, vol 6, (1), pp. 223-228.

Davis, P. E., Warwick, M. E., and Muckett, S. J., *Plating and Surface Finishing*, 1983, vol. 70, pp. 49-53.

Davis, P. E., Warwick, M. E., and Kay, P. J., *Plating and Surface Finishing*, 1982, vol. 69, pp 72-76.

DeHaven, P. W., *Proc. of the Materials Research Society*, 1985, p 40.

Dirnfeld, S. F., and Ramon, J. J., *Welding Journal*, 1990, October, pp. 373s-377s.

Dunn, D. S., Marinis, T. F., Sherry, W. M., and Williams, C. J., *Proc. of the Materials Research Society*, 1985, p. 40.

Dunn, B. D., *Metallurgical Assessment of Spacecraft Parts and Materials*, 1989, Ellis Horwood Ltd., Chichester, England, p. 287.

Eddington, J. W., *Practical Electron Microscopy in Materials Science*, 1976, Van Nostrand Reinhold, New York.

Fenton, E. A., *Soldering Manual*, American Welding Society, 1959, New York, Chapter 1.

Fields, R. J., Low, S. R., III, and Lucey, G. K., Jr., "Physical and Mechanical Properties of Intermetallic Compounds Commonly Found in Solder Joints," in *The Metal Science of Joining*, Cieslak, M. J., Perepezko, J. H., Kang, S., and Glicksman, M. E., eds., 1991, The Minerals, Metals, and Materials Society, Warrendale, Pennsylvania 15086, October 1991 pp. 165-174.

Frear, D. R., "Microstructural observations of the Sn/Pb solder/Cu system and thermal fatigue of the solder joint," Ph.D. Thesis, 1987, Univ. of California, Berkeley, CA.

Frederikse, H. P. R., Fields, R. J., and Feldman, A., "Thermal and electrical properties of copper-tin and nickel-tin intermetallics," *J. Appl. Phys.*, vol. 72, (7), October (1991) pp. 2879-2882.

Gangulee, A., Das, G. C., and Bever, M. B., *Met. Trans*. 1973, vol. 4, p. 2063.

Haimovich, J., *Welding Journal*, 1989, vol 8, (3), Research Supplement, p. 102s.

Hansen, M., and Anderko, K., *Constitution of Binary Alloys*, 1958, McGraw-Hill, New York.

Heinrich, F. J., and Newbury, D. E., "Electron Probe X-ray Microanalysis," *Metals Handbook Ninth Edition Volume 10 Materials Characterization*, 1986, American Society for Metals, Metals Park, OH, pp. 516-535; Leyden, D. E., "X-Ray Spectrometry," *Metals Handbook Ninth Edition Volume 10 Materials Characterization*, 1986, American Society for Metals, Metals Park, OH, pp. 82-101.

Ho, C. T., and Chung, D. D. L., "Carbon fiber reinforced tin/lead alloy as low thermal expansion solder preform," *J. Mater. Res.*, January, 1990, vol. 5, (6), p. 1266.

Howes, M. A., and Saperstein, Z. P., *Welding Journal, Research Suppl.*, 1969, vol. 48, (2), p. 80.

Hoyt, J., "Influence of Leg Shape and Solder Joint Metallurgy on Surface Mount Solder Joint Strength," 1987, Navy Weapons Center Technical Paper #6789, p 409.

Jacobs, E. G., Foster, L. A., Wu, Y., Wilson, A. R., and Pinizzotto, R. F., "Ultramicrotomy: A Unique Method for preparation of composite solder for transmission electron microscopy," *J. Mater. Res.*, 1993, vol. 8, (1) January, pp. 87-94.

Kang, S., and Glicksman, M. E., eds., 1991, The Minerals, Metals, and Materials Society, Warrendale, Pennsylvania 15086, October, pp. 183-190.

Kay, P. J., and MacKay, C. A., *Transactions of the Institute of Metal Finishing*, 1976, vol. 54, pp. 68-74.

Kay, P. J., and MacKay, C. A., *Transactions of the Institute of Metal Finishing*, 1973, vol. 51, p. 85.

Kay, P. J., and MacKay, C. A., *Transactions of the Institute of Metal Finishing*, 1979, vol. 57, p. 169.

Keeler, R., *Electr. Pack. & Prod.*, vol 27, (7), pp. 45-47.

Keller, H. N., *IEEE Transactions on Components, Hybrids, and Manufacturing Technology*, 1986, CHMT-9, (4), p. 433.

Lynch, C. T., ed., *Practical Handbook of Materials Science*, 1989, CRC Press, Inc., Boca Raton, FL.

Malis, T. F., and Steele, D., "Ultramicrotomy for Materials Science" in "Workshop on Specimen Preparation for TEM of Materials II," Anderson, R., ed., 1992, vol. 199, *Materials Research Symposium Proceedings*, Materials Research Society, Pittsburgh, PA.

Manko, H. H., *Solders and Soldering*, 2nd ed., 1979, McGraw-Hill, New York, Chapter 1.

Marshall, J. L., "A Microstructural Study of the Action of Flux on Copper-Tin Intermetallics," *Electrecon '91 Proceedings*, October, 1991, pp. 130-141.

Marshall, J. L., Calderon, J., Sees, J., Lucey, G., and Hwang, J. S., "Composite Solders," *IEEE Trans CHMT*, December, 1991, vol. 14, (4), pp. 698-702.

Marshall, J. L., "Gold Embrittlement in Soldering Processes," *Tex. Soc. Electron. Mic.*, 1985, vol. 6, p. 10.

Marshall, J. L., and Walter, S. R., "Fatigue of Solders," *Int. J. Hybrid Microelectr.*, 1987, vol. 10, p. 11.

Marshall, J. L., Lucey, G., Handwerker, C., Tench, D., and Sunwoo, A., "An Hypothesis for the Mechanism of Solderability Failure," *Proceedings NEPCON 91*, February, 1991, Anaheim, CA, pp. 3-10.

Marshall, J. L., "Solder Fatigue in Electronics," *Brazing and Soldering*, 1988 (15), Autumn, pp. 4-9.

Marshall, J. L., "The Mechanism of Solder Fatigue in Electronic Packaging," *Proc. Electr. Manuf. Symp.*, 1988, Naval Weapons Center, China Lake, CA, February, pp. 389-409.

Marshall, J. L., "Scanning electron microscopy and energy dispersive X-ray (SEM/EDX) characterization of solder -- solderability and reliability," in Lau, J. H., ed., *Solder Joint Reliability*, 1991, Van Nostrand Reinhold, New York, pp. 173-225.

Metals Handbook, 8th ed., 1973, vol. 8, *Metallography, Structures and Phase Diagrams*, American Society for Metals, Metals Park, Ohio 44073.

Morris, J. W., Jr., Tribula, D., Summers, T. S. E., and Grivas, D., "The role of microstructure in thermal fatigue of Pb-Sn solder joints," in John H. Lau, ed., *Solder Joint Reliability*, 1991, Van Nostrand Reinhold, New York, pp. 225-265.

Okamoto, I., and Yasuda, T., *Trans. JWRI*, 1986, vol. 15 (2), p. 73.

Olsen, D., Wright, R., and Berg, H., "Effects of Intermetallics on the Reliability of Tin Coated Cu, Ag, and Ni Parts," *Proc. 13th Annual Conf. on Reliability Physics*, 1975, Las Vegas, pp. 80-86.

Ostrovskaya, L. M., Rodin, V. N., and Kunznetsov, A. I., "Elastic Properties of Intermetallic Compounds Produced by Vacuum Deposition," *Soviet J. of Nonferrous Metallurgy (TSVETNYE METALLY)*, 1985, vol. 26, p 90.

Pinizzotto, R. F., Wu, Y., Jacobs, E. G., and Foster, L. A., "Microstructural development in composite solders caused by long time high temperature annealing," *Proceedings NEPCON '92*, Anaheim, CA, February, 1992, vol. 3, pp. 1284-1293.

Pinizzotto, R. F., Jacobs, E. G., Wu, Y., Sees, J. A., Foster, L. A., and Pouraghabagher, C., *Int. Reliability Physics, Sympos.*, March, 1993.

Purdy, G. R., and Kirkaldy, J. S., *Trans. TMS-ASM*, 1972, vol. 2, (2), p. 371.

Reichenecker, W. J., "Thermal Expansion of the Copper-Tin Intermetallic Compound Cu_6Sn_5 in the Temperature Range -180 to +200° C," *Insulation/Circuits*, 1981, vol. 27, #27, p. 104.

Revay, L., "Interdiffusion and Formation of Intermetallic Compounds in Tin-Copper Alloy Surface Coatings," *Surface Technology*, 1977, vol. 5, p. 57.

Romig, A. D., Jr., Chang, Y. A., Stephens, J. J., Frear, D. R., Marcotte, V. and Lea, C., "Physical metallurgy of solder-substrate reactions," in Frear, D. R., Jones, W. B., and Kinsman, K. R., eds., *Solder Mechanics: A State of the Art Assessment*, 1991, The Minerals, Metals, and Materials Society, Publishers, Warrendale, PA 15086, Chapter 2.

Saperstein, Z. P., and Howes, M. A., *Welding Journal, Research Suppl.*, 1968, vol. 47, (4), p. 162.

Stromswold, E. I., University of Rochester, to be published. The fracture tests were performed by the four-point bend Chevron notch technique as described in Stromswold, E. I., and Quesnel, D. J., *Engineering Fracture Mechanics*, March, 1992, vol. 4, (3), pp. 309-320.

Subrahmanyan, B., "Elastic Moduli of Some Complicated Binary Alloy Systems," *Trans. Jap. Inst. Metals*, 1972, vol. 130, p. 93.

Thwaites, C. J., *Soft-Soldering Manual*, 1982, International Tin Research Institute, Middlesex, UK.

Tu, K. N., *Acta Met.*, 1973, vol. 21, p. 347.

Tu, K. N., and Thompson, R. D., *Acta Met.*, 1982, vol 30, pp. 947-952.

Verhoeven, J. D., "Scanning Electron Microscopy," *Metals Handbook Ninth Edition Volume 10 Materials Characterization*, 1986, American Society for Metals, Metals Park, OH, pp. 491-515.

Walker, G. A., DeHaven, P. W., and Goldsmith, C. C., *IEEE*, 1984, p 125.

Warwick, M. E., and Muckett, S. J., *Circuit World*, 1983, vol 9, (44), p. 5.

Wassink, R. J. Klein, *Soldering in Electronics*, 2nd ed., 1989, Electrochemical Publications Limited, Ayr, Scotland, Chapter 7.

Wassink, R. J. Klein, *Soldering in Electronics*, 2nd ed., 1989, Electrochemical Publications Limited, Ayr, Scotland.

Woodgate, R. W. *Handbook of Machine Soldering*, 1983, John Wiley & Sons, New York, Chapter 2.

Zakraysek, L., *Welding Journal*, 1971, vol. 36,(12), p. 522s.

Zakraysek, L., *Welding Journal*, 1972, vol. 51, p. 536s.

4

Constitutive Models

Donald S. Stone and Mark M. Rashid

4.1 INTRODUCTION

A constitutive model is the mathematical representation of material response to one or more arbitrary, time-varying stimuli. Mechanical constitutive models serve as input into larger computational models, frameworks for stress analysis employed to simulate the distortions and stresses experienced within complicated engineering structures (e.g. solder joints) due to thermal and mechanical loading. Beyond narrow, stress-vs.-deformation descriptions of material behavior, mechanical constitutive models can also be concerned with deformation mechanisms and evolution of microstructure. Some of these concerns can be illustrated in conjunction with Figure 4-1, where scanning electron micrographs show solder joints after thermal cycling, some of them in cross-section. The solder material, nominally Pb-Sn eutectic, consists of a two-phase aggregate in which are dispersed precipitates resulting from the interaction between solder and substrate metallurgies during solidification. In many cases the microstructure has resulted from cooperative growth of the Pb and Sn phases during solidification: their crystallographic orientations bear correlations over extended regions within the solder joint. In terms of the rheology, these extended regions ("colonies" consisting of more than one phase) behave like individual grains within a single-phase aggregate, and, as can be seen in the figure, they slide with respect to one another during deformation. The composition and phase distribution vary throughout the joint.

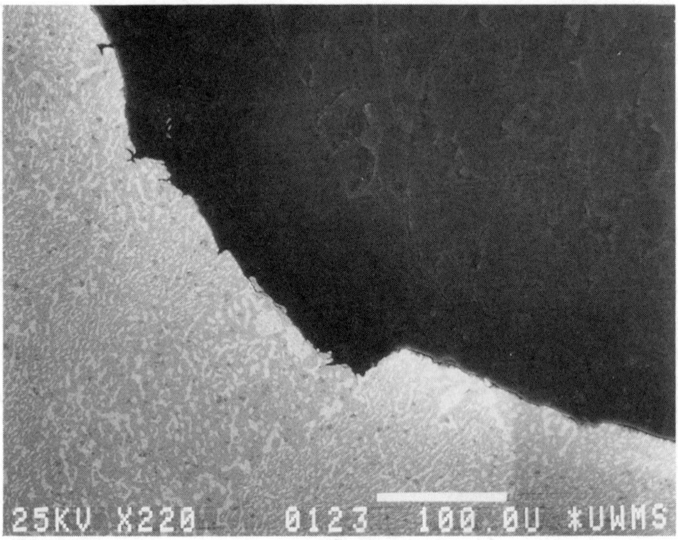

FIGURE 4-1a Scanning electron micrographs of solder joint after thermal cycling:Cross section of a leadless solder joint with wavy microstructure, intermediate between lamellar and globular. Sliding between colonies of the eutectic has resulted in offsets at the surface and some crack initiation [Lee, 1993].

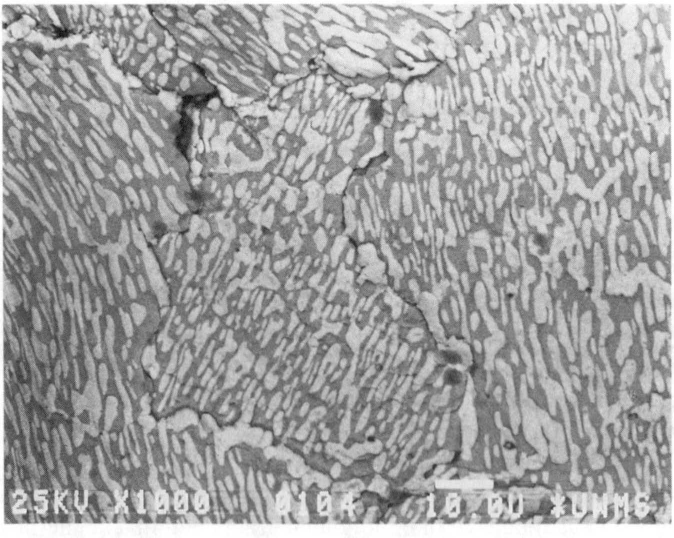

FIGURE 4-1b Backscattered and secondary electron images of one location on the surface of the solder joint, showing sliding and cracking between colonies [Lee, 1993].

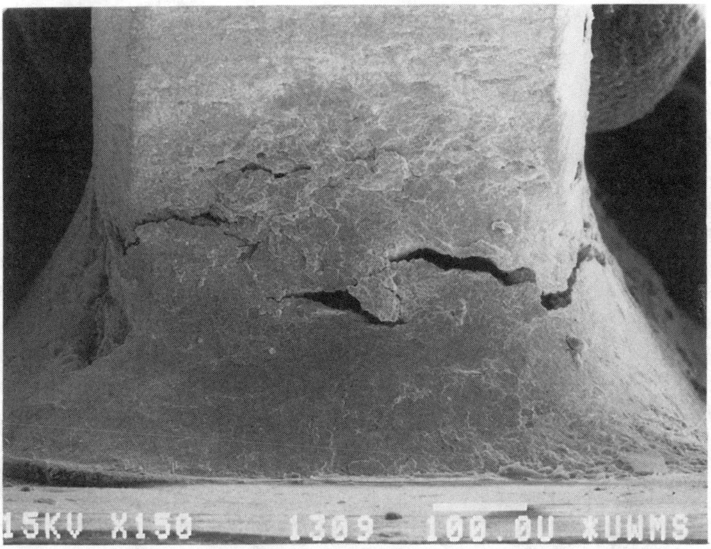

FIGURE 4-1c Intergranular crack initiation in a J-leaded solder joint [Lee, 1993].

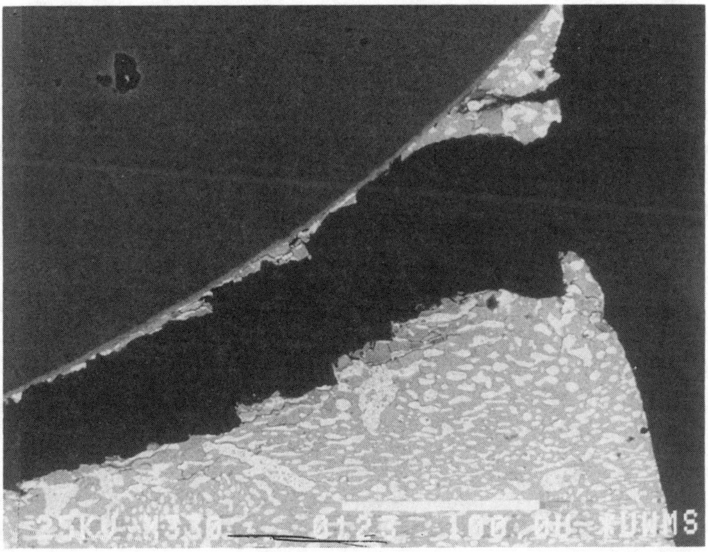

FIGURE 4-1d The same as figure 4-1c, but in cross section, showing how the crack travels adjacent to the J-lead. In this case, the stress supported by the solder, not the strain range, controls fatigue life. Note that this solder joint demonstrates more of a globular microstructure [Lee, 1993].

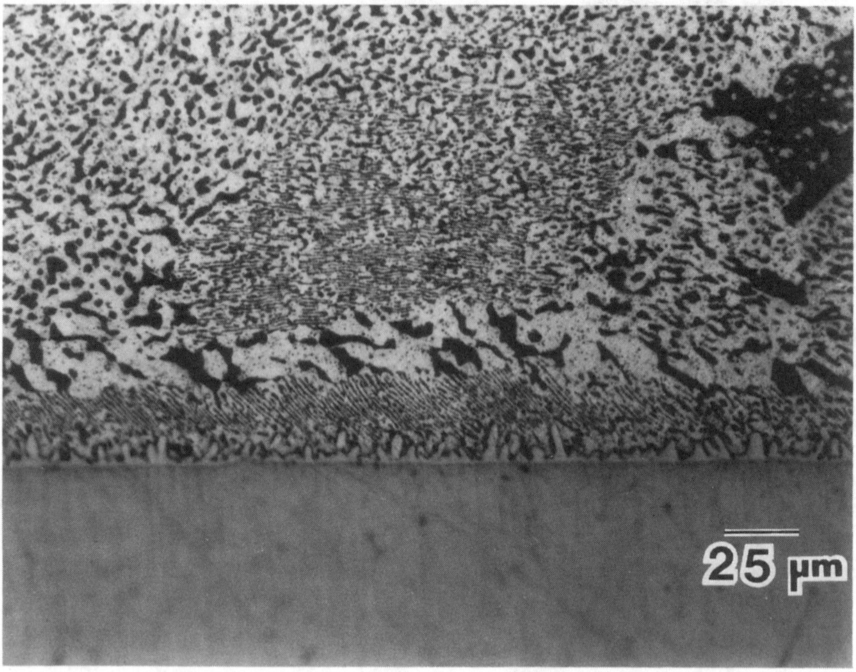

FIGURE 4-1e A coarsened band of the eutectic resulting from a microstructural instability.

In view of the micrographs of Figure 4-1, a number of questions arise: What are the effects of composition and phase morphology (determined by chemistry and rate of cooling) on the flow behavior of solder alloys? Do the properties of a multiphase aggregate reflect simple averages of the properties of its components? How and why do phase distribution and morphology evolve during thermal cycling, and what are the effects of this evolution on flow behavior? What are the mechanisms of deformation, how do these change with temperature and stress, and what is the relation between creep and fatigue crack growth, giving rise to the so-called "creep-fatigue interaction"? What is the importance of solder joint size relative to the scale of the solder microstructure (in particular "grain" size), and how small does a solder joint have to be before treating it as having uniform or continuously-varying properties becomes no longer possible?

The goals of constitutive modeling vary from one application to the next. In the microelectronics context, these goals might include an optimization of solder joint shape, the calculation of strain ranges for a Coffin-Manson based analysis, an extrapolation of laboratory data outside the capabilities of experiment, and the simulation of crack growth taking into account the creep-fatigue interaction. Given this range of goals and the range of experimental, theoretical, and computational tools for achieving them, one should neither expect nor desire that a single, "best," all-encompassing constitutive model exists. Rather, what constitutes an adequate or "good" model is somewhat predicated on the specific goals of the modeler and the willingness to optimize precision and accuracy (achieving detailed, meaningful results) against efficiency (avoiding inessential constitutive parameters).

In recognition of the variety of possible goals and inevitable differences in level of sophistication of constitutive models, this chapter views the construction and application of constitutive models as a flexible undertaking. In fact, only three central ingredients to a constitutive model are postulated: a well defined phenomenology, a basis in empirical data, and a basis in mechanism. Clearly, then, a "good" model somehow optimizes these three ingredients, given the objectives and the experimental and computational tools available for achieving them.

The phenomenology – represented, for instance, by a standard linear solid – provides an efficient means of taking experimental data and placing them within a mathematically rigorous context demanded by the stress analysis. The phenomenological framework guides measurement of constitutive parameters and interpretation of experimental results.

Experiments serve as a means for obtaining the constitutive parameters; they can also be used to test the appropriateness of the model phenomenology and to adjust it to conform to empirical observation, not the other way around. Given that we are concerned with microscopic models, experimental methods employing microscopic examination of the solid serve as an important part of the experimental program.

An understanding of mechanisms helps to ensure that the model extrapolates outside the range of measurement. It also helps to guide and interpret experiments, and can be used in some instances to examine whether the constitutive parameters are physical or not. However, rarely, if ever, are deformation phenomena simple enough that constitutive models can be constructed based entirely on mechanistic considerations. It is more often the case that a single constitutive phenomenon can be explained by more than one physical model.

As is the case in any mathematical model of physical phenomena, the formulation of a constitutive model involves the introduction of a number of idealizations. Chief among the assumptions adopted here is the concept of state. The crucial aspect of the concept of state is that any influence of the past history on the current material behavior must be expressed through the current values of the state variables. Therefore, two material elements that possess identical state variable values are identical in response to future stimuli inputs, regardless of the deformation histories that resulted in their current state. It is recognized that if sufficient numbers of parameters can be incorporated into a model, the model can come arbitrarily close to simulating the actual behavior of a solder. The issue is not, therefore, whether a mechanical equation of state actually exists, but rather how accurately one can portray the flow behavior of a solder with as few (hopefully physical) material parameters and state variables as possible.

4.1.1 Organization of the Chapter

The bulk of this chapter is divided into two parts. Part I approaches constitutive modeling from a materials scientist's perspective, where identifying and measuring constitutive parameters and understanding constitutive behavior are important. Part II adds to this perspective by using finite element analysis to simulate the growth of a shear band across the solder joint.

In Part I, it is shown how microscopic processes (mechanisms) can be incorporated within a phenomenological framework. It is shown how this framework can be approached from the perspective of experiment and how the two approaches can be integrated. Not surprisingly, these demonstrations lead to the discovery that, although many of the observations (pertaining to a Pb-Sn eutectic) can be rationalized based on physical grounds, the construction of the constitutive model must rely most heavily on phenomenology and experiment. Why, then, is an understanding (or at least partial understanding) of the mechanisms so important? Apart from the previously stated, and commonly held, argument that knowledge of mechanisms aids in the ability to construct models that extrapolate outside the range of experiment, the most compelling reason for consideration of deformation mechanisms is in the potential impact these mechanisms might have on the creep-fatigue interaction. For this reason the last section of Part I deals with some of the potential roles played by creep mechanisms in the creep-fatigue interaction.

Part II of this chapter concerns the implementation of a microstructural constitutive model into a computational framework, alluded to earlier, for the stress analysis of a solder joint. The model allows the two-phase microstructure to coarsen as a direct consequence of straining; combined with the way the solder joint is loaded and the effects of solder joint shape, coarsening leads to a characteristic microstructural instability often seen in solder joints after thermal cycling. The constitutive model used in Part II, like all of the ones in Part I, is in many respects an oversimplification; yet it is sufficiently realistic to be used as a predictive tool. Models such as this can be tested with experiment and modified accordingly.

4.2 CONSTITUTIVE MODELING FROM A MATERIALS PERSPECTIVE

4.2.1 Microstructural Models of Constitutive Behavior

Within the following pages, physical models are placed within phenomenological frameworks made available, for the most part, through application of rheological diagrams, spring-and-dashpot analogs. The strength of these models lies in their ability to represent, in pictorial form, the concepts of microscopic compatibility and equilibrium. Unfortunately, the fact that the models are one-dimensional ultimately limits their utility because the manifestations of equilibrium and compatibility can be far more subtle in three dimensions than one. Despite this limitation, no apologies are made for their extensive use throughout this chapter. They serve adequately to express many of the concepts presented below.

This portion of the chapter treats stress, strain, and strain rate as if they were scalar quantities. Extension to three dimensions is left to the reader with the understanding that shear stresses drive plastic deformation in crystalline solids. While it is clear from some of the microstructures shown in Figure 4-1 that the properties of solders are unlikely to be isotropic, none of the models shown below warrant extension into anisotropy at this time.

For historical reasons as well as for clarity in organization, the deformation phenomena described below will be divided into four classes: elasticity, anelasticity, microplasticity, and plasticity. This division is somewhat arbitrary because a single mechanism can give rise to different phenomena depending on the magnitude of the stress and how long it is held.

Elasticity

Elastic deformation is the fully recoverable, time-independent (apart from inertial effects) distortion of a body resulting from the stretching and shearing of atomic bonds. If anelastic effects are to be avoided, elastic moduli should be measured based on phonon scattering and velocity-of-sound techniques. In practice, however, the slopes of stress-strain curves near zero stress, if carefully measured, often fall to within about 10% of the high frequency moduli.

Most solders are multiphase aggregates, for which elastic modulus data do not exist. It is possible, therefore, to make upper- and lower-bound estimates of the Young's Moduli based on Voigt and Reuss averages, respectively, of the elastic moduli of the individual components:

$$E_u = V_1 E_1 + V_2 E_2 \tag{4.1}$$

$$\frac{1}{E_l} = \frac{V_1}{E_1} + \frac{V_2}{E_2} \tag{4.2}$$

where V_1 and V_2 are the volume fractions and E_1 and E_2 are the elastic moduli of the respective phases. These averages correspond to the rheological diagrams shown in Figure 4-2. More-restrictive upper and lower bounds have been determined by Hashin and Shtrikman [1963] for the bulk and shear moduli (B and K).

Anelasticity

Anelasticity involves time-dependent, recoverable strain. Linear at small strains, most anelastic phenomena become nonlinear at large strains [Findley et al.,1976]. A variety of physical mechanisms can give rise to anelasticity, including: grain

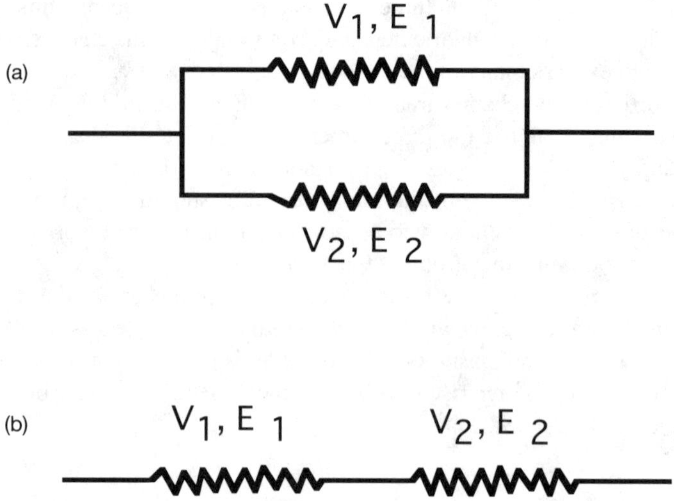

FIGURE 4-2 Rheological diagrams establishing (a) upper-bound (Voight) and (b) lower-bound (Reuss) averages of the Young's Modulus for a two-phase material.

boundary and interface sliding, point defect motion, thermal currents, and dislocation motion below the macro- flow stress [Zener 1948; McClintock 1966]. In Part II of this chapter, a rheological model for the solder will be presented, in which a mismatch in (modulus)/(flow strength) ratio between phases in a multiphase system can give rise to an anelastic-like phenomenon. Not the subject of Part I, this "anelasticity" will be treated in the subsection on microplasticity.

Even though linear anelasticity is probably not relevant to solders except at very low levels of strain (say below 10^{-4}), many of the same physical processes that underlie it will also give rise to microplasticity and creep. Also, the constructions of phenomenological models for linear anelasticity, Figure 4-3, are similar to those for large deformations, as shown later.

Two familiar rheological diagrams for linear anelastic solids are shown in Figure 4-3 [e.g., Findley et al.,1976]. These diagrams are formally equivalent because, apart from differences in the definitions of parameters, they represent a single constitutive equation:

$$\sigma + \tau_A \dot{\sigma} = E_B (\varepsilon + \tau_B \dot{\varepsilon}) \tag{4.3}$$

where:

$$\tau_A = \frac{\eta}{E_1 + E_2} = \frac{\eta'}{E'_2} \tag{4.4}$$

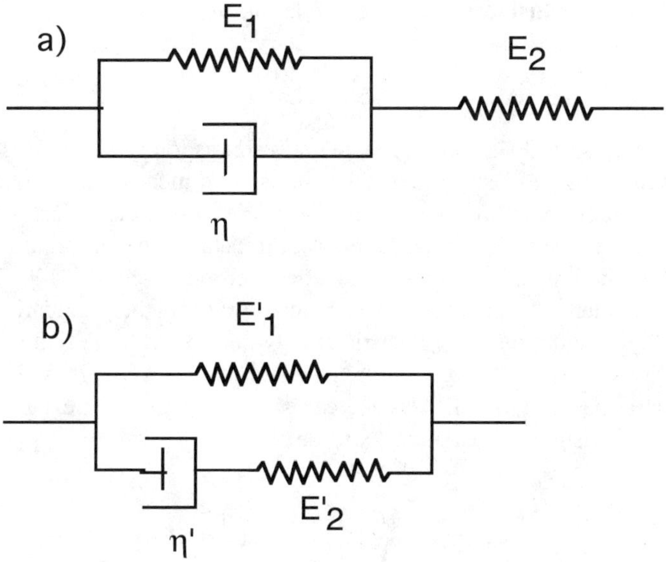

FIGURE 4-3 Standard linear solids.

$$E_B = \frac{E_1 E_2}{E_1 + E_2} = E' \tag{4.5}$$

and

$$\tau_B = \frac{\eta}{E_1} = \frac{\eta' (E'_1 + E'_2)}{E_1 E'_2} \tag{4.6}$$

Below, microscopic models of anelasticity are presented. The physical processes considered include grain boundary sliding and the reversible motions of dislocations well below the macroscopic flow stress. In the case of grain boundary sliding, both linear and nonlinear dashpots are considered.

Grain Boundary Anelasticity

Like most anelastic phenomena in crystalline solids, grain boundary sliding is driven by shear stresses. At low frequencies where the grain boundaries become relaxed, they act, effectively, as shear cracks, which means that grain boundary anelasticity manifests as a "defect" (or reduction) in the measured shear modulus but not the bulk modulus [Zener 1941; Ghahremani 1980a]. Boundary sliding viscosity derives from the motion of grain boundary dislocations and the creep (either diffusion or dislocation) accommodation of matter around particles and inhomogeneities distributed along the boundary [Ashby 1972].

Anelastic Modulus from Grain Boundary Sliding

The relaxed modulus does not depend on grain size as long as the grains are much smaller than the overall specimen dimensions. A dimensional analysis reveals why: even though there are more grain boundaries when the grains are smaller, each boundary has a smaller effect. Suppose, for instance, the grains are polyhedra of characteristic dimension d (e.g. tetrakaidekahedrons in Figure 4-4). Now suppose the specimen is loaded under fixed shear stress. If the boundaries are held fixed, unable to slide, the stored elastic energy per unit volume will be $\tau^2/2G$. If the boundaries are then allowed to relax, the stored elastic energy will increase at constant stress because the stiffness decreases. The increase per hexagonal boundary (of characteristic dimension d/2) will be about:

$$\Delta U \approx \frac{1}{2} \bullet \frac{4\pi}{3} \left(\frac{d}{4}\right)^3 \frac{\tau^2}{2G} g\left(\frac{d'}{d}\right) \tag{4.7}$$

where G is the shear modulus, $(4\pi/3)(d/4)^3$ is the volume of material unloaded due to sliding, and g is a function of the ratio of distance between boundaries (d') to

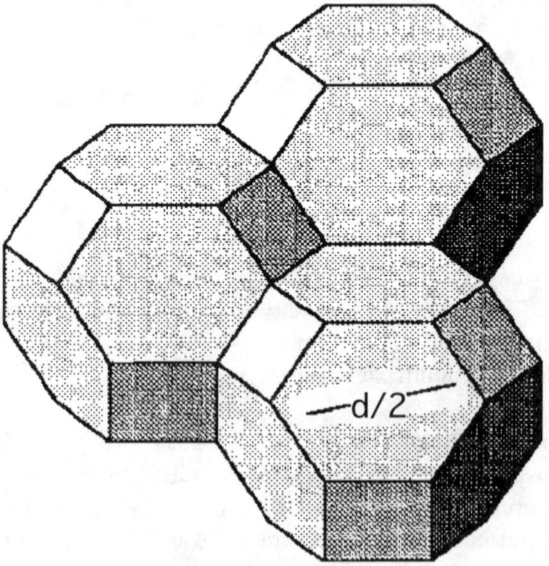

FIGURE 4-4 Grain structure consisting of tetrakaidekahedrons. The boundaries between the grains can act as shear cracks if the sliding viscosity is low.

grain size. This ratio depends, of course, on grain shape but not size, and a reasonable value for the reduced modulus is reached below by assuming g(d'/d) ~2.

There are $4/d^3$ boundaries per unit volume (we ignore contributions by the small, square boundaries in Figure 4-4); hence, when the boundaries are allowed to fully relax, the stored elastic energy per unit volume is:

$$U' = U + \Delta U \approx \left(1 + \frac{\pi}{12}\right)\frac{\tau^2}{2G} \tag{4.8}$$

Based on this, the reduced modulus is about 0.8 G. In Zener's original [1941] treatment based on a spherical grain model, the reduced modulus ranges from 0.4 G to 0.75 G, depending on Poisson's ratio. More recently, Ghahremani [1980] estimated a reduced modulus ranging from 0.75 to 0.85 G, also depending on Poisson's ratio, for two-dimensional hexagonal grains.

4.2.2 Anelastic Viscosity Parameter and Characteristic Relaxation Time Based on Grain Boundary Sliding

The viscosity parameters in Figure 4-3 can also be estimated based on physical grounds, in the present case by assuming grain boundary precipitates inhibit sliding. Whether such estimates are accurate is, as always, a matter to be tested experimentally.

Sliding can be by accommodated by diffusion of matter around the particle as indicated in Figure 4-5b, or by dislocation creep of the matrix. The diffusional mechanism, linear with stress, dominates at low stresses, while the dislocation mechanism dominates at high stresses.

Consider first the case of diffusional accommodation [Raj and Ashby, 1972; Mori et al., 1983]. The precipitates are cubes of dimensions a and separation on the grain boundary as shown in Figure 4-5a. Across the boundary a stress acts, of which only the shear component (τ) in the plane of the boundary is important to sliding.

To estimate the rate of sliding, unconstrained by triple points, we consider the stress-directed diffusion of matter surrounding the particles. For the boundary to slide a distance δx, $\delta N = a^2\delta x/2\Omega$ atoms must remove themselves from the particle/matrix interface located on one side of the boundary, then diffuse to the same interface on the other side of the boundary where they reattach themselves. Diffusion takes place along the interface and through the crystal lattice (of particle or matrix), and the faster of these processes will govern the kinetics of diffusional accommodation. The diffusion flux is indicated as arrows in Figure 4-5b, assuming the matrix atoms are the diffusing species. Meanwhile, the work done by the stress, per particle, is $\delta W = \tau\lambda^2\delta x$. If diffusion is rate-limiting (and not detachment and reattachment of atoms at the interface), a steady state chemical potential gra-

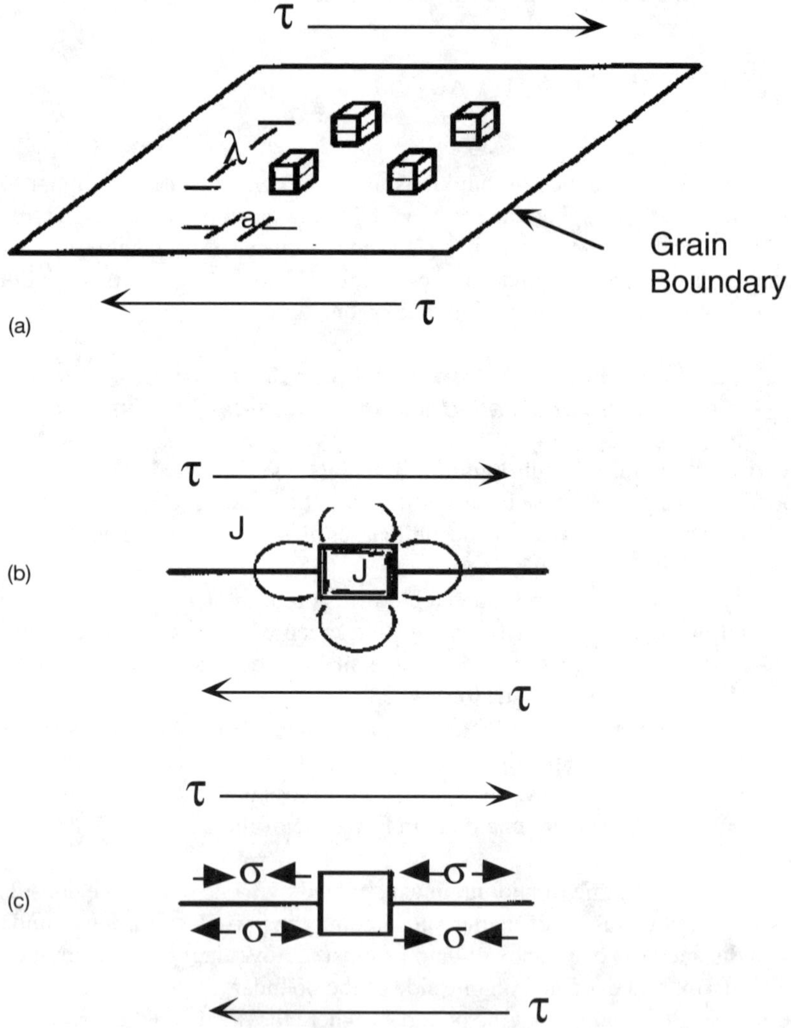

FIGURE 4-5 Grain boundary sliding inhibited by particles: (a) the component of shear stress acting parallel to the boundary provides the driving-force for sliding; (b) in the case of diffusional accommodation, material is transported through the matrix (larger particles) or along the interface (smaller particles); (c) stresses in matrix surrounding the particles. Clearly, this system of stresses is an approximation (it violates equilibrium due to non- zero torque on the particle), but it is useful for estimating the stresses that drive diffusion and dislocation creep accommodation.

dient is established given by $\nabla\mu=(\delta W/\delta N)/(a/2)=4\tau\Omega\lambda^2/a^3$. Here, a/2 is the distance over which the difference in chemical potential persists. The steady state gradient is associated with a flux of atoms given by $J=(D/kT\Omega)\cdot\nabla\mu=4\tau\lambda^2D/kTa^3$, where D is the diffusion coefficient, k is Boltzman's constant, and T is the temperature in Kelvin. Supposing diffusion is fastest along the interface between particle and matrix, then $D = D_i$ where D_i is the interfacial diffusion coefficient. In this case the total number of atoms leaving one side of the interface and arriving at the other per unit time is $dN/dt = ahJ$ where h is the interface thickness. The rate of sliding is then given by $(dN/dt)\cdot(x/N)$:

$$\dot{x} = \frac{8\tau\Omega\lambda^2 hD_i}{kTa^4} \tag{4.9}$$

The contribution of sliding to the specimen strain rate is proportional to $\frac{\dot{x}}{d}$ where, again, d is the grain size. Hence, the viscosity parameter in Figure 4-3 is:

$$\eta_{dif} \propto \frac{kTa^4 d}{\Omega\lambda^2 hD_i} \tag{4.10}$$

The characteristic time τ_0 for relaxation of the anelastic spring is about:

$$\tau \propto \frac{kTa^4 d}{\Omega G\lambda^2 hD_i} \tag{4.11}$$

The constant of proportionality has been determined by Mori et al. [1983] to be about 2. If lattice diffusion dominates, then hD_i can be replaced with $aD_l/2$, where D_l is the lattice diffusivity in Equations (4.9) to (4.11).

Dislocation creep can also accommodate sliding. The power law exponent, n, does not necessarily have to correspond to the power law exponent for creep of a single crystal, because the dislocation structures and diffusion fluxes that generated within the centers of grains might be different from those near particles on the grain boundaries. In any case, another dimensional analysis reveals the form of the sliding viscosity: like the mass-diffusion accommodation mechanism discussed above, dislocation creep is driven by stresses $\pm\sigma_p$ in front of and behind the particle. Indicated in Figure 4-5c, the stresses can be estimated based on a force balance: $4\sigma_p a^2/2 = \tau\lambda^2$, where $a^2/2$ is approximately the area over which σ_p acts and the factor of 4 takes into account the four regions in the matrix near the particle over which a stress $\pm\sigma_p$ acts. Each of these regions has a volume of order $a^3/2$.

Within each region, dislocation creep accommodation will be governed by an empirical, power-law expression:

$$\dot{\varepsilon}_{acc} = A\frac{D_1 G b}{kT}\left(\frac{\sigma_p}{G}\right)^n \tag{4.12}$$

where b is the Burgers vector and A is a dimensionless constant. The rate of sliding, $\dot{x}$, is given by $a \cdot \dot{\varepsilon}_{acc}$. Noting again that the specimen strain rate is proportional to the sliding rate divided by the grain size, we can estimate a stress-dependent sliding viscosity (stress divided by strain rate) as:

$$\eta_{dis} = \frac{2^n a^{2n-1} dkT}{\lambda^{2n} A D_1 b}\left(\frac{G}{\tau}\right)^{n-1} \tag{4.13}$$

Note that, kinetically, dislocation creep and diffusional accommodation act in parallel: the fastest one dominates. Thus, the net viscosity is the inverse of the sum of the inverses of the individual viscosities (i.e. the lesser of the two viscosities). Additionally, the grain boundary will have an intrinsic viscosity even in the absence of particles. As mentioned earlier, this viscosity might arise due to the structure of the boundary, meaning that sliding involves the motion of grain boundary dislocations, or it might arise from the waviness of the boundary. This intrinsic viscosity adds to the others.

A rheological diagram illustrating the relationships among these different mechanisms is shown in Figure 4-6. Also included are springs representing the fact that the boundary must slide a small distance before the stresses build up sufficiently around the particles to drive diffusion and dislocation creep.

Anelasticity due to Bowing of Dislocations at Frank-Read Sources

Segments of dislocations bowing between obstacles can give rise to anelastic strains [Gittus 1975]. If the bow-outs are sufficiently small, the strain will be approximately linear with stress. The linear approximation only serves properly at stresses well below the stress required to operate the dislocation source. Also, even at low stresses, some of the dislocation sources (those having large spacing between pinning points) will operate and emit dislocation segments, only to have the new dislocations impeded at other, nearby obstacles. Thus, the model described below is only a limiting behavior at small stresses.

Anelastic Modulus due to Dislocation Bowing

If the resolved shear stress is below the value needed to operate the dislocation segment as a Frank-Read source ($\tau_{critical} = \alpha Gb/L$; $\alpha \approx 1$), the segment will

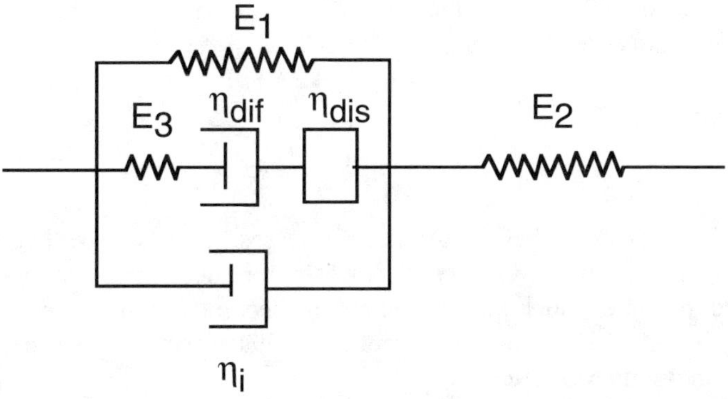

FIGURE 4-6 Rheological diagram for grain boundary anelasticity. Accommodation of sliding at particles involves mass transport by diffusion (η_{dif}) and dislocation creep (η_{dis}). These are parallel kinetic processes; hence the fastest one controls accommodation and they appear in series in the diagram. The grain boundary has an intrinsic sliding viscosity (η_i) representing a series kinetic process and a parallel element in the rheological diagram. A third spring (E_3) reflects the fact that elastic stresses must build up around particles in order to drive the accommodation mechanisms. The box used to represent dislocation accommodation emphasizes this element as a nonlinear dashpot.

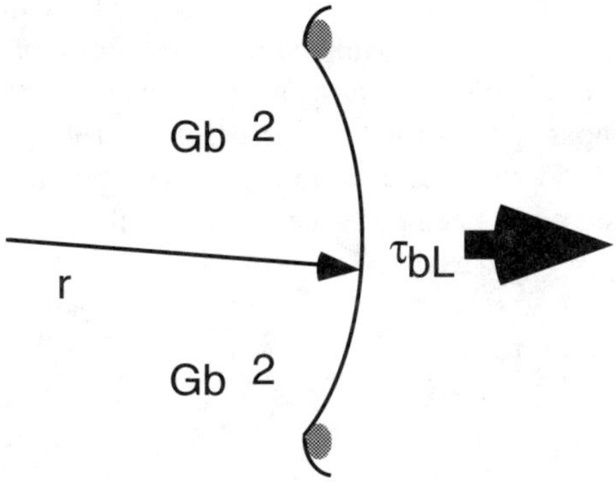

FIGURE 4-7 Schematic of a pinned dislocation which, bowing due to resolved shear stress in its plane, gives rise to anelasticity. For low stresses (small bow-outs or large radii of curvature) the area swept out is approximately proportional to the stress, giving rise to a linear anelastic strain.

assume an arc-like shape with radius of curvature $r = Gb/\tau$ to balance force per unit length b with line tension Gb^2. For small bow-outs, the total area, A, swept out by the dislocation, is approximately $L^3/16r = L^3\ \tau/16Gb$ (Figure 4-7). If the total number of dislocation segments per unit volume is N, then the strain caused by the bow-outs of these dislocations is about NbA/2. At most, N is of order $1/L^3$, meaning that the relaxed modulus for this mechanism will differ from the high frequency modulus (i.e. G) by at most:

$$\Delta G \approx \frac{G}{32} \tag{4.14}$$

This is a rather small effect. At higher stresses, where sources begin to operate and dislocations can travel large distances before becoming temporarily pinned, the effect becomes much larger. The strains become nonlinear and begin to lose reversibility. A phenomenological model for this process is described in the next subsection on microplasticity.

Characteristic Relaxation Time due to Dislocation Bowout

Generally, there will be a viscous retarding force against the motions of dislocations due, for instance, to interactions between the dislocation and solute atoms or (if the dislocation is nearly screw) the drag of jogs. For the former case, the force per unit length per unit velocity acting on a dislocation is approximately $f=(C_0\beta^2/D_skT)\ln(R/b)$ where C_o is the solute concentration, β is a quantity measuring the strength of interaction between dislocation and solute atom (typically $\beta/b \approx 0.01Gb^3$, which has units of energy), D_s is the diffusivity of the solute atom, and R is a cutoff radius (which can, for instance, be taken as L; with dislocations, such logarithmic terms are usually approximated as being constant) [Hirth and Lothe, 1982]. The dislocation can be viewed as a string of tension Gb^2 stretched between points a distance L apart [Weertman and Weertman, 1992]; by analogy to the string, and in absence of inertial effects, the characteristic time for decay in position of the string (once it has been perturbed from equilibrium) is given by:

$$\tau_0 = \frac{fL^2}{8\pi^2 Gb^2}$$

$$\tau_0 = \frac{C_0\beta^2L^2\ln\left(\frac{R}{b}\right)}{8\pi^2 D_s kTGb^2} \tag{4.15}$$

4.2.3 Microplasticity

Microplasticity is a term used to describe inelastic flow at small strains, typically 0.1% to 1%. Microplasticity has in common with anelasticity the recovery or partial recovery of small reversible strains, but the reversion might in practice take forever unless the loading direction is reversed [Jackson et al., 1981]. The Bauschinger effect (kinematic hardening), permanent softening, and the backstress are all observable phenomena associated with microplasticity.

Any microscopic picture of deformation quickly establishes that on a local scale it is a heterogeneous process: dislocations are unevenly distributed [Mughrabi, 1983; Laird et al., 1986]; dislocations pile up against different-strength barriers during yielding [Jackson, 1981]; non-deforming particles give rise to a backstress [Pedersen, et al., 1981; Brown and Stobbs, 1971]; two or more phases with different flow strengths may be present; and individual grains are oriented differently [Hutchinson, 1964; Ono, et al. 1983]. These and other considerations give rise to models of deformation in which microplasticity is a direct consequence of inelastic flow on the microscopic level.

Phenomenological Models of Microplasticity

Models for microplasticity are illustrated in Figures 4-8 and 4-9. Figure 4-8 is a Masing-type model [Sowerby et al., 1979] representing the material as a collection of parallel "phases." In turn, each "phase" consists of a spring in series with an inelastic element. This model clearly resembles a multiphase material such as the Pb-Sn eutectic, but the resemblance should not be carried too far. Apparently, there are other effects, discussed in the next section, that can make a real two-phase material with fine-scale microstructure stronger (or, for that matter, weaker) than a simple average would imply.

A model due to Jackson et al. [1981] and applied to eutectic solder by Wilcox [Wilcox 1990; Subrahmanyan et al., 1989] is shown in Figure 4-9. In this model, barriers against dislocation motion consist of two types: weak and strong. The parameter a_1, a linear spring, represents the buildup of dislocations against strong barriers while a_2, another linear spring, represents the pileup against weak barriers; $\dot{\alpha}_2$ represents the leakage of dislocations past weak barriers and $\dot{\alpha}_1$, giving

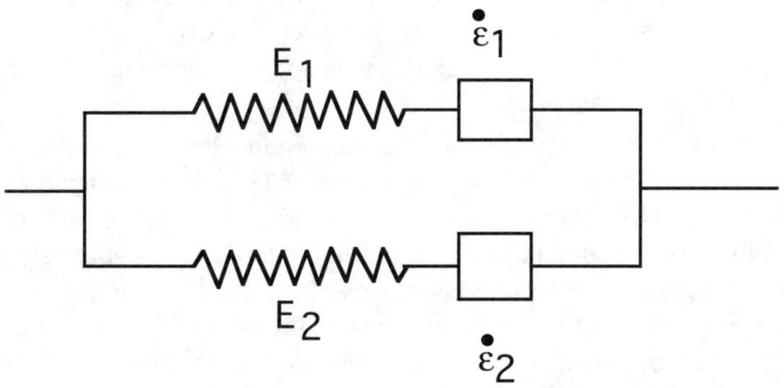

FIGURE 4-8 A Masing-type of model for the Bauschinger Effect. The boxes represent nonlinear dashpots.

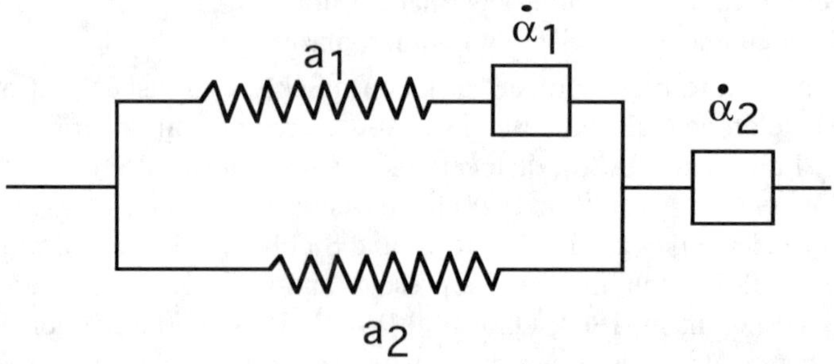

FIGURE 4-9 Microplastic model of Jackson et al. [1981] for the Bauschinger effect. In the model, spring a_1 represents the pileup of dislocations against weak barriers and a_2 represents the pileup of dislocations past strong barriers. $\dot{\alpha}_1$ represents the leakage of dislocations past weak barriers and $\dot{\alpha}_2$ their leakage past strong barriers.

rise to irreversible deformation, represents the leakage of dislocations past strong barriers. During straining in one direction, say, shortly after a stress reversal, the microplastic element "hardens" as progressively "stronger" weak obstacles are encountered.

4.2.4 Creep (Plasticity)

The term "creep" implies time- and temperature-dependent deformation. Because all inelastic deformation phenomena involve kinetic processes on the microscale, creep takes place at all temperatures. Creep itself is therefore not a mechanism but rather a collection of phenomena including, at low temperatures, logarithmic creep. For that matter, the creep test, which involves the application of a constant load or stress, does not measure "creep" any better than does a displacement-controlled experiment; rather, different types of experiments emphasize different aspects of creep phenomena. From this perspective the division of stress-strain curves into creep and plastic components, as is done with strain range partitioning, has little basis in mechanism.

There are fundamental differences in rate-controlling mechanisms between high temperature "creep" and low temperature "plasticity," particularly in terms of the roles played by vacancies in dislocation climb at high temperatures; but as a matter of practical concern, low temperature "plasticity" is not ordinarily relevant to the deformation of solders because solders are low melting-point alloys. Room temperature is 1/2 the liquidus temperature of a 95Pb-5Sn solder and 2/3 the melting point of the eutectic. A model that distinguishes between "plasticity" and "creep" for these materials, even at such low temperatures as –40°C, contains what would appear from a materials science perspective to be an awkward and unnecessary distinction. This is not to say, however, that the processes underlying low temperature "plasticity" are irrelevant to high temperature "creep." Primary creep at elevated temperatures is related to workhardening at low temperatures because both result from the accumulation of dislocations.

There are only a few situations in which creep phenomena are understood in any great detail. There are only a few classes of problems, such as in diffusional creep [Nabarro, 1948; Herring, 1949; Coble, 1963] and the creep of class-A solid solution alloys [Weertman, 1957; Bird, et al. 1972; Mohamed and Langdon, 1974], in which a first-principles approach will predict both the power law exponent and, to within an order of magnitude, value of the creep rate. As a consequence of this limitation we shall have to rely extensively on the phenomenological construction of the constitutive model.

Phenomenological Considerations

Efficiency and simplicity in the stress analysis demand that the inelastic deformation response of a material depends on the values of a few internal parameters η_i, representative of the microstructure and history of deformation. Moreover, plastic flow is manifestly kinetic and demands a rate formulation:

$$\dot{\varepsilon} = \dot{\varepsilon}(\sigma, T, \eta_i) \tag{4.16}$$

In general, the η_i evolve according to:

$$\frac{d\eta_i}{dt} = A(\eta_i) + B(\eta_i)\,\dot{\varepsilon} \tag{4.17}$$

The first term on the right of the evolution equation (4.17) corresponds to processes that in order to occur do not require that deformation be simultaneously taking place. These include static recovery and static recrystallization. The second term corresponds to processes that can only occur while the material is deforming ("dynamic" phenomena, e.g. dynamic recovery, dynamic recrystallization, workhardening).

These equations are very general, and could, with minor modifications such as incorporating $\dot{\sigma}$ on the right-hand side of Equation (4.16), be made to accommodate elastic, anelastic, and microplastic phenomena.

Equation (4.17) presupposes dynamic and static processes to be of the same kind or at least lead to the same types of effects. Likely, this is an oversimplification. For instance, static recrystallization and rotation recrystallization (a type of dynamic recrystallization), both of which tend to soften a material, do so in different ways and evidently lead to different looking microstructures [Poirier, 1985]. Likewise, the supposition that effects of microstructure and microstructural evolution can be handled in terms of smoothly varying functions must be an oversimplification. It lies in conflict with evidence that microstructures, when they evolve, often do so toward radically different-looking microstructures with (presumably) different deformation mechanisms. An example is the as-cast Pb-Sn eutectic vs. the same material after rolling. Despite the possible oversimplifications implied by Equation (4.17), it serves as a starting point for exploring the roles of dynamic and static processes, including evolution of microstructure, on the mechanical properties of the solid.

Steady State Creep

The majority of creep models concern the stress- and temperature-dependence of steady state creep [Weertman and Weertman, 1983]. The familiar power law, or Dorn equation, can frequently be fit to secondary creep data over at least four or five decades of strain rate [Bird et al., 1969]:

$$\varepsilon = A \frac{D D_{eff} b}{kT} \left(\frac{\sigma}{G}\right)^{n} \tag{4.18}$$

Here, A is an empirical constant and D_{eff} is an effective diffusion coefficient. To take into account power law breakdown, Garafalo's [1963] hyperbolic sine equation is sometimes used. It has the form:

$$\dot{\varepsilon} = C' \left(\sinh \left(B\sigma \right) \right)^{n} \tag{4.19}$$

where A, B, and C' are empirical constants and n is a power, typically about 4 or 5. C' is proportional to D_{eff}. At low stress levels, the hyperbolic sine law reverts to conventional power law creep, Equation (4.18).

Temperature Dependence of Power Law Exponent and Activation Energy

It should be emphasized that both the power law exponent, n, and activation enthalpy in the diffusion coefficient, D_{eff}, depend on temperature and level of stress [Weertman and Weertman, 1983; Frost and Ashby 1982]. A semi-empirical expression for the effects of temperature and stress on D_{eff} is:

$$D_{eff} \approx D_{l} + \frac{10 a_{c}}{b^{2}} D_{d} \left(\frac{\sigma}{G}\right)^{2} \tag{4.20}$$

Here, D_l is the coefficient of self-diffusion along dislocations, a_c is an effective cross-sectional area of a dislocation core ($a_c \approx b^2$), and $(10/b^2)\times(/G)^2 \approx \rho$ is the dislocation density.

A physical argument leading to the low-temperature activation enthalpy recognizes dislocations as fast paths for vacancy diffusion [Robinson and Sherby,

1969]. This means the effective diffusion coefficient will be a weighted average of the matrix and dislocation core diffusion coefficients [Hart, 1957]. The diffusion of vacancies along dislocation cores dominates at lower temperatures (yet typically above 0.5 T_m) and high stresses, in which case the activation energy decreases to about one-half of the high-temperature value.

An alternative justification for the reduced activation enthalpy at lower temperatures is the generation of vacancies by deformation [Nabarro, 1985, 1989], typically in concentrations of order 10^{-4} per unit strain, far exceeding intrinsic concentrations [Lawson and Meshii, 1991; Cottrell, 1953]. According to this argument, the self-diffusion coefficient, which is proportional to the product of vacancy concentration with vacancy mobility, should have its activation enthalpy reduced by about one-half. The controversy over this mechanism concerns whether plasticity-induced vacancies should be important at temperatures above 0.3 of the melting point [Nabarro, 1985; Mecking and Estrin, 1980].

Creep of Solid Solution Alloys

One of the more successful applications of dislocation theory has been in the area of solid solution hardening at elevated temperatures. Solid solution impurities can introduce a linear viscous drag on the motion of dislocations [Cottrell and Jawson, 1949; Hirth and Lothe, 1982] (see discussion in conjunction with Equation (4.15), leading to power law creep with an exponent of 3 [Weertman, 1957]). In such alloys (sometimes called class-A or class-I alloys) the strain rate, as a function of creep stress, shows a sigmoidal behavior as shown in Figure 4-10. At both low and high stresses the power law exponent is about 5, while at intermediate stresses the power law exponent is about 3. An expression for the secondary creep rate in class-A alloys in the intermediate regime of stresses is [Mohamed and Langdon, 1974]:

$$\varepsilon = \frac{\pi(1-\nu)\,kT\tilde{D}}{6e^2Cb^5G}\left(\frac{\sigma}{G}\right)^3 \tag{4.21}$$

where ν is the Poisson's ratio, $\tilde{D}$ is the diffusion coefficient of Darken, C is the concentration of solute, and ε, typically about 0.01 to 0.03, is a dimensionless measure of the interaction between dislocations and solute atoms. The term ε is usually taken as the fractional change in lattice constant per unit solute concentration.

Equation (4.21) is included in this chapter in recognition of solid solution strengthening as an important phenomenon in Pb-Sn alloys and other solder alloys [Fang et al., 1986; Stone et al., 1986; Weertman, 1960]. However, potential limitations of the applicability of Equation (4.21) should be noted. Most of the theoretical and experimental studies on solid solution strengthening at high temperatures concern the regime where diffusion is accomplished through motion of thermal vacancies transported through the lattice. Few fundamental, systematic studies have been conducted on solid solution creep in the regime of moderate temperatures (e.g. 1/2 T_m) where the activation enthalpy for creep is about 1/2 that for lattice diffusion. In the latter case, it is unlikely that Equation (4.21) applies, although the effect might be as straightforward as adding 2 to each of the stress exponents in Figure 4-10.

Evolution of Subgrain Size away from Steady State Creep

Even in constitutive modeling of large-strain deformations it is often necessary to be concerned with more than just steady state creep. Below, we address some observations and models of dislocation creep under conditions outside the steady state. The terminology "steady state" implies a dynamic stability in strain rate and microstructure; it should come as no surprise, then, that phenomenologies of creep away from conditions of steady state should emphasize the evolution of strain rate and microstructure during deformation.

In complex alloys, phase distribution, morphology, and dislocation structure will evolve during deformation. While phase distribution (i.e. the phases present and their respective volume fractions) and morphology are at least as important to mechanical properties as is dislocation structure, and their evolutions are likely to have a heavy impact on the evolution of dislocation structure, we only concern

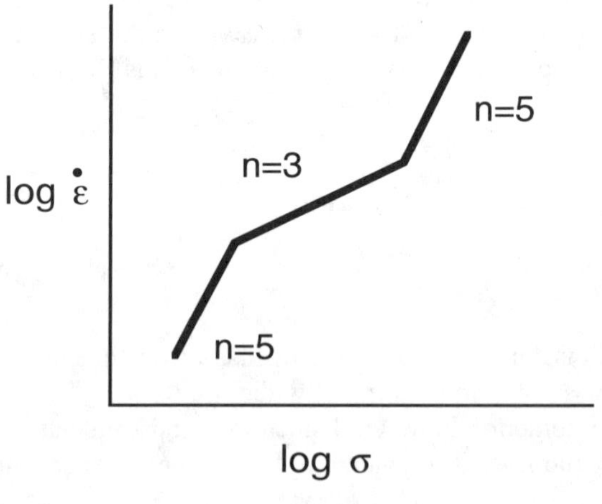

FIGURE 4-10 S-shaped locus of creep data typical of class-A solid solution alloys, in which there is a strong interaction between solute impurities and dislocations. The intermediate (n = 3) regime, perhaps the best understood of the three, is governed by the motion of dislocations under viscous drag by the solute. The other two regimes are considered as climb-controlled [Mohamed and Langdon, 1974; Oikawa et al., 1985]. Other deformation phenomena, such as superplasticity, can give rise to similar s-shaped curves.

ourselves with evolution of dislocation structure at this point, leaving for Part II of the chapter the evolution of phase size. It will be noted in this section, however, how the distributions of phases or precipitates can impact upon dislocation structure during creep.

It is common to picture the dislocation structure consisting of two components, “soft” and “hard.” The “soft” component consists of mobile dislocations distributed relatively uniformly throughout the crystal. The density of these dislocations adjusts relatively rapidly to changes in the level of applied stress; hence the terminology “soft.” The “hard” dislocation structure consists of dislocations trapped within subgrain walls. Because the average subgrain diameter appears to evolve more slowly with straining during primary creep, “hard” is used. Some of the most compelling evidence for this picture is provided in the investigations of Blum et al. [Blum et al., 1983; Eggeler and Blum, 1981] who suggest the evolution of the two structures is coupled. Gibeling and Nix [1981] argue for the subgrain walls (hard structure) as sources and sinks for the mobile dislocations (soft structure). In addition to Blum et al., other researchers have fully documented the connection between evolution of subgrain size and evolution of strain rate during primary creep and after stress reductions and/or increases [Ferreira and Stang, 1983, 1979; Goel et al., 1983].

Subgrains form in pure metals and to a lesser extent in solid solution alloys during creep. They consist of regions of crystal (in which are found the mobile dislocations) separated by low angle grain boundaries ($\approx$ 1°). The boundaries themselves consist of regular networks or arrays of dislocations. Sherby et al. [1977] pointed toward evidence that subgrains strengthen a material against creep and speculated that precipitation hardening stabilizes the subgrain size and therefore the creep strength. They examined a number of metals and found that the strain rate could be expressed as a function of stress and subgrain diameter:

$$\dot{\varepsilon} = \dot{\varepsilon}_0 \sigma^N D_A^P \tag{4.22}$$

where $\dot{\varepsilon}_0$ depends upon temperature, and $N \approx 8$ and $P \approx 3$. Actually, as has been pointed out by Gibeling and Nix [1982], the power law exponent N is probably not constant (at fixed D_A), but depends on stress, as illustrated schematically in Figure 4-11.

During creep, D_A evolves until a steady value is reached in steady state:

$$(D_A)_{ss} = 20G\frac{b}{\sigma} \tag{4.23}$$

where σ is the stress in tension. Meanwhile the strain rate evolves, also approaching a steady value [Ferreora and Stang, 1983]. The locus of steady state strain rates is shown in Figure 4-11 as a line with slope $\approx$ 5.

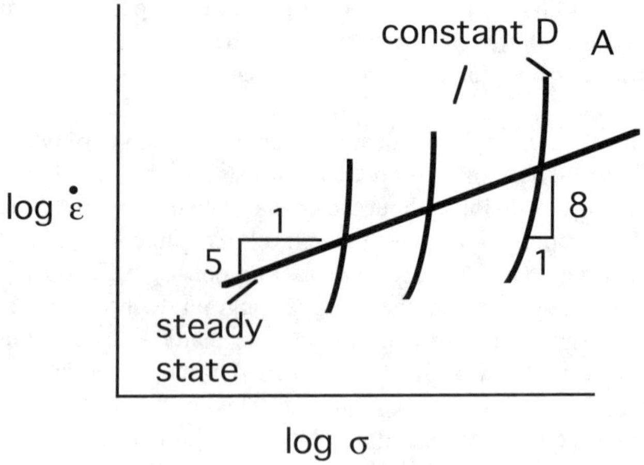

FIGURE 4-11 Strain rate as a function of stress at constant "structure" (subgrain size, D_A) compared with locus of steady state data. If, in a pure metal, the stress is kept constant, D_A will evolve until a steady state $\dot{\varepsilon}$ and D_A are reached. Sherby et al. [1977] postulate that precipitation hardening stabilizes a maximum subgrain size and therefore a minimum creep strength. This is why the power law exponent in precipitation hardened alloys is often high,~ 8. The figure is adopted from Sherby et al. [1977] except the constant lines, instead of being straight, are drawn as curved, more in line with experimental evidence.

Stone [1991] has proposed a physical model for the type of behavior exhibited in Equations (4.22) and (4.23) and illustrated in Figure 4-11. An essential point of the model is that dislocations encounter a distribution of obstacle spacings within the subgrain structure. Where the spacings are large, it is easy for dislocations to glide past obstacles, but dislocation climb in these regions is slow because the distances between sources and sinks for vacancies are large. Where the spacings are small, it is difficult for dislocations to glide past obstacles, but climb is rapid.

At a given level of applied stress (and fixed D_A), straining is accomplished through a mixture of two processes: glide past obstacles in regions where the obstacle spacing is large, and climb around obstacles where the spacing is small enough to impede glide. What determines how the two processes are partitioned throughout the crystal is the level of externally applied stress as well as subgrain size: if stress is high or subgrain size large, then most obstacles in the material are transparent to dislocation motion and the stress exponent is potentially very high (e.g. 50 or 100), characteristic of low temperatures and/or high strain rates. If, on the other hand, the stress is low (or the subgrain size small), then most of the obstacles are too close for the dislocations to glide past them; in this case dislocation climb dominates flow behavior and the stress exponent approaches 1. Throughout the regime of intermediate stress levels, the stress exponent varies continuously between the two extremes, which accounts for the curvature in the constant-D_A curves in Figure 4-11.

In this model a material tends to workharden at stresses where dislocation glide dominates, and it tends to soften (dynamically recover) at stresses where climb dominates. Therefore, if a material with specified subgrain size is crept at high stresses, it tends to workharden (the subgrains become smaller) as manifested in primary creep. If it is crept at low stress, it softens (larger subgrains). Because the subgrain size is a measure of what constitutes "high" stress vs. "low" stress, the ability of the subgrain size to evolve at constant stress serves as a type of negative feedback loop, allowing at any level of stress (or applied strain rate) a steady state to be achieved, in which workhardening and dynamic recovery cancel each other out.

This picture of structural evolution is interrupted when precipitates stabilize the subgrain size. In this case, if the precipitate size and spacing remain stable, to workharden the material (introduce smaller subgrains) will always be possible given a sufficiently high strain rate or low enough temperature. But to soften the material beyond a lower limit might not be very easy because the precipitates pin dislocations (of which subgrain boundaries consist) or at least slow them down. If the spacing between precipitates increases with time due to Ostwald ripening or deformation-induced coarsening (as appears to take place in the Pb-Sn eutectic [Frear, 1991]), the maximum subgrain size will also increase. Formally, such a process might be incorporated into a constitutive model through the second term on the right-hand side of Equation (4.17).

Hart's Model

This model owes its origins to Hart's analysis [1970] of the experimental criteria necessary for rigorously establishing whether the plastic equation of state (i.e. Equation (4.16)) can be satisfied in terms of a single internal state variable $\dot{\varepsilon} = \dot{\varepsilon}(\sigma, \eta_i, T)$. Indeed, there is evidence that, apart from anelastic or microplastic transients, a single-parameter description does exist, at least in some cases. Such evidence is found in the demonstration that strain rate sensitivity, $\nu = \partial \ln\sigma / \partial \ln\dot{\varepsilon}|_{\varepsilon}$, is solely a function of strain rate and stress (independent of deformation and thermal history) and that the workhardening parameter, $\gamma = \delta \ln\sigma / \delta\varepsilon_{\dot{\varepsilon}}$ is also a function solely of strain rate and stress. For instance, Hart and Solomon [1973] demonstrated $\nu = (\sigma, \dot{\varepsilon})\,\nu$ for high purity Al, while evidence for $\gamma = \gamma(\sigma, \dot{\varepsilon})$ in Al is found in the works of Li and coworkers [Korhonen, Hannula, and Li, 1985c; Wire, Ellis, and Li, 1976]. Excellent discussions of the underlying mathematics, formally analogous to the Second Law of Thermodynamics, along with possible physical ramifications of Hart's model behavior are provided by Nabarro [1990].

In its more familiar form, Hart's model is several steps removed from the original, more rigorous considerations. For instance, in the application of Hart's model to solders [Stone et al., 1986; Wilcox ,1990], little emphasis has been placed on the path-independent integrability criteria necessary to establish a single-parameter equation of state. Instead, the lambda law and scaling equations (shown below) are fit to stress relaxation data much the same way Equations (4.18) or (4.19) might be fit to secondary creep data without first establishing whether the secondary creep data actually reflect the steady state.

At elevated homologous temperatures, the internal state variable (σ^*; the "hardness" state) is identified as the high strain-rate-limit of stress based on a lambda-law extrapolation of stress relaxation data. The stress relaxation data look something like the constant-D_A curves in Figure 4-11, and it has been suggested that the stress-dip technique, usually used to obtain constant data, generate equivalent data to stress relaxation apart from the effects of transients [Korhonen et al., 1985a; Stone, 1991]. At constant σ^* the empirical relationship between flow stress and plastic strain rate is:

$$\ln\left(\frac{\sigma^*}{\sigma}\right) = \left(\frac{\dot{\varepsilon}^*}{e}\right)^{\lambda} \tag{4.24}$$

where σ^* is the high strain rate limit of the flow stress, $\dot{\varepsilon}^*$ is a rate parameter, and $\lambda \approx 0.1$ to 0.2 [Hart et al., 1975; Hart, 1976]. This equation is the "lambda law." The parameter $\dot{\varepsilon}^*$ is not independent of σ^*; rather it depends on σ^* and temperature according to:

$$\dot{\varepsilon}^* = \dot{\varepsilon}^*_0\left(\frac{\sigma^*}{G}\right)^m \exp\left(-\frac{Q}{RT}\right) \tag{4.25}$$

where $\dot{\varepsilon}^*_0$ is a constant, m can range anywhere from 1 to 20 (but is more typically 3 to 8 [Korhonen et al., 1987]), and Q is the activation energy for secondary creep. During deformation and in the absence of static recovery effects, it has been proposed that σ^* evolves according to [Korhonen et al., 1985b]:

$$\Gamma \equiv \frac{d\ln\sigma^*}{d\varepsilon} = \Gamma^*(\sigma^*)\exp\left(-\left(\frac{\dot{\alpha}^*}{\dot{\varepsilon}}\right)^{\Lambda}\right) \tag{4.26}$$

where

$$\Gamma^* \propto \sigma^{*\,-B} \tag{4.27}$$

Λ and B are additional fitting parameters, and $\dot{\alpha}^*$ obeys a scaling law similar to Equation (4.25).

Note that the form of the evolution law Equation (4.25) does not allow for the possibility of a steady state because Γ can never become zero or negative. It should be cautioned, however, that measurements giving credence to Equation (4.26) have always been undertaken at a combination of temperature and strain rate far from those needed to detect softening during deformation [Goel et al.,

1983; Ferreira and Stang, 1979, 1983]; therefore, there is no overriding reason to expect that Equation (4.26) can be extrapolated to conditions where the steady state is observed.

Grain Boundary Sliding During Creep

High temperature deformation generally proceeds by a combination of two phenomena, one in which all of the grains shear uniformly, elongating in the direction of applied stress, and another in which they slide relative to each other. Grain boundary sliding is not independent of matrix creep because, unless the boundaries migrate, displacements of grains can not be made compatible at triple points except by creep within the grain matrix [Brunner and Grant, 1959]. The accommodation process therefore becomes rate-limiting.

Migration readily takes place in pure metals at sufficiently elevated temperatures and may, in many cases, be an intrinsic part of sliding related to the dislocation makeup of the boundary [Ashby, 1972; Yavari and Langdon, 1983; Langdon et al., 1983]. Migration also takes place in solid solution alloys [Betrabet and Raman, 1988], but in most engineering alloys grain boundary precipitates inhibit it. In specimens with small grain size (say, below 10 μm in diameter), diffusional accommodation can take place [Raj and Ashby, 1971]. But in most alloys with grain sizes, say, in the 50 μm range, dislocation creep within the grain matrix is likely to become the predominant accommodation mechanism.

Based on the work of Hart [1967], rheological diagrams for matrix-accommodated grain boundary sliding are shown in Figure 4-12. Not surprisingly, these mechanical analogs show a close resemblance to the models of grain boundary anelasticity discussed earlier, the major difference being that the matrix springs in the latter are now replaced by nonlinear dashpot elements. The weighting factors (x) and (y) can be derived from mechanics-based analyses [Crossman and Ashby, 1975; Ghahremani, 1980b; Speight, 1976].

In Hart's and the other models, it is assumed that the rate of sliding is, in the absence of triple point constraints, governed by a power law expression:

$$\dot{\varepsilon}_{gb} = \sigma^m \tag{4.28}$$

where m is less than n, the power law exponent for matrix creep. The justification seems to be dictated by experiment and is rationalized based on sliding models in which diffusional processes play strong roles in accommodating local incompatibilities brought about by sliding; for an example see Equation (4.10) above. Because $m < n$, the grain boundary sliding elements in Figure 4-12 contribute negligibly at high stresses. For the same reason, at low stresses the grain boundaries become relatively slippery (they act as shear cracks) and the stress becomes transferred to the matrix. The width of the intermediate regime is a measure of the difference between the exponents n and m [Speight ,1976].

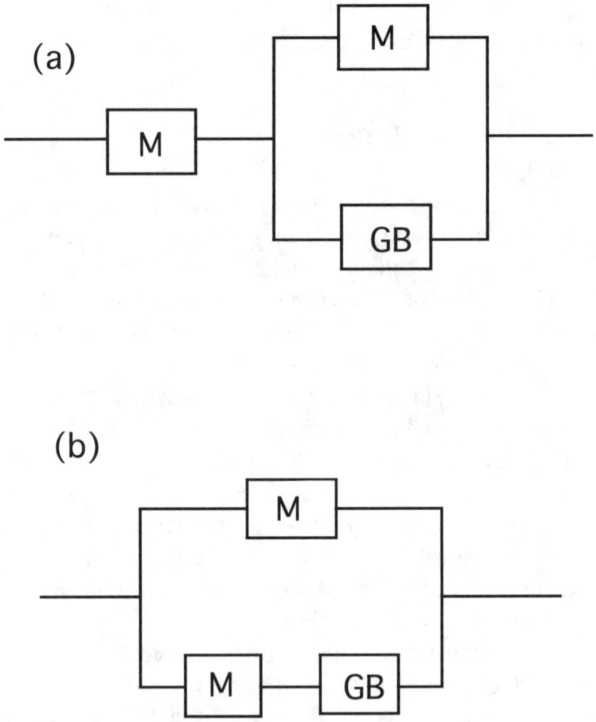

FIGURE 4-12 Rheological models for grain boundary sliding based on the work of Hart [1967]. The models are similar to those for grain boundary anelasticity, with the linear springs replaced with nonlinear dashpots.

4.2.5 Experiments to Examine Constitutive Behavior and Measure Constitutive Parameters

General Considerations

How does one experimentally probe for constitutive behavior? Without recourse to a physical or phenomenological model, one has little hope of addressing this problem. Other than having this model to serve as a guideline, one must recognize the most basic objective of constitutive modelling: simulating the deformation response of a solid under a variety of loading paths. This objective should be reflected in an experimental program.

If solders were to behave nicely like the standard linear solid shown in Figure 4-3a, the identification and measurement of constitutive parameters would be a simple matter. To characterize E_1, E_2, and $\eta=\eta(T)$ would require internal friction experiments in which the frequency and temperature were varied. Unfortunately, the standard linear elastic solid is probably not a very appropriate constitutive model for solders even for fatigue loading at strain ranges below 1%. Moreover, except for very universal behavior like the Dorn equation (Equation (4.18)), one is not likely to have a clear picture of what the deformation response will be. Therefore, it is perhaps better to establish an experimental program based on the most general description of constitutive behavior, Equations (4.16) and (4.17), and then work toward a more specific description given that the physical processes similar to those discussed above might contribute to the flow behavior.

Philosophical Underpinnings

Again, it should be reiterated that, although Equations (4.16) and (4.17) are located in the section on macroplastic (large-strain) deformation, they are very general and include, for instance, the possibility of encompassing anelastic and microplastic phenomena as well. In fact, they might seem to be so general as to be useless for designing a set of experiments. Fortunately, this is not the case; in fact, Equation (4.17) holds a clue as to how to probe for constitutive behavior: identify experiments in which certain of the internal state variables are constant and identify other experiments in which they evolve. The strain, time, and microstructure are all variables of relevance to this evolution.

Now, there are a number of ways in which the internal state variables can be held "constant" during an experiment. Equation (4.17) implies that if strain is held constant, those parameters sensitive to further straining will not change during the experiment. It implies that if time is held constant, variables that are explicitly time-dependent will not change during the experiment. Microstructure-dependent variables will remain constant if experiments are performed where the microstructure is held constant. Finally, if conditions can be established where the sum on the right-hand side of Equation (4.16) is identically zero, then the internal state variables remain constant with time and straining.

It is of course impossible to measure the strain rate as a function of stress under conditions of either constant strain or constant time. Thus, what is meant by "constant" must be taken to be that: (1) if the strain increment over which the stress-strain rate behavior is measured is small enough that dynamic processes $B(\eta_i)$ can be neglected, then the experiment is performed at "constant" strain; (2) if the time increment over which the stress-strain rate behavior is measured is small enough that static processes $A(\eta_i)$ can be neglected, then the experiment is measured at "constant" time. Such operational definitions of "constant" are central to certain types of tests including stress relaxation, strain rate-change, and stress change. Similarly, the possibilities of $A(\eta_i)$ and $B(\eta_i)\dot{\varepsilon}$ either adding to zero or being separately zero is central to the concept of the steady state. Usually, "steady state" creep is something one characterizes by using constant stress experiments to measure the rate of secondary creep. There are some problems with this approach insofar as secondary creep is not guaranteed to imply that a steady state has truly been reached [Hart ,1980]. On the other hand, none of the experiments discussed in this chapter are, by themselves, highly reliable for identifying the η_i and monitoring their evolution. It is only through a combination of experimental techniques that one can probe the applicability of concepts in Equations (4.16) and (4.17).

We start first with experiments to measure flow properties at constant based on "constant" strain and time. The considerations held below can be generalized to microstructures as well.

Stress, Load Relaxation, Rate Change, and Stress Change Experiments

The stress-change and strain rate-change tests are in principle similar, differing only in the variables of experimental control. In the stress-change experiment, the specimen is crept at one level of stress until a predetermined condition (such as strain) is reached, at which point the stress is rapidly changed. The specimen responds with an immediate elastic contraction (or extension, depending on the direction of stress change) followed by a series of transients throughout which the strain rate and dislocation structure change with time. In the strain rate-change test, the specimen is strained under displacement (or strain) control until a predetermined stress or strain is reached, then the displacement rate (or strain rate) is rapidly changed. The resulting transient response is in the stress required to maintain deformation at the new strain rate, and the information contained within this transient is similar to that contained within a strain transient in a stress-change test. There are some practical differences, however. Because strain rate is much more sensitive to stress than vice versa, it is easier to achieve large stress reductions (strain rate reductions) in a stress-change experiment than a strain rate-change experiment.

Following a stress change, the different η_i's will evolve at different rates. Taking the example of a pure metal such as Al, it is possible to describe the response of the material in terms of a two-parameter model, the mobile dislocation density and subgrain size D_A corresponding to the "soft" and "hard" dislocation structures discussed above. The first transient involves reorganization of mobile dislocations in response to the new level of stress. It is variously called anelastic or microplastic backflow [Gibeling and Nix, 1981; Korhonnen et al., 1985a] and may even cause the strain rate to become negative immediately after the stress change if the latter is negative and sufficiently large. Gibeling and Nix [1981] point out that the "constant structure" or constant subgrain size D_A—see Equation (4.22)—experiments of Sherby et al. [1977] correspond to the point where the first transient has decayed. The second transient, of longer duration, involves the slow evolution of D_A to a new value predicated by the new stress. How these types of transients affect the strain vs. time curve are illustrated in Figure 4-13 for a pure metal. Analogous phenomena can be seen in strain rate change tests in which the displacement, not the stress, is controlled. The results of a strain-rate change experiment are illustrated schematically in Figure 4-14. It should be noted that tran-

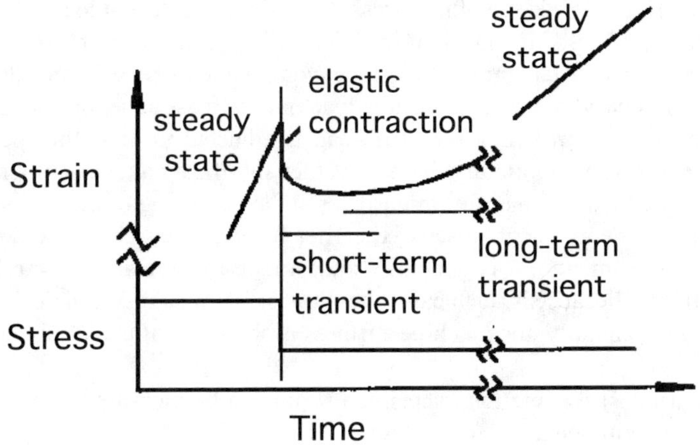

FIGURE 4-13 Strain vs. time behavior of a metal before and after a reduction in stress in a stress change experiment.

sients in class-A solid solution alloys (solid solution alloys in which solid solution strengthening is important) differ systematically from those in pure metals [Yaney et al., 1987].

Stress relaxation is another means of obtaining data at (nearly) constant strain and time. It is often stated that stress relaxation experiments take place under constant total strain, a statement which leads to much confusion because it is generally untrue. Instead, in most stress relaxation experiments the crosshead displacement

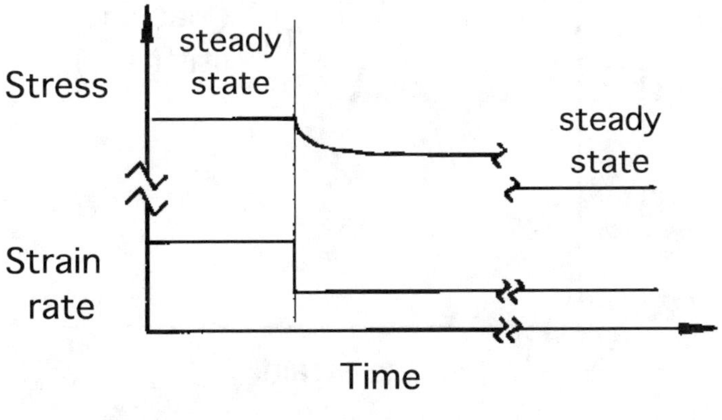

FIGURE 4-14 Stress vs. time behavior of a metal before and after a reduction in imposed strain rate during a strain rate change experiment.

is maintained fixed. The slope of the stress vs. strain curve during load relaxation is the machine stiffness, as indicated in Figure 4-15. If the stress relaxation test were performed at constant strain, this slope would have to be infinitely steep. In practice, it is not easy to obtain an infinite machine stiffness except through some feedback mechanism in which the crosshead is adjusted to keep the specimen strain constant. More importantly, because rather significant anelastic/microplastic transients are likely to manifest themselves at the beginning of the stress relaxation test (the same transients discussed earlier for the stress-change and strain-rate change experiments), it is generally undesirable to have an extremely high machine stiffness because in such cases these transients would dominate throughout relaxation. Generally, the machine stiffness is about 1/3 of the specimen stiffness.

Given that the stress and total specimen strain can be measured, the inelastic strain rate is determined as:

$$\dot{\varepsilon}_{in} = \dot{\varepsilon}_{meas} - \frac{\dot{\sigma}}{E} \tag{4.29}$$

where $\dot{\varepsilon}_{meas}$ is the measured strain rate. Stress relaxation data are illustrated in Figure 4-16 in the form of log σ vs. log $\dot{\varepsilon}$.

It is common, after an initial stress relaxation experiment, to perform what is called a reloading relaxation test in which the specimen is reloaded to below the

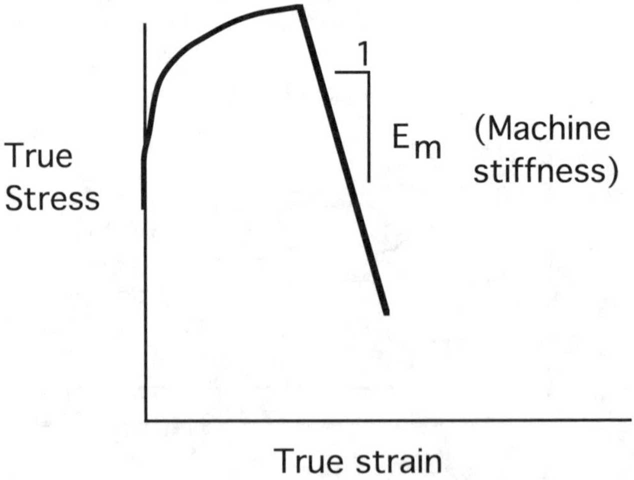

FIGURE 4-15 Tensile data from a displacement-controlled test that includes stress or load relaxation. The slope of the stress vs. strain curve during relaxation corresponds to the machine stiffness.

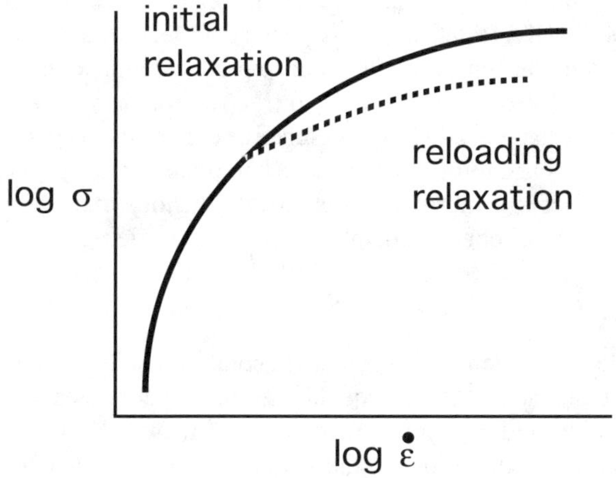

FIGURE 4-16 Stress relaxation and reloading relaxation data where the reloading data converge on the (initial) stress relaxation data.

previous stress level immediately preceding the first stress relaxation, then allowed to relax a second time [Hart and Solomon, 1973]. The reason for the second relaxation is to identify whether the structure has changed during the first experiment by determining whether the two sets of experimental data converge at low strain rates. The reloading relaxation test can be a powerful tool to help in the interpretation of stress relaxation data and construction of constitutive models. One should approach interpretation of the data with some caution, however, because what constitutes "constant structure" is predicated partly on the experiment. In other words, the constitutive behavior derives from certain internal parameters η_1, η_2, η_3,..., η_k. Some of these (e.g. the mobile dislocation density) adjust very rapidly during the relaxation in response, say, to the smooth variation in the level of stress. Others, such as the grain size, will adjust very slowly and may therefore evolve imperceptibly during a relaxation. Clearly, then, the concept of "constant structure" applies only to some of the internal parameters. If one or more of the parameters evolves incompletely during the initial relaxation, the data of the reloading relaxation will not converge with data of the initial relaxation. On the other hand, if all of the η_i evolve so rapidly that they could be considered to be functions solely of the current level of stress at any point during relaxation, the concept of constant structure would clearly not apply even though the reloading relaxation data would quickly converge to the initial relaxation data.

Stress relaxation, stress change, and strain rate change experiments each hold advantages and disadvantages over the others. In practice, therefore, it might be useful to combine two or more experiments. With stress relaxation data it is possible in bulk specimens to sweep out a spectrum of strain rates between, say, 10^{-2}/

s and 10^{-9}/s in tests that last 24 hours given that excellent temperature stability is needed to achieve reliable data at 10^{-9}/s (data at 10^{-7}/s are easy to obtain, however). The low strain rate achievement is not so practical with stress change experiments, where over 100 days are required to creep a specimen 1% at 10^{-9}/s. On the other hand, stress-change and strain rate-change experiments are useful for establishing laws governing evolution of the state parameters. They provide a well-controlled, yet significantly more complex loading history than is provided under constant stress creep or constant displacement rate tensile tests.

Steady State Creep

In principle, the steady state corresponds to a condition, under fixed stress or strain rate, where all of the $d\eta_i/dt$ are equal to zero, due for instance, to a balance between hardening and recovery processes, and in which the system is stable against momentary perturbations in the stress or strain rate. Strictly speaking, the steady state is not identified based on secondary creep under constant stress, but rather experiments in which both the microstructure (including dislocation structure) and strain rate (or stress) are monitored with time subject to perturbations in the stress or strain rate. Stress-change and strain rate-change experiments can be employed in this regard.

Evolution of the Constitutive Parameters

Evolution laws for the internal variables may be obtained based on the transient behaviors, e.g., the primary transient in the creep test or the workhardening transient in a constant displacement-rate test. A number of papers dealing with fatigue loading can be found in a recent conference proceeding [Freed and Walker, 1991].

Cyclic Loading

Finally, we need to note the importance of obtaining cyclic stress-strain data as a means of examining cyclic hardening, softening, and the Bauschinger effect.

4.2.6 Synthesis of a Constitutive Model from Experiment, Phenomenology, and Microscopic Theory

In this section, data of the Pb-Sn eutectic [Lee, 1993] are used as a vehicle to illustrate and expand upon some of the foregoing concepts.

As an aside, we should note the importance of experimenting on materials having microstructures similar to the system being modeled. Stress relaxation data of wavy (as-cast; moderate cooling rates), globular (as-cast; fast cooling), and superplastic (recrystallized) solders are shown in Figure 4-17. Here, at intermediate stress levels, there are almost 3 orders of magnitude distribution in strain rates, in-

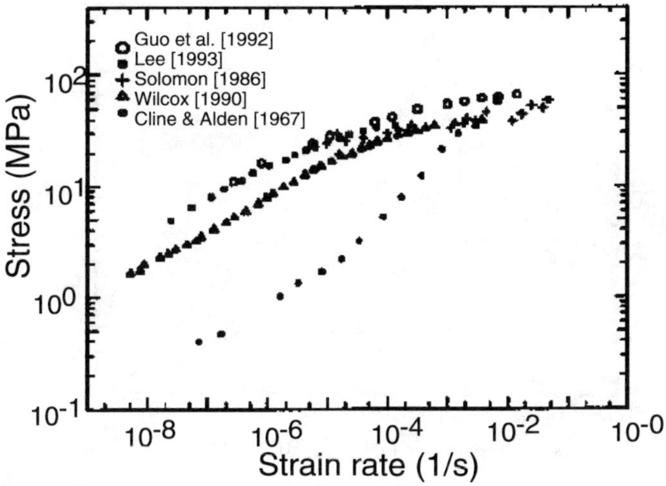

FIGURE 4-17 Room temperature creep, stress relaxation, and fatigue (stress amplitude vs. applied strain rate) data of eutectic and near-eutectic Pb-Sn solders. Data from bulk solders and shear-lap solder joints are included. The differences result from differences in microstructure.

dicating that the creep strength is heavily dependent on microstructure. This is not to say that one microstructure is always necessarily stronger than another; some specimens with globular microstructures have been found to be stronger than others with lamellar microstructures, possibly due directly to the influence of impurities on creep strength or indirectly due to the effects the impurities have on solidification [Lee, 1993]. Literature data obtained from fatigue loading in shear are also shown in Figure 4-17 [Guo et al., 1992; Solomon, 1986], and have been converted by factors of $\sqrt{3}$ for the stress and $1/\sqrt{3}$ in the strain rate to take into account von Mises equivalence. It is seen that they fall along the same general trends as the data obtained under monotonic loading.

Constitutive Phenomena: Experimental Data

Tensile Data

Stress-strain data of the wavy material are shown in Figure 4-18 [Lee, 1993]. In the wavy material (as well as the globular [Wilcox, 1990]) it is seen that the primary creep transient is relatively short-lived (<1% strain) except at higher strain rates where a weak type of workhardening manifests.

Comparison among Data Obtained from Different Types of Experiments

The lack of workhardening at low strain rates is emphasized in a striking way in Figure 4-19, where creep data (fixed stress), tensile data (fixed displacement rate),

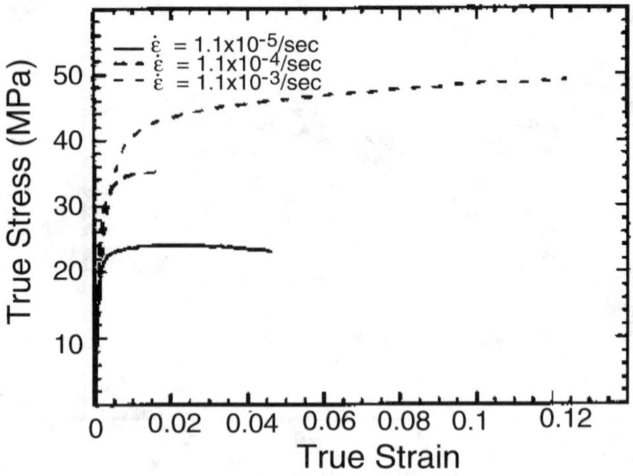

FIGURE 4-18 Tensile data of wavy Pb-Sn eutectic.

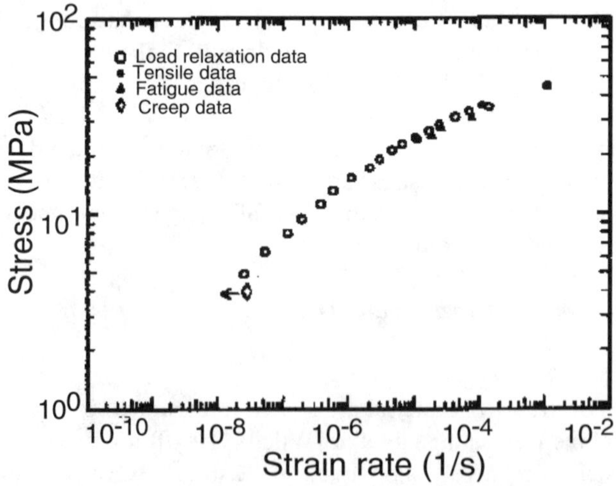

FIGURE 4-19 Comparison among stress relaxation, tensile and fatigue-loading data from specimens of a single wavy microstructure [Lee, 1993]. The data suggest that, to first order (and apart from the Bauschinger effect) the creep of the solder along any loading path can be treated with a steady state expression in which the power law exponent is a function of stress. There is about 10% cyclic softening of the Pb-Sn eutectic during the course of fatigue life apart from the softening due to cracking.

stress relaxation data, and fatigue data (stroke control) all fall approximately along a single curve. There is a slight cyclic softening during fatigue, which would account for the systematic placement of the fatigue data at 1-2 MPa below data of the stress relaxation and tensile tests.The close correspondence among the different experimental data means that, to a first approximation, they can be modeled in a quasi-steady-state approximation; that is, as if the internal state variables did not change with time: $\dot{\varepsilon} = \dot{\varepsilon}(\sigma, \eta_i, T); \eta_i = \text{const}$.

The power law exponent in Figure 4-19 depends upon stress, making the Dorn equation (Equation (4.18)) inappropriate unless fitted piecewise. The hyperbolic sine law, Equation (4.19), and Hart and Li's law, Equation (4.24), can be fit equally well to the experimental data as shown below.

Evidence of Evolution in the Dislocation Structure and Microstructure

Despite the quasi-steady-state description providing a good first-order approximation there is ample evidence that the dislocation substructure and two-phase microstructure both evolve during deformation. Figures 4-20a and 4-20b show a series of relaxation data obtained by successive load relaxation and reloading relaxation runs in which the specimen was strained by about 7% over a period of five days [Lee, 1993]. The sequence of experiments is: 1st load relaxation; reloading relaxation (#1); reloading relaxation (#2); and 2nd load relaxation. For the sake of clarity, the first relaxation data and two sets of reloading data are shown in 4-20a, while the second load relaxation run are included in 4-20b. An interesting feature of the data is that, apart from the initial transients, the first and second sets of reloading-relaxation curves fall at higher levels of stress than the initial load relaxation curve. The second load relaxation curve, taken after the material has been workhardened a second time at a high strain rate, follows the trend of the first load relaxation curve; that is, the data fall at stresses lower than those in the preceding reloading relaxation curves (arrows). These data show that workhardening the material at high strain rates causes a temporary softening to take place during the relaxation at low strain rates. Additionally, a type of permanent softening takes place during relaxation, in which the workhardening introduced at a high strain rate is lost during load relaxation.

To gain possible insight into microscopic phenomena that might underlie this anomalous behavior, the microstructure of the material is examined after deformation, the eutectic colonies have begun to coarsen at their boundaries, Figure 4-21. Such coarsening has been observed by others and has been associated with sliding between colonies. The colony (or grain) boundaries are indeed seen to slide, as shown in Figure 4-22.

Temperature Dependence of Creep

The temperature dependence of stress relaxation data of the globular eutectic is shown in Figure 4-23 [Wilcox, 1990]. Wilcox observed that the data in Figure 4-23

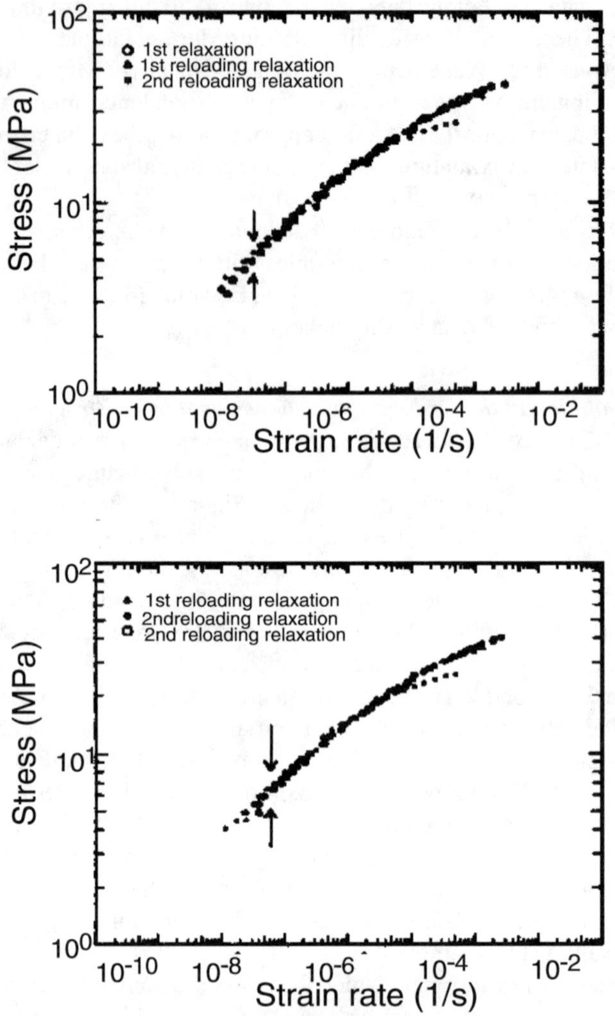

FIGURE 4-20 Load relaxation and reloading relaxation data showing a temporary softening in the data at low strain rate caused by workhardening the material at high strain rate. Permanent softening, witnessed at high strain rate, occurs during relaxation [Lee, 1993].

can be translated to overlap each other based on an activation energy of 46 kJ/mol. By deconvoluting the S-shaped curves based on a grain boundary sliding model (Figure 4-12a above), it was also possible to obtain activation enthalpies of about 30 kJ/mol and 82 kJ/mol for grain boundary sliding and matrix deformation, respectively.

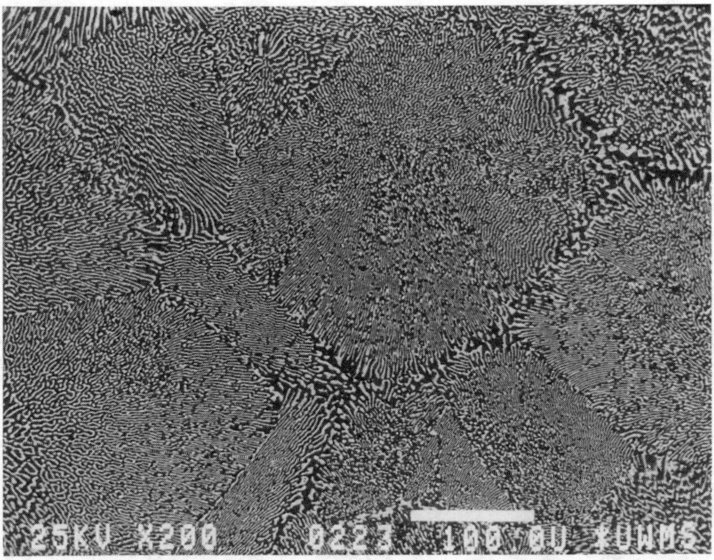

FIGURE 4-21 Coarsening in the solder, induced as a result of the deformation experiments in Figure 4-20 [Lee, 1993]. The coarsening might be related to the temporary softening at low strain rates, permanent softening at high strain rates, or both (see Figure 4-20)

The Underlying Mechanisms

This subsection examines the following constitutive phenomena: the temperature dependence of the strain rate; the effect of the two-phase microstructure on flow strength; the evolution of the flow strength during deformation; grain boundary sliding; and the Bauschinger effect.

Activation Energy for Creep

The temperature dependence of the strain rate is consistent with an activation enthalpy of about 46 kJ/mol, compared to 100-110 kJ/mol for bulk diffusion in Pb and Sn. This contrast does not constitute a discrepancy; in fact, in pure Pb and pure Sn the activation enthalpies for dislocation creep only begin to approach those of bulk diffusion at temperatures near the melting point. For pure Sn, the activation enthalpy is about 40-50 kJ/mol below 0.8 T_m [Suh et al., 1983], whereas in pure Pb, the activation enthalpy is about 65 kJ/mol at temperatures below 0.65 T_m [Frost and Ashby, 1982]. The melting points of these two materials fall such that, in each case, the transition temperature is about 400K. Grain boundary sliding and perhaps even interphase sliding are also likely to have activation energies about half of those for the bulk diffusivities; it is not an easy matter, therefore, to separate deformation mechanisms based on activation enthalpies alone.

FIGURE 4-22a Offset in a scratch mark due to grain boundary sliding.

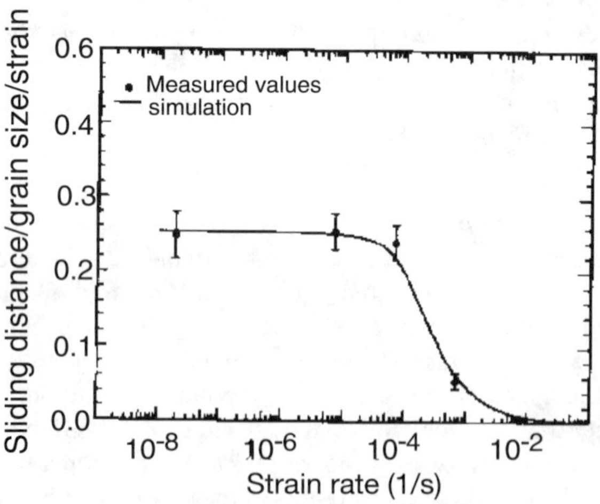

FIGURE 4-22b Sliding distance normalized against grain size, as a percentage of the strain during deformation at different strain rates. A transition is seen between no observable sliding at high strain rates and saturation in observed sliding at low strain rates. The transition is simulated based on a model in Figure 4-25.

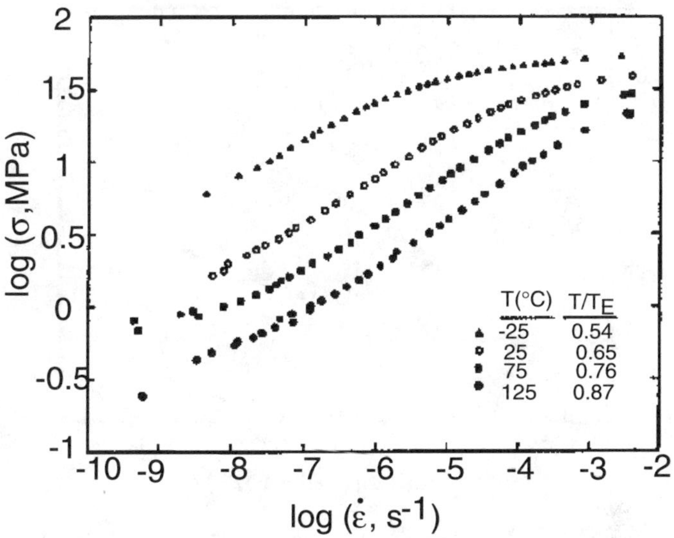

FIGURE 4-23 Temperature dependence of the flow strength of a globular eutectic. [Wilcox, 1990]

Comparison Between Creep Strength of Pb-Sn Eutectic and other Pb-Sn Alloys: Interpretation in Terms of Strengthening Mechanisms

In single-crystal, eutectic Pb-Sn alloys with well-aligned lamellae, sliding of Pb-Sn interfaces easily dominates over plastic deformation of the phases themselves unless the lamella are aligned perpendicular to the stress axis [Cagnon et al., 1977; Eberhardt and Baudelet, 1974]. The same type of sliding probably takes place within wavy, polycrystalline Pb-Sn eutectics, but it is not as likely to govern the flow of the polycrystalline material for two reasons: first, interphase sliding alone is incapable of maintaining compatibility between grains when a polycrystal deforms. Second, the phase boundaries in a polycrystalline, wavy eutectic are not very planar nor are the lamellae parallel over large distances. Thus, we can conclude that sliding, if it is present, must be accommodated by dislocation (or at low stresses, diffusional) mechanisms. For these reasons, it is likely that dislocation mechanisms control the flow of the polycrystal at high stresses, even in the presence of phase boundary sliding.

At high strain rates, the Pb-Sn eutectic is considerably stronger than either pure Pb or pure Sn. This assertion is demonstrated in Figure 4-24, where stress relaxation data of the eutectic [Lee, 1993] are compared with creep data of the pure metals [Cline and Alden, 1967] and a 95Pb-5Sn alloy [Stone et al., 1986]. The latter alloy combines both solid solution and precipitation-strengthening and perhaps, therefore, offers a better basis for comparison with the eutectic than does pure Pb. Because Sn will not hold large concentrations of Pb in solid solution, sol-

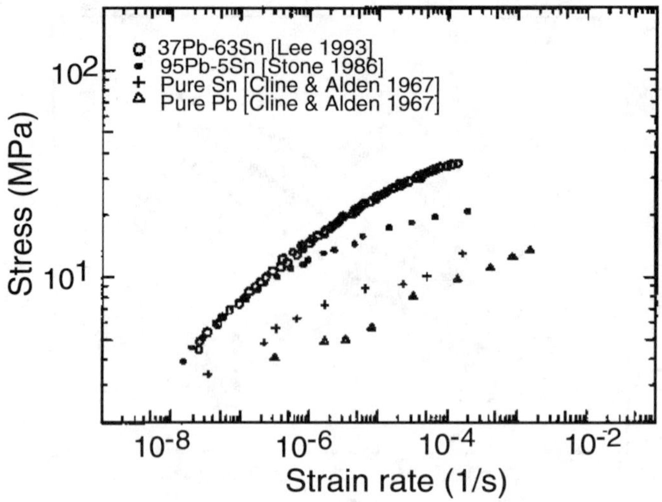

FIGURE 4-24 Comparison between the flow strength of the wavy Pb-Sn eutectic and those of its constituents.

id solution strengthening is not likely to be as important for Sn as it is for Pb. It is clear, then, that at high strain rates the eutectic is stronger than would be rationalized based on a volume fraction composite rule involving any of the other three materials in Figure 4-24. At low strain rates, it is questionable whether the eutectic material holds a great improvement in creep strength over the Pb-rich alloy.

Why is the wavy eutectic comparatively so strong? The effect of the Orowan mechanism – dislocations having to bow between particles – can be estimated as contributing at most:

$$\sigma \approx \frac{4Gb}{r} \tag{4.30}$$

to the flow stress at high strain rates, where r is the spacing between obstacles [Frost et al., 1987]. For Sn, G ≈16 GPa and b ≈ 3 Å; also, r ≈ 2 μm, representing the separation between Pb layers in this eutectic alloy. At most, therefore, the Orowan mechanism can account for about 10 MPa, a factor of 5 below the actual strengthening effect observed at high strain rates.

A more likely explanation for the high strength of the eutectic lies in the possibility that dislocation structures—if not subgrains then structures analogous to subgrains—are stabilized by this microstructure. Thus, using Equation (4.23) as a rough guideline, the eutectic microstructure locks in a dislocation structure of similar scale to that present in Sn crept at 50 MPa stress. Such a structure would be highly unstable in pure Sn; for instance, if crept for a short amount of time at this stress at room temperature, pure Sn would recrystallize. But in a two-phase alloy such as the Pb-Sn eutectic, the boundary migration processes responsible for recrystallization in the pure material would become inhibited. Finally, it should be noted that an external stress is not needed in order for such dislocation structures to come into existence; the Sn-rich phase is known to polygonize prior to application of external stresses due to thermal expansion mismatch between the Pb and Sn phases [Frear, 1992].

Evolution of the Dislocation Structure and Microstructure

The main point of the foregoing argument is that the stress-strain rate data in Figure 4-19 constitute constant "structure" behavior similar to the constant-D_A curves in Figure 4-11. This analogy should not, perhaps, be taken too far given the possibility of deformation mechanisms unique to the wavy microstructure, such as phase boundary sliding. Nevertheless, there is one further extension of the analogy that bears mention: at stresses below 50 MPa, a rough estimate based on Equation (4.23), the material will not workharden (i.e. exhibit a prolonged primary creep transient) because dynamic recovery dominates over workhardening. Yet, under the same conditions there is a limit to how much the material can soften because of its microstructure. In order for softening to take place the microstructure must itself evolve. At higher stresses, the material will workharden because of the enhanced role of dislocation glide; the strengthening effects of workhardening at high stress or strain rate are then quickly removed; for instance, when the stress is relaxed (Figure 4-18). The softening that takes place during relaxation might involve both static and dynamic processes.

Slow evolution in microstructure is evidenced in Figure 4-21, which shows coarsening at boundaries. Although coarsening is enhanced by deformation and accumulates during fatigue loading, the rate at which it accumulates is a strong function of strain rate and possibly strain range. Room temperature experiments at strain ranges below 1% and strain rates of 10^{-4} do not exhibit this type of coarsening except after hundreds of cycles [Lee, 1993]. Also, even though coarsening is enhanced, if not initiated, by straining, it continues to occur long after straining is stopped and is enhanced by subsequent annealing at elevated temperatures.

Apparently, to workharden the material at high strain rates temporarily makes it softer (compared to the unworked material) at low strain rates, as shown in Figure 4-20. The softening effect might be due to a type of dynamic recrystallization apparent at low strain rates; or, softening at low strain rates might be intrinsic to a workhardened structure as discussed by Stone [1991]. Regardless of the mechanism, a micromechanical constitutive model that relates workhardening to the low strain rate softening and to the coarsening in Figure 4-21 would present a significant improvement in the understanding of creep in the eutectic. This is an area that needs to be researched further.

Grain Boundary Sliding

The localization of coarsening at colony boundaries is consistent with the fact that they slide and that localized deformation takes place along them. When, on the other hand, the strain rate is high and sliding minimal, similar coarsening takes place except across colonies [Frear, 1991].

Let us speculate on the rate-controlling mechanism of sliding. Much of this speculation must rest on the power law exponent of the grain boundary element, Figure 4-12, which can only be estimated based on the width of the sliding/no sliding transition in Figure 4-22. The width of the transition is a qualitative measure of difference in stress exponents between matrix deformation and grain boundary sliding [Speight, 1976]. The sharper the transition, the larger will be the difference between exponents. In the present case, the difference is large because the transition takes place within about 1.5 orders of magnitude in strain rate; therefore, the stress exponent for the grain boundary is at most about 2 based on a straightforward analysis employing the rheological model shown in Figure 4-25. The qualifier "at most" means this rheological model does not account for the difference between individual boundaries within a polycrystal. For instance, grain boundaries at 45° to the stress axis generally have a higher driving force for sliding than do boundaries at, say, 85° to the stress axis. Additionally, boundaries will have a distribution of sliding viscosities caused by differences in grain boundary structure and the density of the precipitates on the boundary. These factors should make the apparent power law exponent somewhat larger than the value for a single, isolated boundary in a bicrystal.

Noting the likelihood that the stress exponent for sliding is lower than 2, we can further investigate the plausibility that sliding is governed by diffusion of matter around Pb precipitates on the boundary. First, we note the contribution to the strain rate due to grain boundary sliding is about x/2d where x is the rate of sliding. The average shear stress, τ, on the boundary is roughly $\sigma/2$ where σ is the uniaxial stress. Substitution into Equation (4.9) for the sliding rate gives the contribution of sliding to the strain rate:

$$\dot{\varepsilon}_{gb} = \frac{2\sigma\Omega\lambda^2 hD_i}{kTa^4 d} \tag{4.31}$$

In Figure 4-24 the transition from no sliding to sliding takes place at about 10^{-4}/s strain rate and 30 MPa stress. These values can be substituted in Equation 4-30, allowing us to solve for the value of hD_i based on knowledge of other parameters: $\Omega = 3\times10^{-29}$ m^3 (Pb), $\lambda \approx 5$ μm, a ≈ 2 μm, and d ≈ 200 μm. The calculation gives $hD_i \approx 3\times10^{-20}$ m^3/s, which is only reliable to within about an order of magnitude. Existing data for diffusion in alloys of Sn and Pb suggest hD_i should be roughly 10^{-21} m^3/s at room temperature [Kaur et al., 1989]. This value is close enough to our estimate that the diffusional accommodation mechanism cannot be

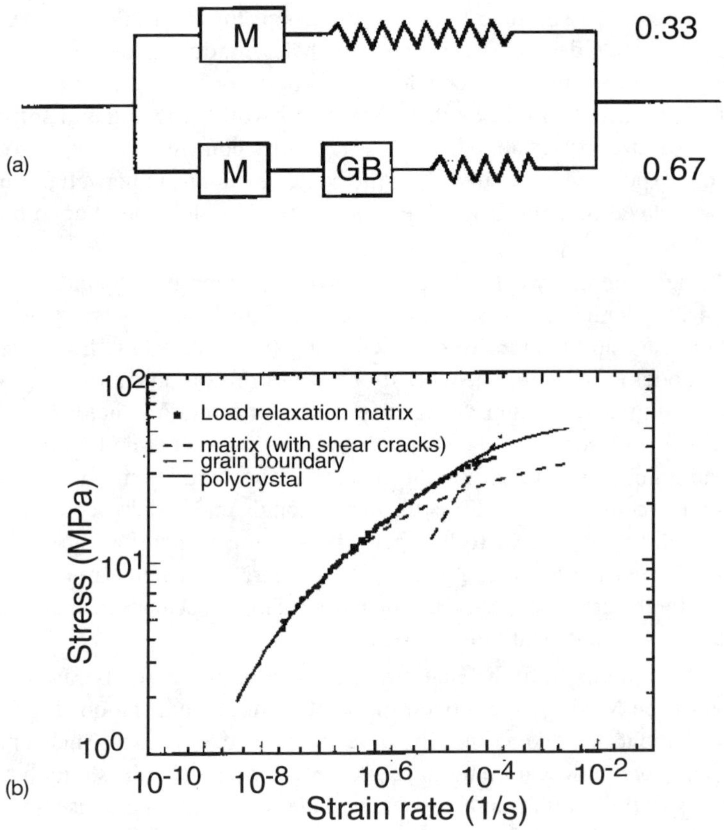

FIGURE 4-25 Simulation of data based on model incorporating grain boundary sliding: a) rheological model; b) simulation of stress relaxation data.

ruled out; however, evidence for dislocation creep-accommodation is suggested by the fact that the microstructure begins to coarsen adjacent to the boundaries under deformation at higher strains. The accumulation of dislocations and/or breaking up of lamella accompanying creep might be responsible for the initiation of coarsening at the boundaries.

Microplasticity and the Bauschinger Effect

We now turn to the subject of microplasticity. To do so, it is best to begin with an attempt to simulate stress-strain hysteresis loops based on the stress relaxation data in Figure 4-20. To facilitate the simulations, stress relaxation data are fit to both the hyperbolic sine law, Equation (4.19), and the lambda law, Equation (4.24), with parameters shown in Table 4-1. It is seen that both of the equations adequately describe the relaxation data, Figure 4-26a.

In an attempt to simulate the hysteresis loop, Figure 4-26b, we see that the models adequately describe the steady state behavior during loading in one direction, but they fail to predict the yielding transients, or Bauschinger effect. A portion of the transients might be due to repeated workhardening and softening as the level of stress is changed. However, if this contribution were strong, we would not expect the peak stresses in the hysteresis loops to be well predicted by the stress relaxation data; instead, the flow stress would depend upon history of deformation.

In the next attempt, we use the constitutive model for grain boundary sliding in Figure 4-25a, which gives a transient due to grain boundary anelasticity. Both grain boundary and matrix are modeled using the lambda law, Equation (4.24), and the model is fit simultaneously to the stress relaxation data in Figure 4-25b and the sliding data in Figure 4-22. The stress concentration factor Y is equal to 0.67, which is reasonable based on earlier discussions on grain boundary anelasticity and grain boundary sliding during creep. The simulation of the stress-strain data is shown in Figure 4-27. Still, even though under these conditions grain boundary sliding is known to be operative and must, therefore, give rise to an anelastic or microplastic-like transient, the measured transient is much larger than predicted by the grain boundary sliding model. This is because the modulus defect introduced by sliding is at most only E/3.

The final attempt to fit the Bauschinger effect in this solder is based on an adaptation of the Masing-type model, Figure 4-8. In this adaptation (Figure 4-28), the two elements are taken, literally, to be the Pb and Sn phases. Such a picture is clearly an oversimplification because, as shown previously, the strength of the alloy is not strictly a volume average. Nevertheless, a part of the Bauschinger effect in this alloy is likely to derive from a disparity between properties of Pb and Sn.

TABLE 4.1 Parameters Used in Simulations

Constitutive Law	*Parameter Values*
λ-Law (Figures 4-26a,b)	$\sigma^* = 137$MPa $\dot{\varepsilon}^* = 1.46\times10^{-7}s^{-1}$ $\lambda = 0.11$
Hyperbolic Sine Law (Figures 4-26a,b)	$C' = 3.78\times10^{-7}s^{-1}$ $B = 7.63\times10^{-2}MPa^{-1}$ $n = 2.8$
λ-Law (with g.b.s.) (Figures 4-25b and 4-27)	$\sigma^*_m = 66$MPa $\dot{\varepsilon}^*_m = 2.19\times10^{-6}s^{-1}$ $\lambda_m = 0.18$ $\sigma^*_{gb} = 295$MPa $\dot{\varepsilon}^*_{gb} = 5.80\times10^{-3}s^{-1}$ $\lambda_{gb} = 0.18$ $y = 0.67$

The volume fractions of Pb and Sn in the eutectic material are about 0.29 and 0.71, respectively. Based on these averages, based on the elastic moduli of Pb and Sn (16 GPa and 41 GPa, respectively), the data can be simulated by assuming that the Pb has twice the flow strength of the Sn. Whether such an assumption is warranted based on physical grounds can be questioned. At this point the simulations have become an exercise in curve-fitting.

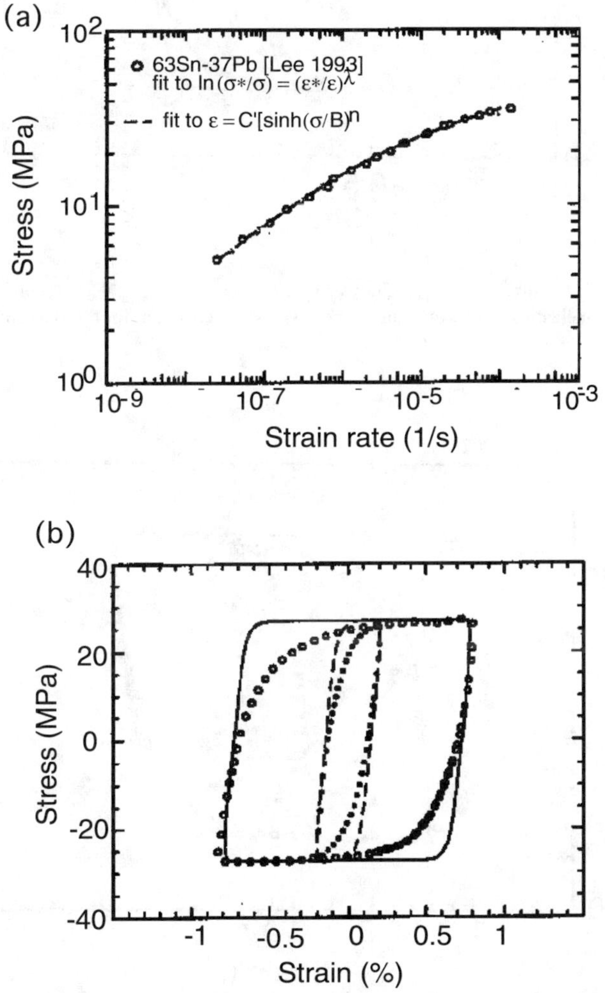

FIGURE 4-26 Simulations of flow behavior: a) fits of stress relaxation data to lambda law Equation (4.23) and hyperbolic sine law Equation (4.18), assuming that the initial part of the data contain no transient; b) simulations of hysteresis loops (there is some cyclic softening).

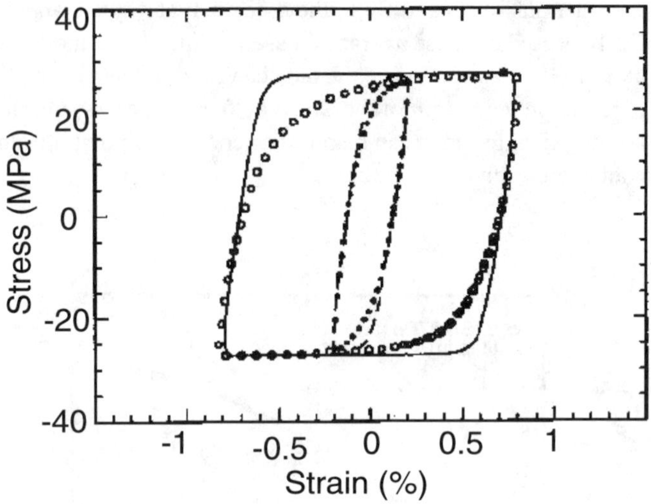

FIGURE 4-27 Simulation of hysteresis loops based on model in Figure 4-25. Grain boundary sliding introduces an anelastic effect which, nevertheless, does not account for the Bauschinger effect.

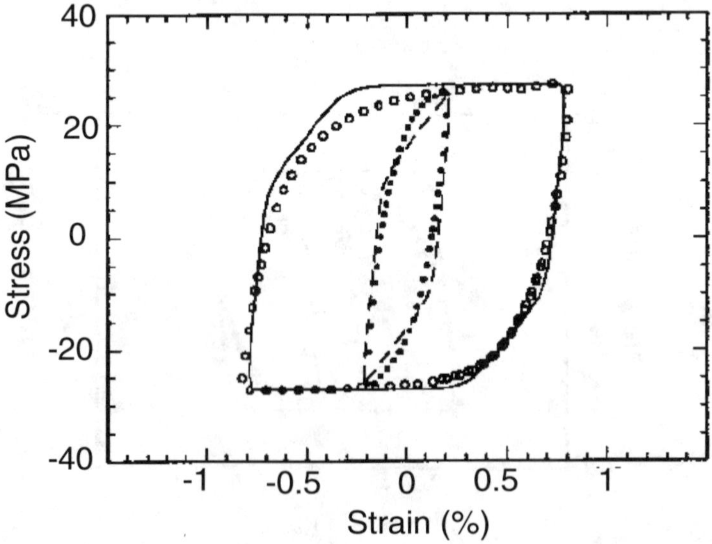

FIGURE 4-28 Simulation of hysteresis loops based on model in Figure 4-8. To account for the magnitude of the Bauschinger effect it is assumed that the Pb-phase has twice the strength of the Sn phase. (Perhaps this is too literal an interpretation of Figure 4-8).

The third microplastic model, based on the work of Jackson [1981], has been adapted to the globular Pb-Sn eutectic by Wilcox [1990], Figure 4-9. The model (Figure 4-29) is seen to fit the experimental data of Wilcox well, in part because the anelastic modulus is so low, approximately G/25.

We conclude by stating that all three of the microscopic phenomena discussed above—grain boundary sliding, different (flow strength)/(modulus) ratios of the phases, and transients in the dislocation structures—are likely to contribute to the Bauschinger effect in this material. Other microscopic processes not treated here, such as phase boundary sliding, might also contribute to the effect.

4.2.7 Application of the Constitutive Models

We have now seen how it is possible to implement varying levels of sophistication into the models to represent different microscopic processes. As a general rule, the effort into greater detail and greater levels of sophistication gives rise to improved accuracy but at the cost of greater numbers of constitutive parameters. Now that we have explored some of these options, the question arises: for what purpose and at what expense does greater detail get incorporated into the model?

Elastic-Steady State (Elastic-Viscoplastic) Model

Perhaps the simplest model capable of treating the creep of a solder is a type of elastic-steady state creep (often called elastic-viscoplastic) model shown in Fig-

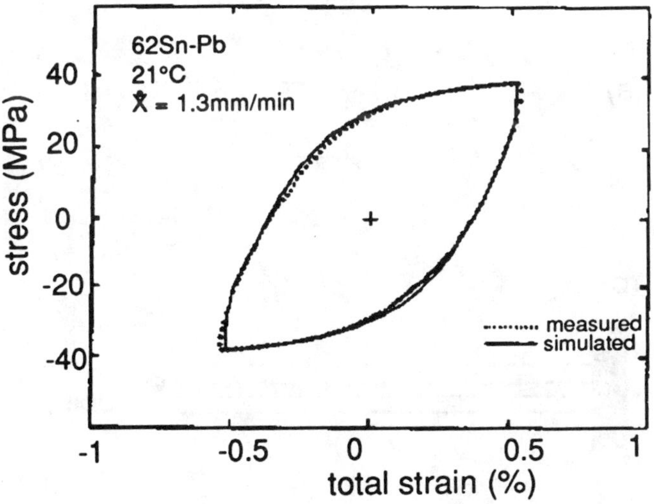

FIGURE 4-29 Simulation due to Wilcox [1990] of a hysteresis loop for the Pb-Sn eutectic. The simulation is based on the microplastic model of Jackson et al. [1981]. (See Figure 4-9.)

ure 4-30a. An example of such a model has been used in a simple one-dimensional analog of leadless solder joints between a ceramic chip carrier and the printed circuit board, 30b [Stone et al., 1985]. The model was used to simulate Hall's data [1984] on thermal stresses and strains induced in pillar-shaped, leadless solder joints. The governing equation relating changes in stress ($\dot{\tau}$) to changes in temperature (T) and inelastic strain rate of the solder ($\dot{\gamma}$) is:

$$\dot{\tau} = GS\left[\left(\frac{l_a}{h_s}\right)\Delta\alpha\dot{T} - \dot{\gamma}\right] + \tau S\frac{d\ln G}{dT}\dot{T} \tag{4.32}$$

where G is the shear modulus of the solder, S is the ratio of assembly stiffness to solder joint stiffness, l_a is the length of the soldered assembly, h_s is the height of the solder joint, and $\Delta\alpha$ is the difference in thermal expansion mismatch between chip carrier and printed circuit board. The simulation of Hall's data is shown in Figure 4-31. Qualitatively, the simulation behaves similar to the experimental data; however, there were probably two major shortcomings of the model at the time it was constructed that gave rise to the discrepancy: the wrong assembly stiffness, S, was used. Also, the creep strength of the solder at low temperatures was underestimated.

Wilcox [1990] has recently characterized the properties of solders and solder joints more carefully, and Subrahmanyan, Wilcox and Li [1990] generated a better simulation of the stress vs. temperature data, Figure 4-32. Their model includes a microplastic element as well, Figure 4-9, but the microplastic element is probably not as important in the improvement of Figure 4-32 over 4-31 as is getting the low

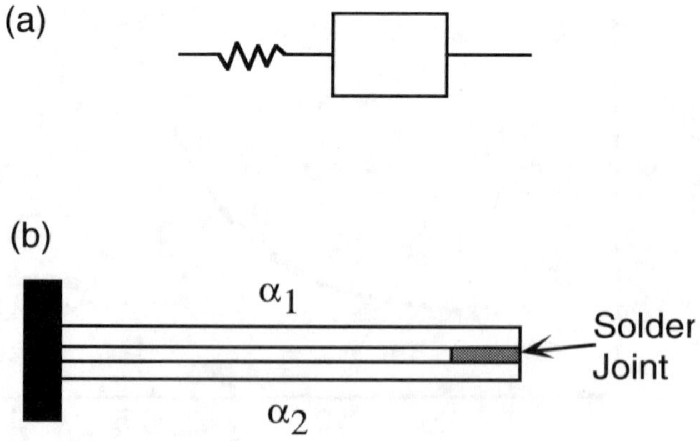

FIGURE 4-30 Model for simulation of thermal mismatch-induced stresses and strains in solder joints: a) rheological model; b) one-dimensional analog of Hall's [1984] pillar-type solder joints placed between printed circuit board and ceramic chip carrier [after Stone, Hannula, and Li (1984)].

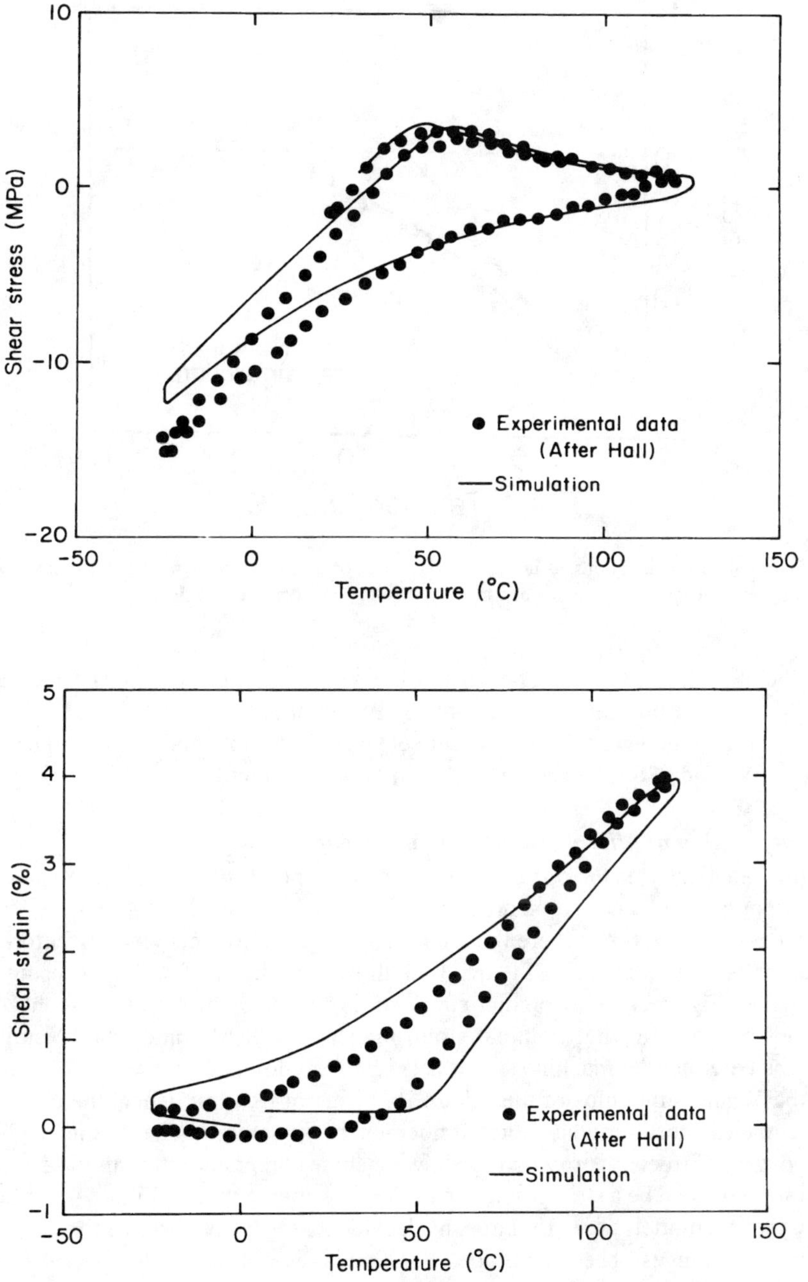

FIGURE 4-31 Stress vs. temperature and strain vs. temperature in simulations of Stone et al. [1986] compared with data of Hall [1984].

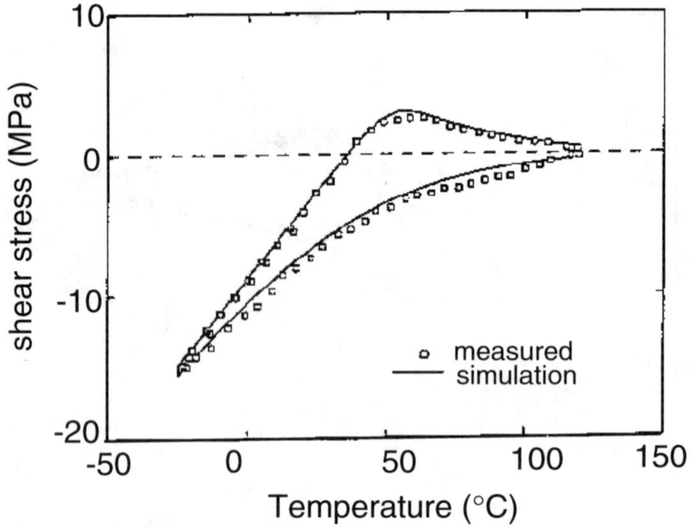

FIGURE 4-32 Simulations, as in Figure 4-31, based on a more accurate model by Wilcox [1990], that incorporates greater accuracy into the temperature dependence of the flow

temperature flow stress and power law exponent right because the shear strain is rather large, approximately 4%. Others [Pao et al., 1992, 1993] have employed steady state creep models within finite element codes to simulate the response of 95Pb-5Sn and 97Sn-3Cu solders to thermal displacement-controlled loading.

Incorporation of Microplasticity into the Model

What, practically, is the difference between incorporating microplasticity and not incorporating it? The major difference comes at low levels of stress, i.e. small ranges of temperature, where the microplastic model predicts yielding before the macroplastic model does. Figure 4-33 illustrates this difference by using the model in Figure 4-28 to predict stress-strain loops for displacement-controlled cycles at the following nominal strain ranges: 3%, 1.5%, and 1%. "Nominal" means here that the machine stiffness is relatively low, effectively E/10 where E is the Young's modulus of the solder. By virtue of its compliance, therefore, the machine can accommodate much if not most of the imposed displacement. Given the degree of freedom provided by a low machine compliance, the simulated strain in the solder can be a strong function of the constitutive model. Figure 4-33 shows that the two models agree for large displacement ranges but disagree for small displacement ranges. The magnitude of the stress is not as much affected by the model as is the predicted strain range. If one is using, therefore, the strain range to predict fatigue life, one achieves a more conservative estimate of fatigue life based on the microplastic model of deformation [Subrahmanyan et al., 1988].

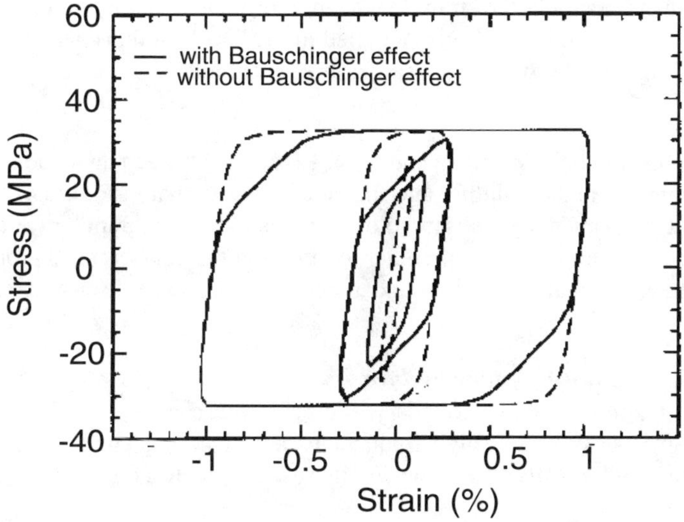

FIGURE 4-33 Simulations of hysteresis loops based on displacement-controlled loading, with low machine stiffness. The model incorporating microplasticity allows the material to flow at lower stresses than does the model without microplasticity; for this reason, when the displacement range, much of which can be accommodated by the machine, is low, the microplastic model predicts larger strain ranges than does the model without microplasticity.

The Creep-Fatigue Interaction

The two foregoing models weigh heavily on the phenomenological approach. Combined with an experimental program to obtain constitutive parameters, such models are in principle capable of accurately simulating constitutive behavior of solders. On the other hand, given our limited knowledge of the underlying mechanisms, considerable effort must be placed into constructing physically based, microstructural constitutive models. What then, would be the advantage of such models?

One advantage would be the ability to extrapolate constitutive models outside the range of experimental verification. But as far as temperature and strain rate are concerned, this ability is not as important in solder joints as it is in other applications; for instance, the power industry where engineers need to predict the performance of high-temperature components over periods approaching 30 years. Instead, it is probably more important to be able to extrapolate based on differences in microstructure than on temperature or strain rate because, while the temperatures and strain rates relevant to solder joint fatigue can be achieved in the laboratory, it is more difficult to duplicate, in bulk specimens, the microstructures relevant to solder joints.

An even more important reason to use physically based constitutive models concerns the creep-fatigue interaction at elevated temperatures. An illustration of one part of this interaction, the strain localization brought about by deformation-induced microstructural instability, is detailed in the Part II of this chapter. In the remainder of this section, we discuss some other phenomena.

Coarsening

One such connection between creep and fatigue is found between the incidence of grain (or colony) boundary sliding and the coarsening shown in Figure 4-21. At high strain rates, coarsening is observed but it tends to be transgranular (or trans-colony). The transition from transgranular to intergranular coarsening is probably related to the grain matrix-grain boundary sliding transition discussed earlier [Frear, 1991].

Intergranular vs. Transgranular Failure

Another ramification of grain boundary sliding is the change in failure mode between fatigue performed at high frequencies and low frequencies. A fatigue mechanism map of Lee [1993] is shown in Figure 4-34, where a transition between

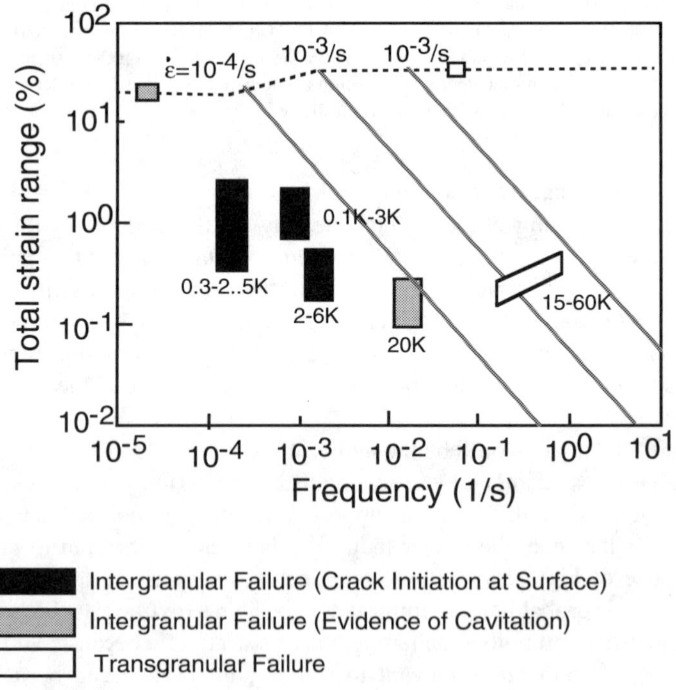

FIGURE 4-34 Fatigue mechanism map illustrating a change in failure mode from transgranular at high frequencies and strain ranges to intergranular at low frequencies and low strain ranges. The transition corresponds to line strain rates of roughly 10^{-3}/s to 10^{-4}/s. According to Figure 4-22, the same transition distinguishes between grain boundary sliding and no grain boundary sliding.

intergranular and transgranular fatigue is observed. Within the intergranular regime, there is evidence to suggest two types of failure under tension/compression loading: one in which fatigue cracks penetrate along grain boundaries from the surface, and the other, which involves internal cavitation. The surface cracking is observed at lower stress and strain ranges than the internal cavitation failure.

The transition region between transgranular and intergranular failure is roughly a line of constant strain rate, between 10^{-4}/s and 10^{-3}/s. This regime is the same one across which a transition takes place between grain boundary sliding and no grain boundary sliding, Figure 4-22. Evidently, then, the stress and strain concentrations generated by sliding promote intergranular crack propagation.

To nucleate intergranular cavities by intergranular sliding requires both large stress concentrations (of order E/100) and large sliding displacements [Nix, 1983]. This is a rationale for why the cavitation regime of fatigue behavior falls generally at higher frequencies and strain ranges than the surface cracking regime [Lee, 1993]. The stress concentrations—which only occur when the boundary slides—were estimated in connection with Equation (4.12) to be about σ_{gb}/f, where f is the fraction of boundary covered by Pb precipitates and σ_{gb} is the stress in the grain boundary element in Figure 4-26a. (Actually, due to the intrinsic viscosity of the boundary, the concentrations are likely to be lower than σ_{gb}/f because the boundary partly supports the shear stress itself). Likewise, the amount of sliding can be estimated based on the same model, using data such as those in Figure 4-22 as a guideline.

The surface cracking regime of fatigue failure appears to involve oxidation of the grain boundary, a process which is accelerated by sliding [Lee, 1993]. Again, the sliding introduces stress concentrations at the Pb precipitates on the boundary. Beginning where the boundary intersects the surface, Pb/Sn interfaces oxidize and separate under the action of the stress on the grain boundary. The sliding displacement incurred during cycling aid oxygen penetration because they open up holes. Combined together, sliding and oxidation introduce a type of stress-corrosion in which cracks penetrate along grain boundaries. As in the case of grain boundary cavitation, the modelling of the stress concentrations on the boundary is central to understanding the mechanism of fatigue.

A number of authors have reported intergranular failure and/or transitions between transgranular and intergranular failure as functions of frequency and/or temperature and/or environment. These transitions are observed not only in the Pb-Sn eutectic but also in Pb-rich alloys and, for that matter, a number of other engineering alloys [Gohn and Ellis, 1951, 1964; Rathore et al., 1973; Vaynman, et al., 1988; Berriche et al., 1987; Raman and Reiley, 1987, 1988; Wild, 1974; Betrabet and Raman, 1988; Solomon, 1973, 1986, 1988, 1990; Summers and Morris, 1990; Majumdar and Jones, 1991; Min and Raj, 1977, 1979; Baik and Raj, 1982a, 1982b). In most if not all of these cases involving fatigue at elevated homologous temperatures, grain boundary sliding is at least partly responsible for the intergranular failures.

4.3 THEORETICAL ASPECTS OF CONSTITUTIVE MODELLING

4.3.1 Definitions and Role

As mentioned at the beginning of this chapter, the term "constitutive model" is taken to refer to a mathematical description of the response of a material to arbitrary, time-varying stimuli. For a solid material such as solder, the "response" is often simply the Cauchy stress, although other relevant information such as the microstructural state may be considered. The "stimuli" will here be regarded as the time-histories of deformation (as quantified by an appropriate strain measure), and temperature and its gradient. This viewpoint is opposite from that usually taken in the materials science community, in which stress is regarded as the stimulus and strain (or its rate) as the response. Our current motivation stems from the fact that stress evolution is always regarded as strain-driven in finite element and other approximate solution schemes. It bears emphasis that a constitutive model must be meaningful for arbitrary stimuli histories if it is to be useful as a predictive tool. This necessity implies that the task of designing a suitable constitutive model for any given material is neither trivial nor precise, since the response of the material can be experimentally probed with only a limited number of deformation histories.

In formulating a constitutive model for any material, it is useful to bear in mind the intended use of the model. Complex constitutive models for the inelastic response of solids are typically used in a computational setting (e.g. finite element codes) to help make engineering predictions of the behavior of structures. In the context of solder joints, inelastic finite element analyses are of use in assessing the relative performance of candidate joint designs, and are therefore useful in the design process. Computational simulation may also be used as an important partner alongside experiment in understanding the microstructural mechanisms leading to fracture of solder joints. In particular, microstructural models, i.e. mathematical models that include features that are intended to represent material behavior and evolution on the microscale, may be useful in assessing the correctness of mechanisms of deformation and failure that are postulated based on experimental evidence. Such models are potentially effective in this role, since different physical mechanisms can be studied computationally, either alone or in combination, in a way that is not accessible to physical experimentation. This line of research has as its ultimate goals both contribution to the understanding of deformation and failure mechanisms in solders, and synthesis of a predictive tool of sufficient reliability to impact the material selection and joint design process.

As is the case for any mathematical model of physical phenomena, the formulation of a material model necessarily involves the introduction of a number of idealizations, as discussed previously. Chief among these is the concept that the state of a material element is taken to be the collection of all information, as expressed by the values of a set of state variables, such that, once known at a time t_1, the state at a later time t_2 may be computed from the constitutive model given the stimuli history from t_1 and t_2. The response (e.g. the stress) is taken to be a direct function of the state. The crucial aspect of the concept of material state is that any influence of the past history on the current material behavior must be expressed through the current values of the state variables.

In the context of solder joints, what is of primary practical interest is not the load-carrying capability of the body, but rather the resistance to fracture under imposed (and usually cyclic) deformations. The common eutectic 60Sn-40Pb solder alloy is a two-phase mixture of Sn-rich and Pb-rich phase regions, in which the characteristic phase-region size is highly dependent on the conditions of solidification [Morris et al., 1991]. Through experimental observation, several researchers have identified the progression of failure in 60Sn-40Pb solder joints to involve coarsening of the microstructure in a localized band, leading to softening, concentrated deformation, and eventual fracture in this band [Frear, 1991, 1989; Morris and Mei, 1991]. On the other hand, based on other experimental results, Lee [1993] does not observe this strict, sequential relationship between coarsening and fracture: in his experiments, cracks appear to propagate along the boundaries between lamellar colonies during most of the fatigue life, with coarsening occurring simultaneously and most strongly at the triple points between colonies. In any event, it appears that a dominant feature of the microstructure of 60Sn-40Pb solder as it relates to deformation resistance is the phase-region size. In the next subsection, a constitutive model for the Sn-Pb eutectic alloy is postulated in which an attempt is made to represent the evolution and effects of the characteristic phase-region size. The intended use of this model is to help assess the extent to which microstructural coarsening plays the same dominant role in actual solder-joint geometries as it appears to in simple-shear experiments. Before proceeding to the model, however, a brief description (omitting the mathematical details) of a three-dimensional constitutive framework suitable for inelastic solids is presented.

For simplicity, thermomechanical coupling is ignored in what follows, so that temperature appears only as a (generally time-varying) parameter in the constitutive equations. Roughly, this assumption amounts to the neglect of transient heat conduction and heat generation due to plastic work. The ideas concerning the concept of material state discussed above can be cast in a mathematical form with few additional assumptions and specifications. This mathematical expression takes the form of a functional that relates the current state to the previous state history as well as to the imposed deformation and temperature histories. This basis was used by Dashner [1986] and Rashid [1991] in developing state-variable formalisms for inelastic solids. Proceeding beyond this starting point, the requirement of invariance under superposed rigid body motions is imposed on the constitutive functional. This condition expresses the requirement that responses generated by the constitutive equations for two deformations that differ by only a rigid rotation, should themselves differ by only a rigid rotation. In order to proceed further, it must be assumed that the constitutive functional can be represented by a set of rate equations, i.e. by a set of (in general nonlinear) ordinary differential equations in time, that involve the state variables. This assumption amounts to a very mild restriction on the types of material behavior that are representable. That only first derivatives with respect to time of the state variables appear in these rate equations can be deduced by prescribing that the response is "solids like"; i.e., that abrupt deformations are accompanied by abrupt, but finite, changes in stress.

It is not possible to proceed further without introducing some postulates concerning the nature of the material response. In particular, for metals it seems appropriate to suppose that inelastic flow ceases in the absence of "elastic" deformation, since the defect motion that is responsible for inelasticity in metals is driven by the stress that accompanies elastic distortion. In this case, it turns out that a suitable representation of the elastic deformation is [Rashid 1991]:

$$\dot{\mathrm{H}} - \mathrm{WH} = \mathrm{DH} - \beta\mathrm{H} \tag{4.33}$$

in which D and W are the symmetric and antisymmetric parts, respectively, of the spatial velocity gradient, and β is a second-rank tensor function of the state. The second-rank tensor H characterizes the elastic deformation in the following sense: if a material element is made to undergo an abrupt unloading to zero stress from a stressed state, then under a mild set of conditions on β, H^{-1} is the deformation gradient associated with the unloading (to within a rigid rotation). Thus, H is analogous to the elastic deformation gradient employed by many authors [Lubarda and Lee, 1981; Dafalias, 1987, 1988; Bammann and Johnson, 1987] in the context of the multiplicative decomposition of the total deformation gradient into elastic and plastic parts as $\mathrm{F} = \mathrm{F}^{e}\,\mathrm{F}^{p}$. In this connection, it is also noted that β is analogous to the plastic velocity gradient (or plastic strain rate) in classical theories. Equation (4.33) is analogous to Equation (4.17), which was given in a one-dimensional context.

Three additional hypotheses are now introduced: (1) the elastic strain is assumed to be small (in a certain sense); (2) the material state is assumed to be characterized completely by the elastic strain and a set of scalar state parameters (to be discussed in the next subsection); and (3) the plastic velocity gradient is taken to depend directly on the Cauchy or "true" stress T and the set of scalar state variables. These three prescriptions seem reasonable in relation to the present intended application. In particular, exclusion of tensorial state variables can be shown to preclude the representation of both initial and evolving anisotropies of the type associated with, for example, crystallographic texture. Although it seems likely that initial texture is present in solder alloys due to the directionality of the solidification process, its effects are here ignored in the interest of simplicity. More will be said in the next subsection concerning the merits of including such features as anisotropy in relation to the nature and intended purpose of the model.

Using Equation (4.33) and the additional hypotheses mentioned above, and also assuming that the Helmholtz free energy depends only on the elastic strain measure H and temperature (a well-founded assumption for metals up to moderate strains), it can be shown that the Cauchy stress T is governed by the following rate equation:

$$T^0 = C \bullet D - (\mu_0 1 + \mu_1 T + \mu_2 T^2) \tag{4.34}$$

In Equation (4.34), $T^0 = \dot{T} + TW - WT$ is the Jaummann rate of Cauchy stress, C is the rank-four isotropic elastic modulus tensor, 1 is the rank-two identity, and μ_0, μ_1, and μ_2 are scalar functions of the state variables and of the invariants of T. The term in parentheses in Equation (4.34) is just the plastic velocity gradient; that it takes the form shown is a result of objectivity requirements and the Cayley-Hamilton theorem. Since the defect motion that is responsible for plastic flow in metals involves negligible volume change, it is typically assumed that tr $\beta = 0$; i.e., that the volumetric part of the plastic deformation rate is zero. Also, in light of the usual notions of defect motion in metals, the plastic deformation rate is taken to be independent of the pressure. Finally, it is nearly always assumed in models of isotropic inelasticity that the plastic deformation rate is independent of the third invariant det T. (This assumption appears to stem more from historical rather than physical motivation. Indeed, the experimentally observed axial effects in tube torsion tests are excluded by this assumption [Rashid, 1991; Hill, 1950].) Applying all of these assumptions to Equation (4.34) results in the evolution equations:

$$T^{0,} = 2G(D' - \mu T') \tag{4.35}$$

$$\dot{p} = -B \mathrm{tr} D$$

in which T' indicates deviatoric part; e.g. T' = T- 1/3(tr D)1, p = -1/3 tr T is the pressure, B and G are the (possibly temperature-dependent) elastic bulk and shear moduli, respectively, and μ is a scalar function of $\tau = \mathrm{tr}(T'^2)$ (the second invariant of the Cauchy stress deviator or "equivalent stress") and the scalar state variables.

Equation (4.35) represents the standard blueprint for finite-deformation, isotropic inelasticity. In the case of rate-independent plasticity, the function μ is proportional to a loading function which, in turn, is determined by the stress trajectory relative to the yield surface and which is a discontinuous function of stress. In rate-dependent theories (e.g. secondary creep), on the other hand, μ may be a continuous function of τ and the (scalar) state variables. In view of the fact that all mechanisms responsible for inelastic deformation in metals (e.g. dislocation glide and grain-boundary sliding) must proceed at a finite rate, rate-independent theories of plasticity are to be considered mathematical idealizations that are useful for some purposes. In any case, it is always possible to approach the rate-independent limit with a rate-dependent theory simply by including terms in the function μ that are very strong functions of the effective stress over some range of values of τ.

The point of giving this brief sketch of the reduction of the general constitutive formalism to Equation (4.34) is to illustrate the assumptions and specifications inherent in the usual constitutive framework for inelasticity in metals. Forms other than Equation (4.34) are of course possible (e.g., a very different evolution equation for the stress is obtained in the case of crystal plasticity [Asaro, 1983; Rashid, 1992]) in which inelastic deformation is assumed to occur exclusively by crystallographic slip on slip planes and in slip directions. If one accepts Equation (4.35) as being appropriate for modeling some aspects of the behavior of the material at hand, then the tasks of the modeler are reduced to two: first, to decide on a suitable set of state parameters along with their evolution equations; and second, to decide on the dependence of the constitutive function μ on the effective stress τ and the state parameters. These issues will be addressed in relation to solder in the following subsection.

4.3.2 Application to the Mechanics of Solder Joints

In this subsection, the considerations of the previous subsection will be made specific for 60Sn-40Pb eutectic solder. The development of this specific model will serve as a vehicle with which to address some issues common to all physically based constitutive models. As mentioned previously, the phase-region size appears to be an important microstructural parameter in eutectic solder, with coarser microstructures generally associated with lower flow resistance. Indeed, solders are materially heterogeneous on a range of size scales, from the atomic scale up to the scale of typical solder joints (1.0×10^{-3} m or more). However, for the purpose of analysis, the highly heterogeneous physical reality is replaced by a mathematical model in which some aspects of the small-scale heterogeneity are represented by smoothly varying state variables. For example, in analyses of fiber-reinforced composite structures, average stresses in each of the phases are often retained, while the actual geometry and interfaces of the fibers and matrix are not taken into account. The problem is thereby made tractable by idealizing the material as two homogeneous, interacting components. Such an approach is capable of representing the large-scale behavior, which is often all that is of interest.

In the case of solders, the size scale of a continuum element is chosen such that a large number of individual phase regions is contained in each element. However, the representative volume element (RVE) must not be so large that larger-scale heterogeneities, e.g. entire lamellar colonies, are comparable in size to the RVE. Although the RVE is used only as a conceptual guide (rather than an analytical ingredient) in the model formulation, it is an essential concept with regard to differentiating between features of the material and features of the body. In the present context, the heterogeneity associated with the individual phase regions is regarded to be a feature of the material, and is therefore represented by appropriate state variables; whereas that associated with the larger-scale variation in phase-region size is taken to be a feature of the body, and is therefore represented by smooth variations of the state variables in space.

Specific model assumptions are now introduced. An RVE containing a large number of individual phase regions of 60Sn-40Pb eutectic solder is considered. "Average" Cauchy stress tensors T^1 and T^2 are retained for the lead-rich and tin-rich phases, respectively; and the inelastic flow in each phase is assumed to respond only to the corresponding stress. The central idealization is that the deformation suffered by each phase region is equal to the overall deformation of the RVE. This idealization is familiar in the context of the theory of polycrystalline metals, in which it is referred to as the Voigt or Taylor assumption (Van Houtte and Wagner 1985). Under the Taylor assumption, it is clear that compatibility of deformation is satisfied on the microscale, although equilibrium is, in general, violated across phase-region boundaries. The appeal of this idealization rests on the enormous simplification it affords: the phase stresses T^{α} are uniform on the microscale under the Taylor assumption, and the overall stress T is given simply by a volume-weighted average of the phase stresses. Thus:

$$T = fT^1 + (1 - f)\,T^2 \tag{4.36}$$

in which f is the volume fraction of the lead-rich phase. Departure from the Taylor assumption, e.g., using a self-consistent method [Nemat-Nasser and Obata, 1986], involves enormous complexity (and therefore computational expense) for little gain in performance in this context. In any case, application of the Taylor assumption to metallic polycrystals is known to give reasonable results for the overall response [Asaro and Needleman, 1985].

The stress response in each phase region is assumed to be governed by the isotropic viscoplastic evolution Equation (4.34), so that:

$$\Gamma^{0\alpha'} = 2G_{\alpha}(D' - \mu_{\alpha}T^{\alpha'}),\; p'_{\alpha} = B_{\alpha}\mathrm{tr}D \tag{4.37}$$

in which $\alpha = 1$ or 2, and is not summed. The flow-rate parameter is taken to be given by the power-law expression:

$$\mu_{\alpha} = \frac{\dot{\varepsilon}_0}{\tau_{\alpha}}\left(\frac{\tau_{\alpha}}{\sigma_{\alpha}}\right)^{\frac{1}{M_{\alpha}}} \tag{4.38}$$

in each phase, where M_α ($0 < M_\alpha < 1$) is the rate-dependence exponent, σ_α is the (as-yet unspecified) flow resistance, and $\dot{\varepsilon}_0$ is a constant; recall that the stress magnitude is given by $\tau_\alpha = (T^\alpha_{ij}T^\alpha_{ij})^{0.5}$. The flow resistance in each phase is taken to evolve in time according to:

$$\sigma_\alpha = (\bar{\sigma}_\alpha + \sigma_\alpha)\,\sigma^0_\alpha \tag{4-39a}$$

$$\dot{\bar{\sigma}}_\alpha = C^\alpha_1 \frac{\dot{\gamma}_\alpha}{\dot{\varepsilon}_0} - C^\alpha_2 \left(\frac{\dot{\gamma}_\alpha}{\dot{\varepsilon}_0}\right)^{\eta\alpha} \bar{\sigma}_\alpha \tag{4-39b}$$

$$\dot{\sigma}_\alpha = C^\alpha_3 \left(\frac{\lambda}{\lambda_0}\right)^{-\frac{1}{2}} \tag{4-39c}$$

$$\dot{\lambda} = \lambda_0 f C^1_4 \dot{\gamma}_1 + (1 - f)\, C^2_4 \dot{\gamma}_2 \tag{4-39d}$$

in which $\dot{\gamma}_\alpha \equiv \mu_\alpha \tau_\alpha$ (equivalent plastic strain rate in each phase). In Equations (4.38) and (4.39), C^α_1, C^α_2, C^α_3, $C^\alpha_{4(1)}$, M_α, σ^0_α, n_α, $\dot{\varepsilon}_0$ and λ_0 are material constants, whereas $\lambda, \bar{\sigma}_\alpha$, and σ_α are state variables. The term is intended to quantify the contribution of the dislocation network to flow resistance, and is therefore made to evolve with inelastic deformation; its evolution equation is written in hardening-recovery format, as in Rohde and Swearengen [1980] and Busso et al. [1992]. The term λ is a characteristic length associated with the phase-region size; its evolution equation is also related to a scalar measure of inelastic deformation rate as suggested by Arrowood et al. [1991]. The parameter σ_α is defined directly by the Hall-Petch relationship equation (4.39c), and is intended to quantify the effects of disruption of dislocation glide due to finite crystallite size.

Equations (4.36) to (4.39) completely specify the proposed constitutive model for two-phase solder alloys. From these equations it may be seen that the model consists of a mixture of the responses of two separate, viscoplastic phases, with a common state variable (the phase-region size λ) affecting the behavior of each phase. The hardening-recovery format in Equation 4-39b was suggested by the work of Rohde and Swearengen [1980], who obtained good results with an expression analogous to Equation (4.39b) in the context of a single-component model for solder. The expression (4.39c), on the other hand, follows along the lines of the classical Hall-Petch model for the effects of grain size on flow resistance. The role of experimental observation and measurement in models such as the present one is not only to motivate the form of the model, but also to assign values to the constants appearing therein. It is clear that, despite the significant idealizations that have been made in arriving at Equations (4.36) to (4.39) (e.g. isotropy, Taylor's assumption), assigning accurate values to the material constants would be a major undertaking. Estimates of the parameter values can be made from uniaxial tests on single-phase, solid-solution alloys of lead and tin, along with supporting tests using the eutectic alloy and involving phase-region size measurements. Alternatively, very approximate values might be obtained by choosing values such that the overall response corresponds to that of the eutectic alloy. In any case, even without precise parameter values, a model of this type is of some utility in helping to assess the validity of various failure-sequence hypotheses through numerical experimentation.

As mentioned previously, the primary issue of interest in the analysis of a solder joint is not the load-bearing capability of the joint, but rather the life-to-failure under a given temperature and deformation history. Also, as discussed above, the progression to failure in solder joints can involve strain localization due to localized coarsening, subcritical crack propagation, or some combination, depending on the geometry, imposed deformation rate, and temperature. In the preliminary model described by Equations (4.36) to (4.39), an attempt has been made to represent the effects of microstructural coarsening, although in a very simplified manner. Also, microplastic effects are exhibited by the model, which correspond physically to the mismatch in elastic strains in the two phases, which in turn arises from the mismatch in inelastic flow rate in the two phases.

4.3.3 Numerical Implementation

The model discussed above was implemented in the quasistatic, three-dimensional finite element code JAC3D [Biffle, 1993]. The role of a constitutive model in a finite element code is to provide the code with the updated stress corresponding to a given increment in deformation, and, in the case of implicit finite element formulations, to supply the tangent moduli for the material. The updated stress is used by the code to enforce equilibrium at the end of each time step, whereas the tangent moduli are used in the iterative scheme by which the program obtains the equilibrium solution. The accuracy of the finite element solution depends strongly on the accuracy of the numerical integration of the constitutive equations, whereas the quality of the tangent moduli only affects the rate of convergence to equilibrium at the end of each time step.

In this implementation of the material model, the constitutive routine performs two passes (one for each phase) at an algorithm that integrates the evolution equations (4.35) for isotropic viscoplasticity. The difficulties that arise in numerically integrating inelastic constitutive relations such as Equation (4.35) are primarily related to the drastic differences between the elastic and inelastic components of the overall behavior. In particular, such evolution equations are mathematically stiff [Lambert, 1970], implying, among other things, that stability requirements impose severe restrictions on the choice of numerical integration scheme. A specialized integration method has been developed for evolution equations of the form of Equation (4.35). The essence of the technique is the decomposition of Equation (4.35) into a scalar ordinary differential equation that governs the stress magnitude, and a second ODE that governs the stress "direction" T'/τ. An approximate analytic solution to the stress direction equation is obtained by means of an asymptotic expansion. The essential problem is thus reduced to the update of a simple scalar equation, which is carried out with an implicit technique. This integration scheme is unconditionally stable and extremely accurate for arbitrary time increments.

With regard to the tangent moduli, what is required is the rank-four tensor that relates small variations in prescribed D (see Equation 4-34) to the corresponding variation in T at the end of the time step. In the present implementation, the tangent modulus tensor is obtained by applying a perturbation technique to the evolution equations 4-34 for each phase. The perturbed equation gives rise to a first-order, linear ordinary differential equation for the variation δT in stress. This equation is solved approximately using a single backward-Euler step, giving an approximation to the tangent modulus tensor.

4.3.4 Example Calculation

To illustrate the capabilities of the foregoing constitutive model, an example calculation was performed using an idealized solder lap-joint geometry (Figure 4-35). The analysis was carried out with a fully three-dimensional finite element formulation, although boundary conditions were imposed such that the deformation was constrained to be planar. The top and bottom adherends were modeled as stiff elastic bodies, and the deformation developed by thermal cycling was approximated by a horizontal back-and-forth displacement boundary condition on the upper adherend. The nominal shear strain range in the gap part of the joint was 6.7%. The imposed displacement history for the upper block is shown in Figure 4-35. From this figure, it may be seen that small time increments were taken near the points of load reversal to ensure convergence of the equilibrium iterations. The material properties that were used in the analysis were as follows (note that the "1" and "2" subscripts correspond to the lead-rich and tin-rich phases, respectively):

$$G_1 = 6.9\text{GPa},\ \ B_1 = 15.2\text{ GPa},\ \ M_1 = 0.3,\ \ \sigma^0{}_1 = 6.9\text{MPa}$$

$$C^1{}_1 = 20.0,\ \ C^1{}_2 = 100.0,\ \ C^1{}_3 = 1.0,\ \ C^1{}_4 = 10.0$$

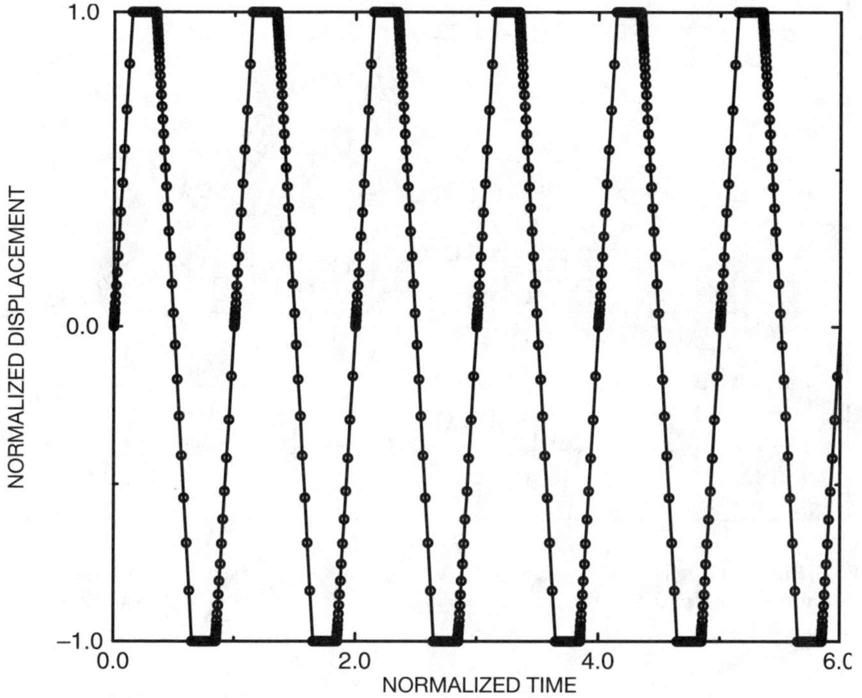

FIGURE 4-35 Normalized imposed displacement history of the upper block, indicating the time increment size used in the computation (open circles).

$$G_2 = 13.8\ \text{GPa},\quad B_2 = 30.3\ \text{GPa},\quad M_2 = 0.1,\quad \sigma^0{}_2 = 13.8\ \text{MPa}$$

$$C^2{}_1 = 20.0,\quad C^2{}_2 = 100.0,\quad C^2{}_3 = 1.0,\quad C^2{}_4 = 10.0$$

$$\lambda_0 = 1.0\ \mu\text{m},\quad \dot{\varepsilon}_0 = 1.0\ \text{min}^{-1},\quad f = 0.4.$$

A grayscale magnitude plot of the phase-size parameter λ is shown in Figure 4-36b. This figure indicates that λ has reached a value of approximately 19 μm after six deformation cycles of one-minute duration each. The concentration of coarsening in a narrow band is evident in Figure 4-36b. However, that the level of coarsening shown has accumulated after only six cycles strongly suggests that the values C_4^{α} used in the simulation are unrealistically high. In any case, it seems probable that the overall approach used to arrive at this result could be used both to help assess proposed coarsening models, and to aid in the design of solder-joint geometries that are resistant to the development of coarsened bands.

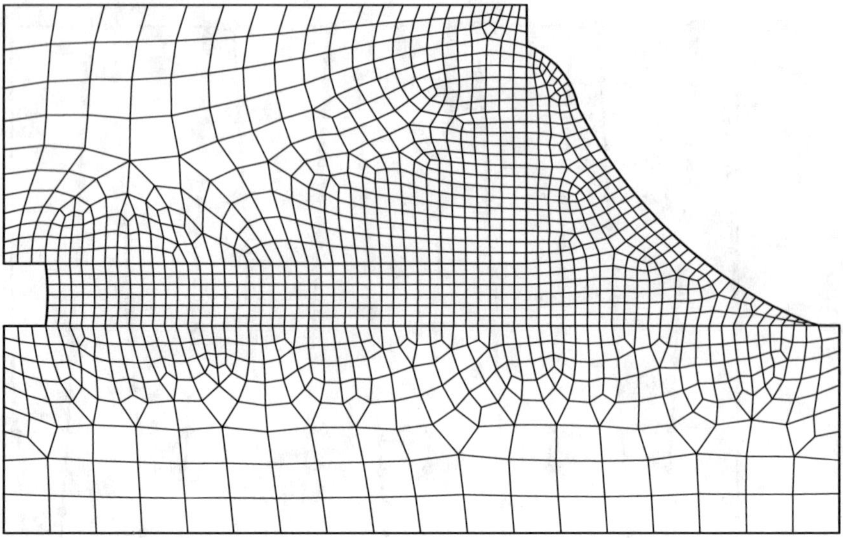

FIGURE 4-36a Finite element mesh of an idealized solder surface mount joint. Twelve equally-spaced levels are shown, with the maximum value of about 19 μm indicated by the star.

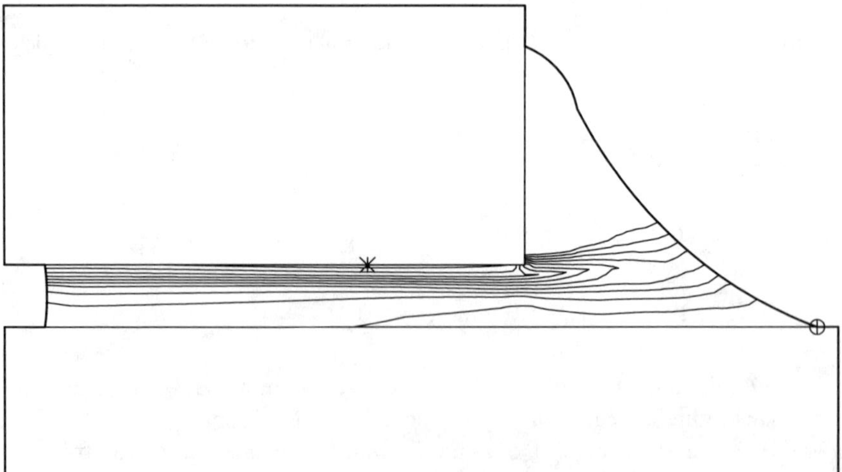

FIGURE 4-36b Grayscale magnitude plot of the phase-size parameter λ after six deformation cycles of one-minute duration each. Twelve equally-spaced levels are shown, with the maximum value of about 19 μm indicated by the star.

4.3.5 Conclusions

Both the model and the analysis discussed above are very preliminary, but they serve to bring into focus the avenues of research that will ultimately lead to enhanced usefulness of computational mechanics as regards solder-joint reliability. Above all, it seems that work of the type described in the first half of this chapter is crucial to both the design of constitutive models with physically meaningful features and to the assignment of reasonable material parameter values. Conversely, once the form of a constitutive model has been decided upon, the model serves as a tool that may be used to guide the design of an experimental program for the determination of the relevant parameters. In addition to this synergism between the materials-science and mechanics aspects of the problem, it appears that much work yet remains in relating analysis results to the actual event of failure in solder joints. Specifically, just as it is often impractical to test solder joints through many thousands of cycles to failure, so it is with computational simulation of solder joints. In addition to this difficulty, accounting for moving cracks with complicated shapes is a difficult task, and the physical relevance of such simulations is uncertain at present. It therefore seems that examination and extrapolation of trends of the type shown in Figure 4-36 will ultimately form the basic ingredients of any computationally based failure prediction methodology for solder joints.

4.4 CONCLUDING REMARKS

In this chapter we have explored constitutive models from three complementary perspectives. They include application of the constitutive model, construction of the constitutive model, and relation of the underlying constitutive phenomena to performance, in particular through the creep-fatigue interaction. The first of these perspectives reflects the driving force for constructing and using constitutive models. The second recognizes three central ingredients to a successful constitutive model: phenomenology, experiment, and mechanism. The last perspective points to the future of constitutive modeling, where the term "creep-fatigue interaction" will assume a more precise meaning, and where physical models will replace such artificial distinctions as "creep strain" vs. "plastic strain."

These perspectives have been illustrated by means of the Pb-Sn eutectic, an alloy that has been around for a very long time and will continue to serve even as new, "leadless" solders find application to address environmental concerns. The Pb-Sn eutectic is, perhaps, the best understood of all of the solder materials, and even then our understanding—the "state-of-the-art"—is in many ways crude relative to what we hope to achieve in the future. For the solders that we have not yet studied in any great detail, it is believed that many of the approaches to modeling and experimentation outlined above will apply, although the details of the constitutive behavior will differ.

REFERENCES

Arrowood, R., A. Mukherjee, and W.B. Jones,1991, *Solder Mechanics, a State of the Art Assessment*, D.R. Frear, W.B. Jones, and K.R. Kinsman Eds., The Minerals, Metals, and Materials Society, Warrendale PA, 107.

Asaro, R.J., 1983, *J. Appl. Mech.* 50:921.

Asaro, R.J. and A. Needleman, 1985, *Acta Metall.* 33: 923.

Ashby, M.F., 1972, *Surface Sci.*, 31:498.

Baik, S., and R. Raj,1982, *Metall. Trans. A*, 13A:1207.

Baik, S., and R. Raj, 1982, *Metall. Trans. A*,13A:1215.

Bammann, D.J. and G. C. Johnson, 1987, *Acta Mech.* 70:1.

Berriche, R., S. Vaynman, M.E. Fine, and D.A. Jeannotte, 1987, *Proc. 3rd Conf. on Electronic Packaging: Materials and Processes in Corrosion in Microelectronics*, M.E. Nicholson, Ed., ASM International, IEEE, NACE, Minneapolis, 169.

Betrabet, H.S., and V. Raman, 1988, *Metall. Trans. A*, 19A:1437.

Biffle, J.H., 1993, "JAC3D—a three-dimensional finite element computer code for the nonlinear quasistatic response of solids with the conjugate gradient method," Sandia National Laboratories report SAND87-1305, Albuquerque, NM.

Bird, J.E., A.K. Mukherjee, and J.E. Dorn, 1969, *Quantitative Relations Between Properties and Microstructure,* D.G. Brandon and A. Rosen, Eds. Israel University Press, 255.

Blum, W., A. Absenger, and R. Feilhauer, 1979, *Proc. 5th International Conference on the Strength of Metals and Alloys* 1:265.

Brown, L.M., and W.M. Stobbs, 1971, *Phil. Mag.* 23:1185.

Brunner, H., and N.J. Grant, 1959, *Trans. Metall. Soc. AIME*, 215:48.

Busso, E.P., M. Kitano, and T. Kumazawa, 1992, *J. Eng. Mat. and Tech.*114:331.

Cagnon, M., M. Suery, A. Eberhardt, and B. Baudelet,1977, *Acta Metall.*, 25: 71.

Cline, H.E., and T. H. Alden, 1967, Trans. TMS-AIME, 239:710.

Coble, R.L., 1963, J. Appl. Phys., 34:1679.

Conrad, H.,1965, *High Strength Materials* V.F. Zackay, Ed., Wiley, New York, 436.

Cottrell, A.H., 1953, *Phil. Mag. Ser* 744:31.

Crossman, F.W., and M.F. Ashby, 1975, *Acta Metall.*, 21:149.

Dafalias, Y.F., 1987, *Acta Mech.* 69:119.

Dafalius, Y.F., 1988, *Acta Mech.*69:121.

Dashner, P.A., 1986, *Int. J. Solids Struct.* 22:571.

Eberhardt, A., and B. Baudelet, 1974, J. Mat. Sci., 9:865.

Eggeler, G., and W. Blum, 1981, *Phil. Mag. A*, 44:1065.

Fang, T.-T., R.R. Kola, and K.L. Murty, 1986, *Metall. Trans. A*, 17A:1447.

Ferreira, I., and R.G. Stang, 1979, *Mat. Sci. Engr.*, 38:169.

Ferreira, I., and R.G. Stang, 1983, *Acta Metall.*, 31:585.

Findley, W.N., J.S. Lai, and K. Onaran, 1976, *Creep and Relaxation of Nonlinear Viscoelastic Materials* North-Holland (Dover, New York, 1989) 6.

Frear, D.R., 1991, *Solder Mechanics, a State of the Art Assessment*, D.R. Frear, W.B. Jones, and K.R. Kinsman Eds., The Minerals, Metals, and Materials Society, Warrendale PA, 191.

Frear, D.R., 1989, *IEEE Trans. on Components, Hybrids, and Manufacturing Technology,* 12:492-501.

Frear, D.R., and J.W. Morris, Jr., 1992, personal communication.

Freed, A.D., and K.P. Walker, Eds., 1991, High Temperature Constitutive Modelling—Theory and Application, The American Society of Engineers, New York.

Frost, H.J., and M.F. Ashby, 1982, *Deformation Mechanism Maps* 12.

Frost, H.J., P.R. Lavery, and S.D. Lutender, 1987, *Proc. 3rd Conf. on Electronic Packaging: Materials and Processes in Corrosion in Microelectronics*, M.E. Nicholson, Ed., ASM International, IEEE, NACE, Minneapolis, 259.

Garafalo, F., C. Richmond, W.F. Domis, and F. von Gemmingen, 1963, in *Proc. Joint Int. Conf. on Creep*, Inst. Mech. Engr., London, 1-31.

Gemmell, G.D., and N.J. Grant, 1957, *Trans. AIME, J. Metals*, 417.

Ghahremani, F., 1980a, *Int. J. Solid Struct.*, 16:825.

Ghahremani, F., 1980b, Int. J. Solid Struct., 16:847.

Gibeling, J.C., and W.D. Nix, 1981, *Acta Metall.* 29:1769.

Gittus, J., 1975, *Creep, Viscoelasticity, and Creep Fracture in Solids Applied Science Publishers LTD*, London, 312.

Goel, A., T.J. Ginter, and F.A. Mohamed, 1983, *Metall. Trans. A*, 14A: 2309.

Gohn, G.R., and W.C. Ellis, 1951, *Proc. ASTM*, 51:721.

Guo, Q., E.C. Cutiongo, L.M. Keer, and M.E. Fine, 1992, Trans. ASME, J. Electr. Pack., 114:145.

Guo, Z., A.F. Sprecher, and H. Conrad, 1992, Trans. ASME, J. Electr. Pack., 114:112.

Hall, P.M., 1984, *IEEE Trans. Comp. Hybrids, Manufact. Techn.*, 7:314.

Hart, E.W., 1957, *Acta Metall.* 5:597.

Hart, E.W., 1970, *Acta Metall.* 18:599.

Hart, E. W., and H. D. Solomon, 1973, Acta Metall., 21: 295.

Hart, E.W., C.-Y. Li, H. Yamada, and G.L. Wire, 1975, *Constitutive Equations in Plasticity*, A.S. Argon Ed., MIT Press, Cambridge Mass, 149.

Hart, E.W., 1976, *J. Engr. Mater. Techn.*, 98:193.

Hart, E. W., 1981, in Creep in Structures, A. R. S. Ponter and D. R. Hayhurst, Eds., Springer, Berlin, 90.

Hashin, Z., and S. Shtrikman, 1963, *J. Mech. Phys. Solids*, 11:127.

Herring, C., 1950, *J. Appl. Phys.* 21:437.

Herring, C., 1963, *J. Appl. Phys.*, 34: 437.

Hill, R., 1950, *The Mathematical Theory of Plasticity*, Clarendon Press, Oxford, 37.

Hirth, J.P., and J. Lothe, 1982, *Theory of Dislocations*, 2nd Ed. John Wiley and Sons, New York, 650.

Hutchinson, J.W., 1970, *J. Mech. Phys. Solids* 12:11.

Jackson, M.S., G.W. Cho, P. Alexopoulos, and C.-Y. Li, 1981, *J. Eng. Mater. Technol.*, 103:314.

Kaur, I., W. Gust, and L. Kozma, 1989, *Handbook of Grain and Interphase Boundary Diffusion Data*, Vol. 2, Ziegler Press, Stuggart, 1179,1185,1277.

Korhonen, M.A, C.-Y. Li, and J.C. Gibeling, 1985a, *Mat. Sci. Engr.*, 74:179.

Korhonen, M.A., S.-P. Hannula, and C.-Y. Li, 1985b, *Metall. Trans. A.*, 16A:411.

Korhonen, M.A., S.-P. Hannula, and C.-Y. Li, 1987, *Unified Constitutive Equations for Creep and Plasticity*, A.K. Miller, Ed., Elsevier, London, 89.

Laird, C., P. Charsley, and H. Mughrabi, 1986, *Mat. Sci. Engr.* 81:433.

Lambert, J.D., 1970, *Computational Methods in Ordinary Differential Equations,* Wiley, London.

Langdon, T.G., D. Simpson, and R.C. Gifkins, 1983, *Acta Metall.*, 31:939.

Lawson, L., and M. Meshii, 1991, *Scr. Metall. et Mat.*, 25(7):1713.

Lee, S.M.,1993, *Creep and the Creep-Fatigue Interaction in Pb-Sn Eutectic and Eutectic Solder Joints*, PhD Thesis, The University of Wisconsin, Madison.

Levine, E., and J. Ordonez, 1981, *IEEE Trans. Comp. Hybrids, Manufact. Techn.*, 4:515.

Lubarda, V.A. and E.H. Lee, 1981, *J. Appl. Mech.* 48:35.

Majumdar, S., and W.B. Jones, 1991, *Solder Mechanics, a State of the Art Assessment*, D.R. Frear, W.B. Jones, and K.R. Kinsman Eds., The Minerals, Metals, and Materials Society, Warrendale PA, 273.

McClintock, F.A., 1966, *Mechanical Behavior of Materials*, F.A. McClintock and A.S. Argon Eds. Addison-Wesley, Reading Mass. 471-483.

Mecking, H. and Y. Estrin, 1980, *Scr. Metall.*, 14:815.

Min, B.K., and R.Raj, 1977, *Acta Metall.*, 25:1007.

Min, B.K., and R.Raj, 1979, *Canad. Metall. Quarterly*, 18:171.

Mohamed, F.A., and T.G. Langdon, 1974, *Acta Metall.*, 22:779.

Mori, T., M. Koda, R. Monzen, and T. Mura, 1983, *Acta Metall.*,31:275.

Morris, J.W. Jr.; D. Tribula, T. S. E. Summers, and D. Grivas, 1991, in *Solder Joint Reliability: Theory and Applications* (J. H. Lau, ed.), Van Nostrand Reinhold, New York, 225.

Mughrabi, H., 1983, *Acta Metall.*, 31:1367.

Nabarro, F.R.N., 1948, in *Report of a Conference on the Strength of Solids, Physical Society*, London, 75.

Nabarro, F.R.N., 1985, in *Strength of Metals and Alloys* (ICSMA 7) H.J. McQueen, J.-P. Bailon, J.I. Dickson, J.J. Jonas, and M.G. Akben Eds., Pergamon, Oxford, 1667.

Nabarro, F.R.N., 1989, *Acta Metall.*, 37:1521.

Nemat-Nasser, S. and M. Obata, 1986, *Proc. R. Soc. Lond. A* 407:343.

Nix, W. D., 1983, Scr. Metall., 17:1.

Oikawa, H., H. Sato, and K. Maruyama, 1985, *Mat. Sci. Engr.*, 75:21.

Ono, N., T. Tsuchikawa, S. Nishimura, and S. Karashima, 1983, *Mat. Sci. Engr.*, 59, 223.

Pao, Y.-H., E. Jih, S. Badgley, and J. Browning, 1993, to appear in Trans. ASME, J. Electr. Pack.

Pao, Y.-H., S. Badgley, R. Govila, L. Baumgartner, R. Allor, and R. Cooper, 1992, Trans. ASME, J. Electr. Pack., 114:135.

Pedersen, O.B., L.M. Brown, and W.M. Stobbs, 1981, *Acta Metall.* 29:1843.

Poirier, J.-P., 1985, *Creep of Crystals,* Cambridge University Press, Cambridge 73:181.

Raj, R., and M.F. Ashby, 1971, *Metall. Trans.*, 2:1113.

Raj, R., and M.F. Ashby, 1972, *Metall. Trans A*, 3A:1937.

Raman, V., and T.C. Reiley, 1987, *J. Mat. Sci. Lett.*, 6:549.

Raman, V., and T.C. Reiley, 1988, *Metall. Trans. A*, 19A:1533.

Rashid, M.M., 1991, in *Proc. 12th Army Symposium on Solid Mechanics* (S.-C. Chou, ed.), 4-7 November, Plymouth, MA; 521.

Rashid, M.M., 1992, *Computer Methods Appl. Mech. Eng.* 94: 201.

Rathore, H.S., R.C. Yih, A.R. Edenfeld, 1973, *J. Test. Eval.*, 1:170.

Rohde, R.W., and J. C. Swearengen, 1980, *J. Eng. Mat. and Tech.* 102:207.

Robinson, S.L., and O.D. Sherby, 1969, *Acta Metall.* 17:109.

Sherby, O.D., R.H. Klundt, and A.K. Miller, 1977, *Metall. Trans. A*, 8A:843.

Snowden, K.U., 1964, *Acta Metall.*, 12:295.

Solomon, H.D., 1973, *Metall. Trans.*, 4:341.

Solomon, H.D., 1986, *IEEE Trans. Comp. Hybr. Manf. Techn.* CHMT:423.
Solomon, H.D., 1988, in *Low Cycle Fatigue*, ASTM STP 942,
H.D. Solomon, G.R. Halford, L.R. Kaisand, and B.N. Leis, Eds., Americal Society for Testing and Materials, Philadelphia PA, 342.
Solomon, H.D., 1990, *Trans. ASME J. Electr. Pack.,* 112:123.
Sowerby, R., D.K. Uko, and Y. Tomita, 1979, *Mat. Sci. Engr.* 41:43.
Subrahmanyan, R., D. Stone, and C.-Y. Li, 1988, in *Proc. Electr. Pack. Symp.,* 108 MRS, 381.
Subrahmanyan, R., J. R. Wilcox and C.-Y. Li, 1989, *IEEE Trans. Comp. Hybr.,* Manuf. Techn., CHMT-12:480.
Summers, T.S.E., and Morris, J.W. Jr., 1990, *Trans. ASME J. Electr. Pack.*, 112: 94.
Speight, M.V., 1976, *Acta Metall.*, 24:725.
Stone, D., S.-P. Hannula, and C.-Y. Li, 1985, *Proc. 35th ECC, IEEE*, New York NY 46.
Stone, D., H. Wilson, R. Subrahmanyan, and C.-Y. Li., 1986, *Proc. Elctr. Pack. Symp.* 74: MRS 175.
Stone, D., 1991, *Acta Metall. Mater.*, 39:599.
Suh, Cohen, and Weertman, 1983, *Metall. Trans. A*, 14A:117.
Van Houtte, P. and F. Wagner, 1985, in *Preferred Orientation in Deformed Metals and Rocks: an Introduction to Modern Texture Analysis* (H.-R. Wenk, ed.), Academic Press, Orlando, FL, 233.
Vaynman, S., M.E. Fine, and D.A. Jeannotte, 1988, *Metall. Trans. A*, 19A:1051.
Weertman, J., 1957, *J. Appl. Phys.* 28:1185.
Weertman, J., 1960, *Trans. AIME*, 218: 207.
Weertman, J., and J.R. Weertman, 1983, in *Physical Metallurgy*, 3rd Ed., R.W. Cahn and P. Haasen Eds., Elsvier, London, 1310.
Weertman, J., and J. R. Weertman, 1992, *Elementary Dislocation Theory,* Oxford University Press, New York, 52.
Wilcox, J.R., R. Subrahmanyan, and C.-Y. Li, 1989, in *Proc. 2nd ASM Electr. Mat. Proc. Congr.*, Philadelphia, PA, 203.
Wilcox, J.R., 1990, *Inelastic Deformation and Fatigue Damage in Metals at High Homologous Temperatures*, PhD Thesis, Cornell University, 119.
Wild, R.N., 1974, *Proc. of the Technical Program, National Electronic Packaging Conference*, Industrial and Scientific Conf. Management Inc., Chicago, IL, 105.
Yaney, D. L., J. C. Gibeling, and W. D. Nix, 1987, *Acta Metall,* 36:1391.
Yavari, P., and T.G. Langdon, 1983, *Acta Metall.*, 31:1595.
Zener, C., 1941, *Phys. Rev.*, 60:906.
Zener, C., 1948, *Elasticity and Anelasticity of Metals* University of Chicago Press, Chicago, 69-159.

5

Prediction of Solder Joint Geometry

S. M. Heinrich

5.1 INTRODUCTION

When an analysis of a solder joint is to be performed for the purpose of estimating stress and strain fields (and ultimately joint fatigue life and reliability), several input quantities must be specified. These include the thermomechanical loading conditions to which the joint will be subjected, the material properties of the solder alloy and associated microstructure, and the geometry, i.e. size, shape, and position of the solder interconnect. All of these factors will influence the results of the analysis as well as the performance of the actual joint. Thus, any hope of accurately predicting joint life and reliability is dependent upon the ability to accurately characterize the material, loading, and geometric parameters associated with the solder joint. Usually the loading conditions considered by the analyst are dictated by design specifications, while the description of material behavior is experimentally based and the subject of current research, as discussed in Chapter 2. Since the joint geometry is not known prior to the soldering process, the need arises for methods of predicting the final joint geometry which results from the soldering operation. This need provides a major motivation for the present chapter, the purpose of which is to review and assess existing models for predicting the geometry of solder joints in electronic packaging applications. These models relate joint geometry to the design parameters of the component and substrate, and to the soldering process parameters, such as solder paste volume in a reflow operation. By using models of this type in conjunction with appropriate constitutive models (Chapter 4), analysis tools (Chapter 7), and lifetime prediction methods (Chapter 6), the optimization of solder joints and electronic assemblies can be achieved at the design and manufacturing stages.

In addition to being a significant factor in determining joint reliability, the ability to accurately model joint formation (as it relates to joint geometry) is of paramount importance in maximizing the yields of soldering processes. For example, low yields and high "infant mortality" rates of electronic assemblies are often dictated by geometry-related defects, such as insufficient solder ("skips"), excessive solder ("bridges"), undesirable fillet shapes, and large contact angles. Thus, models for predicting joint geometry can also be implemented to establish "windows" for design and process parameters, i.e., to specify tolerances for manufacturing variability in order to minimize the number of solder-related defects.

5.2 OBJECTIVES

The remainder of this chapter has been organized into four main sections, each of which aims to achieve a particular objective. A brief description of each section and its objective is presented below:

Section	*Objective*
Soldering Methods	To provide a brief introduction to common soldering methods used in electronic packaging, including reflow and wave soldering processes.
Surface Tension Theory	To introduce the reader to surface tension theory, upon which most of the existing models for predicting joint geometry are based; two alternate formulations of the joint formation problem (energy minimization and local force balance) are presented.
Models for Predicting Solder Joint Geometry	To review and assess the models that are currently available for predicting solder joint geometry in electronic packaging applications, and to summarize those results that are conducive to practical implementation.
Future Research Needs	To identify the shortcomings of the existing models of joint geometry prediction and to suggest those areas in which additional research is required.

5.3 SOLDERING METHODS

As with the development of any theoretical model, a practical understanding of the physical process to be modeled is prerequisite to the modeling phase. Therefore, prior to describing the theoretical models that are currently available for predicting joint geometry, a brief discussion of soldering methods commonly used in electronic packaging will be presented. These include reflow soldering and wave soldering.

5.3.1 *Reflow Soldering*

According to the ANSI/IPC-T-50E Standard [1992], reflow soldering is defined as "the joining of mating surfaces that have been tinned and/or have solder between them, placing them together, heating them until the solder fuses, and allowing them to cool in the joined position." Thus, reflow processes are characterized by a separate application of the solder and the heat used to melt, or reflow, the solder. In some reflow processes, such as when a molten solder is used to coat the mating surfaces prior to fusion, the solder is truly "reflowed," or "melted a second time." However, in other cases the term "reflow" is a misnomer. For example, when solder paste is used (described below), the solder liquefies only once during the soldering process.

In general the sequence of steps involved in a reflow process may be listed as follows:

1. Solder/Flux Application: application of solder and flux to the metallizations on the components and/or substrate;
2. Component Placement: positioning of components prior to reflow;
3. Heating: resulting in melting of solder and wetting of metallizations (promoted by flux action);
4. Cooling: causing solidification of solder joints; and
5. Cleaning: removal of flux residues, if necessary.

Since a wide variety of techniques can be implemented in practice to apply solder, flux, and heat, a description of the various options will now be given.

Techniques for applying solder onto component and/or board metallizations include dipping, electroplating, solder preforms, and solder paste [Klein Wassink, 1989]. Once layers of solder have been applied to one or both of the mating surfaces by dipping or electroplating, flux is usually applied prior to reflow to remove surface oxides and promote wetting and spreading of the reflowed solder. When solder preforms or solder paste is used, the flux is usually included in the preform and always included in the paste, so that additional fluxing is unnecessary. A solder preform is simply a small mass of solder alloy of a particular shape (disc, sphere, cylinder, etc.) that may contain flux in the form of an inner core or a coating [Johnson, 1982]. Solder paste, or solder cream, is a suspension of solder powder particles in a cream flux, in which additives have been included to promote wetting and control paste properties (viscosity, tackiness, slump, etc.) [Klein Wassink, 1989; ANSI/IPC 1992]. The paste is usually applied to the metallizations on the circuit board (called footprints, solder lands, or solder pads) using a screen printing or stenciling operation, or it may be dispensed by syringe. Once the paste has been applied, components are placed on the board and held in position by the tackiness of the paste. The assembly is then conveyed through a reflow oven and the joints are formed. This sequence is depicted in Figure 5-1 for a leadless chip component. For a more detailed treatment of solder paste applications the reader is referred to the book by Hwang [1989].

Solder reflow methods are often classified according to the heating mechanism that is used. The most frequently used methods of supplying heat for reflow soldering are thermal conduction, thermal convection, laser, infrared radiation, and vapor phase, with infrared (IR) and vapor phase being the most popular [Hinch, 1988]. An IR reflow oven contains a conveyor belt which transports the assembly through various heating zones, in which lamps or panels emit infrared energy. The temperature profile experienced by the assembly is a function of conveyor speed, emitter-conveyor spacings, and individual temperature settings for each emitter. In a vapor phase reflow system the assembly is conveyed through a chamber which contains saturated vapor, created by a heater which boils a liquid at the bottom of the chamber. As the relatively cool assembly enters the chamber, the vapor condenses on the assembly and releases the latent heat of vaporization. This heat transfer continues until the system achieves thermal equilibrium. Solder reflow

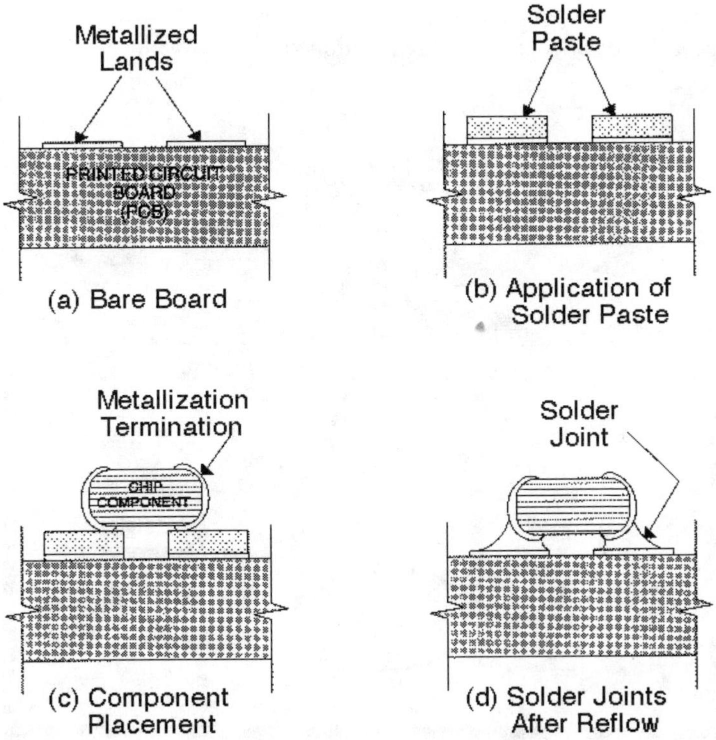

FIGURE 5-1 Steps involved in reflow process using solder paste.

will occur provided that the boiling temperature of the liquid exceeds the melting temperature of the solder [Hinch, 1988; Klein Wassink, 1989].

5.3.2 Wave Soldering

In a wave soldering process, a printed circuit board assembly is passed over, and in contact with, the crest of a continuously flowing fountain of molten solder (Figure 5-2). Upon exiting the "solder wave," the metallized portions of the assembly will have captured small amounts of solder from the wave which, upon cooling, will form solder joints (Figure 5-3). Thus, a wave soldering process differs from a reflow process in the respect that the application of solder and heat (via the hot liquid solder) to the assembly occurs simultaneously.

The sequence of steps that must be carried out in association with a wave soldering operation are the following [Klein Wassink 1989]:

1. Automatic Through-Hole Insertion: automated placement of through-hole components by inserting leads through holes in circuit board;
2. Adhesive Application: dispensing of adhesive onto board at locations of surface mount components (SMCs);
3. Placement of SMCs: SMCs are positioned at adhesive locations;

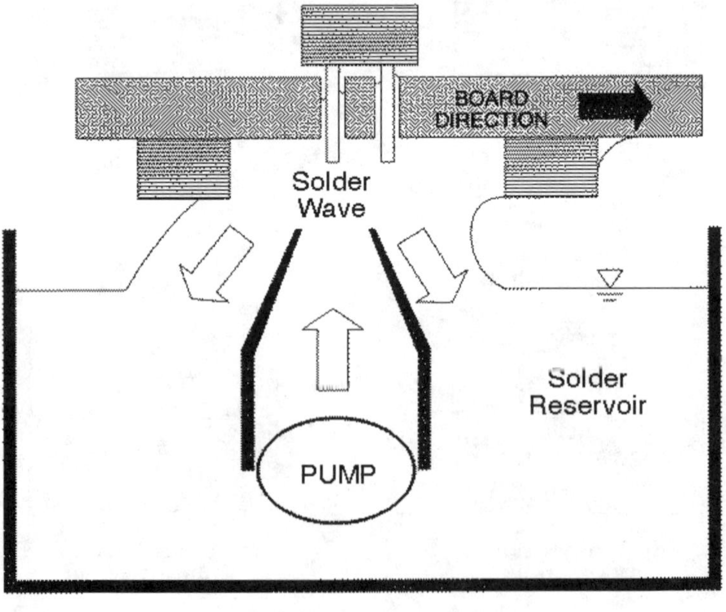

FIGURE 5-2 The wave soldering process.

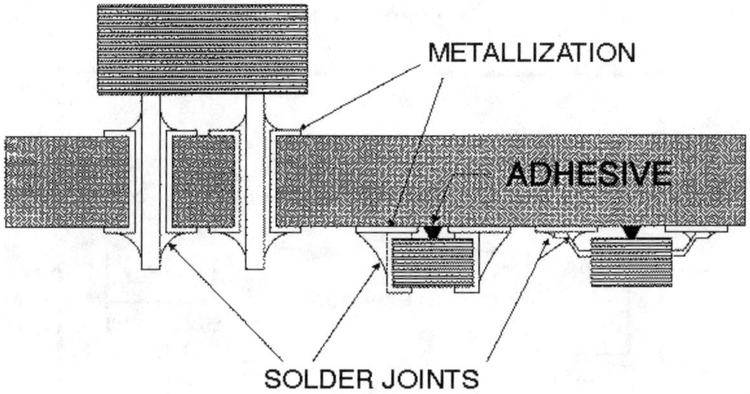

FIGURE 5-3 Joints resulting from wave soldering process (from left): leaded through-hole component, leadless surface mount component, and leaded surface mount component.

4. Curing of Adhesive: temporary bond created between SMCs and board, permitting component to be supported on underside of board;
5. Manual Through-Hole Insertion: manual placement of through-hole components by inserting leads through holes in circuit board;
6. Flux Application: by foam, wave, or spray;
7. Preheating: to evaporate solvents in flux, minimize thermal shock, and activate fluxes;
8. Wave Soldering: assembly passes over solder wave, capturing molten solder from wave;
9. Cooling: solder solidifies and joints are formed; and
10. Cleaning: removal of flux residues, if necessary.

Steps 6-8 all occur within the wave soldering machine, as shown in Figure 5-4.

Two fundamental requirements must be satisfied by a wave soldering process: all terminations must be wet by solder (i.e. no skips), yet excessive amounts of solder resulting in the formation of bridges must be avoided. To satisfy these somewhat opposing requirements, the solder wave is adjusted so that at the entry point the solder flow is turbulent, while at the exit a laminar flow exists. In many cases multiple waves are incorporated in the machine, each having different flow characteristics. In fine-pitch applications (center-to-center lead spacings $\leq$ 0.625 mm (0.025 in)) a hot-air knife may also be employed at the exit point to direct streams of hot air at the relatively unstable molten bridges between neighboring joints, thus causing separation and the elimination of the unwanted bridges. Although the wave soldering process was first introduced at a time when through-hole technology was the dominant interconnect technology, many facilities still utilize this soldering method for surface mount applications. To do so successfully, several

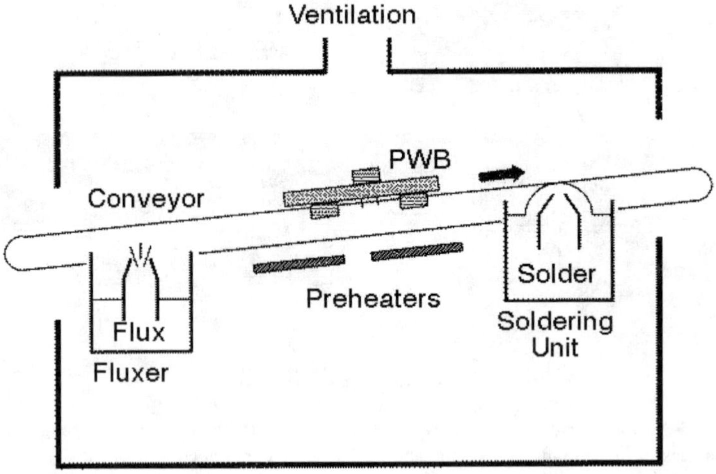

FIGURE 5-4 Schematic of wave soldering machine.

improvements have had to be made to the original process (e.g. hot-air knife). As lead spacings continue to decrease, however, the demands placed on wave soldering become greater, perhaps approaching the limit beyond which wave soldering might no longer provide a feasible means of interconnection.

5.4 SURFACE TENSION THEORY

When attempting to model the flow or predict the static equilibrium configuration of a molten solder interconnect, one usually resorts to a theoretical model that incorporates the concept of "surface tension" (or "surface energy"), since surface effects are significant in joints as small as those encountered in electronics applications. Since most of the models surveyed in this chapter are based upon surface tension theory, the purpose of this section is twofold: (1) to introduce the reader to the fundamental concepts of the theory; and (2) to summarize the two alternative problem formulations that have been used in the literature to model solder joint formation.

5.4.1 Fundamental Concepts

The surface energy per unit area associated with an interface between a liquid (e.g. molten solder) and another medium (solid, liquid, or gas) is defined as the amount of work required to isothermally enlarge the interface by one unit of area. This quantity, denoted here by γ, has dimensions of energy/area or, equivalently, force/length. The origin of this energy is the difference in electrochemical forces acting on a liquid molecule near the interface, due to the different nature of the neighboring molecules (Figure 5-5). Rather than using this work-related definition, the surface energy is often defined (or interpreted) as a uniform internal tensile force per unit length, acting within an imaginary interfacial membrane (Figure 5-6); hence, the alternate term surface tension is often used. Although the concept of surface tension involves several physical inconsistencies (e.g., the interfacial membrane does not actually exist), it does lead to models that are mathematically equivalent to those based on the surface-energy concept, and it results in the accurate prediction of liquid surface geometries [Rouse, 1978].

Intimately associated with the concept of surface tension is the contact angle θ that exists at the juncture of a solid, liquid, and gas (Figure 5-7a). The value of θ at the triple point P can be related to the surface tensions of the three interfaces by

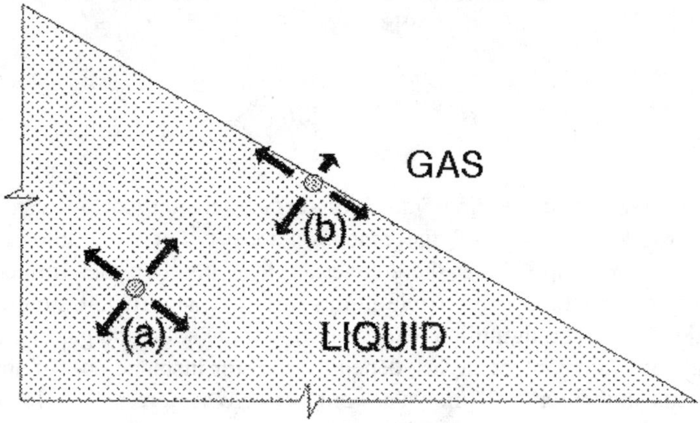

FIGURE 5-5 Intermolecular forces acting on molecules (a) away from surface and (b) near surface.

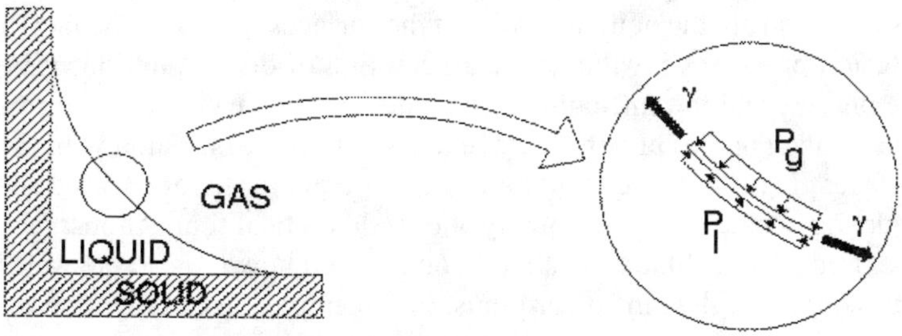

FIGURE 5-6 Interfacial "membrane" model of liquid/gas interface.

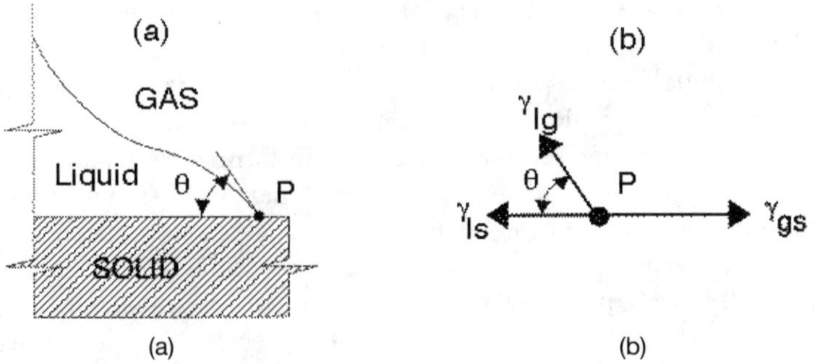

FIGURE 5-7 (a) Definition of contact angle θ at triple point *P* of solid/liquid/gas system and (b) Force diagram showing surface tensions at triple point.

minimizing the free energy or by considering "equilibrium" of the triple point along the solid's tangential direction (Figure 5-7b). This results in Young's Equation [e.g., Johnson and Dettre, 1969]:

$$\gamma_{ls} + \gamma_{lg}\cos\theta = \gamma_{gs} \tag{5.1}$$

If one assumes that the three values of surface tension (for a given constant temperature) depend only on the materials meeting at each interface, then Young's equation implies that the contact angle is uniquely determined by the three materials meeting at the triple point. This is a significant simplification, since real droplets of a liquid in contact with the same solid and gas can exhibit a wide range of static contact angles, dependent on, for instance, the volume of the droplet. This phenomenon is known as contact angle hysteresis [e.g. Johnson and Dettre, 1969; Dussan V., E.B., 1985]. Attempts to reconcile this phenomenon with Young's Equation have been explored by considering microscopic roughness of the solid surface and distinguishing between the microscopic and macroscopic contact angles. Other problems associated with the practical utilization of Young's Equation are the nonuniformity of the surfaces involved, the fact that surface diffusion processes invalidate the assumption of distinct interfaces between the three phases, and the difficulties encountered when trying to experimentally determine contact angles and the surface tension values associated with solid/liquid interfaces. Despite these shortcomings, the equation is often used in solder joint modeling because of its simplicity and its theoretical foundations; moreover, it may be used in a qualitative fashion to understand how to promote wetting (attain small contact angles) in solder joints. For example, an examination of Equation (5-1) or Figure 5-7b leads to the conclusion that θ may be decreased (cosθ increased) by decreasing γ_{lg} or γ_{ls}, or by increasing γ_{gs}. In practice, flux is used to remove oxides from the free solid surface (the metallization), thereby increasing γ_{gs} and thus promoting wetting [Klein Wassink, 1989].

5.4.2 Problem Formulations

To mathematically formulate the problem of predicting the geometry of a molten solder joint, one has a choice between two different, yet entirely equivalent, options: the variational formulation and the boundary value problem (BVP) formulation. In the variational approach the problem is cast as one of minimizing the total energy functional over the family of all admissible joint geometries, i.e. those geometries satisfying the geometric constraints of the problem. The total energy functional includes surface energies, gravitational potential energy of the solder (often negligible for small joints), and potential energy associated with other external forces (e.g. the self-weight of a settling component). The constraints are related to the fixed volume of solder in the joint, the impenetrable nature of the bounding solid surfaces (e.g. solder lands and component leads), and the particular wettability conditions that exist where the free solder surface contacts the solid surfaces. The BVP approach incorporates a governing differential equation that may be derived from a force balance on a differential element of the interfacial membrane between the liquid solder and the surrounding atmosphere (or flux). (This differential equation may also be derived directly by using variational calculus to set the first variation of the above-mentioned total energy equal to zero [e.g. Weinstock, 1974].) Also included in the BVP formulation are boundary conditions (BCs), which are related to physical constraints (impenetrability and wettability conditions) at the edges of the free solder surface, and the solder volume constraint. The two formulations are equivalent in the sense that a solution for one is also a solution for the other. More specifically, an admissible solder joint geometry that minimizes the total energy of the system also corresponds to a (stable) solution of the BVP (differential equation, BCs, and volume constraint), and vice versa.

The two alternate mathematical formulations are summarized below. The motivation for choosing one formulation over the other is usually dictated by the type of results desired from the analysis. The variational formulation tends to be more conducive to obtaining approximate solutions, as are often necessitated by complex 3-D joint geometries, while the BVP approach is preferable in generating closed-form or semi-analytical "exact" solutions to problems involving relatively simple geometries.

Variational Formulation

Consider a conservative (zero energy loss) three-phase system consisting of a volume V of molten solder in contact with several solid bodies, as shown in Figure 5-8. Let γ denote the surface tension (surface energy per unit area) of the "free solder surface" S, i.e., the solder/atmosphere (or solder/flux) interface. The surface tensions associated with the solid/solder and solid/atmosphere interfaces

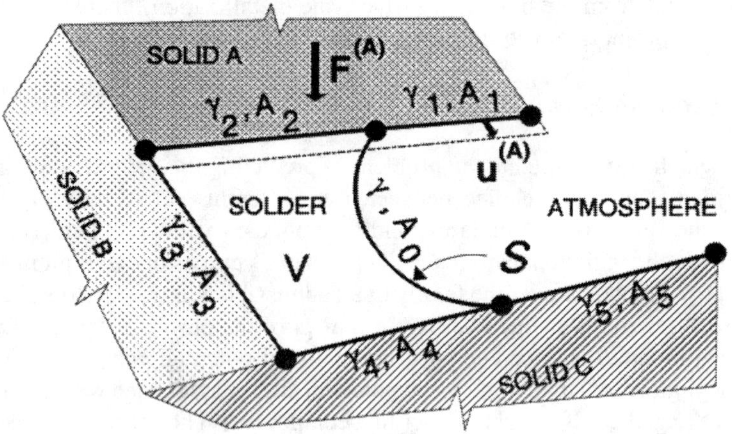

FIGURE 5-8 Definition of quantities appearing in total energy expression for three-phase system.

will be represented by γ_i, i=1,2,3,..., and the surface areas of these interfaces by Ai. The surface area of the free solder surface is A0. The total energy functional E for this three-phase system is defined as follows:

$$E \equiv E_s + E_b + E_f \tag{5.2}$$

where

E_s = surface energy associated with those interfaces whose area may vary

$$= E_s^{(0)} + \sum_{i=1,2,\ldots} E_s^{(i)};$$

$E_s^{(0)}$ = surface energy of free solder surface S = γA_0;

$E_s^{(i)}$ = surface energy of ith interface = $\gamma_i A_i$;

E_b = potential energy of body forces (gravitational, electromagnetic, etc.) acting throughout solder volume;

$$= \int_V \rho g z dV \text{ if only gravitational body forces are present;}$$

ρ = density (mass per unit volume) of molten solder;

g = gravitational acceleration;

z = vertical coordinate, measured positive upward from an arbitrary datum;

E_f = potential energy of all conservative external forces corresponding to movement of solid bodies;

$$= \sum_{i = A, B, \ldots} E_f^{(i)};$$

$E_f^{(i)}$ = potential energy of the conservative external force $F^{(i)}$ corresponding to displacement $u^{(i)}$ of the ith solid body, i = A,B,...;

$$= -F^{(i)} \bullet u^{(i)}$$

Having defined the total energy, the problem of determining a stable equilibrium configuration for the three-phase system may be stated as follows:

Determine a surface S^* that makes the total energy E a (local) minimum over the family of all surfaces S satisfying the geometric constraints of the problem.

Two obvious geometric constraints that must be met by any admissible surface S are that the molten solder cannot penetrate a solid surface and that the volume bounded by surface S and the solids is equal to the volume of solder V that is available to form the joint. Other physical constraints that must be satisfied include those related to assumptions on the wettability of the solid surfaces. Two common wettability assumptions are (a) zero wettability and (b) specified contact angle. When a solid surface is assumed to have zero wettability, the solder cannot advance onto that surface; thus, once the solder spreads to the edge of a non-wettable surface, the contact area cannot increase along that surface, so that a portion of the contact curve (bounding curve of S) is fixed. (In a two-dimensional view the contact curve reduces to a triple point, as indicated in Figure 5-7a.) Although the position of this part of the contact curve is fixed for this case, the distribution of the contact angle along this part is unknown and must be determined by solving the variational (energy minimization) problem. On the other hand, when a specific value of contact angle is assumed along a contact curve (or any portion thereof), the position of the contact curve must be determined as part of the solution.

Regardless of which type (or combination) of wettability condition(s) is assumed, the need to specify values of surface tension corresponding to the solid surfaces (γ_1,γ_2,γ_3 etc.) can be eliminated. If, for instance, area A_1 of Figure 5-8 is non-wettable, areas A_1 and A_2 will be fixed. (The solder cannot spread onto A_1.) Thus, it is unnecessary to include the corresponding surface energy terms $\gamma_1 A_1$ and $\gamma_2 A_2$ in the total energy expression, since they are constant for all admissible S. However, if the contact angle is specified along the contact curve separating areas A_1 and A_2 in Figure 5-8, these energy terms cannot be dropped. In this case, Young's Equation (Equation (5-1)) is usually employed to express these surface energies in terms of γ (surface tension of the free surface) and θ (contact angle) rather than γ_1 and γ_2, which are much more difficult to measure experimentally.

In many cases the volume constraint is treated most conveniently by introducing a Lagrange multiplier term to form a modified total energy functional $\tilde{E}$ as follows [e.g., Finn, 1986]:

$$\tilde{E} = E + E_c \tag{5-3}$$

where

E_c = energy term associated with volume constraint = $\lambda \left(\int dV - V \right)$;

$\int dV$ = volume enclosed by admissible free surface S and the solids;

V = prescribed volume of solder available to form the joint;

λ = unknown Lagrange multiplier.

The problem then becomes one of minimizing $\tilde{E}$ with respect to changes in S and λ, where S is required to meet only the geometric constraints imposed by the solid bounding surfaces (impenetrability and wettability). The volume constraint will automatically be satisfied, since the minimization of $\tilde{E}$ requires that:

$$\frac{\delta E''}{\delta \lambda} = 0$$

which implies

$$\int dV = V \cdot$$

The value of the Lagrange multiplier, which is determined by the minimization process, is physically related to the pressure within the equilibrated molten solder joint.

Boundary Value Problem (BVP) Formulation

Again consider the three-phase system shown in Figure 5-8. Instead of viewing the system equilibrium in a global manner as was done in the variational formulation, one may examine the local equilibrium of any small portion of the free surface. For example, a force balance on a differential element of the free surface "membrane" (Figure 5-9) will result in the Equation of Laplace [e.g. Klein Wassink, 1989]:

$$p_a - p = \lambda \left(\frac{1}{R_1} + \frac{1}{R_2}\right) \tag{5.4}$$

where:

p_a = (uniform) atmospheric pressure;

p = pressure within molten solder;

R_1, R_2 = principal radii of curvature of free surface.

Assuming the pressure within the solder varies hydrostatically, i.e.,

$$p = p_0 - \rho g z \tag{5.5}$$

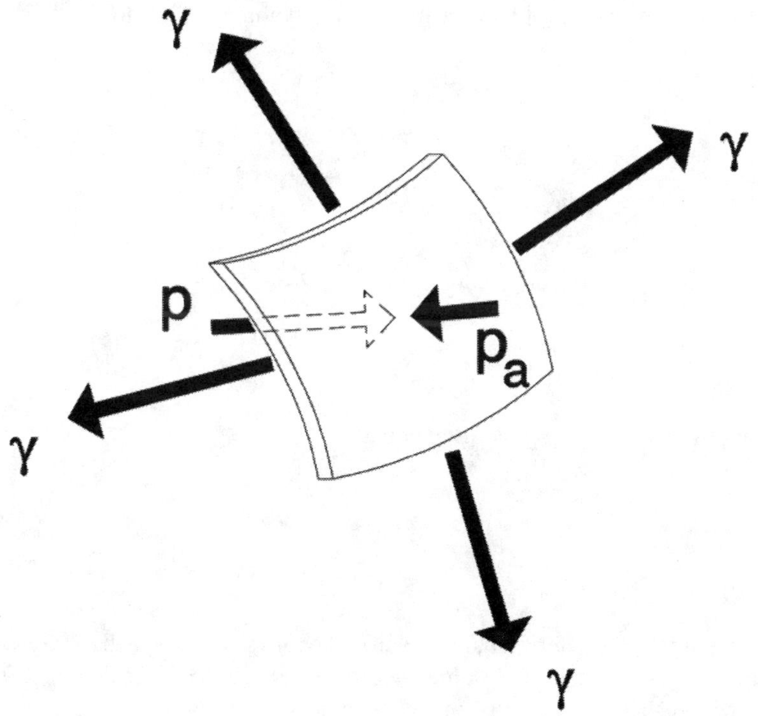

FIGURE 5-9 Differential element of free surface "membrane"

Equation (5-4) becomes

$$(p_a - p_0) + \rho g z = \gamma\left(\frac{1}{R_1} + \frac{1}{R_2}\right) \tag{5.6}$$

The constant p_0 represents the unknown reference pressure corresponding to the elevation z=0. Equation (5-6) is called the governing equation. If the unknown equilibrium surface is represented parametrically, the governing equation may be written explicitly as a second-order nonlinear partial differential equation. However, for the general three-dimensional case, the expressions for the principal radii of curvature are rather cumbersome; for this reason, general 3-D models are usually based on the variational formulation. For 2-D and axisymmetric cases, R_1 and R_2 (and thus the governing equation) take on relatively simple forms. Therefore, the BVP formulations (including governing equation, boundary conditions, and volume constraint) for these two special cases will be explicitly listed below.

2-D Boundary Value Problem

For a two-dimensional problem the principal curvatures reduce to:

$$\frac{1}{R_1} = \frac{y''(x)}{\{1 + [y'(x)]^2\}^{\frac{3}{2}}}$$

$$\frac{1}{R_2} = 0$$

resulting in the following governing equation:

$$(p_a - p_0) + \rho g z = \gamma \frac{y''(x)}{\{1 + [y'(x)]^2\}^{\frac{3}{2}}} \tag{5.7}$$

where x and y are the horizontal and vertical coordinates, respectively, y(x) represents the equation of the free solder surface (curve), and primes denote differentiation with respect to the argument. The boundary conditions associated with the unknown function y(x) are typically of two types: prescribed contact angle or prescribed joint dimension at the two ends of the profile curve. For example, for a fillet-type joint as depicted in Figure 5-10, the boundary conditions would take on the following forms:

$$y(0) = h \tag{5.8}$$

$$y'(0) = -\cot\theta_1 \tag{5.9}$$

$$y(l) = 0 \tag{5.10}$$

$$y'(l) = -\tan\theta_2 \tag{5.11}$$

where h, l, θ_1, θ_2 are the fillet height, length, and contact angles, respectively. The volume constraint would reduce to a constraint on the available solder cross-sectional area A:

$$\int_0^l y(x)\,dx = A \tag{5.12}$$

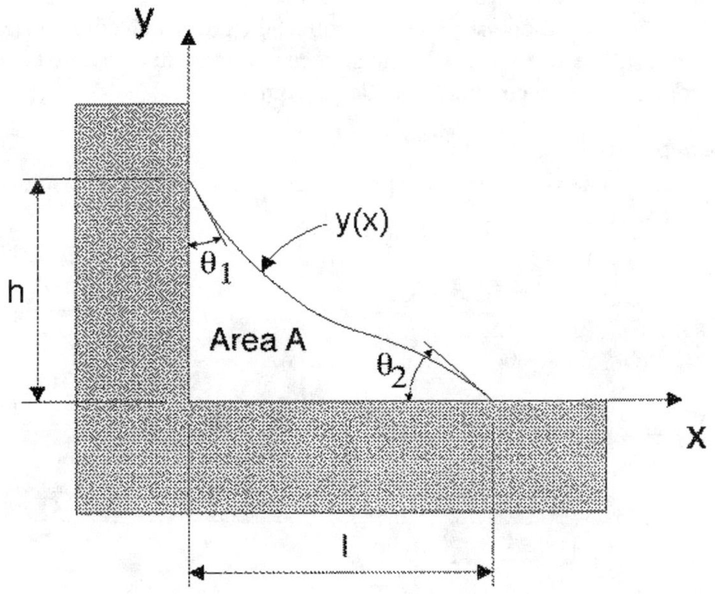

FIGURE 5-10 2-D solder fillet in orthogonal corner.

(The area A could be estimated by dividing the solder volume by the width of the land metallization.) Thus, for the fillet joint of Figure 5-10, the BVP is given by Equations (5-7), (5-8), and (5-9). This set of equations must be solved for the unknown profile shape y(x), given values of h or θ_1, and l or θ_2.

Usually a BVP involving a second-order differential equation (such as Equation (5-7)) will require only two boundary conditions. However, in this particular problem two additional BCs and the area constraint are necessary due to the presence of three unknown parameters in the model. For example, two practical problems to which the above formulation applies include the following:

1. Given solder properties (ρg, γ), joint area (A), and contact angles (θ_1,θ_2), determine the joint shape y(x) and the unknown parameters p_0, h, and l. This problem corresponds to a case in which the solder does not flow to the end of either metallization, such that the values of the contact angles may be estimated *a priori* (Figure 5-11a);
2. Given solder properties (ρg, γ), joint area (A), and the overall joint dimensions (h, l), determine the joint shape y(x) and the unknown parameters p_0, θ_1, and θ_2. This problem corresponds to a situation in which the solder flows to the ends of both metallizations, enabling one to specify the overall fillet dimensions a priori (Figure 5-11b).

In addition to these two types of problems, "mixed problems," in which θ_1 and l, or θ_2 and h are prescribed, are also possibilities. For 2-D problems involving solder flow along non-orthogonal or curved solids, the governing equation (Equation (5-7)) would still be valid; the BCs and area constraint would have to be modified to incorporate the more complex boundary geometries.

Axisymmetric BVP

For an axisymmetric problem the principal curvatures of the free solder surface can be expressed as:

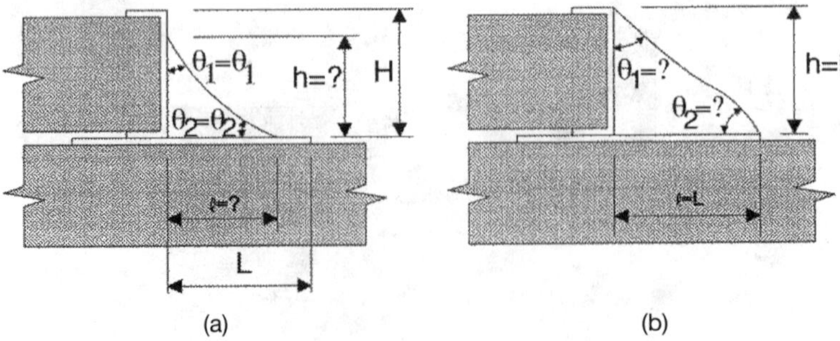

FIGURE 5-11 2-D problems involving (a) prescribed contact angles and unknown joint dimensions and (b) prescribed joint dimensions and unknown contact angles.

$$\frac{1}{R_1} = \frac{z''(r)}{\{1 + [z'(r)]^2\}^{\frac{3}{2}}}$$

$$\frac{1}{R_2} = \frac{z'(r)}{r\{1 + [z'(r)]^2\}^{\frac{1}{2}}}$$

where r and z are the radial and vertical (circular cylindrical) coordinates, respectively, and z(r) describes the meridional profile of the surface of revolution. Substitution of these expressions into Equation (5-6)) yields the governing equation for axisymmetric joint surfaces:

$$(p_a - p_0) + \rho g z = \gamma \left\{ \frac{z''(r)}{\{1 + [z'(r)]^2\}^{\frac{3}{2}}} + \frac{z'(r)}{r\{1 + [z'(r)]^2\}^{\frac{1}{2}}} \right\} \tag{5-13}$$

As with the 2-D case, the values of the profile function and its first derivative at the end points of the meridian can be related to the contact angles and overall joint dimensions. For definiteness consider an axisymmetric fillet joint connecting a pin to a substrate (Figure 5-12). The boundary conditions for this type of problem are the following:

$$z(a) = h \tag{5-14}$$

$$z'(a) = -\cot\theta_1 \tag{5-15}$$

$$z(a+l) = 0 \tag{5-16}$$

$$z'(a+l) = -\tan\theta_2 \tag{5-17}$$

where a is the radius of the pin. The volume constraint is obtained by setting the volume of solder beneath the free surface equal to the available solder volume V:

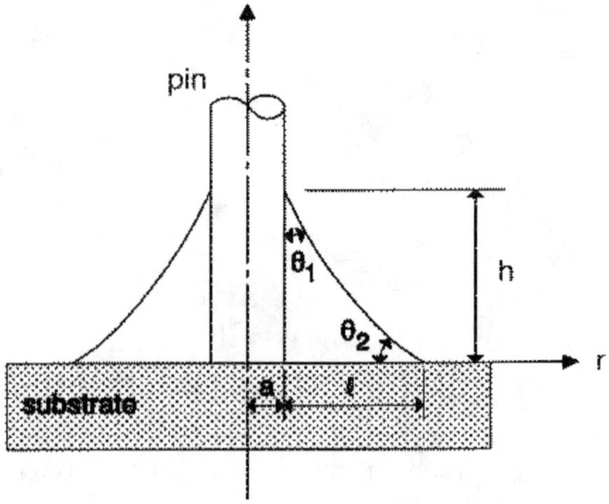

FIGURE 5-12 Axisymmetric solder fillet.

$$2\pi \int_{a}^{(a+l)} rz(r) = V \tag{5-18}$$

Thus, the BVP for the axisymmetric fillet joint of Figure 5-12 consists of determining the unknown function z(r) that satisfies Equations (5-10), (5-11), and (5-12), for given values of h or θ_1, and l or θ_2.

For axisymmetric joints different from the type depicted in Figure 5-12, the governing equation (5-10) would remain unchanged, but the BCs and volume constraint would require appropriate modifications.

5.5 MODELS FOR PREDICTING SOLDER JOINT GEOMETRY

The models that are currently available in the literature may be classified into three categories; in order of increasing complexity, these are: two-dimensional, axisymmetric, and general three-dimensional models. In this section the various models will be reviewed, and the results that are most conducive to practical implementation will be presented here. The reader is, of course, encouraged to consult the original references for details regarding the assumptions and technical derivations associated with these models.

5.5.1 Two-Dimensional Models

The models that are currently available for predicting two-dimensional joint geometries are summarized in Table 5-1. In this table, the area of application, solution type, major results, and restrictions associated with each study are also included. As can be seen from the table, the most prominent use for a 2-D model is to predict the geometry of solder joints connecting leadless (passive) chip components (e.g. capacitors and resistors) to a printed circuit board. This is due to the fact that there is very little variation in the fillet cross-sectional shape along the width of the relatively wide component (Figure 5-13). One of the major assumptions made in the development of these joint geometry models for leadless chip components is that the standoff height between chip and board, as indicated in Figure 5-1d, is either ignored or prescribed rather arbitrarily, as opposed to being predicted. Another assumption made in deriving most of these models is that the chip boundaries are straight and orthogonal, although finite element models, such as that proposed by Nigro et al. [1992], can easily handle more general chip geometries.

The studies by Klein Wassink and Verguld [1986] and by Ellis and Masada [1990] utilized 2-D fillet profile predictors to model the phenomenon of "drawbridging" or "tombstoning," which is characterized by chips rotating about one end during reflow, into the position of a partially raised drawbridge or even a nearly vertical tombstone-like position (Figure 5-14). The former study simply took the fillet shape to be linear, while the latter numerically solved the 2-D BVP for

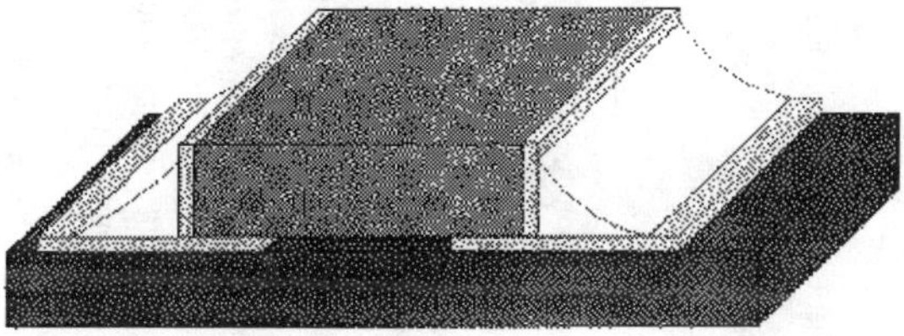

FIGURE 5-13 Leadless chip component with "two-dimensional fillets."

FIGURE 5-14 Examples of tombstone (drawbridge) phenomenon.

TABLE 5.1 Literature Involving Models for Predicting 2-D Joint Geometries

Reference	*Applications*	*Solution Type*	*Main Results*	*Restrictions on Results*
Klein Wassink & Verguld 1986	Straight-line Fillets	Straight-Line Fillets	Gen'l Guidelines for Preventing Tombstoning	Gen'l Guidelines for Preventing Tombstoning
Alexander & Bello 1987	Reflow of Leadless Chips	Elastica Solution	Gen'l Recommendations to Increase Joint Compliance by Controlling Solder Volume	Elastica Solution is Approximate
Klein Wassink 1989	Reflow & Wave Soldering of Leadless Chips	Numerical Solution of BVP: Elastica Solution	Samples of Upright & Inverted Fillet Profiles	Numerical Results are Problem-Specific; Perfect Wettability; Infinite Tinning
Jopek 1989	Reflow of Leaded Fine-Pitch Components	"Piecemeal" Elastica	General Guidelines for "Optimal" Volume	Elastica is Approximate: Theory Underestimates "Optimal" Volume
Ellis & Masada 1990	Reflow of Leadless Chips: Tombstone Prevention and Self-Alignment Enhancement	Numerical Solution of BVP for Fillet Profile; Numerical Solution of Equations of Motion for Chip Dynamics	General Guidelines for Preventing Tombstoning and Promoting Self-Alignment	Specific Contact Angle (12°) Assumed at Chip Metallization
Heinrich et al. 1990a,b	Reflow of Leadless Chips	Explicit Integral Solution of BVP; Circular Arc Approximation (Zero-Gravity Solution)	Fillet Height & Length vs. Solder Area (Volume); Closed-Form Expressions for Small & Large Joint Profiles and for Maximum Fillet Height	Infinite Tinning Assumption; Acute Contact Angles
Rooks et al. 1991	Reflow of Leadless Chips	Explicit Integral Solution of BVP; Circular Arc Approximation (Zero-Gravity Solution)	Samples of Fillet Profiles; Comparison of Exact and Zero-Gravity Solutions	Numerical Results are Problem-Specific; Infinite Tinning
Liedtke et al. 1991	Wave-Soldering of Leadless Chips	Explicit Integral Solution of BVP	Fillet Size & Shape vs. Pad & Component Sizes (Dimensionless)	Perfect Wettability'; (Finite Tinning)
Heinrich et al. 1992	Reflow of Leadless Chips	Explicit Integral Solution of BVP; Circular Arc Approximation (Zero Gravity Solution)	Fillet Profiles as Function of Chip & Pad Geometry (Dimensionless Results); Closed-Form Expressions for Small Joints	Perfect Wettability; (Finite Tinning)
Nigro et al. 1992	Reflow of Leadless Chips	Energy Minimization by Finite Element Method	Samples of Fillet Profiles for Finite & Infinite Tinning	Numerical Results are Problem-Specific
Jahsman 1992	Reflow of Long (Channel-Type) Solder Bumps	Circular Arc Approximation (Zero-Gravity Solution)	Bump Volume and Bump Force vs. Contact Angle and Bump Height (Dimensionless); Effect of "Count Ratio" of Two Bump Sizes on Dimensionless Bump Height	Equal Pad Radii; Count Ratio Results are Problem Specific

the fillet profile and subsequently solved the equations of motion for the chip. Both studies made some general recommendations for eliminating the tombstoning problem by modifying metallization dimensions on the chip and board or changing the amount of solder paste.

Studies by Alexander and Bello [1987] and Jopek [1989] have implemented the classical closed-form solution for the displaced shape of the "elastica," taking advantage of the analogy between the governing equation of 2-D surface tension theory (Equation (5.7)) and the differential equation of equilibrium for a looped wire loaded horizontally at its two ends [e.g., Klein Wassink, 1989]. Jopek actually applied the 2-D elastica solution in a piecewise fashion to model the fillets along each edge of a gullwing lead on a fine-pitch component, for the purpose of estimating "optimal" solder volume. Alexander and Bello performed a similar study for chip capacitors and resistors. In either case the elastica solution must be viewed as approximate, since the fillet will only take this shape if a particular solder volume is present and particular boundary conditions are satisfied at the toe of the fillet [Heinrich et al., 1990a].

A numerical solution to the 2-D BVP was obtained by Klein Wassink, [1989] and by Rooks et al. [1991] for the case of prescribed contact angles and different cross-sectional areas of the fillet. The results of Klein Wassink also include profiles for inverted fillets, such as would be found on wave-soldered leadless surface-mount chip components (Figure 5-3). The difficulty in implementing such results is that, unlike in reflow applications, one does not have direct control over the area (volume) of a wave-soldered joint; it, too, must be predicted. Liedtke et al. [1991] proposed a "dipping model" for estimating the size of wave-soldered fillets. The results are shown in dimensionless form in Figure 5-15a. (Note: The lower four contour lines were incorrectly labeled in the original reference.) The shaded region labeled "0.256" represents a "contour plateau," i.e., a region in which the amount of solder captured by the joint is independent of the metallization dimensions. Using the data of Figure 5-15a in the area constraint (Equation (5.9)), Liedtke et al. solved the BVP using an explicit integral expression for the inverted fillet profile and incorporated the effects of the metallization dimensions. The dimensionless results for fillet shapes are plotted in Figure 5-15b. These results are based on the assumption of zero contact angles, i.e. perfect wettability, provided that the solder does not reach the end of a metallization.

An explicit integral solution for a reflowed fillet on the upper side of a board was derived for acute contact angles [1990a] and for arbitrary contact angles [1992] by Heinrich et al. In those studies closed-form approximate expressions (valid for small joints) were also derived for the cases of infinite tinning (prescribed contact angles) and perfectly wettable finite tinning, respectively. Plots of fillet height and length vs. solder area for various degrees of wettability, presented in the earlier study, are shown in Figures 5-16a and b. These results correspond to a solder having density ρ=8.34 g/cm^3 and surface tension γ=490 dyne/cm. For other solder properties the dimensionless plots of Heinrich et al. [1990b] dis-

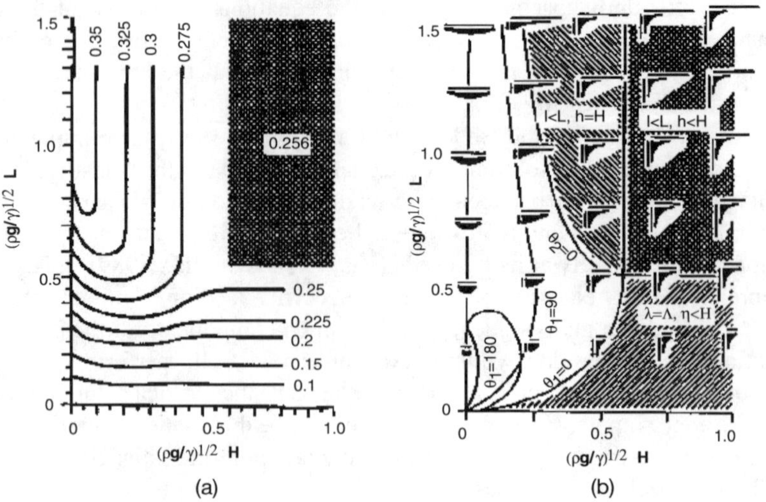

FIGURE 5-15 (a)Dimensionless contour plot of wave-soldered fillet area *A* as a function of tinning lengths *L* and *H;* (b)Array of possible wave-soldered fillet shapes as determined by dimensionless tinning lengths. [Liedtke et al., 1991: *Figure appears courtesy of The American Society of Mechanical Engineers.*]

FIGURE 5-16 Overall joint dimensions for a reflowed 63Sn-37Pb solder fillet and various sets of contact angles: (a) height *h* vs. area *A*; (b) length *ℓ* vs. area *A*. [Heinrich et al., 1990a: *Figure appears courtesy of The American Society of Mechanical Engineers.*]

played in Figure 5-17 may be useful. This plot, corresponding to perfectly wettable metallizations, also includes the results from the closed-form expressions for small joints and a window indicating the accuracy of the approximate formulae. The paper from 1992 considered the effect of chip and pad tinning dimensions on the solder joint shape. The results of that study are summarized in Figure 5-18a, corresponding to a chip height H →0, and in Figure 5-18b, which contains exact results for a chip of finite height. (For a typical solder the chip height associated

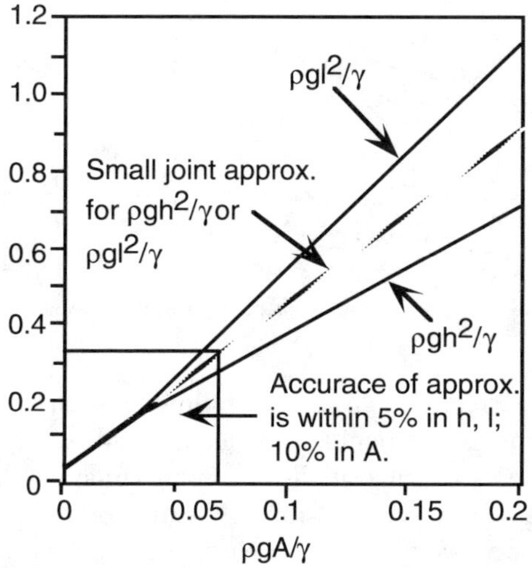

FIGURE 5-17 Dimensionless design curves for predicting overall fillet dimensions _ and h as a function of fillet cross-sectional area A. Perfect wettability ($\theta_1=\theta_2=0$) has been assumed. [Heinrich et al., 1990b: *Figure appears courtesy of The American Society of Mechanical Engineers*.]

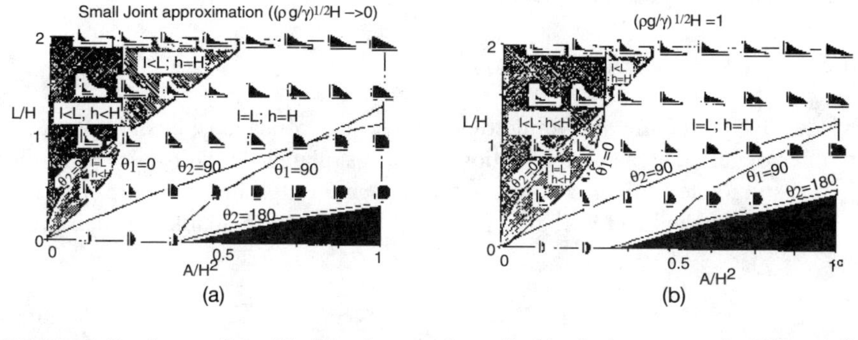

FIGURE 5-18 Arrays of possible fillet shapes as determined by tinning aspect ratio (L/H) and dimensionless solder area (A/H^2): (a) small joint approximation; (b) exact results for. [Heinrich et al., 1992: *Figure appears courtesy of The American Society of Mechanical Engineers*.]

with Figure 5-18b would be approximately 2.5 mm.) Comparison of these two figures gives an indication of the accuracy of the small joint approximation for finite tinning problems.

The only 2-D model listed in Table 5-1 that is not applicable to fillet-type joints is that proposed by Jahsman [1992]. He considered the 2-D analog of the Controlled-Collapse Chip Connection, or C4 joint, used in "flip-chip" attachment. The typical C4 joint is a solder bump connection between circular contact pads on a silicon chip and a ceramic substrate (Figure 5-19), the joint being axisymmetric

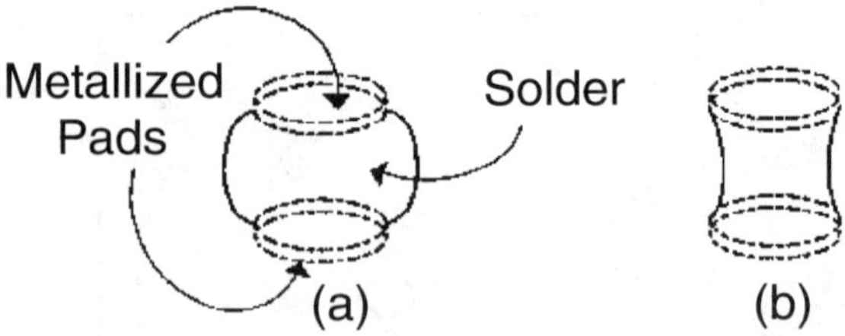

FIGURE 5-19 Typical C4 joints: (a) barrel-shaped; (b) hourglass-shaped.

when the pads are perfectly aligned. (See following section.) Jahsman, however, modeled long "channel-type bumps" connecting pads of equal width (Figure 5-20). His results were based on a small-joint approximation. Although the study is not directly applicable to typical C4 joints, the methodology proposed can be extended to analyze the axisymmetric case. Furthermore, the study by Jahsman gives insight into how stand-off height and joint shapes can be controlled using different solder volumes in the bump array, resulting in both barrel- and hourglass-shaped joints, reduced thermal stresses and strains, and increased fatigue life. This important concept, as it relates to C4 technology, has been investigated by several of the researchers surveyed in the next section.

5.5.2 Axisymmetric Models

Models for predicting axisymmetric solder joint profiles have been implemented in three different areas of application: C4 joints, plated-through-hole (PTH) fillet joints, and melted solder coatings within plated through-holes. These three types of profiles are depicted in Figure 5-21, and the models available for their predic-

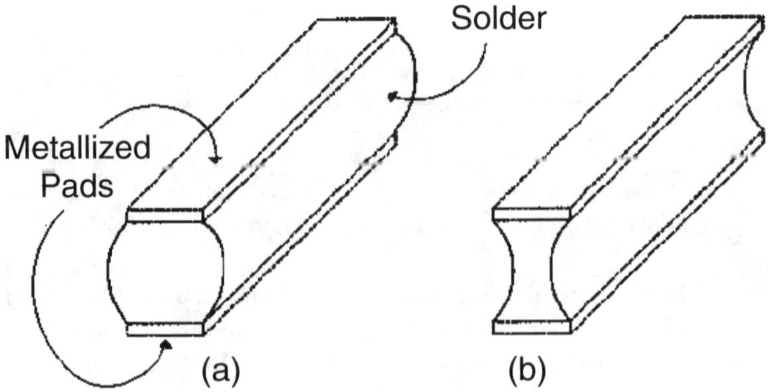

FIGURE 5-20 "Channel-type bumps": (a) barrel-shaped; (b) hourglass-shaped.

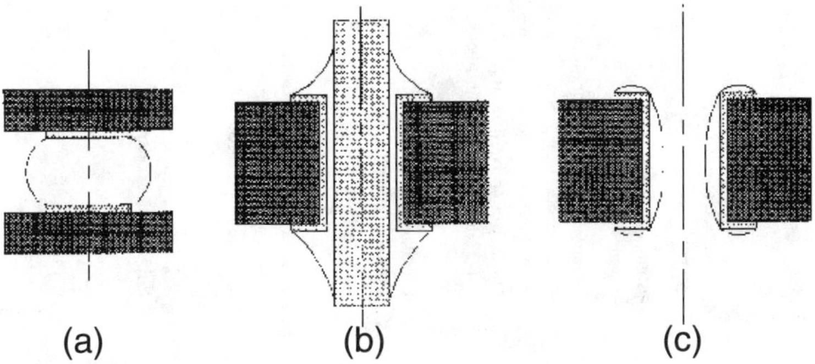

FIGURE 5-21 (a) Controlled-collapse (C4) joint, (b) plated through-hole fillet joint, and (c) fused solder coating in plated through-hole.

tion are listed in Table 5-2. Although the melted solder coatings are not truly "joints" (the fillet joints will be formed subsequently by using a solder preform or by wave soldering), predicting and controlling their geometry is important for the purpose of permitting lead insertion. Since the equilibrium shapes of these fused coatings are governed by the same mathematical formulation as the C4 and PTH fillet joints, they have been included in this discussion.

Several of the references summarized in Table 5-2 are related to C4 joints used in flip-chip technology. In 1969 Goldmann presented the first study that aimed to predict joint geometry and control design and process parameters in an optimal way from the standpoint of maximizing fatigue life. In that investigation a simple truncated sphere model was used to relate joint height, contact pad radii, and bump volume. A series of papers by a group of Japanese researchers [Oshima et al., 1982, 1985; Satoh, et al. 1983, 1987] utilized more general models to predict bump geometry, including a model based on a circular arc profile meridian and a numerical solution of the governing BVP. As in the Goldmann study, these models served as a precursor to optimizing the joint geometry. The major thrust of the Japanese researchers was the development of Self-Stretching Soldering Technology (SST), which utilizes height-control solder bumps (non-loadbearing) to create more desirable hourglass-shaped load-carrying bumps. Recently, Katyl and Pimbley [1991] have described four different methods that can be used to predict C4 joint geometry: elliptic integral and polynomial solutions for zero gravity, an approximate energy minimization technique (Ritz method), and an iterative solution of the exact BVP using Runge-Kutta numerical integration. A new finite element method has been introduced by Nigro et al. [1992] to determine axisymmetric equilibrium shapes of solder joints. The method is based on a parametric polynomial representation of all spatial variables within each element.

Investigators who considered the formation of axisymmetric fillets around through-hole pins include Chu [1975], Klein Wassink [1989], and Rooks et al.

TABLE 5.2 Literature Involving Models for Predicting Axisymmetric Joint Geometries

Reference	*Applications*	*Solution Type*	*Main Results*	*Restrictions on Results*
Goldmann 1969	C4 Joints in Flip-Chip Applications	Truncated Sphere	Approximate Analytical Expression Relating Pad Radii, Bump Height, and Solder Volume; Optimization of Joint Geometry	Truncated Sphere Assumption is Approximate
Chu 1975	Plated-Through-Hole Fillets Using Solder Preforms	Numerical Solution of BVP Coupled with Approximate Energy Minimization	Samples of Top & Bottom Fillet Profiles; Ideal Solder Volume vs. Board Thickness; Fillet Heights vs. Solder Volume	Numerical Results are for Specific Pin and Hole Diameters and Pad Radii; "Operationally Ideal" Solution is Arbitrarily Taken to be Within 0.5% of True Minimum-Energy State
Golachev et al. 1980	Melting of Tin-Lead Coatings in Plated-Through-Hole Apertures	Circular Arc Approximation for Meridian	Change in Hole Diameter vs. Volume of Plated Solder Alloy; Recommended Solder Layer Thickness vs. Pad and Hole Diameters	Circular Arc Approximation is Invalid for Larger Pad Diameters; Results are Problem- Specific
Kudryavtsev et al. 1982	Melting of Tin-Lead Coatings in Plated-Through-Hole Apertures	Circular Arc Approximation for Meridian	Change in Hole Diameter vs. Board & Plating Thicknesses and Pad & Hole Diameters	Circular Arc Approximation is Invalid for Larger Pad Diameters; Results are Problem- Specific
Oshima et al. 1982	C4 Joints in Flip-Chip Applications	Numerical Solution of BVP; Truncated Sphere; Circular Arc Meridian	Exact Results for Bump Height vs. Solder Volume; Optimization of Joint Geometry Using Approximate Solutions	Exact Numerical Results Limited to Equal Pad Diameters

Reference	*Applications*	*Solution Type*	*Main Results*	*Restrictions on Results*
Satoh et al. 1983, 1987; Oshima et al. 1985	C4 Joints in Flip-Chip Applications Using Self-Stretching Technique (SST), i.e., Height-Control Solder Bumps	Circular Arc Meridian	Bump Height vs. Volumes of Joint Solder and Control Solder; Optimization of Joint Geometry	Results are Problem-Specific
Klein Wassink 1985, 1989	Melting of Solder Coatings in Plated-Through-Hole Applications	Numerical Solution of BVP with and without Gravity; Circular Arc Meridian	Sample Profiles of Fused Solder Coatings; Comparisons of Exact and Approximate Profiles	Results are Problem-Specific
Klein Wassink 1989	Plated-Through-Hole Fillets	Numerical Solution of BVP	Samples of Top & Bottom Fillet Profiles	Results are Problem-Specific; Perfect Wettability
Rooks et al. 1991	Top Fillet in Plated-Through- Hole Applications	Numerical Solution of BVP; Elliptic Integral Solution for Zero Gravity	Samples of Top Fillet Profiles; Fillet Volume vs. Height (Dimensionless)	Results are Problem-Specific; Contact Angle at Pin is 15 deg
Katyl & Pimbley 1991	C4 Joints in Flip-Chip Applications	Numerical Solution of BVP; Energy Minimization by Ritz Method; Elliptic Integral and Approximate Polynomial Solutions for Zero Gravity	Sample Profiles of Sessile and Pendant Drops on Finite Pad; Sample Profiles of C4 Joints	Results are Problem-Specific
Nigro et al. 1992	C4 Joints in Flip-Chip Applications	Energy Minimization by Finite Element Method	Sample Profile for C4 Joint; Bump Force vs. Bump Height	Results are Problem-Specific

[1991]. All of these researchers treated the case in which the solder volume is specified (e.g. a solder preform is used). Chu numerically solved for both the upper and lower fillet profiles, until both the volume constraint and pressure constraint were satisfied. He also introduced the concept of an "operationally ideal" solution as one which is "near" the minimum-energy configuration, yet is not able to achieve the true minimum-energy position due to surface roughness and contamination effects. Chu also found that, for the case of square pins, the 2-D solution (revolved around the pin) provided better results than an axisymmetric solution. Results from this "hybrid" solution are illustrated in Figure 5-22, in which top and bottom fillet heights are related to the total solder volume in the fillets (i.e. excluding that in the hole) for a 25-mil square pin. Klein Wassink presented some specific numerical results for fillets on double-sided boards. His results were based on a numerical solution of the axisymmetric BVP under the assumption of 0-degree contact angles. Rooks et al. considered top fillets only, generating "exact" results by solving the BVP numerically and approximate (zero-gravity) results using an elliptic integral expression. Dimensionless results relating solder volume to fillet height are plotted in Figure 5-23, assuming a 15-degree contact angle at the pin and a completely covered contact pad; i.e. the radius of the outer edge of the fillet is prescribed. The results indicate that solder self-weight is negligible in joints within the size range considered, and that the dimensionless volume-height relationship is essentially linear.

The remaining type of axisymmetric solder profile that has received attention is the fused solder coating on a plated through-hole. Investigations by Golachev et al. [1980] and Kudryavtsev et al. [1982] were based on a circular arc meridian within the fused hole. Klein Wassink [1985, 1989] later proposed the following

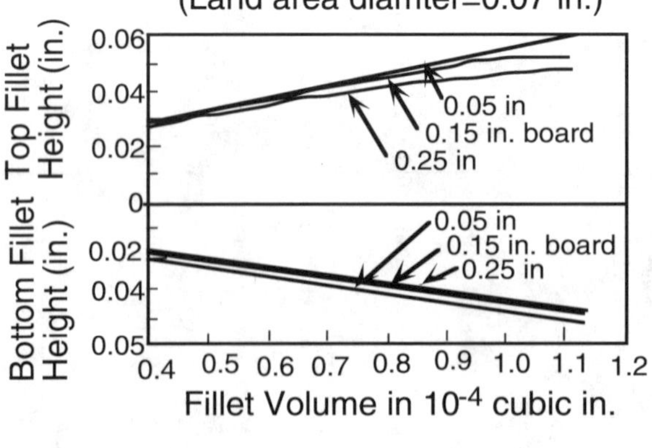

FIGURE 5-22 Predicted heights of through-hole fillet joints on double-sided board, as functions of fillet volume. [Chu, 1975]

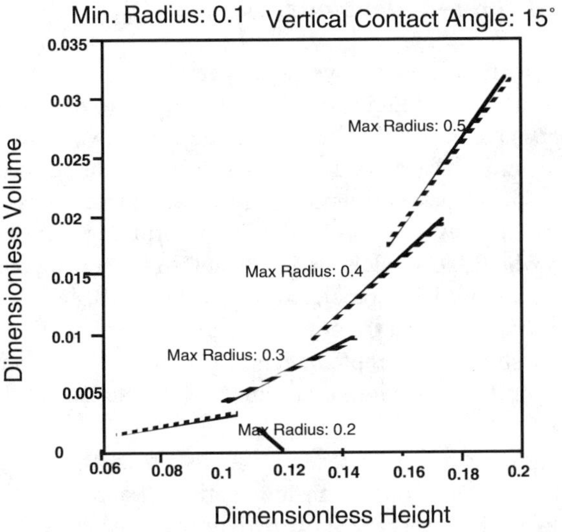

FIGURE 5-23 Relationship between dimensionless height of axisymmetric fillet and dimensionless solder volume for a dimensionless pin radius of 0.1 and for several dimensionless pad radii. (Dotted curves represent the zero-gravity solution. All length parameters have been scaled by the characteristic length parameter defined by $\sqrt{\gamma/(\rho g)}$.) [Rooks, et al. 1991: *Figure appears courtesy of The American Chemical Society.*]

design formula, based on the assumptions that the meridian within the hole is a circular arc and that the solder plating on the circular pads does not flow into the aperture after melting:

$$\delta_m = 1.5\delta_0 \approx 0.5\frac{V}{D_0 t} \tag{5.19}$$

where

δ_m = maximum thickness of the melted solder layer within the hole;

δ_0 = (uniform) thickness of the solder layer within the hole, prior to melting;

V = volume of solder layer within the hole;

D_0 = hole diameter;

t = board thickness.

Klein Wassink [1985, 1989] also performed a more sophisticated analysis of the problem by obtaining a numerical solution to the axisymmetric BVP with and without gravity. Sample profiles and numerical data are included in both references.

5.5.3 General Three-Dimensional Models

Due to the complexity associated with the general three-dimensional mathematical formulation for determining equilibrium shapes of molten solder joints, relatively few papers have been published on the subject. However, as fine-pitch technology becomes more prevalent in microelectronics, the ability to predict the complex 3-D joint shapes surrounding small fine-pitch leads and controlling geometry-related defects such as skipping or bridging will become critical. Another area in which 3-D models are important is in modeling the self-alignment of flip-chip devices used in optoelectronic applications, in which the required alignment accuracy may be on the order of microns. In this section a summary will be given of those studies that implement general 3-D joint geometry predictors to solve problems in these two important areas. The relevant papers are listed in Table 5-3.

A series of articles has been recently published on the realignment of controlled-collapse (C4) joints in flip-chip applications [Landry et al., 1991; Patra and Lee, 1991; Patra, et al. 1992]. The study by Patra and Lee presents an approximate numerical solution technique which neglects the self-weight of the solder and assumes that the horizontal cross-sections of a misaligned joint are circles having centers lying along a single line. They consider the case in which the contact pads on the chip and substrate are circular. Samples of their results are shown in Figure 5-24. In Figure 5-24a the original solid bumps (prior to reflow) are shown on the chip and substrate, in addition to the initial (misaligned) molten configuration at the start of reflow and the final fully aligned joint. Corresponding to this simulated alignment process are plots of the total energy and horizontal restoring force vs. the misalignment level (Figures 5-24b and c). Data of the type shown in Figure 5-24c was used in the investigation by Landry et al. to determine the value of surface tension γ such that the restoring force data is consistent with the measured natural frequency associated with misalignment oscillations. A similar fitting technique was used to estimate a viscous damping coefficient. Having determined values of surface tension and damping coefficient by fitting one set of data, Landry et al. then proceeded to use these values in their geometry prediction model and in a simple dynamic model to predict the oscillatory motion measured in another experiment using smaller joints. The comparison is indicated in Figure 5-25. Models of this type may be useful in characterizing the self-alignment mechanism in scaled-down applications in which experimentally monitoring the small joint motions may be infeasible [Landry et al., 1991]. The paper by Patra et al. [1992] develops a general 3-D finite-difference method for solving the BVP which governs the shape of flip-chip joints spanning circular or non-circular pads. Self-weight of the small solder joints is neglected. Three-dimensional renderings of the model's results are illustrated in Figures 5-26a and b.

The work of Racz et al. [1993] and Racz and Szekely [1993] utilizes a computer program known as the Surface Evolver [Brakke, 1991, 1992] to predict me-

TABLE 5.3 Literature Involving Models for Predicting General 3-D Joint Geometries

Reference	*Applications*	*Solution Type*	*Main Results*	*Restrictions on Results*
Landry et al. 1991	Realignment of C4 Joints in Flip-Chip Applications	Approximate Numerical Solution of BVP Neglecting Solder Self-Weight, Coupled with Curve-Fit of Oscillatory Realignment Motion	Sample Profiles of C4 Joints Before and After Realignment; Restoring Force & Equivalent Stiffness vs. Misalignment; Verification of Predicted Oscillatory Motion	Horizontal Cross-Sections Assumed to be Circular; Numerical Results are Problem-Specific
Patra & Lee 1991	Realignment of C4 Joints in Flip-Chip Applications	Approximate Numerical Solution of BVP Neglecting Solder Self-Weight	Sample Profiles of C4 Joints Before and After Realignment; Restoring Force vs. Misalignment	Horizontal Cross-Sections Assumed to be Circular; Numerical Results are Problem-Specific
Patra et al. 1992	Realignment of C4 Joints on Circular or Non-Circular Pads	Numerical (Finite-Difference) Solution of BVP Neglecting Solder Self-Weight	3-D Graphical Representations of Joints on Fully Wetted and Partially Wetted Circular and Rectangular Pads (Zero Misalignment); Joint Height vs. Rectangular Pad Aspect Ratio; Joint Height vs. Wetted% of Circular Pad Area	Numerical Results are Problem-Specific
Racz & Szekely 1993; Racz et al. 1993	Prediction of Complex Joint Geometries, such as in Gullwing and J-Lead Fine-Pitch Applications	Energy Minimization by Finite Element Program (Surface Evolver)	3-D Graphical Representations of a Variety of Solder Joints; Dimensionless Design Monographs for Selecting "Optimal" Solder Volume as a Function of Wettable Pad/Lead Area for Gullwing & J-Leads	Stand-Off Height Between Lead and Pad is Specified a Priori; Criteria for "Optimal" Joints are Qualitative "Rough Guidelines" Extrapolated from Through-Hole to Surface- Mount Technology

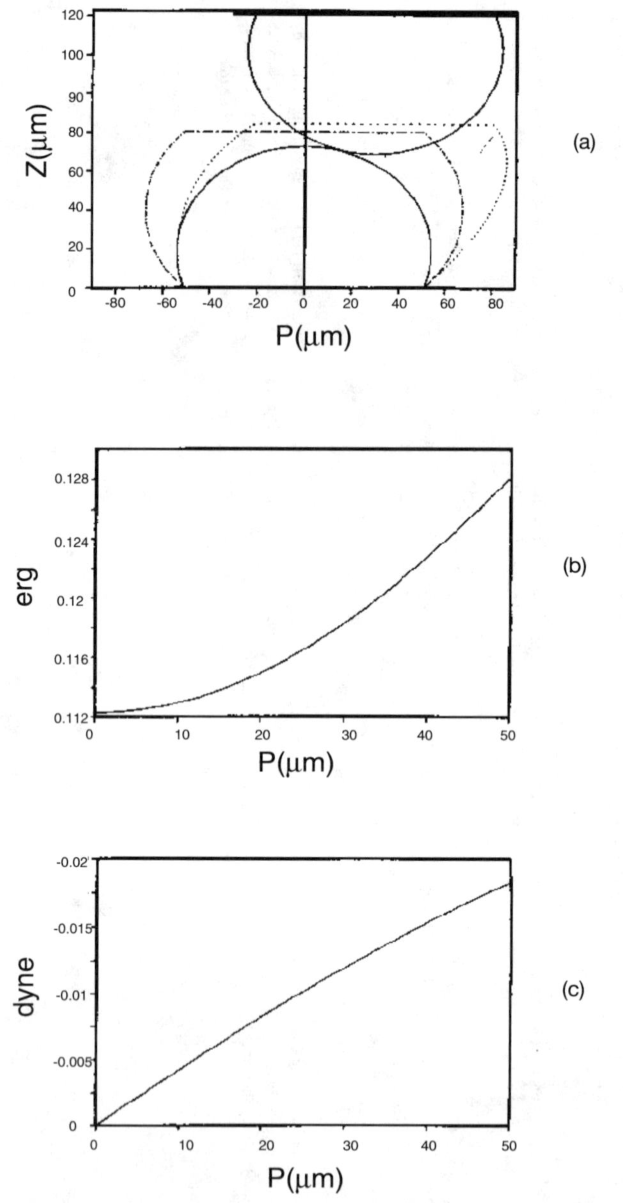

FIGURE 5-24 (a) Solder bump profiles before reflow (solid line), immediately after melting in misaligned configuration, and at the end of reflow in the equilibrated fully aligned position; (b) Total energy vs. misalignment for the predicted movement in (a); (c)Horizontal restoring force vs. misalignment for the predicted movement in (a). [Patra and Lee, 1991: *Figure appears courtesy of The American Society of Mechanical Engineers.*]

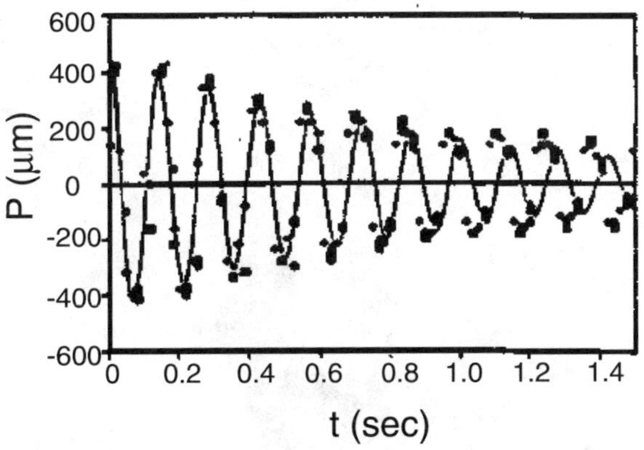

FIGURE 5-25 Comparison of discrete experimental data and predicted results for oscillatory re-alignment of flip-chip solder joint (P = *misalignment*). [Landry et al., 1991: *Figure appears courtesy of The American Society of Mechanical Engineers.*]

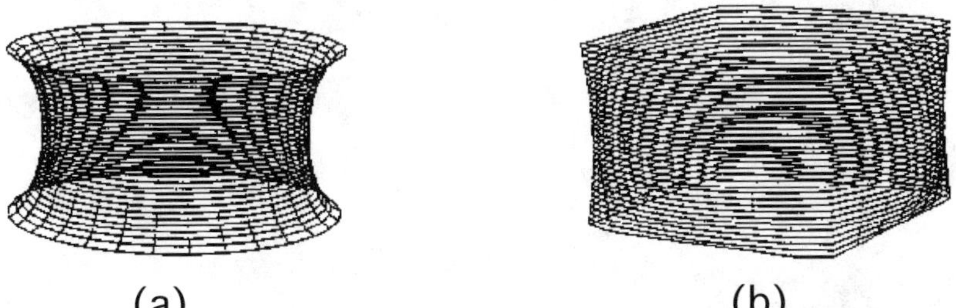

FIGURE 5-26 Examples of predicted joint geometries for (a) circular pads; (b) square pads. [Patra et al., 1992: *Figure appears courtesy of The American Society of Mechanical Engineers.*]

niscus shapes that arise in materials processing, with particular emphasis on soldering in surface-mount microelectronics applications. The Surface Evolver numerically determines a local minimum-energy configuration of a meniscus (solder joint), given an initial free surface. The program approximates the solder surface with triangular facets ("finite elements") and positions these facets until a local minimum of the total energy is achieved. The program is capable of handling complex geometries such as those encountered in fine-pitch surface-mount applications. Examples of predicted joint geometries around gullwing and J-leads are shown in Figures. 5-27a and b. In Racz and Szekely [1993] the Surface Evolver was implemented to create design guides for selecting "optimal" solder volume for gullwing and J-lead joints as a function of wettable surface area on the lead and board. These dimensionless results are indicated in Figures 5.28a

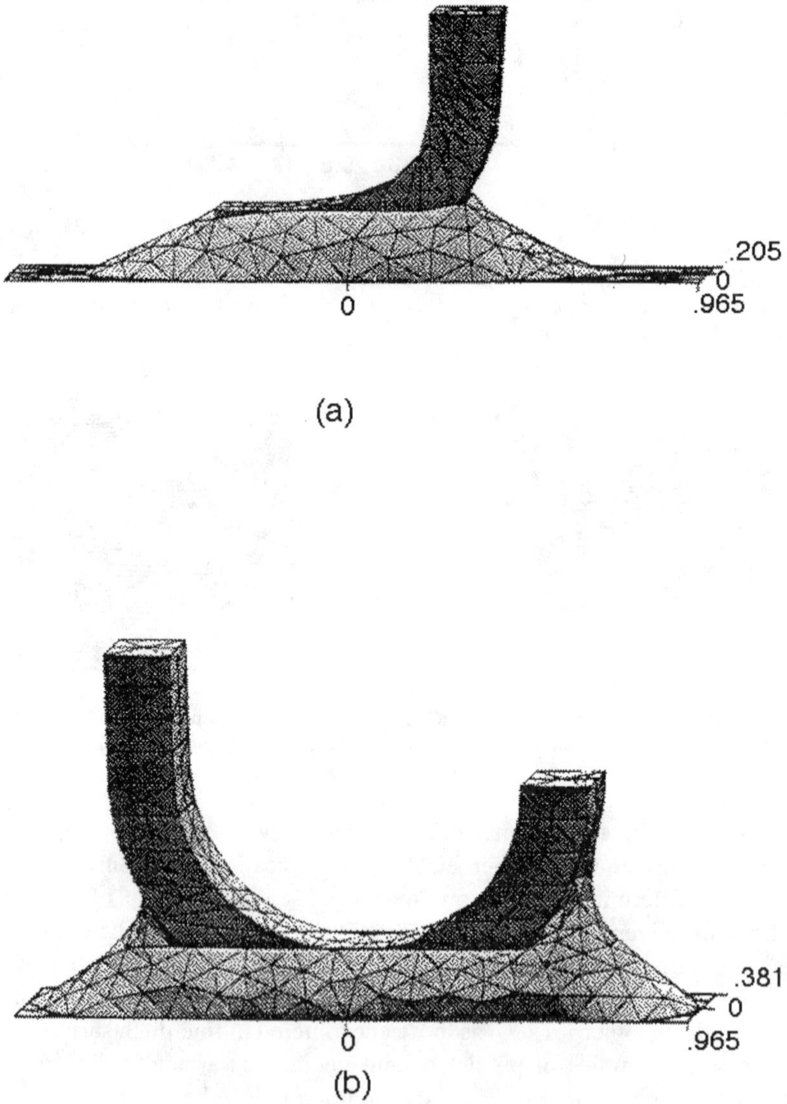

FIGURE 5-27 Examples of fine-pitch joint geometries generated by *Surface Evolver*: (a) gullwing lead; (b) J-lead. [Racz and Szekely, 1993: *Figure appears courtesy of The Iron and Steel Institute of Japan*.]

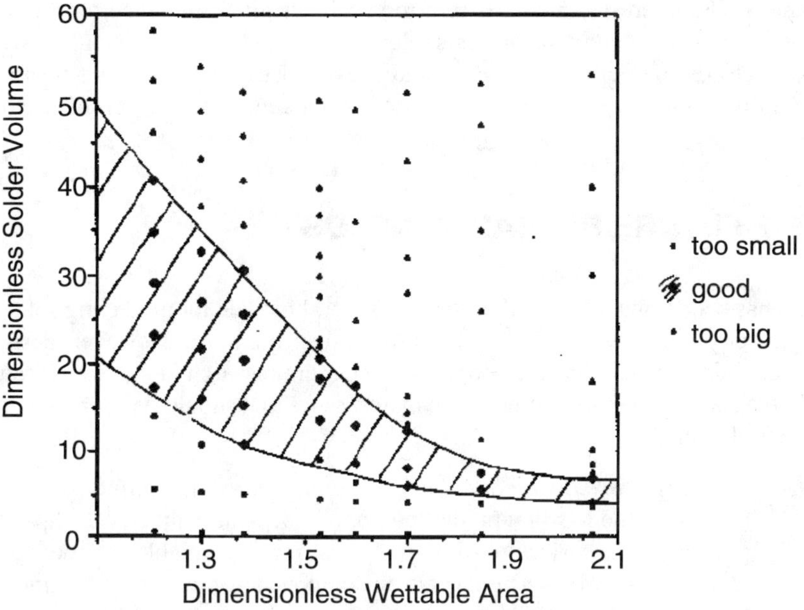

(a)

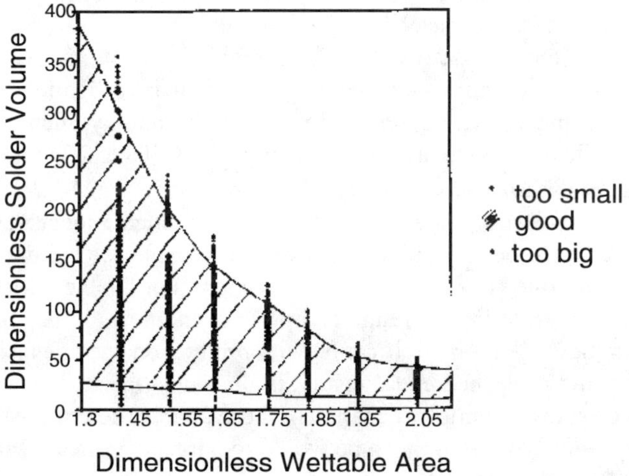

(b)

FIGURE 5-28 Dimensionless design guides for selecting "optimal" solder volume: (a) gullwing leads; (b) J-leads. [Racz and Szekely, 1993: *Figure appears courtesy of The Iron and Steel Institute of Japan.*]

and b. The interested reader is referred to the original reference for the specific definitions of the dimensionless solder volume and the dimensionless wettable surface area. The criteria that Racz and Szekely use to determine whether a solder joint is acceptable or defective are based on the recommendations of Klein Wassink [1989].

5.6 FUTURE RESEARCH NEEDS

As has been pointed out in this chapter, several models for predicting solder joint geometry have been described in the technical literature. However, despite the attention this topic has received in the past, more research in this area is required. In the author's opinion, some of the major issues that should be addressed include the following:

1. *Models for Predicting Stand-Off Height in Fillet-Type Joints:* With the exception of models that predict bump, or C4, joint geometry in flip-chip applications, the surface-tension models currently available are not capable of predicting stand-off height between component (or lead) and substrate. Since this geometric parameter is often intimately associated with thermomechanical strain levels in the solder joint and thus the fatigue life of the joint, the ability to predict and control this parameter is critical;
2. *Three-Dimensional Studies:* Additional work is needed in developing numerically efficient, user-friendly computer programs for determining the complex joint geometries that are present in modern microelectronics applications. In addition, fundamental theoretical investigations are needed to identify under what circumstances a computationally intensive 3-D geometry prediction may be avoided by using the available 2-D results;
3. *Industrial Verification/Implementation of Theoretical Models for Predicting Joint Geometries:* Although most of the theoretical models discussed in this chapter have been verified experimentally, many of the experiments have been controlled in a laboratory environment. For the theoretical results to be of maximum utility to design and process engineers in the electronic packaging industry, more verification with production specimens is required. Incorporating the theoretical results in an industrial setting will most likely require statistical implementation of the available (deterministic) models and their results. Work of this type would require a significant sampling of production joints and characterization of their geometries. Automatic and nondestructive methods for measuring small-scale geometries, such as X-ray laminography [e.g., Read and Bayba, 1991; Read and Lucey, 1991] and laser profilometry, would be critical in generating the required data base;

4. *More User-Friendly Presentation of Results:* The numerical results generated by the currently available models and the models of the future should be presented in the most useful form possible. For instance, use of dimensional analysis may permit a numerical solution (or set of solutions) to be represented in a dimensionless plot (or set of plots) that succinctly describes general results, not simply one set of numerical results for a given set of input parameters. Dimensionless results are of more general use and are more easily converted into practical guidelines. Furthermore, if a model has been validated for a particular type of application and its results can be summarized in, say, a dimensionless graphical form, these results can be programmed (by curve-fitting or storing discrete data), eliminating the need for the original model altogether. This would result in much more efficient computation of joint geometric parameters.

In addition to a continuing research effort in predicting joint geometries, simultaneous progress must be made in understanding and predicting solder joint behavior and performance, specifically joint fatigue life and reliability. Advancement in this area is necessary for the determination of optimal joint geometries, which will be application-dependent. Once accurate models for predicting joint behavior are established, the joint geometry predictors can be interfaced with them (serving as pre-processors). For instance, if a finite element stress analysis is to be performed on a joint, the required geometric data could first be calculated by a joint geometry program, followed by an automatic finite element mesh generation and the subsequent determination of displacement, strain, and stress fields. This output could then be used to estimate joint fatigue life using an appropriate fatigue model. Integration of these various models (geometry, stress analysis, and fatigue) was discussed by Nigro et al. [1991] as it pertains to computer-aided design of solder joints. A flow chart illustrating the building blocks of such a design process is shown in Figure 5-29. Only when the three models are fully developed and integrated can a "concurrent" approach to solder joint design be fully realized, enabling the solder joint to be optimized at the design and manufacturing stages.

5.7 ACKNOWLEDGMENTS

The author would like to express his gratitude to the Allen-Bradley Company, Milwaukee, WI, for technical and financial support; to colleagues A.F. Elkouh, R.A. Fournelle, P.E. Liedtke, and N.J. Nigro; and especially to P.S. Lee, without whom this undertaking would not have been possible.

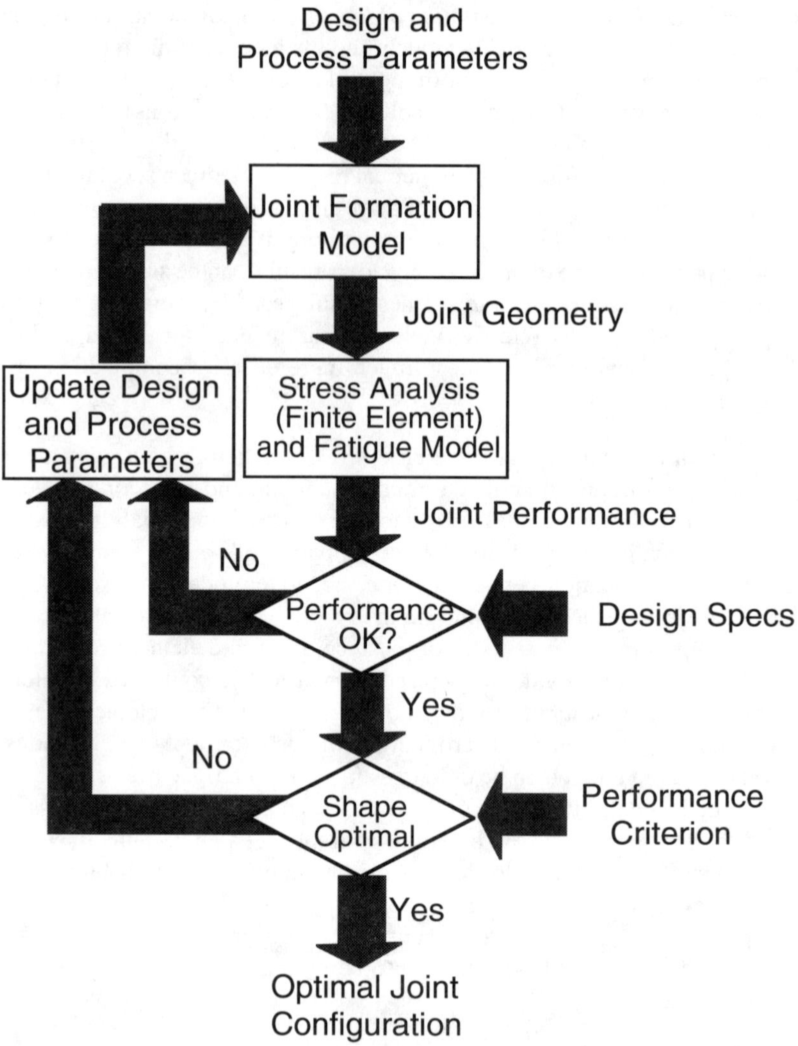

FIGURE 5-29 Flow chart illustrating integration of joint formation model (geometry predictor), stress analysis (finite element) model, and fatigue model, resulting in optimal joint configuration.

REFERENCES

Alexander, P.T., and Bello, D.C.,1987, *Technical Proceedings Expo SMT '87*, Las Vegas, 85-89.

ANSI/IPC-T-50E Standard, 1992, "Terms and Definitions for Interconnecting and Packaging Electronic Circuits," Institute for Interconnecting and Packaging Electronic Circuits.

Brakke, K.A., 1991, *Surface Evolver Manual, version 1.65*, Research Report GCG 31, The Geometry Center, 1300 S. Second St., Minneapolis, MN 55454.

Brakke, K.A., 1993, *Experimental Mathematics*, in press.

Chu, T.Y., 1975, *Western Electric Engineer*, 19:31-42.

Dussan V., E.B., 1985, *J. Fluid Mech.*, 151:1-20.

Ellis, J.R., and Masada, G.Y., 1990, *IEEE Trans. on Components, Hybrids, and Mfg. Technology,* 13:545-552.

Finn, R., 1986, *Equilibrium Capillary Surfaces,* Springer-Verlag, NY.

Golachev, S.M., Kudryavtsev, V.S., and Ovseevich, V.L., 1980, *Welding Prod. (Svar. Proiz.)*, 27:19-21.

Goldmann, L.S., 1969, *IBM J. Res. Develop.*, 13:251-265.

Heinrich, S.M., Elkouh, A.F., Nigro, N.J., and Lee, P.S., 1990a, "Solder Joint Formation in Surface Mount Technology—Part I: Analysis," ASME, *Journal of Electronic Packaging,* 112:210-218.

—., 1990b, "Solder Joint Formation in Surface Mount Technology—Part I: Analysis," ASME, *Journal of Electronic Packaging,* 112:219-222.

Heinrich, S.M., Liedtke, P.E., Nigro, N.J., Elkouh, A.F., and Lee, P.S., 1992, *Proc., ASME/JSME Joint Conference on Electronic Packaging: Advances in Electronic Packaging 1992,* 1:603-611. (Also in press, *Journal of Electronic Packaging*, ASME, 1993.)

Hinch, S.W.,1988, *Handbook of Surface Mount Technology*, Longman Scientific and Technical.

Hwang, J.S.,1989, *Solder Paste in Electronics Packaging,* Van Nostrand Reinhold.

Jahsman, W.E., 1992, "Channel Solder Bump Shape Analysis," Internal Correspondence, Intel Corporation.

Johnson, C., 1982,*Electronic Packaging and Production*, 22:23-24.

Johnson, R.E., and Dettre, R., 1969, *Surface and Colloid Science*, Vol. 2, Matijevic, E. (ed.), Wiley-Interscience, pp. 85-153.

Jopek, S., 1989, *Printed Circuit Assembly*, 3:15-20.

Katyl, R.H., and Pimbley, W.T., 1991, *Mfg. Processes and Materials Challenges in Microelectronic Packaging,* ASME AMD-Vol. 131/EEP-Vol. 1, pp. 57-63.

Klein Wassink, R.J., 1985, *Brazing and Soldering*, 8:5-7.

Klein Wassink, R.J., and Verguld, M.M.F., 1986, *Hybrid Circuits,* 9:18-24.

Klein Wassink, R.J., 1989, *Soldering in Electronics*, 2nd ed., Electrochemical Publications Limited.

Kudryavtsev, V.S., Nikiforov, G.D., Dolgov, Yu. S., Golachev, S.M., and Oparin, M.I., 1982, *Welding Prod. (Svar. Proiz.)*, 29:13-14.

Landry, M., Patra, S.K., and Lee, Y.C., 1991, "Experiment and Modeling of the Self-Alignment Mechanism in Flip-Chip Soldering," *Manufacturing Processes and Materials Challenges in Microelectronic Packaging,* ASME AMD-Vol. 131/EEP-Vol. 1, pp. 49-56.

Liedtke, P.E., Heinrich, S.M., Elkouh, A.F., Nigro, N.J., and Lee, P.S., 1991, "Effect of Chip and Pad Geometry on Solder Joint Formation in SMT" and "Prediction of Wave-Soldered Fillet Geometry in SMT Applications," *Manufacturing Processes and Materials Challenges in Microelectronic Packaging,* ASME AMD-Vol. 131/EEP-Vol. 1, pp. 65-72.

Nigro, N.J., Heinrich, S.M., Elkouh, A.F., Zou, X., Fournelle, R.A., and Lee, P.S., 1992, presented at the *ASME Winter Annual Meeting*, Anaheim, CA.

Nigro, N.J., Elkouh, A.F., Heinrich, S.M., Chang, P., and Lee, P.S., 1991, *Surface Mount Technology,* 5:59-61.

Oshima, M., Kenmotsu, A., and Ishi, I., 1982, *Proc., 2nd Annual Intl. Electronics Packaging Soc. Conf.,* pp. 481-488.

Oshima, M., Satoh, R., Hirota, K., and Ishi, I., 1985, *Proc., 1st IEEE CHMT Symposium: CHMT Tokyo '84,* pp. 165-170.

Patra, S.K., and Lee, Y.C., 1991, *ASME Journal of Electronic Packaging*, 113:337-342.

Patra, S.K., Sritharan, S.S., and Lee, Y.C., "Minimum-Energy Surface Profile of Solder Joints for Non-Circular Pads," *Manufacturing Aspects in Electronic Packaging,* ASME EEP-Vol. 2, PED-Vol. 60, Anaheim, CA, 1992, pp. 197-210.

Racz, L.M., Szekely, J., and Brakke, K.A., 1993, *Trans., Iron and Steel Institute of Japan International*, 33., pp. 336-342.

Racz, L.M., and Szekely, J., 1993, *Trans., Iron and Steel Institute of Japan International,* 33.

Read, D.T., and Bayba, A., 1991, *Paper 91-WA-EEP-17*, ASME Winter Annual Meeting, Atlanta.

Read, D.T., and Lucey, G.K., Jr., 1991, *Journal of Electronic Packaging,* 113:178-185.

Rooks, S., Racz, L.M., Szekely, J., Benhabib, B., and Neumann, A.W., 1991, *Langmuir,* 7:3222-3228.

Rouse, H.,1978, *Elementary Mechanics of Fluids,* Dover Publications, Inc..

Satoh, R., Oshima, M., Komura, H., Ishi, I., and Serizawa, K., 1983, *Proc., 3rd Annual Intl. Electronics Packaging Soc. Conf.*, pp. 455-461.

Satoh, R., Oshima, M., Hirota, K., and Ishi, I., 1987, *J. Japan Inst. Metals*, 51:553-560.

Weinstock, R.,1974, *Calculus of Variations*, Dover Publications, Inc., 1974.

6

Life Prediction and Accelerated Testing

Harvey D. Solomon

6.1 INTRODUCTION

It is 8:00 Monday morning and your boss has just handed you the task of estimating the lives the solder joints present in a product that is being developed. Furthermore, you are also requested to develop an accelerated testing plan that will verify your predictions. For all too many readers the above scenario is a reality. This chapter is intended to present a plan of attack for these problems, the various specific approaches that have been employed and the areas of weakness of these approaches. The questions raised at the recent 3rd International Workshop of Materials and Mechanics Issues of Solder Alloy Applications will be discussed, as will consensus conclusions that were reached.

The problems that are raised in the opening scenario were not as pressing 15 or 20 years ago. This is because the fundamental nature of electrical interconnections has changed. There was a time when electrical interconnections were primarily mechanical, and solder was used only to insure that good electrical contact was maintained. Indeed, the general rule of thumb was that the solder should never be called upon to support a mechanical stress. This has all changed in modern electronics, particularly with the use of surface mounted components. Now the solder is being called upon to act as the source of mechanical as well as electrical interconnection.

The analysis of mechanical connections made with solder is complicated by the low melting point of solder. The commonly used Sn/Pb eutectic has a melting point of 183°C. Room temperature (25°C) is 0.65 of the melting point (based upon degrees Kelvin). This ratio (the temperature divided by the melting point, both in degrees K) is called the homologous temperature. In general, the use temperature of most structural materials is kept well below a homologous temperature of 0.5. For instance, the turbine disk of an aircraft engine is generally not allowed to exceed this homologous temperature. These turbine disk alloys were developed after decades of research and represent what many feel is the zenith of the metallurgists' art. Yet even here, with alloys specially developed for high temperature applications, we are limited to half the melting point. Contrast this to the solder joints utilized in an engine control system, where a temperature of 125°C might be encountered (which is a homologous temperature of 0.87).

Not only are solders, which were developed thousands of years ago, called upon to perform under conditions that are more severe than those encountered by some of our most advanced alloys, but much less attention is given to the prediction of the survivability of mechanical structures held together by solder. Aircraft engine and airframe manufacturers have design groups whose sole function is to consider component life prediction. Decades have been spent in gathering the data necessary for these predictions and developing the methodology for using this data. Furthermore, these data and design methodologies are subjected to extensive review and government regulation. In contrast, the design of electrical interconnections is often left to the end of a design process, and is often just an afterthought, if it is thought of at all. Instead of a design team whose members spend much of their technical career on life prediction problems, solder joints are often designed by engineers who may have little or no experience performing life prediction calculations of the sort described here. It is for such an engineer that this chapter and indeed this whole book is dedicated.

6.2 LIFE PREDICTION

Fortunately, the problem of solder joint reliability is not quite so bleak as might be inferred from the introduction. Solders possess unique properties which make them well suited for their applications, properties which will likely insure their continued use. The low melting point of solders allows joints to be made at temperatures low enough to prevent device damage. Furthermore, solders are very ductile—Sn/Pb eutectic can be superplastic. This allows a solder joint to absorb the thermal strain on cooling from the soldering temperature without failing. The very large ductility also allows solder joints to have an exceptional thermal fatigue resistance, which allows their use under conditions which would cause most other material to fail. It is this thermal fatigue resistance which will be the subject of the life prediction approaches which will now be discussed.

There are four primary steps to determining the life of a solder joint: 1) the determination of the duty cycle that is assumed to exist, providing the boundary condition for the analysis; 2) the calculation of the displacement (or force) that the joint experiences due to the assumed duty cycle; 3) the determination of the strain (or stress) produced by this displacement; and 4) the determination of the life resulting from this strain (or stress). This chapter will concentrate on the fourth step, but first we shall briefly consider the first three steps.

6.2.1 Determination of the Duty Cycle

Before any life prediction can be made, the imposed boundary conditions that the solder joint in question will experience must be determined. Will the joint experience a thermal cycle and, if so, what will be the maximum and minimum temperatures and what will be the cycle time and wave shape (i.e., will the temperature be held at some value, how fast will it change and how often will it change)? The temperature range and cycle wave shape may be fixed by variables such as the maximum and minimum ambient temperatures that can be achieved. This may be a function of possible ambient temperatures and the amount of cooling (or heating) that is to be expected. For military electronics there may be a military specification for a design such as –55°C to + 125°C in a specified time. Perhaps it is not the cycling of the ambient temperature that is concerned, but rather the temperature rise due to the operation of the device and cooling when it is turned off or the power dissipation is reduced. Perhaps there is a problem with vibrations or impact loads. The first problem is to decide exactly what it is being designed for. This may require some testing or assumptions regarding the conditions of use.

This problem often is complicated by the fact that no simple cycle is experienced. The ambient temperature may change and so may the power level. The magnitude of the temperature variation may change with the season (such as what occurs with under the hood automobile components). In the winter, the overnight temperature may be well below freezing and rise when the engine heats up. In the summer the lowest temperature will be much higher, yet the highest operating temperature also increases (but the temperature range may be lower). A cumulative damage rule (to be discussed later) is required to handle such multiple cycling. For now we shall assume that the design will be for a simple thermal cycle with a given maximum temperature, minimum temperature, rate of change of temperature and possibly a hold at any given temperature.

6.2.2 Calculation of the Displacement (or Force) that a Solder Joint Experiences

A solder joint is only part of a very complex electronic assembly. It joins two larger and generally stiffer members (for example, a chip carrier or other component and a PWB). The joint experiences an imposed displacement due to the thermal expansion mismatch between the component and a substrate (generally a printed wiring board (PWB)) and also between the solder and either the component or PWB. Even if there is no expansion mismatch, displacements will be produced by thermal gradients. Joints can also experience mechanical loadings due to vibrations or impact loading. The first step in any life analysis is to determine the displacements or loads that the joint experiences.

For most devices the thermal displacements are the most important, so we shall concentrate on them. We shall also concentrate on the problem of the displacement experienced by the joint between a component and a substrate. Changing the temperature from T1 to T2 produces a thermal displacement which is determined by the relative thermal expansions α_s and α_c, and the distance L between joints. This displacement, Δ, is described by Equation (6.1) and is illustrated in Figure 6-1.

$$\Delta = (\alpha_s - \alpha_c)(T_2 - T_1)L/2 \tag{6.1}$$

Figure 6-1 shows the displacement developed when the temperature is increased from T_o to T_{max} and decreased from T_0 to T_{min}. Equation (6.1) describes the displacement developed between T_{max} and T_{min} (with $T_2 = T_{max}$ and $T_1 = T_{min}$). The factor of 2 assumes that the two joints equally share the displacement that is developed. For conditions where the displacement is not equally shared, the factor of 2 is replaced by a partition coefficient P, which varies between 1 and 2. For P = 1 all the displacement is absorbed by the joints of one side.

Equation (6.1) represents a simplistic approach to the development of thermal displacements. It neglects the thermal gradients that develop in the component and substrate, possible bending of the component or substrate, and the thermal mismatch between the solder and the component of substrate.

Equation (6.1) assumes that there is no bending of the component or substrate, which is not necessarily valid. If the component or substrate bends, then it takes up some of the thermal mismatch, lessening that which the solder experiences. The shear strain that is being imposed is reduced, but a tensile stress component may be added. Equation (6.1) therefore represents a worst-case calculation. Actually it is somewhat better than that. The stored energy that such bending produces is available to the solder and causes the solder to creep during a high temperature hold. In the limit, the solder can creep enough to relax out the thermal stresses developed in the component and substrate. (This assumes that there is no creep of the component or substrate, which, given the very high creep rate of the solder, is an excellent assumption.) If all of the bending is eliminated, then the displacement of the solder will be given by Equation (6.1). Thus, Equation (6.1) describes a limiting condition where the component and substrate are so stiff that they absorb none of the thermal displacement, or the temperature and time are sufficient for the solder to have crept enough to relax out all of the displacements that the component and substrate experience. The rate of this relaxation depends upon the tem-

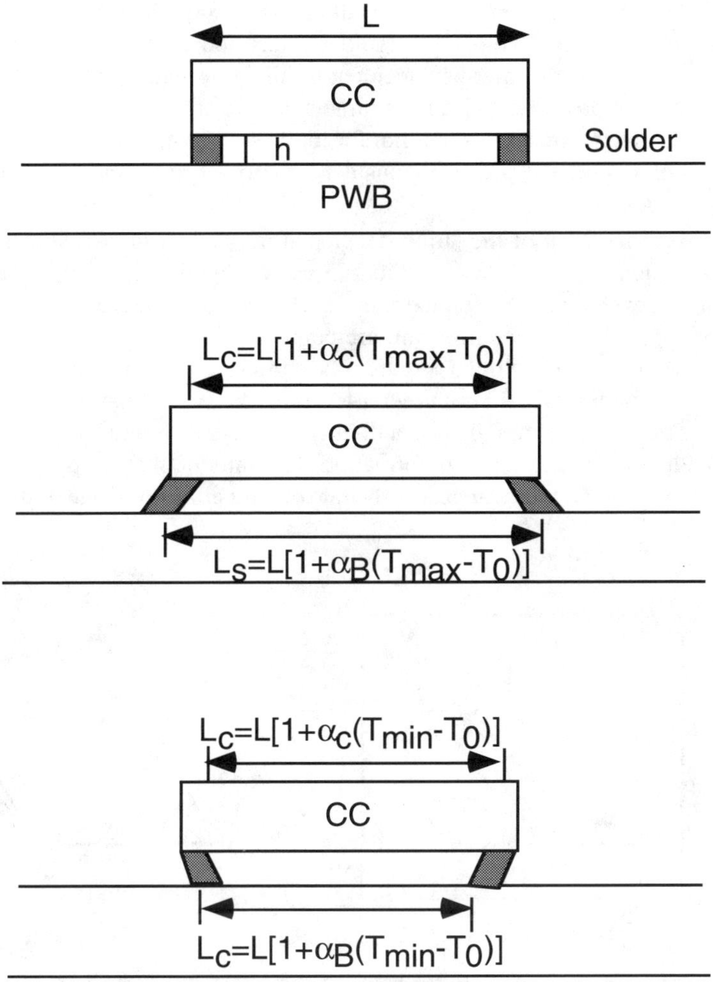

FIGURE 6-1 Development of thermal strains. [Solomon, 1991c]

perature. If these conditions are not met, then Equation (6.1) will overestimate the displacement that the solder experiences. While Equation (6.1) is not completely accurate it still forms the basis for several life prediction approaches described later in this chapter.

The incorporation of creep effects into a life prediction model is complicated, and the result depends upon the choice of constitutive relationship used to describe the rate dependence of the solder deformation. The magnitude of the deformation of solder joint depends upon not only the magnitude of the temperature change, but also upon the peak temperature, the rate of temperature change, and any hold time at a given temperature. Hall [1983; 1984; 1987; 1991] has measured the displacement developed by solder joints and quantitatively described this behavior. He divides the displacement by the joint height to convert it to a strain. Such an approach is valid for simple solder post joints, such as those developed in C4 interconnections, but not for joints with large fillets. We shall first consider Hall's approach and then consider how to approach more complicated joint geometries.

Hall's determination of the strain developed in a solder joint is illustrated in Figure 6-2. Figure 6-2b shows the thermal cycle. Figure 6-2a shows the shear strain that is developed as the temperature is changed and Figure 6-2c shows the force developed. The force and strain are combined in Figure 6-2d to show the hysteresis loop that is developed. Figure 6-2a shows that as the temperature is increased the amount of shear strain increases, but 6-2c shows that less force is required to produce this strain. This is a direct consequence of the temperature and strain rate dependence of solder deformation. The amount of strain developed depends upon the temperature range and the rate of temperature change or the length

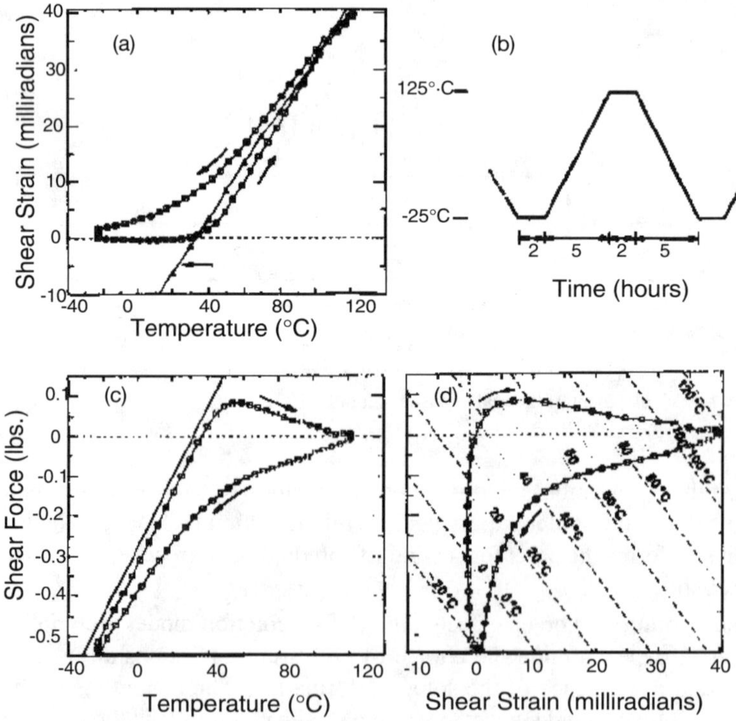

FIGURE 6-2 Temperature cycling with a large ΔT. [Hall, 1991]

of a hold time at any given temperature. At low temperatures and for fast temperature changes, the solder strains may be kept small and elastic.

Figure 6-2 shows that a rather characteristic, "banana" shaped hysteresis loop is produced by the thermal cycling of a solder joint. Such a hysteresis loop is calculated from the constitutive equation of state developed by Stone, Li, and coworkers [1986]. The width and area of the hysteresis loop have both been used to predict the fatigue life.

6.2.3 Determination of the Strains (or Stresses) Developed in a Solder Joint

The simplest approach is to not consider the stresses developed, but rather to keep the problem one of thermal displacements and strains in the joint. The fatigue life can be related to the applied stress, but it is easier to use the strain as the correlating parameter. This is especially true in the case of solder fatigue, where the stress is almost always above yield and creep deformation is an important consideration. There is no easy way of accounting for creep strains in a stress-based life prediction approach. In the preceding discussion, the thermal strain developed in the solder joint was easily determined by dividing Equation (6.1) by the height, h, of the joint to convert the displacement into a shear strain, γ, by:

$$\gamma = (\alpha_S - \alpha_c)(T_2 - T_1)L/2h \tag{6.2}$$

Equation (6.2) is most applicable to the simple solder post illustrated in Figure 6-1.

As has been noted, Equation (6.2) does not completely describe the behavior of filleted joints such as those present in castellated leadless chip carriers or leaded chip carriers. With fillets, a strain distribution develops when a displacement is applied to the joint. This is illustrated in Figure 6-3. Finite element analysis [Solomon et al., 1990] was used to determine the strain distribution. Finite element analysis (FEA) has been used to determine the strain distribution developed in leaded chip carriers, pin grid arrays, TAB packages, ceramic quad flat packages, plastic quad flat packages, as well as in surface mounted packages of the sort shown in Figure 6-3 (a more comprehensive review of these studies is found in [Charles et al., 1990; Sherry et al., 1985; Pao and Jih, 1992; Pan, 1991, Lau et al., 1991a, Barker et al., 1991; 1992; Read and Lucy, 1991; Sharif et al., 1991; Jahsman et al., Lau, 1991a]). It should be noted, however, that in spite of the assumptions made, Equation (6.2) may give a reasonable estimate of the actual strains, especially if the geometry of the joint is simple.

Creep and the strain rate dependence of solder deformation should be included in any analysis of the stresses and strains developed in a solder joint. Other chapters in this book discuss the constitutive relations needed to perform time-dependent analysis.

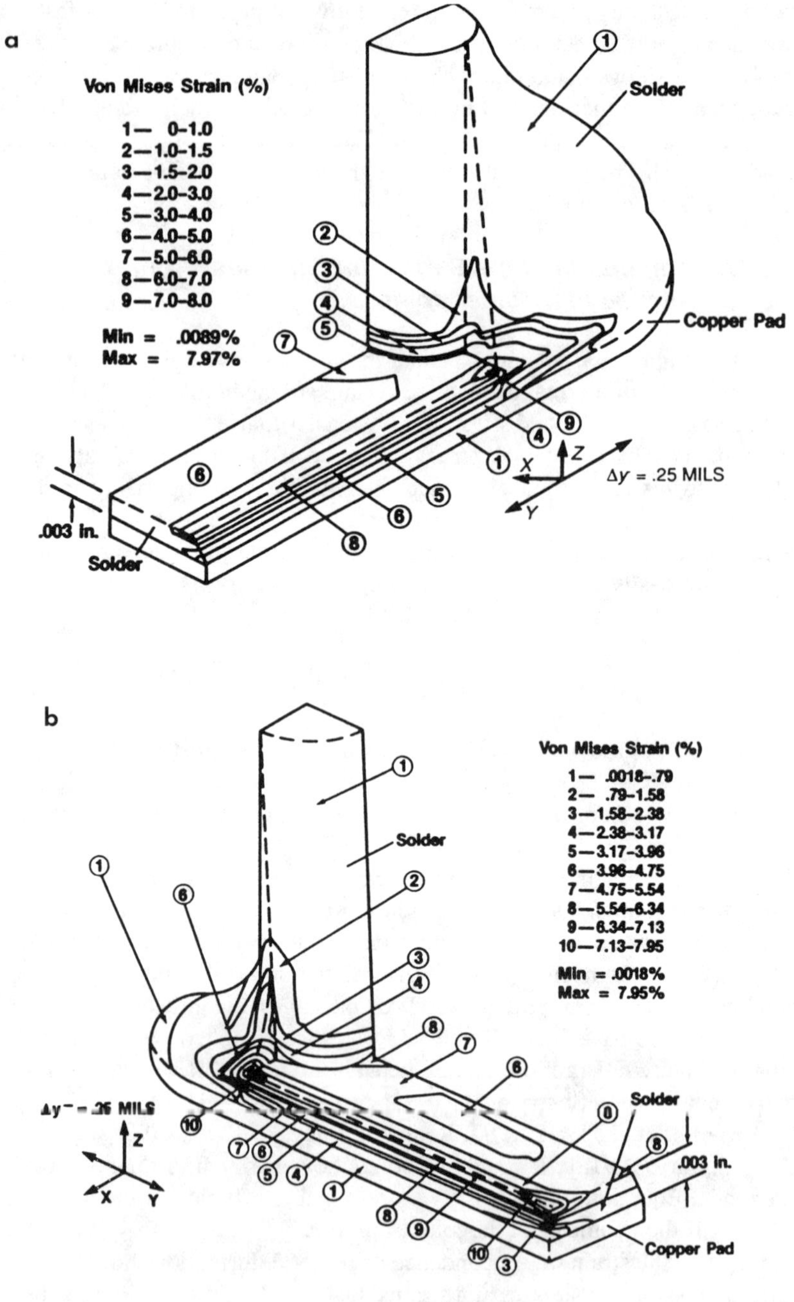

FIGURE 6-3 Calculated von Mises strain for thin (a) and bulbous (b) leadless CC-PWB joints for a castellated chip carrier subjected to a displacement of 0.25 mils (0.006mm). [Solomon et al.,1990]

6.3 LIFE PREDICTION APPROACHES

Let us suppose that the duty cycle has been determined, as have the thermal displacements. These data, the proper constitutive relationship for the solder deformation and a finite element analysis, will yield the strain and possibly also the stress developed by the duty cycle (combining the two gives hysteresis loops of the sort shown in Figure 6-2). Actually, the analysis will yield a distribution of strains and stresses. For simplicity, let us first use just one strain or hysteresis loop and then consider how to incorporate the proper strain/stress distribution in the life prediction analysis. For more information on how to arrive at this point in the analysis, see other chapters in this volume. From here on we shall concentrate of how to utilize the calculated strains and stresses.

Thermal fatigue is a complex process, with the stresses and strains that produce fatigue damage being applied as the result of temperature changes. First we shall consider approaches based upon isothermal fatigue data, with the assumption that the strain or hysteresis energy, produced by a temperature change, can be applied isothermally to model the thermal fatigue process, ignoring the fact that in the actual application the temperature is varying. We shall consider the influence of the mean or peak temperature, but not the varying temperature, per se. In a later section, we shall consider thermal cycling and how the results of thermal cycling tests compare to the predictions made from isothermal data. We shall also briefly consider thermomechanical cycling, where both the temperature and strain can be varied independently.

6.4 COFFIN-MANSON LCF APPROACH

A knowledge of the strain applied to a joint can be used to calculate the fatigue life resulting from this strain. This is done via the Coffin-Manson LCF relationship [Coffin, 1954; 1969a; Manson, 1953; 1960] which relates the plastic strain range, $\Delta\gamma_p$, and the resulting fatigue life, N_f, or the Basquin relation [Basquin, 1910] which relates the elastic strain range with the resulting life. Both relations express the fatigue life as simple power law functions of the elastic or plastic strain. Figure 6-4 is a typical example of a Coffin-Manson plot, where the fatigue life is shown vs. the applied plastic strain range employed in the fatigue cycling [Solomon, 1989a]. In this figure the fatigue life is defined as the number of cycles to reduce the load to half the value of the first cycle. The influence of the definition of failure is discussed in the next subsection. Two slopes are drawn through the data. One (dotted curve) represents the least squares curve fit which minimizes the error with respect to the plastic strain. The other (solid curve) represents the curve fit with respect to the fatigue life. Since the fatigue life is the independent variable the solid curve is preferred. Unfortunately, in fatigue studies it is customary to plot the fatigue life as the X axis, whereas curve fitting routines minimize the error

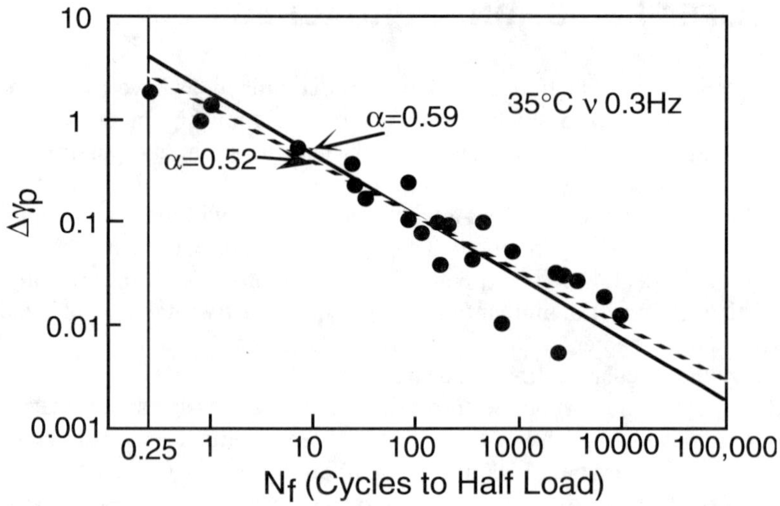

FIGURE 6-4 Low cycle fatigue data for 60Sn-40Pb solder, for failure defined by a 50% load drop. [Solomon et al., ©1990, ASME]

with respect to the Y axis (which is usually the independent variable). To get the proper slope [Solomon, 1989a] it is necessary to invert the axis, determine the slope, and then re-invert the axis to get them in the form shown in Figure 6-4.

Figure 6-5 shows the fatigue life of 60Sn-40Pb solder, tested in shear, as a function of both the elastic and plastic strains [Enke et al., 1989]. (This figure uses the number of strain reversals rather than cycles. There are 2 reversals per cycle. Sometimes the abscissa is thus given as $2N_f$ rather than just N_f. In both cases N_f is the number of cycles to failure.) Using reversals has benefits when considering complex cycling, where determining what constitutes a complete cycle is a problem. Since we will not be considering this sort of cycling, for simplicity we shall stick with cycles.) The sum of the elastic and plastic strain is the total strain, so that expressing the Coffin-Manson and Basquin relationships in terms of the strain and adding these two strain yields a relationship (attributed to [Morrow, 1965]) which correlates the total strain and the fatigue life [Langraf, 1970]. This is shown schematically in Figure 6-6 and for a high Pb solder in Figure 6-7. If the total strain range is calculated then this total strain vs. fatigue life data can be used. The elastic, plastic, and total strains are not independent. A specific elastic and plastic strain will be established for any stress, and the sum of these strains is the total strain. These strains are dependent upon the elastic modulus, yield stress, and work hardening behavior. It is important to remember that, because all these strains are interrelated, either the elastic, plastic, or total strain can be used to predict the fatigue life (providing creep effects do not, with time, decrease the elastic strains and increase the plastic ones).

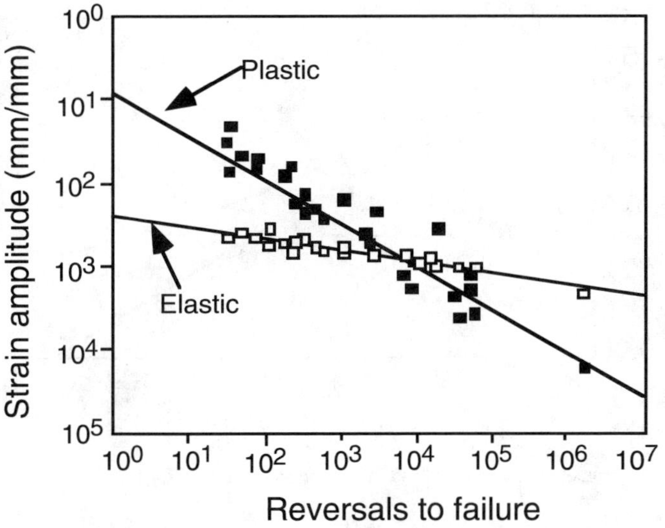

FIGURE 6-5 Elastic and plastic strain-life plots. [Enke et al., ©1989 IEEE]

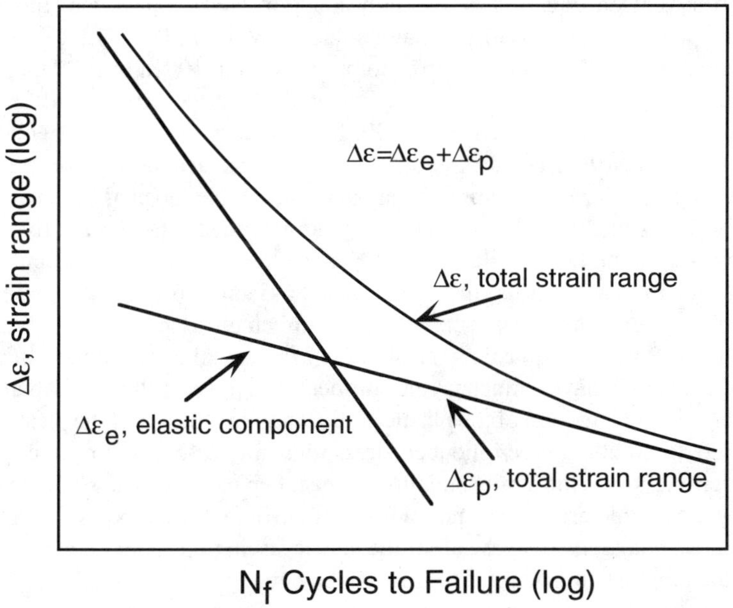

FIGURE 6-6 Schematic fatigue behavior showing the fatigue life as a function of the elastic, plastic, and total strains. [Landgraff, 1970]

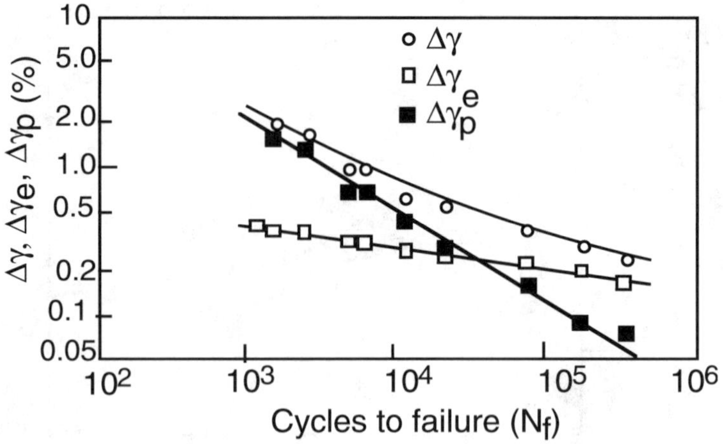

FIGURE 6-7 Fatigue behavior of 95Pb-5Sn at room temperature for tests in torsion. [Kitano, et al.,1987]

Figure 6-8 shows an example where the calculated elastic strain is added to the plastic strain to get a total strain, which was used for the fatigue life prediction [Solomon et al. 1990]. If either the elastic or plastic strain is calculated then either the Coffin-Manson equation or Basquin equation can be used. Coffin-Manson LCF data has been generated for many solders [Wild, 1971; 1972; 1975; 1988; Vaynman et al., 1988; Lawson, 1987; Solomon, 1986a; 1991a,b; 1990a, b; Weinbel, et al., 1987a, b; Shine et al., 1984; Bae et al., 1989; Vaynman and Fine; Cutiongco, 1990; Wilcox et al., 1989, 1990; Engelmaier, 1985a; IPC, 1992; Nix et al., 1991; Stone, 1990; Hendrix et al., 1989; Fine and Jeannotte, 1989; Lau and Jeans, 1989; Oberin, 1989; Subrahamanyan et al., 1989; Guo et al., 1991b; Kinato et al., 1987]; hence, this approach is most recommended. There is relatively little data correlating the fatigue life with the elastic strain; hence, it is difficult to use the Basquin approach. Furthermore, as Figures 6-5 and 6-6 show, the slope of the Basquin equation (elastic strain-fatigue life) is much more shallow that of the Coffin-Manson expression (plastic strain-fatigue life). Small errors in the calculation of the elastic strain have a much greater influence on the calculated life than a similar error in the calculation of the plastic strain, and thus the use of the plastic strain is favored. In addition, creep-induced stress relaxation causes the elastic strain to be converted into plastic strain, and this can be taken into account when the plastic strain is used to determine the fatigue life. If sufficient time exists for creep-induced stress relaxation to cause all of the elastic strain to be converted into plastic strain, then the plastic and total strains will be equal.

The Coffin-Manson and Basquin relationships depend upon empirically derived constants that make the expressions difficult to use from first principles. It is possible, however, as shown in Figure 6-8, to estimate the elastic strain and add

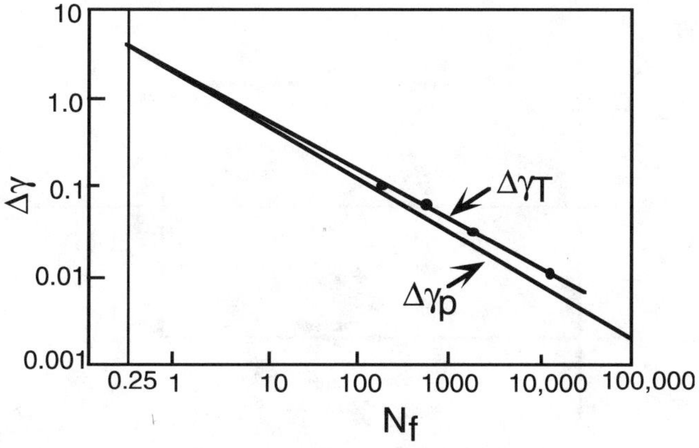

FIGURE 6-8 Fatigue of 60Sn-40Pb in terms of the applied plastic strain and calculated total strain. [Solomon et al., ©1990, IEEE]

this strain to the plastic strain, then construct a total strain vs. fatigue life curve from plastic strain vs. fatigue life data [Solomon et al., 1990]. Generally, a strain analysis calculates the total strain, so it is necessary to do one of three things: extract the plastic strain from this analysis (this requires a knowledge of the temperature and strain rate dependence of the yield strength); convert a plastic strain vs. fatigue life curve to a total strain vs. fatigue life curve and use this altered data; or find the appropriate total strain vs. fatigue life data in the literature. The analysis will generally show that the strain is not uniaxial, so a correction must be made for the stress state. This point will be discussed in more detail later.

For simplicity it will be assumed that the analysis yields the plastic strain, and therefore the Coffin-Manson fatigue life relationship can be used. This relationship, which describes the data of Figure 6-9, is given by Equation (6.3), namely:

$$N_f^{\alpha} \Delta\gamma_p = \theta \qquad (6.3)$$

where α is called the fatigue ductility exponent (this is the slope of the fatigue curve) and θ is called the fatigue ductility coefficient (it defines the position of the fatigue curve and is experimentally found to be proportional to the ductility). Equation (6.3) is quite simple, so it may be reasonably assumed that its application is likewise simple. Unfortunately, there are a number of complicating considerations which much be kept in mind when applying this equation.

6.4.1 Influence of the Definition of Failure

Equation 6-3 relates the plastic strain and the fatigue life. The meaning of plastic strain is straight forward, but the definition of fatigue life is not. Most conven-

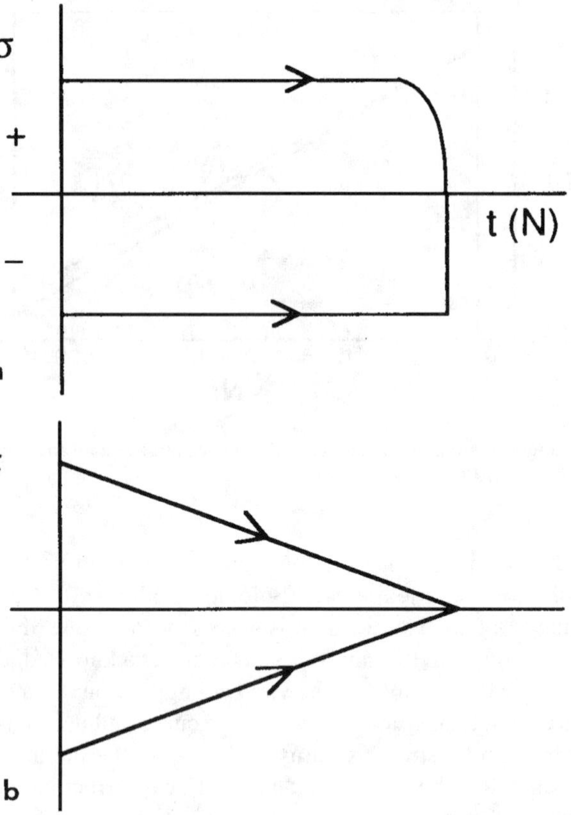

FIGURE 6-9 Schematic showing the drop in load with time (or cycle number N) for (a) tensile loading and (b) shear loading.

tional fatigue tests are run in tension on specimens which are much smaller than the parts upon which the data generated is applied. Solder joints are loaded in shear and the fatigue tests used to predict their lives are also often performed in shear. The joint size is generally much smaller than the test specimen used to predict the fatigue life. Both of these factors make the definition of the fatigue life a critical question when applying fatigue data to solder joints.

Figure 6-9 shows a schematic illustration of the load time behavior of a tensile fatigue test. The locus of the peak tensile and compressive loads is shown vs. time, which is proportional to the cycle number (since these are schematic diagrams, no numbers are included). The load initially changes reflecting cyclic hardening or softening [Langraf, 1970]. For cyclic softening, the load locus would decrease. For cyclic hardening, the load locus would increase. Softening is due to dislocation annihilation and rearrangements. Hardening is due to dislocation multiplication and tangling. Hardening generally occurs in relatively annealed material; softening occurs when the material has experienced prior plastic deformation. After a few cycles, a saturation hysteresis load is reached for a fixed cyclic strain. When cracks develop, they reduce the tensile load-bearing ability and the tensile load decreases. Because the cracks close in compression, the compressive load experiences little or no change. With large cracks there may be some offsets and the cracks may not close completely, resulting in a load change. It is assumed that the stress to produce the control strain does not change so the load drop is directly proportional to the cracked area (i.e., if half the cross section is cracked, the tensile load will drop in half). The onset of a measurable load drop, say 1–5% beyond the load plateau value produced by softening or hardening, could be used as a measure of the fatigue life. Once a measurable load drop is observed the load drop may continue at a relatively rapid rate, making the difference between a load drop of, say, 5% and 95% relatively small, relative to the number of cycles to get to the first 5% load drop. However, if careful metallography is done cracks can be found at still smaller lives, but their influence on the measured load is too small to be resolved. Cracks are present before a 5% load drop, and in some cases cracks may initiate in the first few cycles.

The load drop behavior of Figure 6-9 reflects the fact that in tension the rate of crack growth and hence the rate of load drop varies with the square root of the crack length (i.e., with the stress intensity or J integral). It may take a long time until the load starts to drop, but once it does, the drop in load progresses more rapidly. Because the crack growth rate depends, in part, upon the crack length, as the crack lengthens it grows faster, and the load drops more rapidly. Once the cracks are large enough to produce a measurable load drop they are generally growing fast enough to produce complete failure in relatively few additional cycles. As is discussed below, the load drop is more gradual in tests run in shear, rather than in tension.

The exact nature of the load drop behavior depends upon the solder being tested. Cyclic hardening and some saturation in the hysteresis loop stress was observed in 95Pb-5Sn solder, [Vaynman et al., 1987a] but not in eutectic (63Sn-37Pb) solder [Cutiongco et al., 1990]. In the eutectic solder cyclic softening was observed, and even though tensile deformation was used the load drop was gradual, similar to that shown in Figure 6-9b. With cyclic hardening in 95Pb-5Sn solder, Vaynman used the number of cycles to reach a maximum load as the measure of failure [Vaynman et al., 1987a]. Since cyclic softening was observed in eutectic solders a slightly different definition for failure was used. Since the compressive load does not decrease due to cracking, but does due to softening, the ratio of the tensile load divided by the absolute value of the compressive load corrects for softening and shows just the load drop due to cracking [Cutiongco et al., 1990]. The onset of a drop in this ratio was used to define failure. The most ambiguous definition for failure is complete separation of the test specimen. This definition also yields the longest life, but as Figure 6-9a shows, the fatigue life determined using this definition for failure is only slightly larger than that determined using a small load drop. In terms of the scatter in the data this difference is often unreasonable. Unfortunately, as is shown in Figure 6-9b, this is not the case for tests run on solder tested in shear.

Figure 6-9b shows that in shear the load drop is much more gradual than it is in tension. Figure 6-10a,b shows data which illustrates this gradual drop in load for 60Sn-40Pb and 92.5Pb-5Sn-2.5Ag. The load is correlated by the load drop parameter, ϕ, which is defined by:

$$\phi = 1 - (\Delta P/\Delta P_M) \tag{6.4}$$

where ΔP_M is the maximum load range, and ΔP is the load range after any given number of cycles. ϕ varies from zero to 1 and can be related to the cycle number, N by the general expression:

$$\phi = AN^B \tag{6.5}$$

where A is a constant, and B is a number which can be a constant or in its most general form is a function of the applied plastic strain. Solomon [1986a] has shown that, for 60Sn-40Pb solder, B is about 1 for a plastic strain range of 25%, but decreases with decreases in the plastic strain range, equaling about 0.3 when the plastic strain range decreases to 1%.

Guo and co-workers [1991b], working with the same solder, also found B to be equal to 1 at relatively high strains (6.1%–30.5%), but did not observe a decrease at the lower strains. They did not use very low strains (less than 6%), so the decrease in B at lower strains could not be verified. Their use of total strain limits also tends to keep B constant. With cycling with the total strain being kept constant, as the load dropped the elastic strain decreased and the plastic strain increased. This would increase the load drop rate and increase B, masking the decrease that is observed when plastic strain limits are used.

The load drop can be correlated with an increase in the area that is cracked. This is shown in Figure 6-11, which was developed [Guo et al., 1991b] by fracturing the fatigued specimen in liquid nitrogen to reveal the cracked area. Figure 6-11 shows the variation of the measured remaining area vs. the load drop parameter (Equation (6.4)). The initial area A_o is corrected by the area of the voids A_{void} (produced during the manufacture of the joints) which cannot carry any load. An average linear crack length was determined by taking the square root of the cracked area. Since the load drop parameter is correlated with the cycle number it is possible to extract an average crack growth rate from the load drop. The correlation of the drop in load with the cracked area was also done qualitatively with the aid of an ultrasonic microscopy investigation of samples which were removed from test, observed, and then re-tested [Solomon, 1986a].

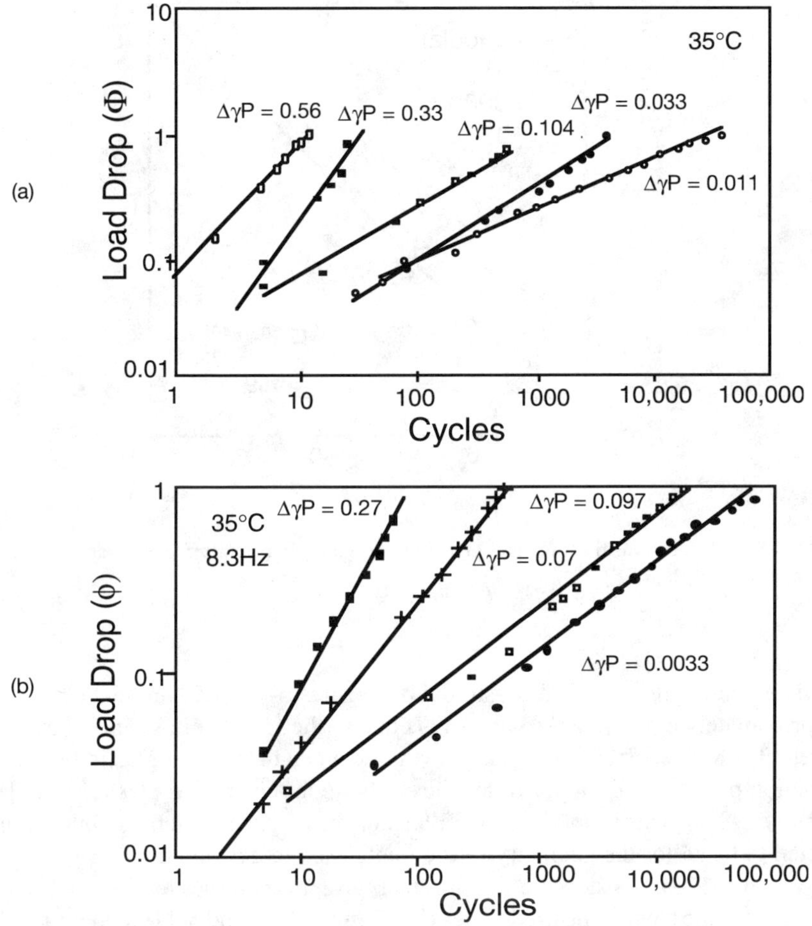

FIGURE 6-10 Load drop parameter vs. cycle number for different applied plastic strains: (a) for 60Sn/40Pb solder. [Solomon, 1991c] and (b) for In151 (92.5Pb-5.0Sn-2.5Ag) solder. [Solomon, 1990]

Care must be taken in applying any literature correlation of the load drop with the area fractured. Wilcox et al. [1990] have calculated the load drop behavior as a function of the joint stiffness. The greater the stiffness, the sooner the load drop begins; the more gradual the load drop and the less the increase in cracked area for a give load drop. Wilcox et al. have also shown that the electrical resistance increase with cycling begins sooner and is more gradual the greater the joint stiffness. Applying literature values for the load drop behavior to a joint with a different stiffness can thus lead to errors.

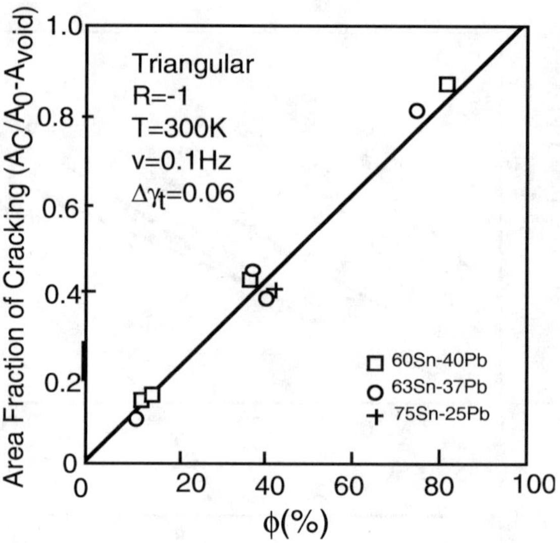

FIGURE 6-11 Area fraction of fatigue cracks vs. the load drop parameter for several Sn-Pb alloys. [Guo et.al., ©1991b, IEEE]

The gradual drop in load illustrated by Figures 6-9b, 6-10 and 6-11 (i.e., B is approximately equal to or less than 1) is due to the fact that in simple shear the J integral, which controls the crack growth rate when there is net plasticity, is not a function of the crack length. It depends only upon the applied plastic strain [Solomon, 1986a]. This is not the case for tension, or for shear in general, but was predicted to be so for the geometry of the joints used to develop the data of Figures 6-10 and 6-11. In tension the rate of crack growth increases as the crack grows longer. For the specimens used to develop Figures 6-10 and 6-11, it does not. This creates a dilemma because different definitions for failure can produce vastly different lives. In this case what definition for failure is appropriate? Conservatively one might choose a very small load drop, say 10%, but as Figure 6-12 shows, this can yield a very short fatigue life, one which is overly conservative. Figure 6-12 (developed from data of the sort shown in Figure 6-10) combines the results for different tests to describe the Coffin-Manson behavior for different load drop definition of failure. For this isothermal data, *the best approach is to treat the load drop data as crack growth data and base the design on this crack growth.* More will be said about the relationship between the load drop and crack growth rate presently, and later in the section on crack growth approaches to life prediction. Load drops measured in thermomechanical fatigue are more complex, with a cyclic hardening and softening behavior that is different from that observed in isothermal tests [Frear and Jones, 1990], and a different approach may be needed.

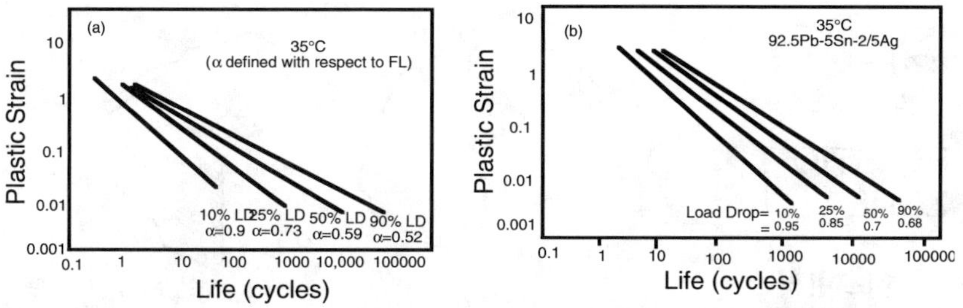

FIGURE 6-12 Fatigue life as a function of the applied plastic strain for different load drops: (a) for 60Sn-40Pb solder. [Solomon, 1989a] and (b) for In151 (92.5Pb-5.0Sn/2.5Ag) solder. [Solomon, 1990]

The second important point to be considered in this section is that solder joints are actually often much smaller than the laboratory specimens used to predict joint behavior. This is exactly the opposite behavior that is observed in most other design situations. It is generally assumed that a LCF test models the behavior at a fillet or other stress concentration. The test specimen can be thought of as being embedded in this region. This approach, attributed to Coffin [1974a], is illustrated in Figure 6-13. The number of cycles to crack this specimen models the number of cycles to start a crack in the structure. Crack propagation approaches are then used to determine the overall life of the complete structure. In the most conservative design it could be assumed that the small crack (modeled by the LCF specimen) is all that is tolerable and the life of this test specimen is used as the structure life.

With solder joints the situation is different from that shown in Figure 6-13. The test specimen cannot be thought of as modeling only a small part of the solder joint because the whole joint is generally much smaller than the test specimen. In shear enlarging the test specimen lengthens the fatigue life (because the crack is not accelerating as it grows longer) and lessens the predictive capability of the test specimen, relative to a small solder joint. To use the load drop defined LCF data, it is necessary to treat it as crack growth data.

If it is assumed that there is no hardening or softening then the load drop is proportional to the cracked area. Furthermore, if it is assumed that a single crack, extends through the width of the shear specimen and grows through the length of the specimen, then the amount of crack growth is equal to the load drop; i.e., a 50% load drop corresponds to a crack growth through 50% of the specimen. Knowing the length D over which the crack is growing enables one to thus determine the crack growth rate $\Delta a/\Delta N$:

$$\Delta a/\Delta N = D/N_f \tag{6.6}$$

Equation (6.6) enables one to relate the LCF data to the behavior of a solder joint. It is assumed that for a given plastic strain the average crack growth rate, $\Delta a/\Delta N$, is

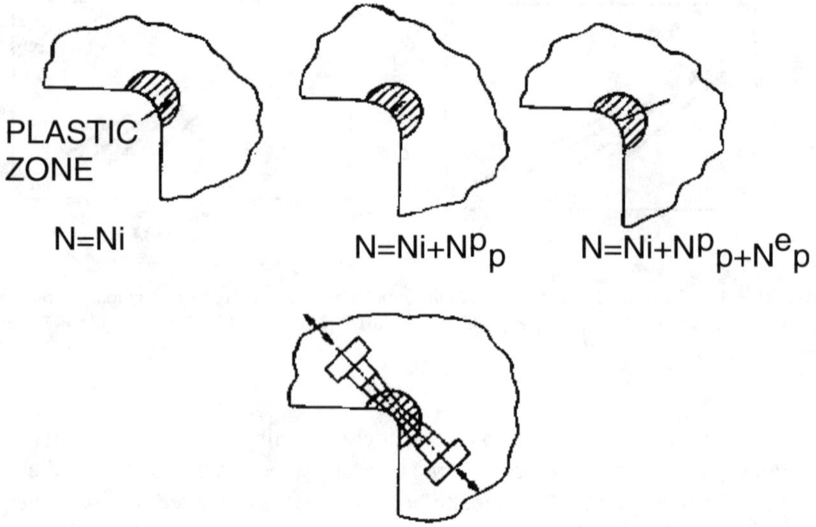

FIGURE 6-13 Schematic showing an LCF specimen simulating the fatigue behavior at a stress concentration. The fatigue life of the specimen corresponds to the number of cycles to initiate a crack and grow it through the region of high strain. [Coffin,1974a]

given by Equation (6.6). N_f is defined for a given load drop and this load drop and the size of the specimen define, D. For a single crack a definition of failure using a 25% load drop requires that the crack must grow 1/4 across the specimen, and D is just 1/4 of the specimen length. D and N_f define the average crack growth rate. Now if the same strain is applied to the solder joint, the $N_{f(joint)}$ is just

$$N_{f(joint)} = N_f L/D \tag{6.7}$$

where L is the length of the joint over which the plastic strain is applied and $N_{f(joint)}$ is the number of cycles to grow the crack through the length L.

The assumption that a single crack is growing is not necessarily a good one. This was corrected for by using a effective value for D [Solomon et al., 1990]. If two cracks were growing instead of one, then for a given load drop each crack would be assumed to be growing on half the distance of a single crack producing the same load drop. In applying the load drop data [Solomon, 1986a] it was assumed that initially there were several cracks, but the number of independent cracks decreased due to coalescence of the cracks.

Figure 6-3 shows that a single strain is not applied to a solder joint; rather, a strain distribution is developed. This requires the sequential application of Equation (6.7) as the crack grows through the whole joint. This process is simplified by the assumption that in shear the crack growth rate depends only upon the applied plastic strain and not also upon the crack length [Solomon et al.,1990]. Approaches which do not require that the crack growth rate be independent of the crack length will be discussed later.

Another parameter which has been used to define failure is the increase in resistance of the joint [Solomon et al., 1990; Wild, 1972; 1975; Wilcox et al., 1990; Engelmaier, 1983; Frear, 1989]. This is quite an attractive choice for defining failure, since it can be applied to thermally cycled joints where the load drop cannot be measured. Furthermore, the resistance of many joints can be measured simultaneously in thermal cycling tests, and statistically significant failure data can be generated. Unfortunately, the joint resistance increase is not very sensitive to joint cracking [Solomon et al., 1990; IPC, 1992]. This is especially true for joints tested in shear. If the crack faces touch, they will conduct electricity even though they may not be able to support a load. If another event, such as a tensile load, causes the joint to separate, then a jump in the resistance will be noted and the apparent failure will be precipitated by this event, even though the joint has actually already failed. This can introduce significant scatter in resistance-defined failures. Rather than use the DC resistance, one should use a high speed event monitor to pick up microsecond resistance fluctuations, which are much more sensitive to the presence of cracks [Engelmaier, 1983; 1985a; Wilcox et al., 1990; Frear, 1989].

6.4.2 Test Temperature vs. Thermal Cycle Range

The Coffin-Manson law (Equation (6.3)) defines the isothermal fatigue life, i.e. it only defines the fatigue life at the temperature at which the tests were run. Tests can be run at different temperatures to determine the temperature dependence of the constants in Equation (6.3). An example of this is shown in Figure 6-14, which shows the behavior of 60Sn-40Pb solder, using a 50% load drop definition for failure. Figure 6-15 shows data for a high Pb solder [Vaynman, et al., 1987a]. Both solders showed an increase in the intercept θ (the fatigue ductility coefficient) with decreasing temperatures. The high Pb solder showed an increase in the Coffin-Manson exponent (fatigue ductility exponent, β in Fig. 6-15a) on reducing the temperature below 80°C. (The different exponents observed at different strain ranges will be discussed in the next section.) The 60Sn-40Pb solder only showed a temperature variation of the exponent when the temperature was increased above 125°C. A decrease in the coefficient decreases the fatigue life. A decrease in both the coefficient and the exponent will decrease the fatigue life at high strains but will increase it at low strains.

Unfortunately, the availability of data such as that shown in Figures 6-14 and 6-15 does not immediately provide fatigue life for a thermally cycled joint. In thermal cycling the joint experiences a range of temperatures and strains. The method to determine the strain was discussed previously. The fatigue life can be estimated

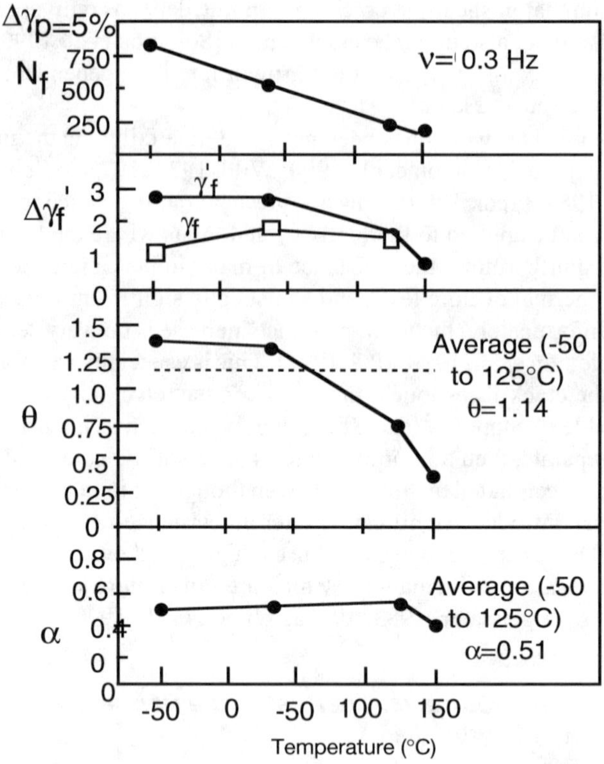

FIGURE 6-14 Temperature dependence of the terms in the Coffin-Manson expression, for 60Sn-40Pb solder tested in simple shear. [Solomon, , ©1986c, IEEE]

using this strain and the Coffin-Manson law (or the total strain approach if the total, rather than the plastic, strain is calculated). However, the temperature data set that should be used has not been considered.

Taira [1973] has considered two possibilities, the use of the mean temperature or the peak temperature. Figures 6-14 and 6-15 shows that it is a conservative approach to use data developed at the peak temperature. This approach is also supported by the fact that most of the strain is actually applied at the peak temperature. Solder deformation (accompanying a temperature change or developed by creep when the temperature is held constant) is highly temperature-dependent, with the flow stress decreasing as the temperature is increased. Since most of the deformation generally occurs at the peak temperature (Figure 6-2) it makes sense to use data taken at this temperature for the fatigue life analysis.

Using just the highest temperature for life prediction assumes that the lowest temperature is only important insofar as it fixes the strain that is applied (the lower

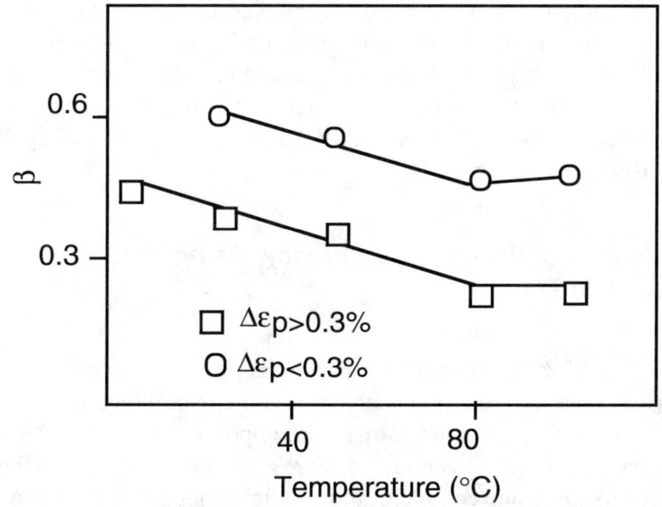

FIGURE 6-15a Temperature dependence of the Coffin-Manson exponent for 96.5Pb-3.5Sn solder tested in tension. The Coffin-Manson curve has been broken into two regions with different exponents in each region. [Vaynman et al.,1987, ©1987, IEEE]

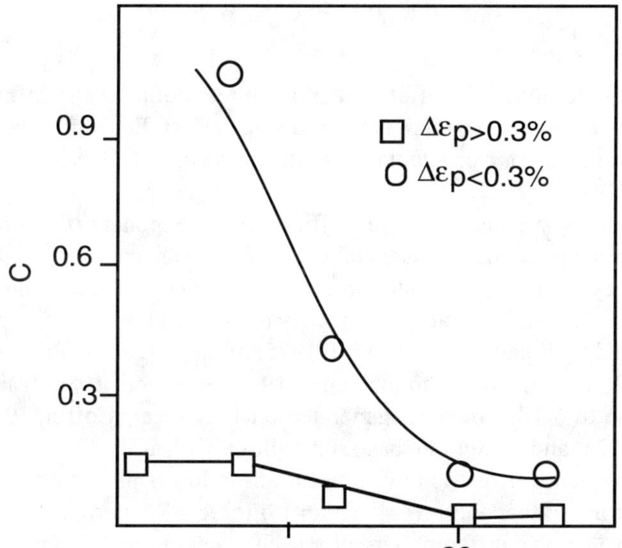

FIGURE 6-15b Temperature dependence of the fatigue ductility coefficient, for 96.5Pb-3.5Sn solder tested in tension. The Coffin-Manson curve has been broken into two regions with different coefficient exponents in each region. [Vaynman et al.,1987, ©1987, IEEE]

the minimum temperature of the cycle, the greater the strain range). This assumption is good if the temperature range is small. Problems can, however, arise if the temperature range is large and different deformation mechanisms are acting at the peak temperature and the minimum temperature [Frear and Jones, 1990]. This lessens the effectiveness of using isothermal data to predict thermal cycling, a subject to be discussed later.

6.4.3 Influence of the Cycle Frequency, Ramp Rate, and Hold Time

Frequency Modified Coffin-Manson Relationship

Equation (6.3) does not take into account the time over which the strain is being applied. At elevated temperatures, this is an important omission. Decreasing the cycle frequency, υ, (cycles/second), by increasing the cycle period (the length of time required to apply one complete strain cycle) generally reduces the number of cycles for failure [Coffin, 1971; 1973]. Coffin described the influence of the cycle frequency by the use of a frequency-modified version of the Coffin-Manson equation, i.e., by:

$$(N_f \upsilon^{(K-1)})^{\alpha} \Delta\gamma_p = \theta \tag{6.8}$$

where K is determined at different temperatures, defining the influence of the cycle frequency. For $K = 1$ there is no frequency effect. For $K = 0$, the time to failure, t_f, is a constant for a constant plastic strain ($N_f/\upsilon = t_f$). Generally K is between 1 and 0.

Equation (6.8) assumes that the Coffin-Manson exponent α is independent of the cycle frequency. This means that the Coffin-Manson curve is just displaced according the cycle frequency, but with the slope unchanged. Unfortunately, experimental data often shows that the frequency effect is much more complicated than that predicted by Equation (6.8) [Wild, 1975; Stone, 1990; Coffin, 1969b; 1972a; 1972b 1973; 1974a; Vaynman and Fine, 1991; Solomon, 1986c]. For example, Figures 6-16 to 6-18 show data generated on 1010 steel [Coffin, 1974a], 304SS [Coffin, 1972a] and A286(a Fe base superalloy) [Coffin, 1972b] where, as can be seen, the frequency effect is only present at the lower plastic strains and in the presence of air (rather than in a vacuum). Figure 6-17 correlates the data with Equation (6.8) with a different curve for each temperature, but with all the curves approaching the same frequency-independent behavior at the highest plastic strains or when a vacuum is used (more will be said about the influence of environment later on). Figure 6-19 [Coffin, 1969b] schematically summarizes the data shown in Figures 6-16 to 6-18. At high frequencies, frequency-independent Coffin-Manson behavior is to be expected, with a Coffin-Manson exponent of about

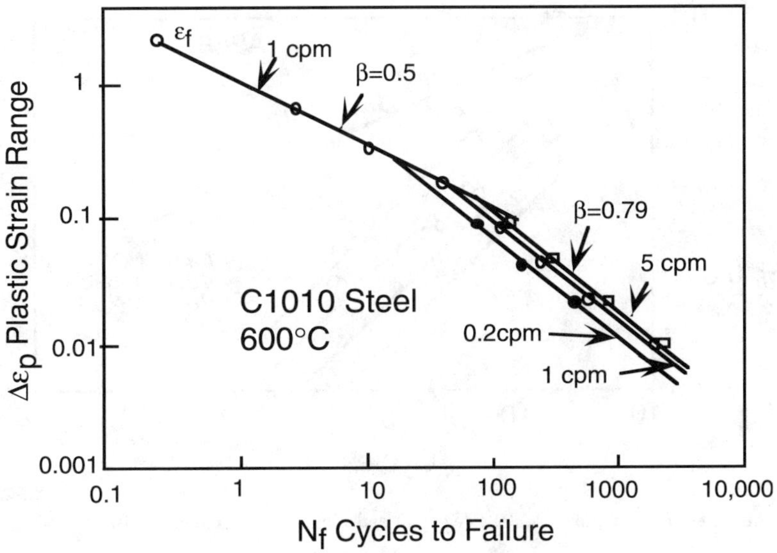

FIGURE 6-16 Effect of cycle frequency on the fatigue behavior of AISI-C1010 steel tested at 600°C. [Coffin,1974b].

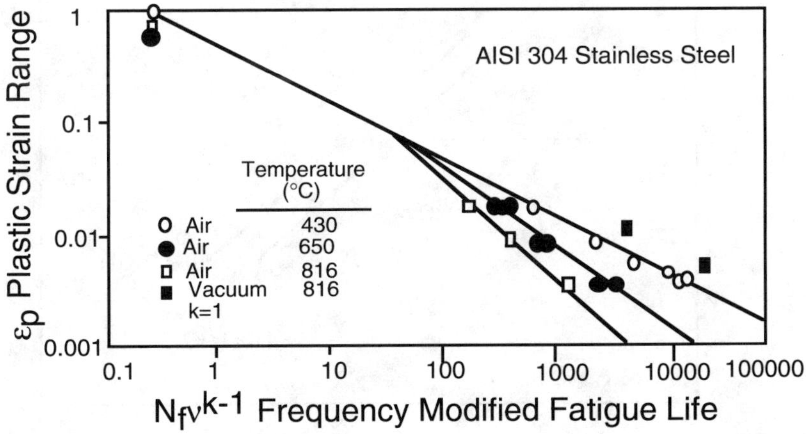

FIGURE 6-17 Frequency modified fatigue life of AISI 304SS tested in air and vacuum. [Coffin, 1972a]

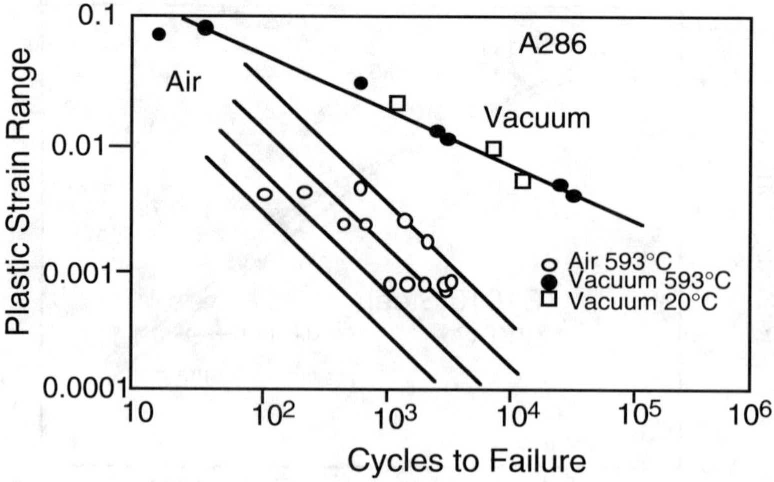

FIGURE 6-18 Fatigue life of A286 tested in air and vacuum and vacuum. [Coffin,1972a]

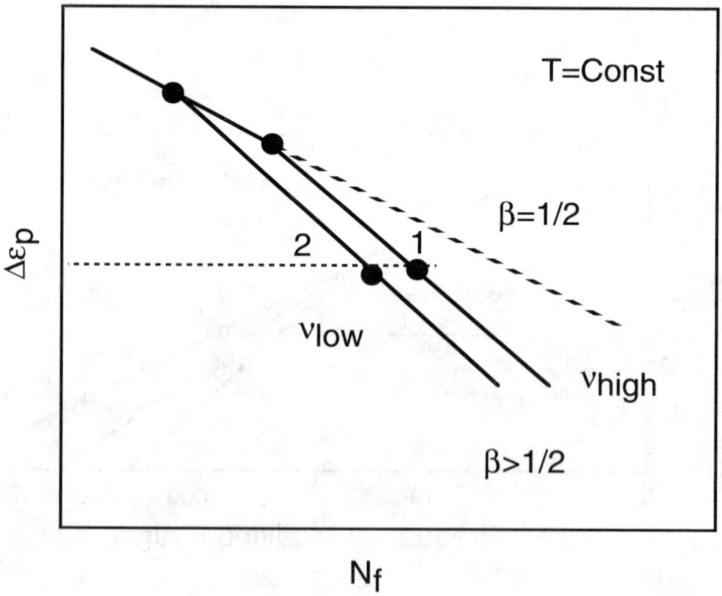

FIGURE 6-19 Schematic view of the effect of frequency on fatigue life. [Coffin, 1969b]

0.5. At low frequencies, frequency-dependent behavior is observed with an exponent that is greater than 0.5. At high strains, the low frequency behavior curve merges with the high frequency curve, i.e., the frequency dependence disappears. The behavior shown in Figure 6-19 is expected at elevated temperatures, which is generally defined as about 1/2 of the melting point (i.e., by a homologous temperature of 0.5). For solders, $T_m/2$ is generally less than room temperature, so frequency effects are generally observed.

Figure 6-20, obtained on a high Pb solder [Vaynman et al., 1988; Vaynman and Fine, 1991] shows the same sort of break in the Coffin-Manson curve as is shown in Figures 6-16 to 6-19. Figure 6-21 shows different Coffin-Manson curves for eutectic solder [Wild, 1972; 1975]. Here, high strain range frequency-independent behavior is not observed, but different Coffin-Manson curves are drawn for different cycle dwell (hold) times and temperatures. These curves appear to merge as predicted in Figure 6-19. Figures 6-17 and 6-20 and 6-21 show that K is temperature-dependent and Figures 6-22 to 6-25 show that K is not constant but has different values in different frequency regimes as well as at different temperatures.

Figure 6-26 [Solomon, 1986c] schematically illustrates several different fatigue life vs. frequency curves of the sort shown in Figures 6-22 to 6-25. Figure 6-27 [Solomon, 1986c] shows the Coffin-Manson curves which are developed from data of the sort illustrated in Figure 6-26. Figure 6-26a shows the case where the break in the curve occurs at the same frequency for all plastic strains. A frequency-independent Coffin-Manson curve at frequencies above the break is shown in Figure 6-26a; below the break there will be series of parallel curves, all with the frequency independent Coffin-Manson exponent, but displaced according to Equation (6.8). This is illustrated in Figure 6-27c.

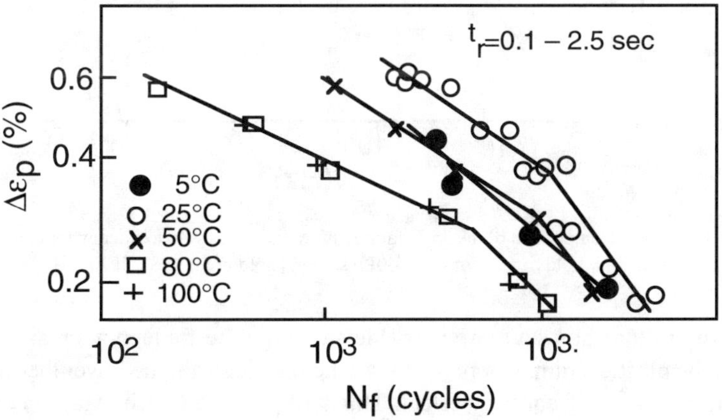

FIGURE 6-20 Fatigue life of 96.5Pb-3.5Sn tested at different temperatures. [Vaynman et al.,1987, ©1987, IEEE]

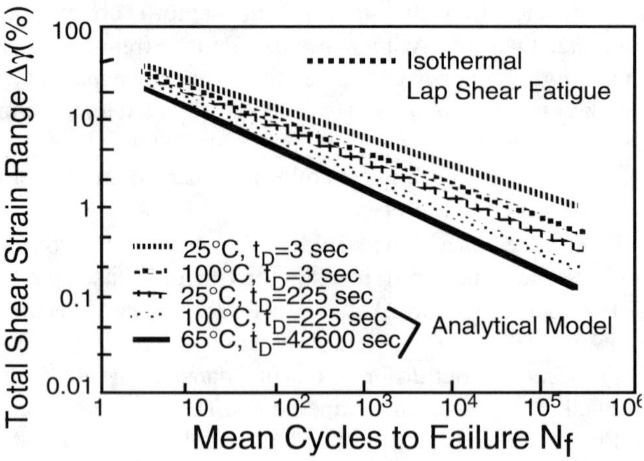

FIGURE 6-21 Fatigue life of eutectic (63Sn-37Pb) solder tested at different temperatures, with different hold times. [Wild,1972] (Re-plotted by Engelmaier)

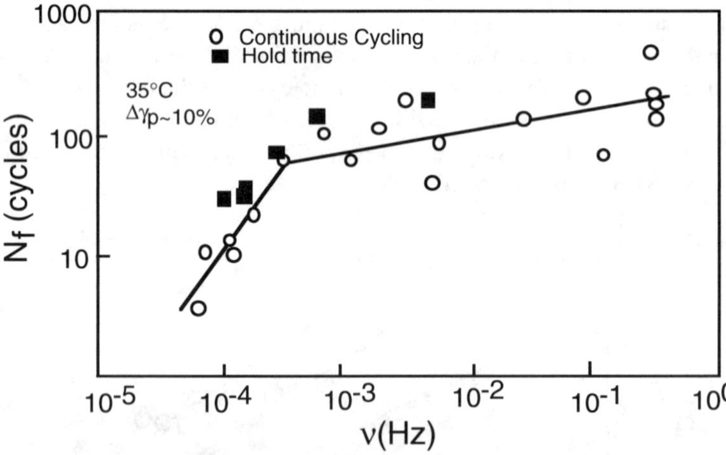

FIGURE 6-22a Comparison of the 35°C fatigue life versus the cycle frequency for continuous cycling and cycling with hold times. For 60Sn/40Pb solder. [Solomon, © 1988, IEEE].

Figure 6-26a shows a more probable situation: the frequency break is a function of the plastic strain, where the lower the plastic strain the lower the frequency at which there is a frequency effect. This will produce Coffin-Manson curves of the sort shown in Figures 6-15 to 6-21; and 6-27a. Figure 6-27a shows that at the highest frequency and plastic strain there will be frequency-independent Coffin-Manson behavior. At lower plastic strain ranges or frequencies there will be a

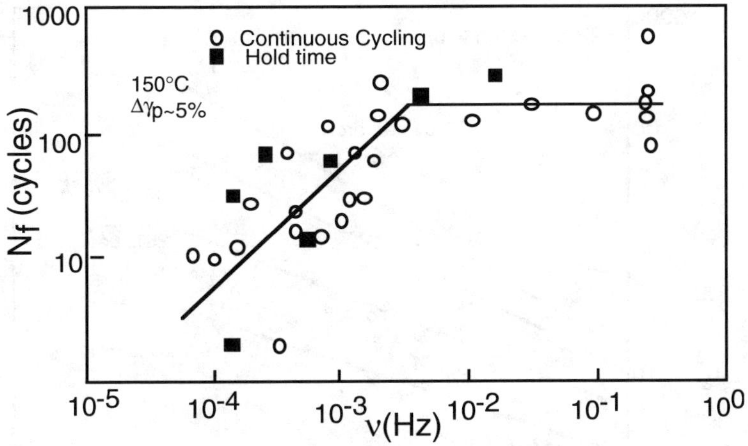

FIGURE 6-22b Comparison of the 150°C fatigue life versus the cycle frequency for continuous cycling and cycling with hold times. For 60Sn/40Pb solder. [Solomon, ©1988, IEEE]

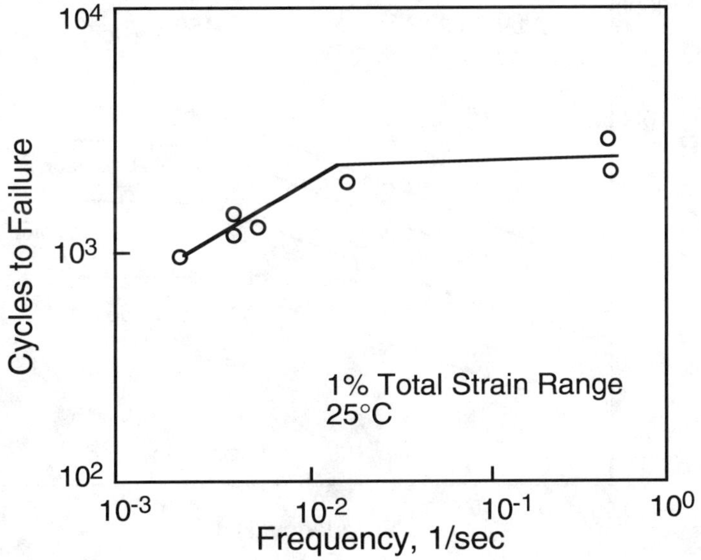

FIGURE 6-23 Effect of cycle frequency on the fatigue life of 63Sn-35Pb-2Ag solder tested at room temperature. [Vaynman and Fine, 1991]

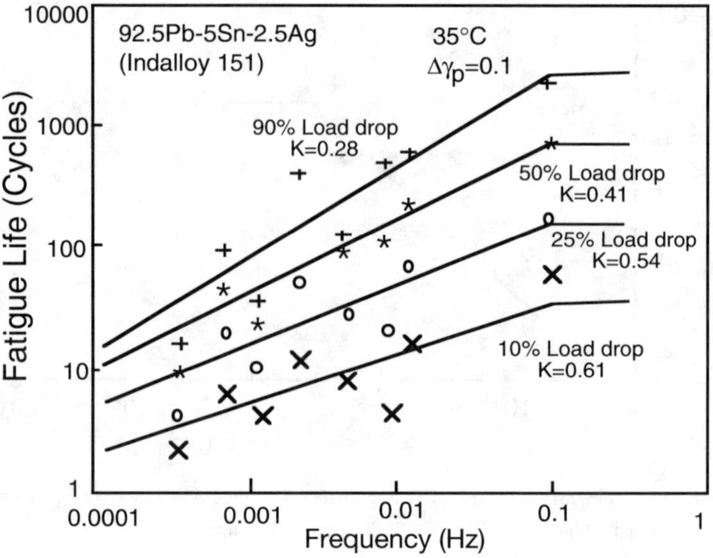

FIGURE 6-24a Fatigue life of In151 solder (92.5Pb-5.0Sn-2.0Ag) vs. cycle frequency for tests at 35°C. [Solomon, 1990]

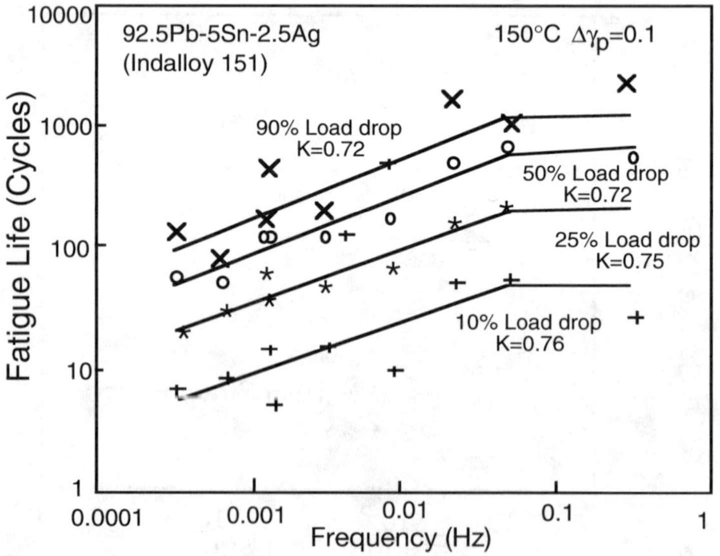

FIGURE 6-24b Fatigue life of In151 solder (92.5Pb-5.0Sn-2.0Ag) vs. cycle frequency for tests at 150°C. [Solomon, 1991a]

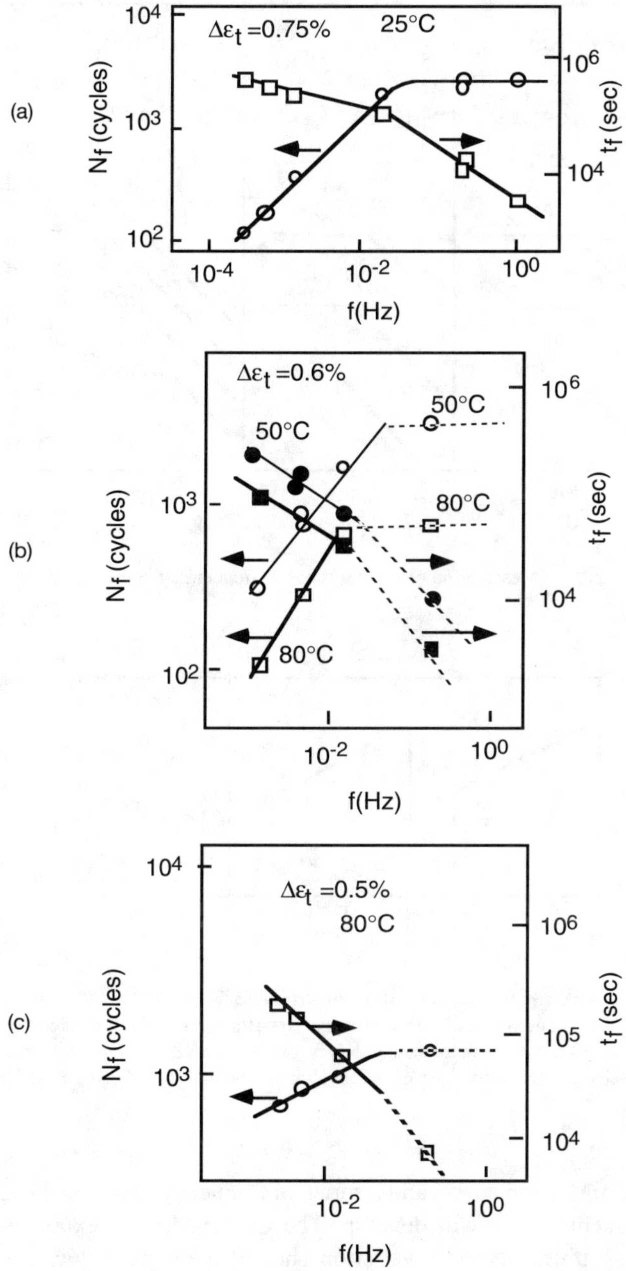

FIGURE 6-25 Effect of cycle frequency on the fatigue life of 96.5Pb-3.5Sn tested at: (a) 25°C, with a total strain of 0.75%, (b) 50°C and 80°C, with a total strain of 0.6%, and (c) 80°C, with a total strain of 0.50%. [Vaynman et al.,1987a]

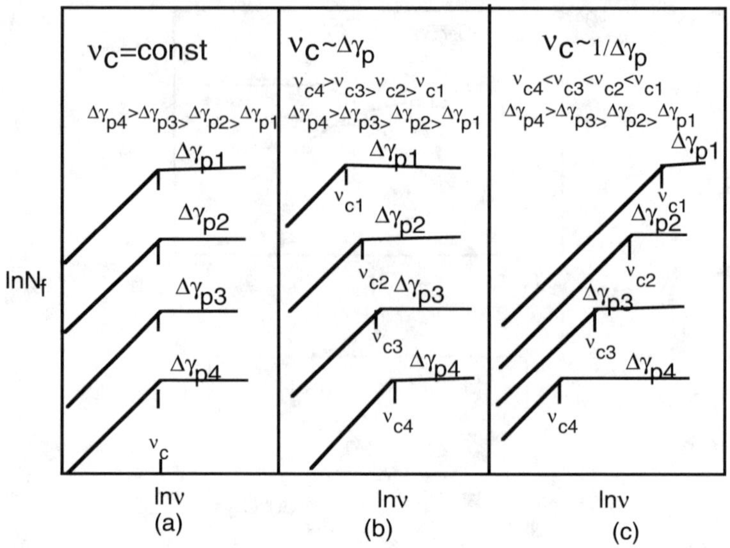

FIGURE 6-26 Schematic representation of several types of frequency-fatigue life diagrams. [Solomon, ©1986c, IEEE]

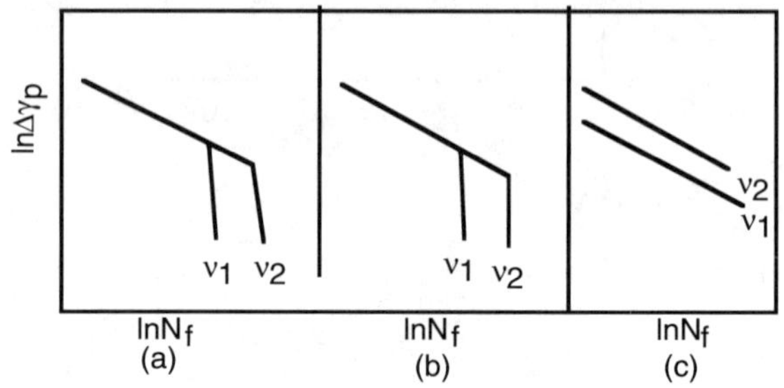

FIGURE 6-27 Schematic representation of different strain-life diagrams: (a) with high strain frequency-independent fatigue and frequency dependence at low strains, with a different curve for each frequency; (b) same as (a) but the low frequency curves showing no dependence of the fatigue life with strain; and (c) Separate curves for each frequency, with no break at low frequencies. [Solomon, ©1986c, IEEE]

break in the Coffin-Manson curve, and a series of frequency-dependent high Coffin-Manson exponent curves will develop. The Coffin-Manson exponent is increased because, with the type of variation shown in Figure 6-26c, the lower frequency portions of the curves move closer together than the frequency independent portions of the curve. In the limit, the curves could come together at low frequencies, producing Coffin-Manson curves of the sort shown in Figure 6-27b.

Figure 6-26b shows the improbable situation where the frequency at which the break occurs is inversely proportional to the plastic strain. This would cause the low frequency Coffin-Manson curves to have a shallower slope than the frequency, independent portion of the curve.

Figures 6-26 and 6-27 are schematic but are based upon experimental observations of the sort shown in Figures 6-16 to 6-25. Figure 6-20 was obtained at a high cycle frequency with no hold time, and here the break is observed at very low plastic strains, less than 0.4%. No break was observed in Figure 6-12b, which shows data for a similar solder, but at higher strains. Figure 6-27a illustrates what is shown in Figure 6-19; namely that the lower the cycle frequency the larger the plastic strain range at which a frequency effect will be noted. Figure 6-27b shows a limiting condition of Figure 6-27a; at low frequencies there is no strain-dependence of the fatigue life. This sort of behavior is shown in Figure 6-28, which combines data taken at a 5% and 10% plastic strain range. At high frequencies there is a considerable difference in the fatigue life, but no frequency dependence. This is what is to be expected at high frequencies and high strains. At low frequencies, Figure 6-28 shows little or no difference in the fatigue life with 5% or 10% plastic strain, i.e. the fatigue life is not a function of the applied plastic strain.

Figures 6-26 and 6-27 summarize data from solders, superalloys, and steels. They show that the fatigue life is a complicated function of the plastic strain and cycle frequency. Equation (6.8) can be used to describe portions of the data but not the whole of the data space. In general, at high plastic strains and cycle fre-

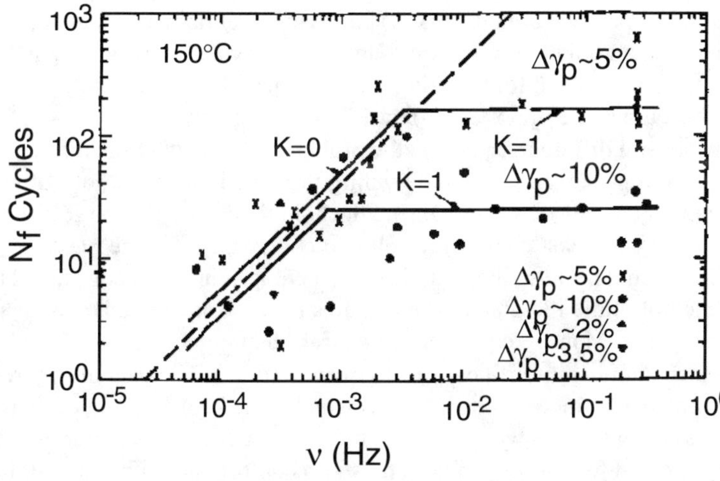

FIGURE 6-28 The fatigue life versus cycle frequency for tests on 60Sn-40Pb tested at 150°C. At low frequencies there is no significant difference between the fatigue life measured when 5 or 10% plastic strain is applied. This is the sort of behavior shown in graph b of the previous figure. [Solomon, ©1986c, IEEE]

quencies K = 1 and standard Coffin-Manson behavior is noted. At very low plastic strains and very low cycle frequencies, there will be a large dependence of the fatigue life with the cycle frequency but a much smaller dependence with the plastic strain. At intermediate ranges of plastic strain and cycle frequency there will be a strong dependency of the fatigue life with both the plastic strain and the cycle frequency. In general, the higher the temperature, the greater the cycle frequency at which a frequency dependency is noted (see Figures 6-22 and 6-24).

Figures 6-12,6-14,6-15, 6-21 and 6-25 give some guidance, for near eutectic and high Pb solders, about the frequency and plastic strain dependence of the fatigue life. Unfortunately, all the plastic strain and frequency regime has not been studied. In particular, there is little or no data at low plastic strains and low cycle frequencies. Such data are hard to obtain because of the long times required to obtain long fatigue lives at low cycle frequencies.

Separation of Hold Time and Ramp Rate Effects

In service cycling, or in thermal cycling tests, the variation in the cycle frequency is not achieved by altering the rate of temperature change, which will change the cycle frequency. What really happens is that the cycle frequency is varied by changing the length of the hold time at a specific temperature or temperatures. Typically these hold times are put in at the maximum and minimum temperature. The rate of temperature change (the temperature ramp rate) is generally fixed, and the time spent during this temperature ramp is usually smaller than the hold time. In isothermal testing the variables are generally the rate at which the strain is ramped and the amount of hold time at a fixed strain or stress. The frequency modified fatigue life approach (Equation (6.8)) lumps together ramp rate and hold time effects and considers only the influence of the overall cycle frequency. While this provides a general approach for handling the influence of both ramp rate and hold time, it can obscure some important effects, described below.

There are several different aspects to the problem of predicting the influence of hold time and ramp rate. In a thermal cycling test, the longer the hold time the greater the amount of plastic strain. During the hold time, creep of the solder joint causes the stress (and elastic strain) to relax to compensate for the plastic strain increase. This occurs because the total displacement is being held constant. Thus, the longer the hold time the greater the plastic strain that is applied; and, as predicted by the Coffin-Manson equation, the shorter the fatigue life.

There is also an inherent influence of frequency (time). The data of Figures 6-22 to 6-24 were all obtained at plastic strains which were kept constant even as the ramp rate was changed. This was done with the aid of an analog plastic strain computer which allowed for cycling under total strain control, but with fixed limits of plastic strain. Figures 6-22a and 6-22b also show data for hold time experiments. Here the limit strains were altered to compensate for the amount of plastic strain developed during the hold time. The amount of plastic strain developed during the first few cycles was used to determine the appropriate cycle strain limits (plastic strain which was used to trigger the start of the hold time). If extensive plastic strain was developed during the hold time, then the cycle limit was reduced. Thus, the sum of the plastic strain developed during the ramp loading and the holds was kept constant. Figures 6-22a and b show that there is not very much difference between this hold time and ramp rate data, although the ramp rate data generally has the lower fatigue life.

In addition to the hold-time induced increases in the plastic strain and the inherent influence of cycle frequency (for a fixed plastic strain), one must also consider the saturation in the hold time effect which occurs at long hold times. At some point a long hold time will have caused virtually all the stress to relax, and further increases in the hold time will cause little or no further straining. The joint will be sitting at close to zero load and deformation induced damage will cease. (There may still be environmental influences and microstructural changes which will be discussed later.) This produces a saturation in the hold-time induced reduction in the fatigue life which is illustrated in Figure 6-29 [Vaynman et al., 1988; Vaynman and Fine, 1989, 1991]. No such saturation occurs when ramp tests are run with the same plastic strain limits with ramp rate variations to produce differences in the overall cycle frequency (see Figures 6-22 and 6-24).

Figure 6-30 combines tests on a 96.5Pb-3.5Sn solder run at different strains so that a strain-fatigue life curve could be drawn for different hold times. At low strains the curves are very steep, close to what is illustrated in Figure 6-27. High strain tests were not run, so it is not possible to determine if the curves would converge at high strains. The jogs in the curves have also been observed by Vaynman et al., in tests run on Sn62 solder. These jogs will be discussed in a subsequent section of this chapter.

There is also an inherent influence of the ramp rate vs. what is produced by a hold time that must be considered. The data of Vaynman and Fine shown in Figure 6-31 indicates that hold time is much more important than the time spent during the ramp loading. Figure 6-31a shows that even a 120 second hold time can outweigh the influence of 240 seconds spent during the ramping. Figure 6-31b shows that the fatigue life with a 120 second ramp time is lower than that observed with a 1 second ramp, but that as little as about a 30 second hold time could outweigh this effect.

When the hold time is longer then the ramp time, Vaynman and Fine [1991] found the number of cycles to failure N_f to be given by:

$$N_f = (C+D(t_{ht}+t_{hc}))/(2t_r+t_{ht}+t_{hc}) \tag{6.9}$$

where C and D are constants and t_{ht}, t_{hc} and t_r are the tensile hold time, compressive hold time, and the ramp time, respectively. This approach requires experiments to determine C and D. With very long hold times, compared to the ramp time, N_f reduces to D, and the value of the fatigue life when the hold time effect saturates.

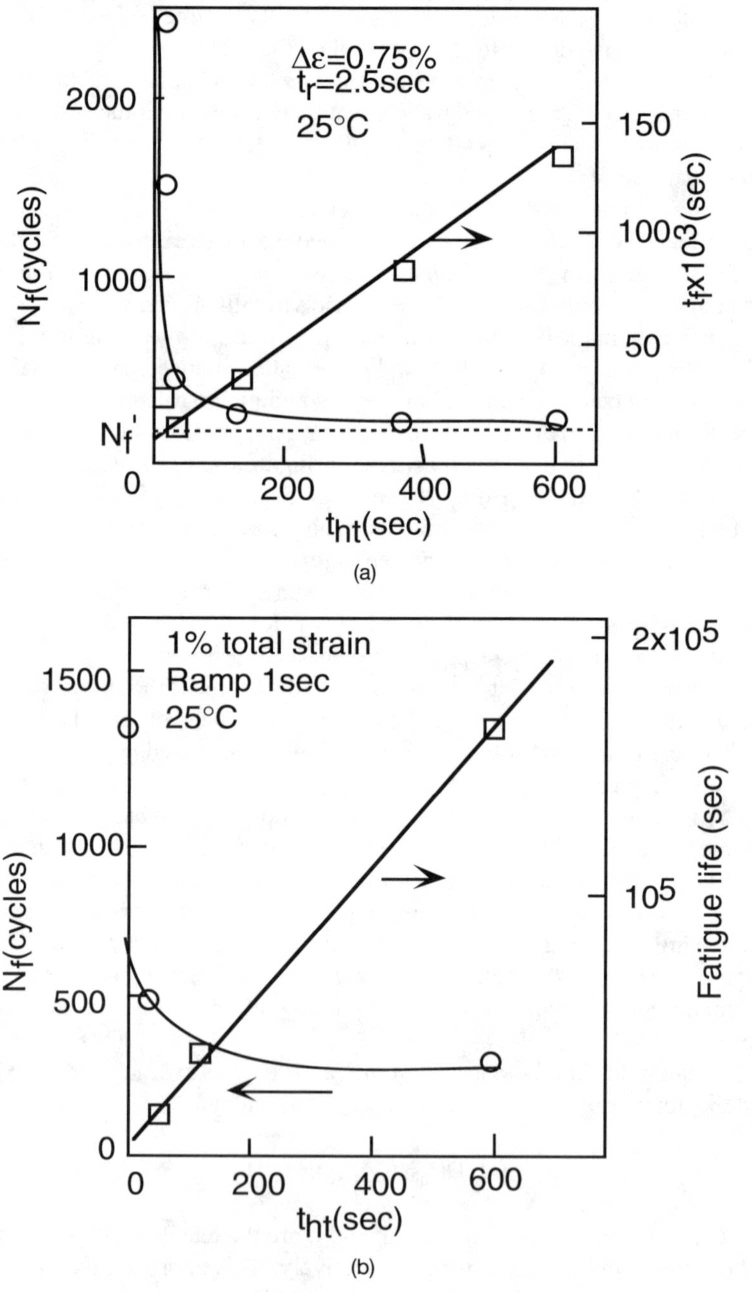

FIGURE 6-29 Effect of tensile hold time on the fatigue life of (a) 96.5Pb-3.5Sn solder [Vaynman et al., 1988] and (b) eutectic (63Sn-37-Pb) solder. [Vaynman et al., 1989]

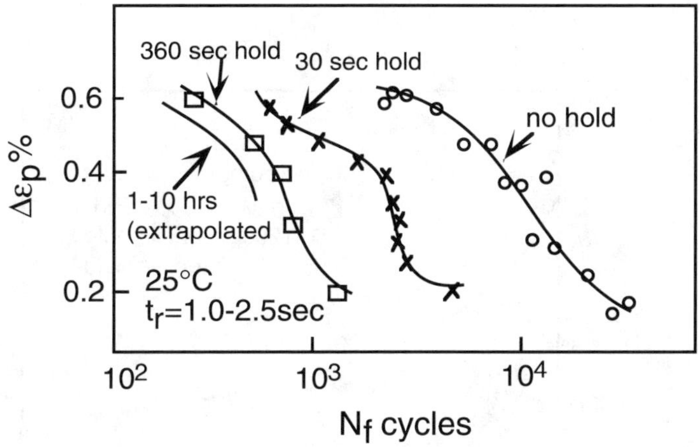

FIGURE 6-30 Strain-fatigue life behavior of 96.5Pb-3.5Sn solder at 25°C for tests run with different hold times. [Vaynman et al.,1988]

The ramp time used in Figure 6-31 extends to only 240 seconds which corresponds to a cycle frequency of 4.2×10^{-3} Hz. The decrease in the fatigue life with this ramp rate is similar to that observed in Figure 6-22a, which was obtained on a similar solder at a similar temperature. In both cases decreasing the cycle frequency to 4.2×10^{-3} Hz decreased the fatigue life by about a factor of 2. However, Figure 6-22a shows that if the ramp rate were decreased still further the influence of the ramp rate would be greatly increased. In contrast to the results of Figure 6-29, Figure 6-22 shows that a hold time produced less of an influence than an equivalent ramp rate. There is, however, a difference in the type of hold time that was used in each case. In the experiments described in Figure 6-22, equal holds were applied at both extremes of shear strain. In the case of the data of Figure 6-29, the hold was initiated only at the maximum tensile displacement. This hold produced asymmetric deformation, which is generally more damaging than symmetric loading (i.e., equal hold times at the limit conditions). This is illustrated in Figure 6-32 [Solomon, 1988], which shows data obtained on the same material used in Figure 6-22a,b (the data of Figure 6-23 was obtained on a similar solder, by different investigators) with the same general testing procedures. Here, ramp loading (the same data as in Figure 6-22a,b) is compared with asymmetric ramp loading, where the ramp rate was altered only for loading in one direction. While this data is for shear and not tension, and for asymmetric ramp loading and not asymmetric hold times, it illustrates that asymmetric loading is more damaging than symmetric loading, at least for the frequency range employed in the studies shown in Figure 6-31. This may explain why there is a greater influence of the hold time in Figure 6-31. There is also a difference between the tensile deformation used in Figure 6-31 and the shear deformation used in Figures 6-22, 6-23, and 6-32. The question

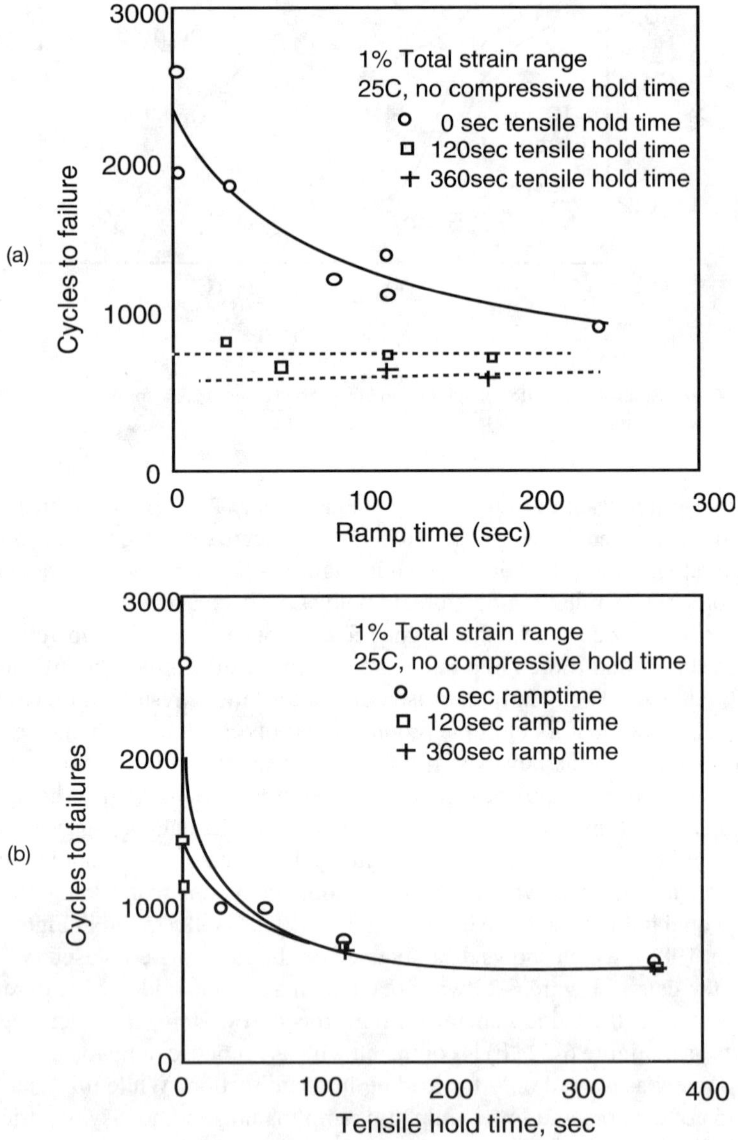

FIGURE 6-31 (a) The effect of the ramp time on the fatigue life of 62Sn-36Pb-2Ag solder, for tests with and without a tensile hold [Vaynman and Fine,1991] and (b) the effect of tensile hold time on the number of cycles to failure of 62Sn-36Pb-2Ag solder in tests with different ramp rates. [Vaynman and Fine,1991]

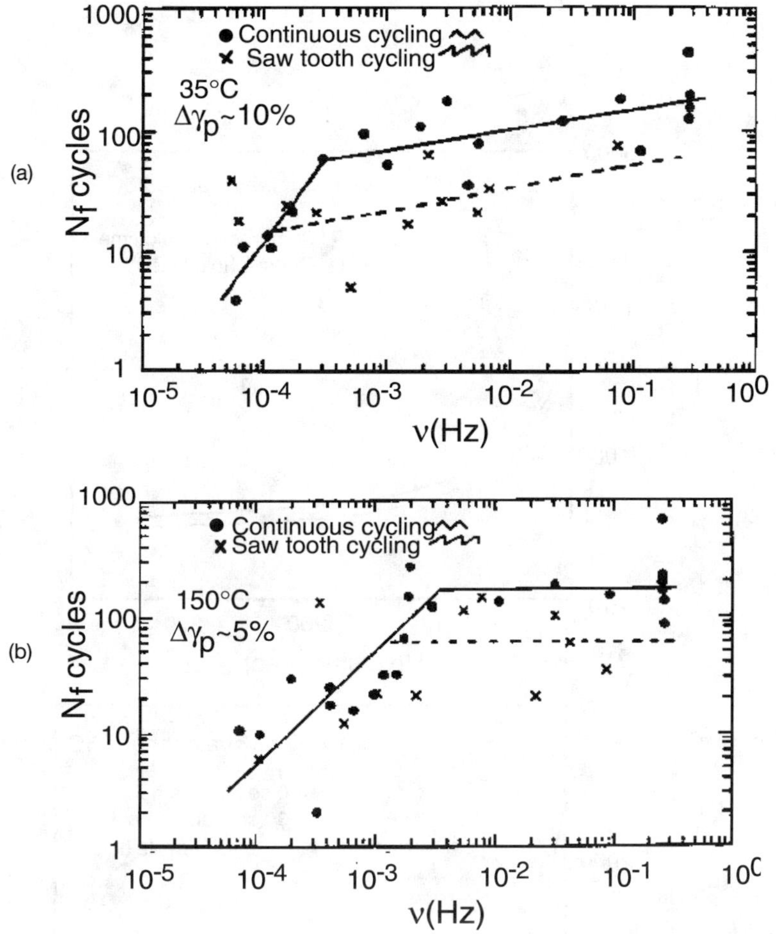

FIGURE 6-32 The cycle frequency dependence of the fatigue life of 60Sn-40Pb solder for tests at (a) 35°C, and (b) 150°C, with symmetric and asymmetric ramp cycles. [Solomon, ©1988, IEEE]

of the influence of the stress state will be covered in a subsequent section. (There is also an effect of the hysteresis loop energy, which will also be discussed later).

Figure 6-33 compares the influence of tensile only vs. compressive only hold times for the same experimental conditions as were used for Figure 6-31. (The compressive holds were established by holding at the zero total strain part of the cycle. This caused compressive stresses to be established). Figure 6-33 shows that there is relatively little difference between the two. Figure 6-33a directly compares tests run with only tensile or only compressive hold times. It shows little difference between the two. There may be a slightly greater influence of the compres-

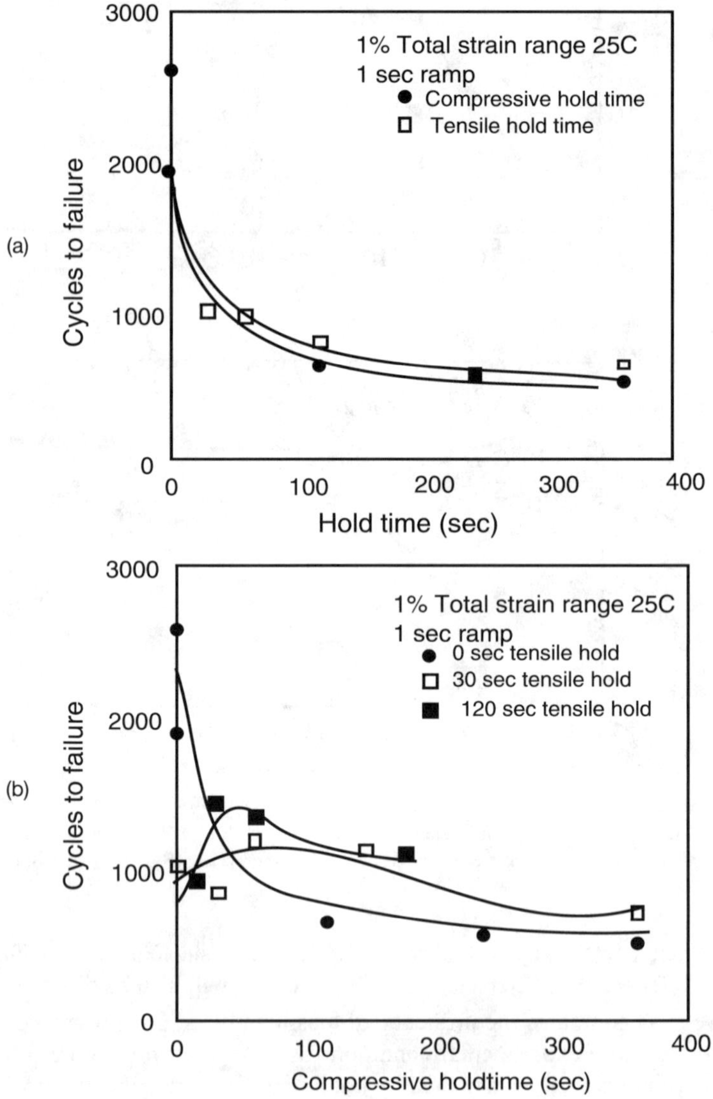

FIGURE 6-33 (a) Effect of Tensile and compressive hold times on the fatigue of 62Sn-36Pb-2Ag and (b) effect of compressive hold time on the number of cycles to failure of 62Sn-36Pb-2Ag, in tests which also experienced a tensile hold.[Vaynman and Fine,1991]

sive hold time, but the difference between the results with compression and tension holds is very small. Figure 6-33b shows what happens when increasingly larger compressive hold times are added to cycles with different amounts of tensile hold time. This figure shows that for more than about a 60 second compressive hold, compressive-only holds are more damaging (lower fatigue life) than cycles which combined tensile and compressive holds. This is another manifestation of the greater damage produced by asymmetric cycling. The behavior observed for short compressive holds is more complicated, since it appears that the fatigue life increases as the hold time increases. Vaynman et al., ascribes this to some sort of recovery process that was occurring in the near eutectic solder they were studying. This sort of recovery was not observed when they studied a high Pb solder [Vaynman et al., 1987a,b].

The problem of symmetric vs. asymmetric loading is not an academic one. In thermal cycle tests or in real life thermal cycling the strains developed at the peak temperature and those developed at the minimum temperature are not the same. A maximum amount of plastic strain is developed at the peak temperature, whereas if the minimum temperature is low enough, little or no plastic strain may be developed during a hold there. This is illustrated in Figure 6-2. Thus, asymmetry should be considered when trying to predict the fatigue life. The use of data developed with symmetric ramp rates or hold times is apt to underestimate the real influence of hold time or ramp rate.

Strain Range Partitioning

All of the preceding used some form of frequency modification to the Coffin-Manson law. There is another approach which also utilizes the Coffin-Manson law but does not use a frequency modification *per se.* This is the technique known as strain range partitioning [Manson et al., 1971; Manson, 1975; Halford and Manson, 1976]. Strain range partitioning (SRP) assumes that there is a characteristic Coffin-Manson curve for time-independent plastic strain, a separate curve to define the fatigue life when the deformation is via creep, and curves which describe asymmetric deformation where the loading in tension is time-dependent and the unloading to compression is via creep, and a curve where the reverse is true. These Coffin-Manson curves are illustrated in Figure 6-34. In general the CP (creep in tension and time independent plastic deformation in compression) curve predicts the shortest fatigue life. This is significant because, as Figure 6-2 shows, this is the sort of deformation that is prevalent in the thermal fatigue of solder joints. An interactive cumulative damage rule is used to determine the fatigue life when there is a combination of PP (no creep in tension or compression), CC (creep deformation in both tension and compression), CP (creep in tension but no creep in compression), and PC (no creep in tension but creep in compression) damage, i.e.:

$$1/N_f = F_{pp}/N_{pp} + F_{cc}/N_{cc} + F_{cp}/N_{cp} + F_{pc}/N_{pc} \qquad (6.10)$$

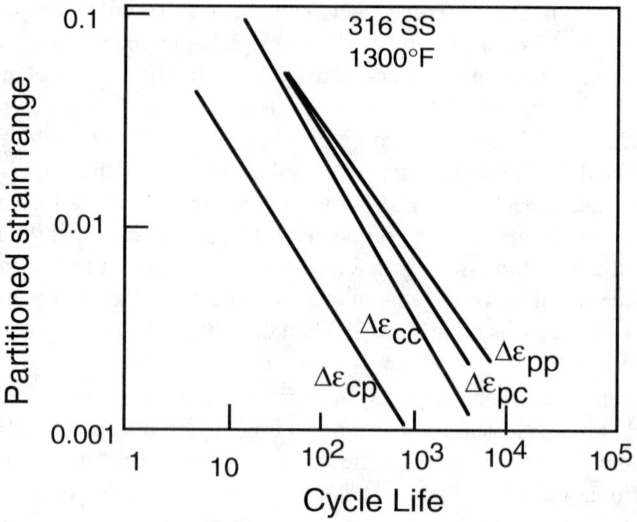

FIGURE 6-34 Partitioned strain-life curves. [Manson,1971]

where N_f is the fatigue life, F_{pp} is the fraction of the deformation which is pp, N_{pp} is the fatigue life when pure pp deformation is used (see Figure 6-34), and F_{cc}, N_{cc}, etc. are similarly defined. The problem of predicting the fatigue life then becomes one of determining the portion of the hysteresis loop that corresponds to each type of deformation and applying the damage rule. This method can be applied to complex thermal cycles [Halford and Manson, 1976] as well as to simple cycles where just the ramp rate is altered. Reducing the cycling frequency increases the amount of creep deformation and reduces the fatigue life. This is illustrated in Figure 6-35. When the frequency is reduced so that all the deformation is via creep, then the fatigue life will be N_{cc}, and further reductions in the frequency will not reduce the fatigue life. There is another frequency-independent regime at high frequencies, where the deformation is entirely time-independent (PP), and increasing the cycling frequency does not alter the type of deformation or the fatigue life. Between these two extremes will be a frequency effect as the deformation transitions from PP to CC behavior.

The observation of a saturation in the fatigue life with hold time (see Figure 6-29) was taken as an indication of the validity of the SRP approach [Shine et al., 1984; Vaynman and Fine, 1991; Kilinski et al., 1991] as applied to solder [Vaynman et al., 1987a]. However, as has been discussed, this saturation is due to the relaxation of the load to zero and a cessation of any additional plastic strain with further hold time. A better approach to seeing if SRP is applicable to solder fatigue is to consider ramp loading at low frequencies, i.e., tests of the sort shown in Figures 6-22 to 6-24. As can be seen in these figures, there is no indication of a ces-

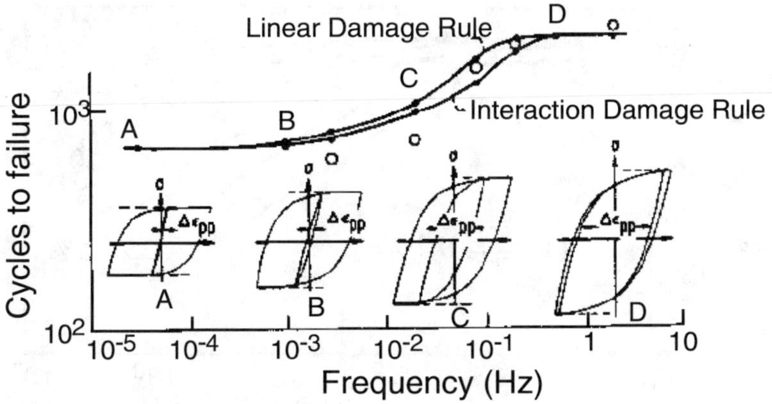

FIGURE 6-35 Variation of the cycles to failure vs. the cycle frequency and a method of determining the time dependent and time independent strain. [Manson,1971]

sation of frequency dependence at low frequencies. It can be argued that further reductions in frequency may be required, but this can be disproved by measuring the creep strains in the hysteresis loop.

The SRP model predicts that when the deformation is all via creep, a frequency-independent fatigue life, N_{cc}, will be measured. The hysteresis loops shown in Figure 6-35 illustrate how to determine the amount of creep that is developed during cycling at any frequency. The amount of creep can be determined by comparing the hysteresis loop drawn at the low frequency with that drawn at a high enough frequency to produce only time-independent deformation. Both loops are drawn for the same stress, with the difference in the plastic strain that is developed being the creep strain. This was done for the material used to develop Figures 6-22 and 6-23. The results of these experiments [Solomon, 1986b; 1988b] are shown in Figure 6-36. The amount of creep strain can also be determined from creep measurements [Halford and Manson, 1976], as is shown in Figure 6-37. Both approaches show that even at 0.3Hz there is a considerable amount of creep strain in the hysteresis loop and that the creep strain increases before there is any decrease in the fatigue life. The calculations made from the hysteresis loops (Figure 6-36) show that for the conditions of Figures 6-22 and 6-23, the creep strain is greater than 80% of the plastic strain before the frequency at which the fatigue life is reduced is reached. Using the creep data yields somewhat lower fractions creep (about 50–60%) before the fatigue life becomes frequency-dependent. Furthermore, the creep strain does not dramatically increase when frequency dependence is observed. Both figures show that an increase in the % creep strain cannot explain the decrease in the fatigue life shown in Figures 6-20 and 6-21. When, as Figures 6-22 and 6-23 shows, the frequency dependence of the fatigue life begins

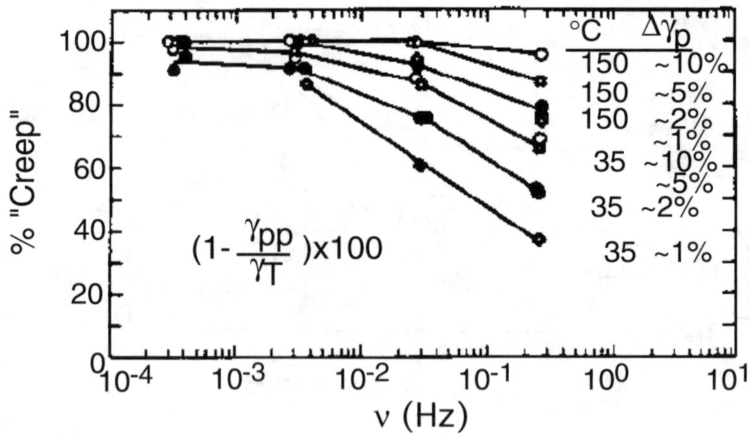

FIGURE 6-36 The cycle frequency dependence of the calculated %-creep, as determined from hysteresis loops using the technique described in the Figure 6-35. [Solomon,1986B]

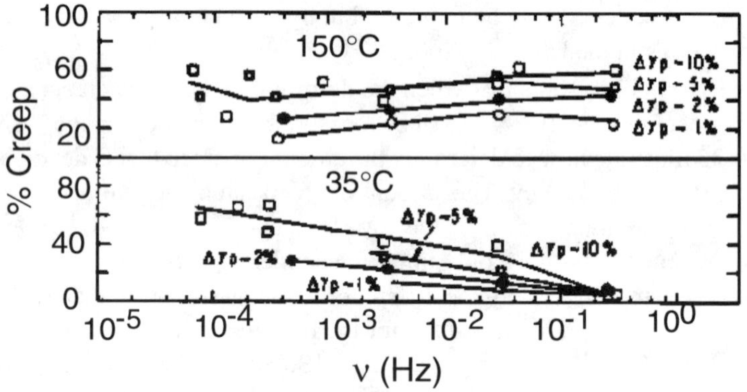

FIGURE 6-37 The cycle frequency dependence of the % creep, as calculated from creep data. [Solomon,1988]

(3×10^{-4} Hz at 35C and 2×10^{-3} Hz at 150°C), the creep strain has already increased to over 50–80%. Further decreases in the cycle frequency produce only a modest increase in the % creep but cause a rapid decrease in the fatigue life.

At low frequencies Figures 6-22 and 6-23 shows that the K value is 0. This means that for a fixed plastic strain, Equation (6.8) predicts that $N_f/\upsilon = t_f$ will be a constant or the time to failure, t_f, is constant. In this frequency regime halving the cycle frequency halves the number of cycles to failure, keeping the time to failure constant. This observation gives rise to a modified version of SRP [Solomon, 1988b]. Instead of using a limiting number of cycles to failure when the deformation is pure creep it is assumed that there will be a limiting time to failure. This approach is illustrated in Figure 6-38, which re-plots the data of Figures 6-22 and 6-23 according to the time modified SRP approach, namely:

$$1/t_f = f_{cc}/t_{cc} + f_{pp}/N_{pp} \tag{6.11}$$

where t_f is the time to failure and t_{cc} is the limiting value when the deformation is all creep. The values for t_{cc} are given in the figure. Multiplying by the cycle frequency converts $1/t_f$ to N_f. Time modified SRP is completely analogous to the frequency modified approach (Equation (6.8)), when K = 0 (f_{cc} = 1, f_{pp} = 0) at low frequencies and K = 1 (f_{cc} = 0, f_{pp} = 1) at high frequencies.

The length of this subsection clearly illustrates the importance and complexity of ramp rate and hold time influences. Unfortunately, there is insufficient data to account for all the important effects, for all conditions. Of most importance is how to handle possible changes in the Coffin-Manson slope with cycle frequency. One limiting condition is to assume that the slope is independent of cycle frequency (see Figure 6-27c). Unfortunately this gives a potentially non-conservative answer. Figure 6-27b shows the other extreme, where the Coffin-Manson curve is vertical at low frequencies and strain ranges. If this assumption is made then one needs experiments at only one strain range and a variety of frequencies. The real solution lies between these limiting cases (see Figure 6-27a). Lacking the data to generate curves of the sort shown in Figure 6-27a, one could bracket the solution using the limiting cases illustrated in Figures 6-27b and 6-27c. Alternatively, the Engelmaier Reliability Model, to be discussed later, gives an estimate of the changes in the Coffin-Manson slope due to temperature and frequency, based upon the data of Figure 6-21.

All of the preceding data was generated in isothermal fatigue. Recent thermomechanical fatigue tests [Frear and Jones, 1990; Frear et al., 1991a] show a different frequency behavior. The evolution of microstructural changes (to be discussed later) were different for thermomechanically cycled samples than for those cycled isothermally. The development of cracks from the formation of a coarsened band was actually retarded in a lower frequency thermomechanical test. There were also differences in the relaxation behavior and the stresses required to develop different strains, compared to those required in isothermal tests. More studies of this type are required, particularly those using low ramp rates and long hold times. Thermomechanical cycling is the subject of another chapter of this book and the reader is directed there for more information.

Regardless of the solder studied or the specific approach taken to describe the frequency, or ramp rate and hold time effects, there is general agreement that, in isothermal tests, the lower the cycle frequency (whether this is produced by lowering the ramp rate or lengthening the hold time) the more the tendency for grain boundary sliding and intergranular cracking. This is reflected in the shorter fatigue lives for cc loading or by a break in the Coffin-Manson curve and the necessity for

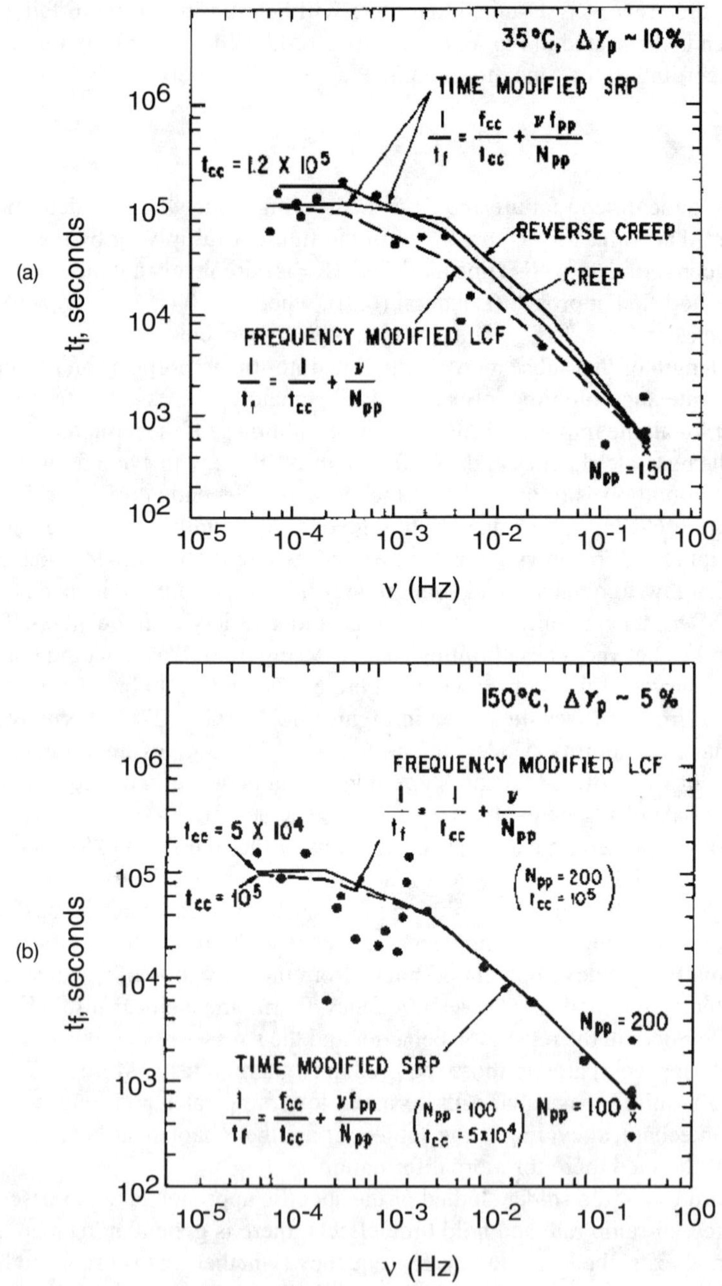

FIGURE 6-38 The frequency-dependence of the time to failure (a) at 35°C and (b) at 150°C, using the time-modified SRP and frequency-modified LCF. [Solomon, ©1988, IEEE]

a frequency modification. As we shall see in the next subsection there is also an influence of environment on the fatigue life, which also influences the degree of intergranular cracking.

Influence of the Environment

Figures 6-17 and 6-18 show that the influence of cycle frequency may not be due to creep; rather, it may be due to the influence of an air environment. The influence of air on the fatigue behavior of Pb has long been known [Snowden, 1964a,b]. Snowden has shown that an oxygen pressure of as little as 3×10^{-2} mm Hg will reduce the fatigue life of Pb by about two orders of magnitude, compared to that measured in vacuum. More recent tests [Berriche, et al., 1987; 1991] run on solders have also showed that the external environment (compared to tests run in vacuum) will reduce the fatigue life of solders.

Berriche et al. [1987; 1991], have done a rather systematic study of the influence of environment (vacuum, air, CO_2, and moisture) on the fatigue life of 96.5Pb-3.5Sn solder at room temperature. These tests employed tensile loading of bulk (6mmx12mm cross section) fatigue specimens (not thin solder joints). The results of these tests are shown in Figures 6-39 to 6-40. Figure 6-39 shows the plastic strain-life behavior to be exactly the same as that observed in the high temperature tests run on 1010 steel stainless steel and a superalloy (see Figures 6-16 and 6-17). At high strains the behavior in air and vacuum is the same, but the Coffin-Manson curve breaks at low strains when the tests are run in air. Figure 6-39

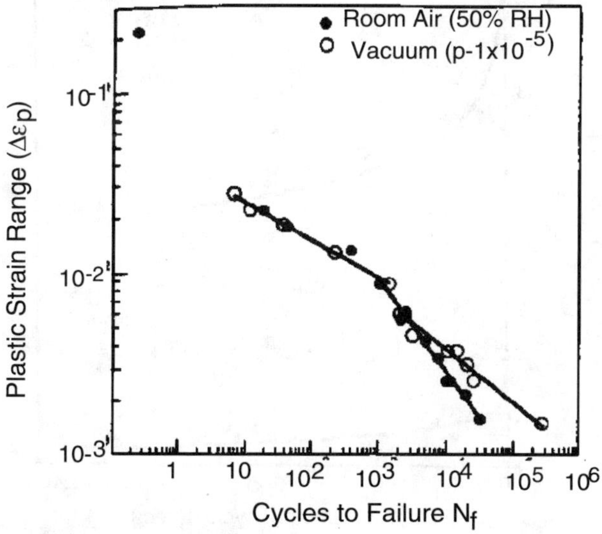

FIGURE 6-39 Strain-life data for 96.5Pb-3.5Sn solder tested in air (50% relative humidity) and vacuum at room temperature. [Berriche et al.,1987,1991]

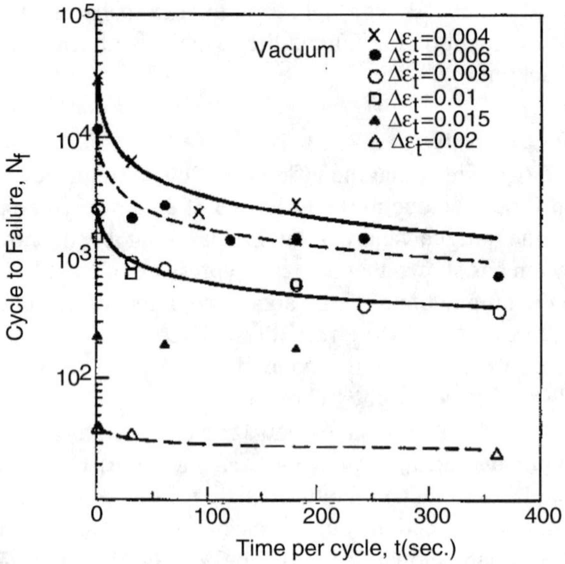

FIGURE 6-40a Variation of the number of cycles to failure as a function of the time per cycle for tests on 96.5Pb-3.5Sn solder tested in vacuum at room temperature. [Berriche et al.,1987,1991]

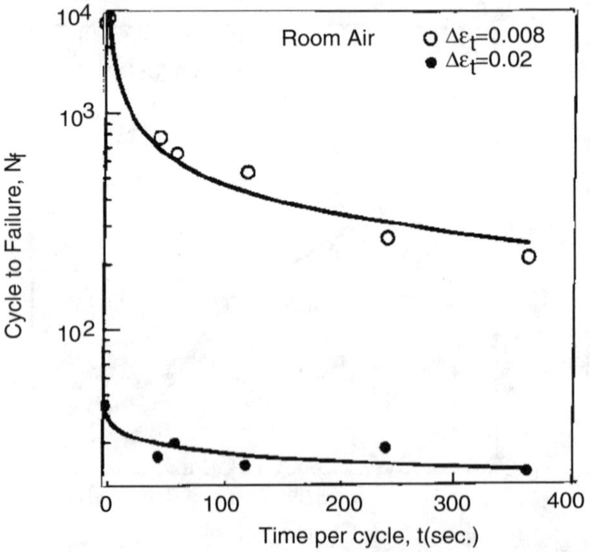

FIGURE 6-40b Variation of the number of cycles to failure as a function of the time per cycle for tests on 96.5Pb-3.5Sn solder tested in air at room temperature. [Berriche et al.,1987,1991]

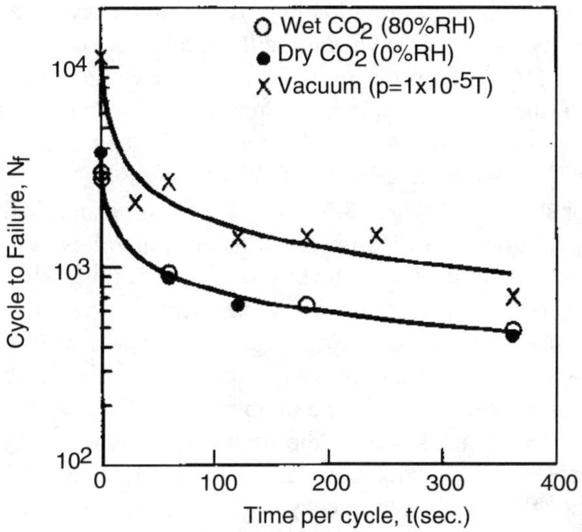

FIGURE 6-40c Variation of the number of cycles to failure as a function of the time per cycle for tests on 96.5Pb-3.5Sn solder tested in wet and dry CO_2, and vacuum, at room temperature. [Berriche et al.,1987,1991]

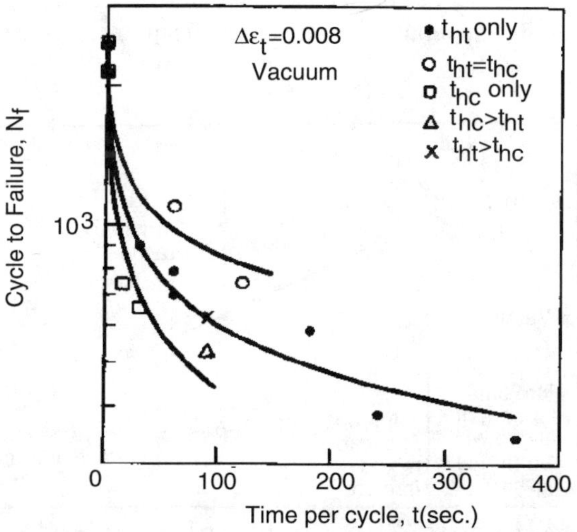

FIGURE 6-40d Variation of the number of cycles to failure as a function of the time per cycle (developed by combinations of tensile and compressive holds with a one second ramp time) for tests on 96.5Pb-3.5Sn solder tested in vacuum, at room temperature. [Berriche et al.,1987,1991]

was obtained using 1 second ramp loading. If the frequency is reduced then the fatigue life is reduced even in a vacuum, although not as much as in air. This is shown in Figure 6-40, where the results using increasingly longer tensile hold times are shown (the time per cycle is the sum of the 1 second ramp loading time and the tensile hold time). Wet and dry CO_2 had about the same effect as air [Berriche et al., 1987] (also compare Figures 6-40b and c).

The behavior shown in Figures 6-39 to 6-42 [Solomon and Coffin, 1973; Coffin, 1975; White, 1969] was also observed in the superalloy A286 and 1/2 Mo steel. Figure 6-41 plots the crack growth rate, which is inversely proportional to the fatigue life (i.e. decreasing the growth rate increases the fatigue life). Figure 6-42 uses the fatigue life and is analogous to Figures 6-22 to 6-24. At high frequencies, time-independent behavior is expected (K = 1). This is what is observed, in the high Pb solder, with the 1 second ramp rate (see Figure 6-39). Figures 6-41 and 6-42 shows that in air, lowering the frequency produces a frequency effect (K<1) and the observed reduction in the fatigue life. However, if the frequency is reduced low enough there is a frequency effect, even in a vacuum. Thus there is an environmental and a creep component to the frequency effect.

Figure 6-40d correlates the influence of tensile vs. compression hold time in a vacuum. In tests run in air (see Figure 6-33a) there was little or no difference between tensile-only or compression-only hold times. There is more of an effect when the tests are run in a vacuum. Compression only, or combined tensile and compression holds with the holds being much longer in compression than in tension, reduce the life by a greater amount than when tensile-only holds of the same length are used. This effect has been observed before [Coffin, 1975; Lord, 1974; Coffin, 1974b] but, in a superalloy, where a significant mean stress could be sup-

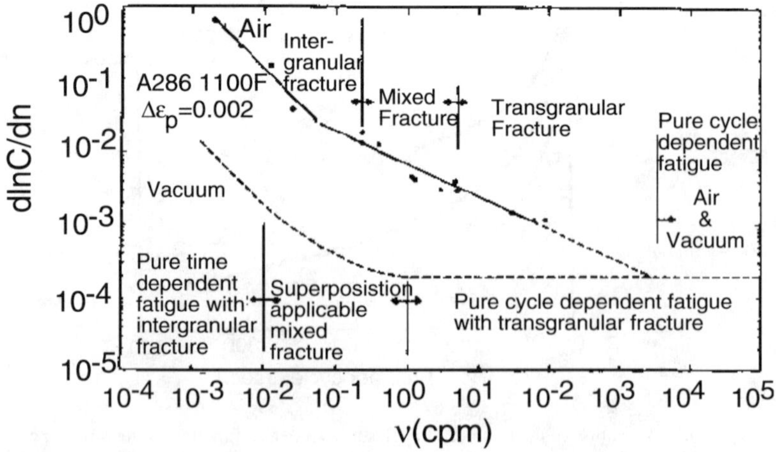

FIGURE 6-41 Comparison of air and vacuum LCF data for A286. [Solomon and Coffin, 1973]

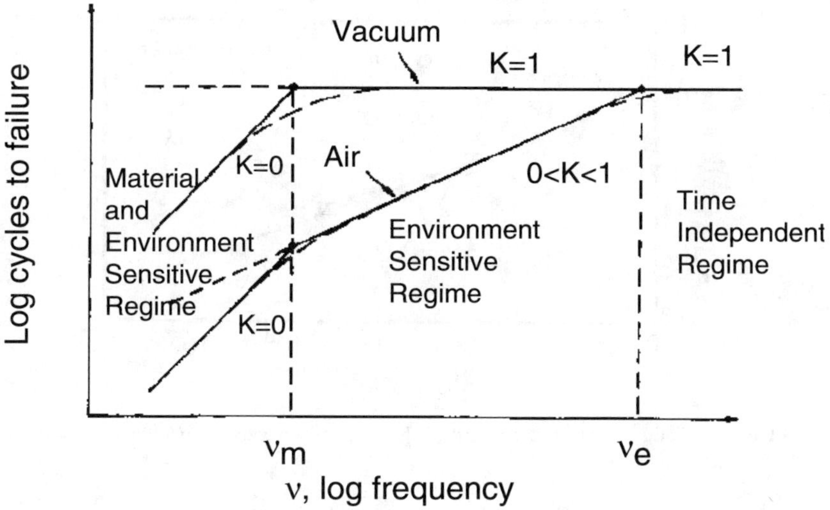

FIGURE 6-42 Schematic showing the effect of cycle frequency on the fatigue life with and without environmental interactions. [White, 1969]

ported. Asymmetric cycling produces complex hysteresis and this influences the fatigue life.

In general, hold times (or low ramp rates) promote intergranular cracking, which is correlated with the reduction in the fatigue life. The intergranular cracking at low frequencies was, however, greatly reduced in vacuum tests, relative to tests run in air [Berriche et al., 1991]. In vacuum, without the promotion of intergranular attack, the variations of the hysteresis loop may dominate and explain why the compressive hold was more damaging. While the amount of intergranular cracking is reduced in vacuum, it is not entirely eliminated. This is in agreement with the results cited in Figure 6-41. At a low enough frequency, even in a vacuum, there will be grain boundary sliding and intergranular cracking, which reduces the fatigue life.

Guo et al. [1991a], also studied the influence of cycle frequency. Their results are shown in Figure 6-43. No environmental effect is noted. Unfortunately, the frequency and strain range employed were too large to expect much of an environmental effect. Figure 6-39 shows that with a 1 second ramp time and a plastic strain of 1% or more, there is no effect of the environment even in a 96.5Pb-3.5Sn solder. The data of Figure 6-43 were obtained with a somewhat lower cycle frequency (0.1 Hz or with a ramp time of 10 seconds) but with total strains of generally greater than 5%. (The elastic strains are quite small so the plastic strain is much greater than the 1% of Figure 6-39).

These vacuum experiments are important because hermetically sealed joints should have a longer fatigue life than ones exposed to air. This has been observed

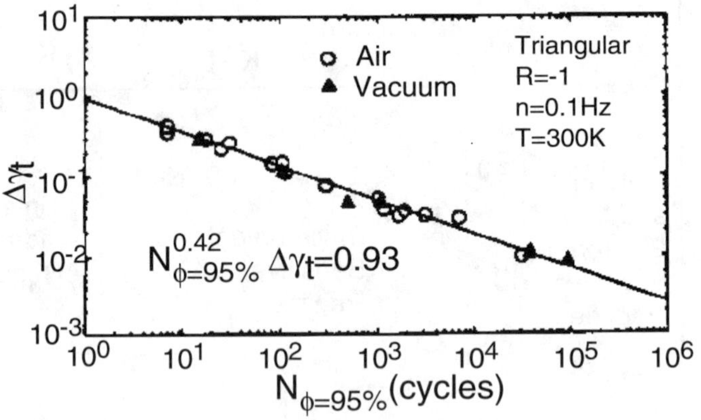

FIGURE 6-43a Fatigue life plots showing the behavior in air and vacuum of 90Pb-10Sn solder tested at 27°C. [Guo et al., ©1991a, IEEE]

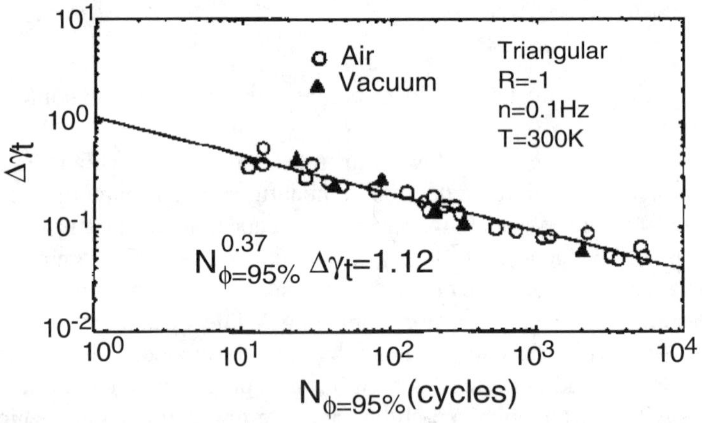

FIGURE 6-43b Fatigue life plots showing the behavior in air and vacuum of 63Sn-37Pb solder tested at 27°C. [Guo et al., ©1991a, IEEE]

by Neugebauer et al. [1984; Burgess et al., 1984]. When the hermiticity was destroyed by drilling a small hole in the enclosure containing the joint, the fatigue life was decreased. Unfortunately, it was found that a significant percentage of supposedly hermetically sealed joints were not so sealed, with an attendant loss in the fatigue life.

Guo et al., [1991a] and Lacey and Woodford [1988] both found an improvement in the fatigue life when vacuum grease was used to cover the joint. This is illustrated in Figure 6-43c. The influence of the grease may not be the result of

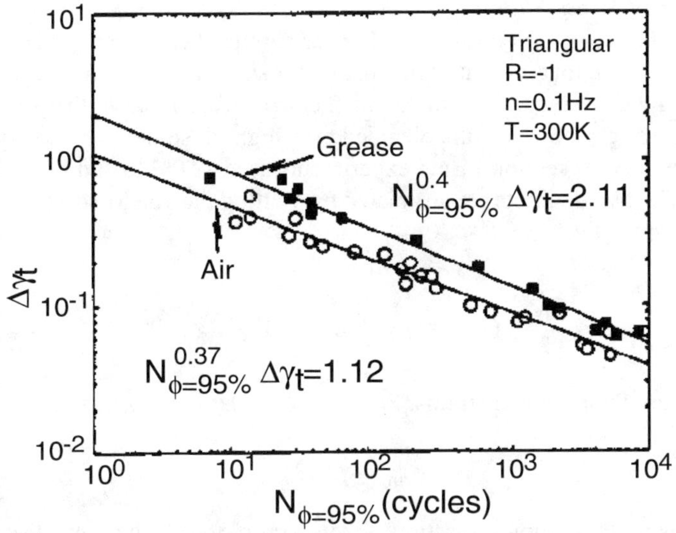

FIGURE 6-43c Fatigue life plots showing the behavior in air and with a grease coating, of 63Sn/37Pb solder tested at 27°C. [Guo et al., ©1991a, IEEE].

preventing oxygen from reaching the solder; instead it may be due to an influence of the grease on the mechanical behavior of the solder joint. Guo et al. found that the flow stress was lower when the grease was used. This was ascribed to a possible interaction of the grease and the oxide on the joint (the Rebinder effect [Lidhman, 1958]). The reduction in the flow stress reduces the energy of the hysteresis loop. As shall be discussed, the fatigue life can be correlated with this energy, in much the same way as the fatigue life is correlated with the applied strain. The lower energy produces a longer fatigue life.

Frear et al. [1991a] have used thermomechanical cycling to study environmental effects. Once again, the thermomechanical results differ from those of isothermal tests. They found that dry oxygen tended to improve the fatigue behavior, by suppressing coarsening and the cracking resulting from this process (to be discussed later). They believe, however, that water vapor will reduce the fatigue life through the influence of corrosion on the fatigue behavior.

The preceding has considered only the gaseous environment. There is also an internal environment associated with the passage of current to consider. Current can cause metallurgical changes due to its influence on diffusion. If some atoms are caused to diffuse at different rates, voids can be produced (this is a form of Kirkendall void formation). This has been observed in Indium solders. Vaynman et al. have recently shown that an electrical current can reduce the fatigue life of 60Sn-40Pb. This effect was most pronounced at the lower strains and when hold times were utilized.

Influence of Stress State

Figure 6-44a compares the fatigue life for simple shear strain (from the data of Figure 6-4) with that measured in tension, for similar solders [Wild, 1972; Vaynman and Fine]. (Note the jog in the Sn62 data curve, similar to that shown in Figure 6-30, which shows results obtained on a high Pb solder. This jog will be discussed in the next section.) Both experiments used a 50% load drop definition for failure. The two data sets are correlated by plotting the von Mises effective strain, which is given by:

$$\varepsilon = \frac{\sqrt{2}}{3}\sqrt{(\varepsilon_1-\varepsilon_2)^2+(\varepsilon_1-\varepsilon_3)^2+(\varepsilon_2-\varepsilon_3)^2} \quad (6.12)$$

where ε_i are the principal strains. For shear $\varepsilon_3 = 0$, $\varepsilon_1 = -\varepsilon_2$, $\gamma = -\varepsilon_1\text{-}\varepsilon_2$ or:

$$\varepsilon_{vm} = \gamma/\sqrt{3} \quad (6.13)$$

In tension (assuming a plastic Poisson's ratio of 1/2) the von Mises effective strain is just the uniaxial tensile strain, ε_0, so $\gamma = \sqrt{3}\,\varepsilon_0$. It is thus simple to correlate simple shear strains and uniaxial tensile strains. Equation (6.13) can also be used to relate any arbitrary stress state to the von Mises effective strain, and to predict the fatigue life with a von Mises effective strain vs. fatigue life plot of the sort

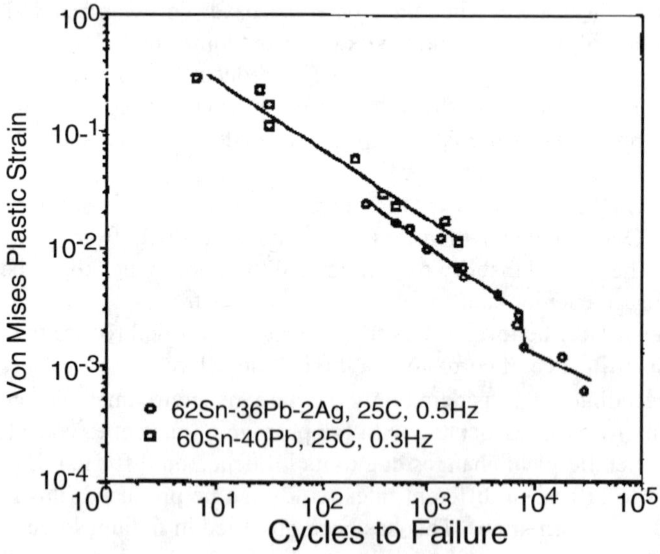

FIGURE 6-44a Fatigue life versus equivalent plastic strain for 62Sn-36Pb-2Ag and 60Sn-40Pb solders. [Cortez et al.,1991,1992]

shown in Figure 6-44a. The validity of this approach is shown in Figure 6-44b, which compares tension, torsion, and tension/torsion fatigue data on the basis of the calculated von Mises effective strain [Cortez et al., 1992].

Cortez, et al. [1991, 1992] have also considered other approaches for correlating multiaxial loadings on eutectic (63Sn-37Pb) solder. In addition to the Von Mises effective strain, they also considered: the maximum principal plastic strain, the maximum plastic shear strain, the Smith, Watson and Topper [1970] relation as extended by Socie [1989] for multiaxial fatigue (which is an energy approach utilizing the maximum principal stress and the maximum principal strain amplitude), and other approaches [Brown and Miller, 1973; Lohr and Ellison, 1980; Kandil et al., 1982] which use the normal stress on the maximum shear planes and the maximum shear strain as correlating parameters.

There was no appreciable difference in the degree to which any of these approaches correlated the data. For all of the models the uniaxial and tension-torsion results differed by about a factor of 3, a small difference for fatigue data. The maximum plastic shear strain approach gave a slightly, but not significantly, better correlation than the von Mises effective strain approach outlined above. All of the data used for these correlations were obtained at ramp loading cycling frequencies of 0.5 Hz. It is therefore not known if low cycling frequencies or hold times might favor one of the approaches considered. This is an area where more work is needed.

Deviations From The Coffin-Manson Relationship

Two types of deviations from the simple power law relationship between the plastic strain and the fatigue life (Equation (6.3)) have been noted. One is associated with tests performed at low cycle frequencies or when long hold times are employed.

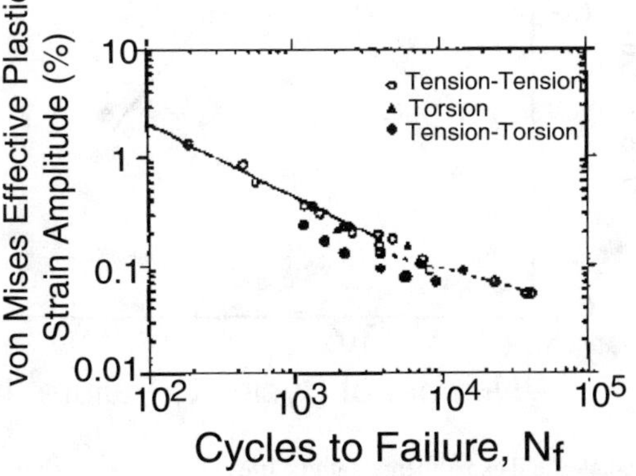

FIGURE 6-44b Fatigue life versus the von Mises effective plastic strain amplitude, for tests run in tension, torsion and tension-torsion. [Cortez et al., ©1991,1992, IEEE]

This behavior is illustrated in Figures 6-16 to 6-19 and, as has been already noted, is observed in many materials when they are tested at elevated temperatures. As has also been noted, this behavior is associated with environmental influences and creep deformation, leading to intergranular cracking at low frequencies and more of a tendency for transgranular cracking at high frequencies. With a difference in the nature of the deformation (more grain boundary sliding at low frequencies) and the resulting change in the nature of cracking, it is not surprising that the Coffin-Manson slope differs in different frequency and strain regimes.

Zubelewicz et al. [1988; 1990] have developed a rather complex model which predicts curved strain life plots of the sort shown in Figure 6-20. Based upon micro-macro correlations of stress and strain, this model takes account of cyclic hardening, recovery and changes in the microstructure. The chief drawback is that it requires high strain fatigue data to determine the necessary parameters; however then it allows extrapolation to lower strains. This is illustrated in Figure 6-45. As is noted, the first 5 (highest strain) data points were used to develop the curve and extrapolate to lower strains. The point of this analysis is that the extrapolation is

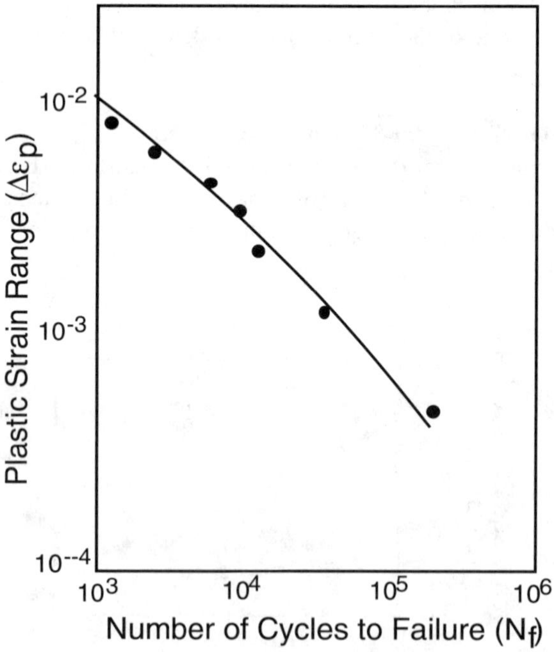

FIGURE 6-45a Predicted and experimental fatigue life curves for Pb-rich solder. The first five data points are used for the prediction. [Zubelewicz et al.,1988,1990] Work performed at Oak Ridge National Laboratory, managed by Martin Marietta Energy Systems, Inc., for U.S. Dept. of Energy under contract DE-AC05-840R21400.

nonlinear (on the log-log scale), whereas the Coffin-Manson relationship (Equation (6.3)) involves a linear extrapolation. While this approach generates the break in the strain-life curves shown in Figure 6-20 (but as a smooth curve instead of as two linear curves), it is not clear that it will handle environmental and cycle frequency effects. For instance, Figure 6-39 shows that at high strains the behavior in air and vacuum is the same, leading to the same parameters and strain life curve. Unfortunately, this predicts that at low strains the behavior in air and vacuum should be the same, whereas Figure 6-39 shows that this is not the case. This could be overcome by using data obtained at lower strains for the determination of the model parameters, yet this eliminates much of the usefulness of this approach because one could no longer just use easily obtained high strain data. There is also a question of how this approach would handle frequency effects. Once again at high strains there may be no frequency effect, with the resulting error when the data is extrapolated to lower strains, where the frequency effect can have a profound effect on the fatigue life.

The second type of deviation from the simple Coffin-Manson behavior shown in Figure 6-4 is the development of a jog in the fatigue life vs. plastic strain curve,

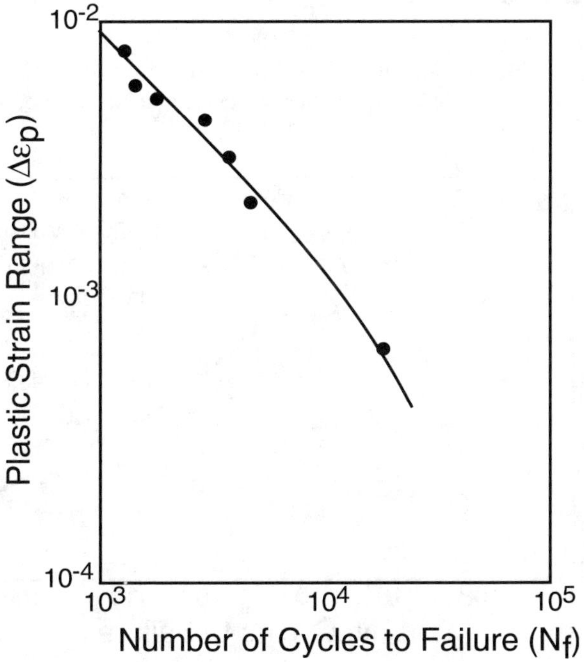

FIGURE 6-45b Predicted and experimental fatigue life curves for 96.5Pb-3.5Sn solder. The first five data points are used for the prediction. [Zubelewicz et al.,1988,1990]

as typified by Figure 6-30 and the Sn62 data of Figure 6-44a. The observation of a jog in the fatigue curve is not limited to tests on solders. It has also been observed in air tests run at 1000˚F on 2 1/4 Cr Mo steel [Coffin, 1975] as is shown in Figure 6-46. As with the solder data, the break shown in Figure 6-46 occurs at the point where the elastic and plastic strains are about equal (which defines the transition from low cycle to high cycle fatigue). In the case of the steel data the jog was ascribed to the influence of strain aging effects [Coffin, 1975]. For the solders it has been suggested [Cationgco and Jeannotte, 1990] that different failure mechanisms are operating in the different regimes of the jog. It is also interesting to note that the jog is present in vacuum for solders (see Figure 6-39), but is not predicted to be present in the steel. (The vacuum curves for the steel were not developed from vacuum data but were extrapolated from the high strain, frequency-independent air data, so the absence of the jog in vacuum has not actually been observed).

The jog in the fatigue life curve is relatively small; one could ignore it and just use a best fit through the data. At low frequencies the jog can be incorporated into the low frequency portion of the fatigue curve, as in Figure 6-20.

6.5 INTEGRATED MATRIX CREEP STRAIN APPROACH

Shine, Fox, and co-workers [Shine et al., 1984; Shine and Fox, 1987; Knecht and Fox, 1991] have developed a life prediction technique based upon the amount of

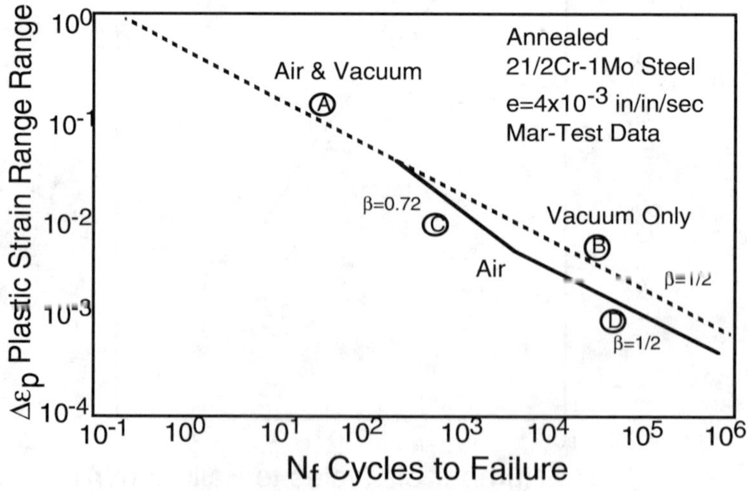

FIGURE 6-46 Fatigue life plots showing a two slope correlation for 2 1/4 Cr-1Mo steel, tested at 538˚C in air and vacuum. [Coffin,1975]

matrix creep developed during cycling. Figure 6-47 shows the measured creep strain rate as a function of the applied shear stress (presented in Knecht and Fox [1991], from the results of several investigators [Grivas et al., 1979; Solomon, 1986b]). Two strain rate regimes are shown. At high stresses and high strain rates the strain rate is a very strong function of the stress, at lower stresses the curve bends and a reduced stress dependence is observed. The high stress and strain rate regime of Figure 6-47 is associated with creep in the matrix and the lower stress regime with grain boundary sliding. (The creep behavior is much more complex [Kashyap and Murty, 1981a,b; Murty and Kashyap, 1984] than that shown in Figure 6-47, with the overall behavior being divided into at least three different regimes. At lower stresses than that shown in Figure 6-47, there is another regime of high stress dependence of the strain rate.)

As shown in Figure 6-48, the integrated matrix creep approach correlates the fatigue life with the matrix creep per cycle. The fatigue data used in this figure comes from Figures 6-4a and 6-21. As Figure 6-48 shows, the matrix creep approach uses a relation similar to the Coffin-Manson relationship, namely:

$$N_f \gamma_{MC} = C \tag{6.14}$$

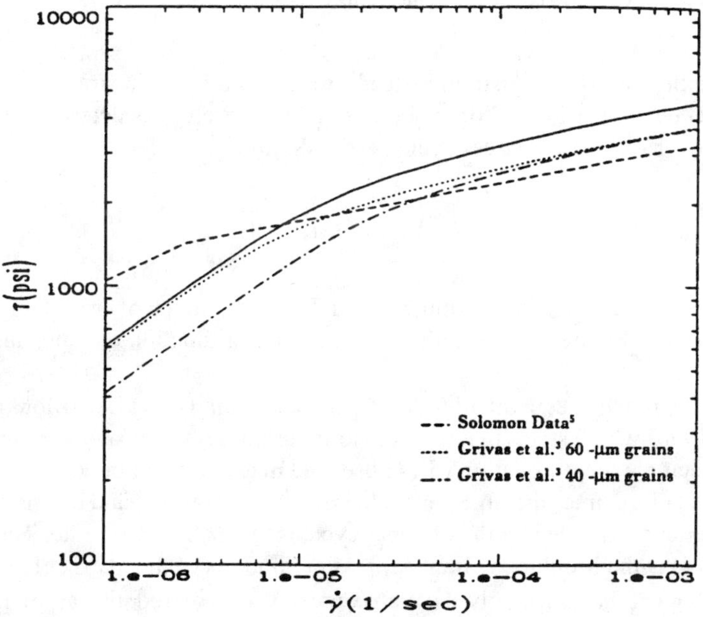

FIGURE 6-47 Creep data for eutectic and near-eutectic solders. [Krecht et al.,1991; data from Solomon,1986b and Grivas et al.,1979]

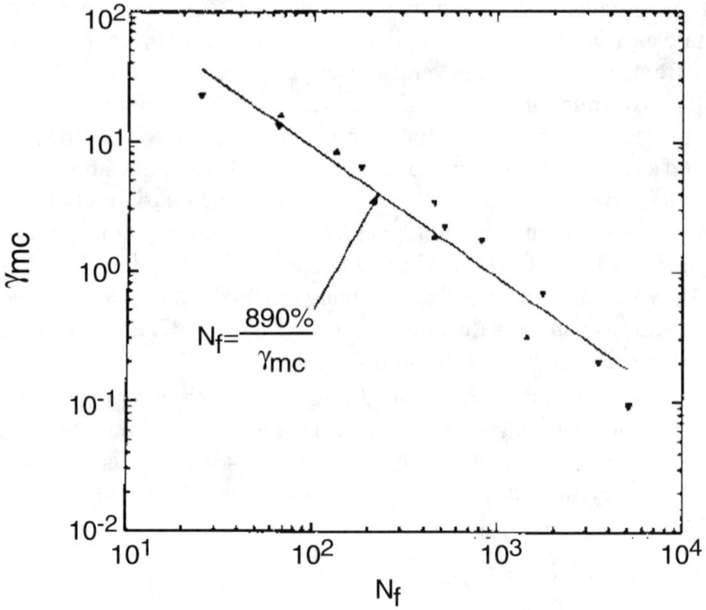

FIGURE 6-48 Cycles to failure versus the matrix creep strain. [Knecht et al.,1991]

but with the matrix creep strain instead of the overall plastic strain, and with an exponent equal to 1 instead of about 0.5. The matrix creep is determined by integrating over the applied stress cycle (stress vs. time) using:

$$\gamma_{MC} = \int C\,(\tau/\tau_0)^{n_{MC}}\,dt \tag{6.15}$$

where the constant n_{mc} is determined from the upper branch of creep curves of the sort shown in Figure 6-47, τ is the shear stress as a function of time, and τ_o is a constant.

This approach is equivalent to the Monkman Grant Creep law [Monkman and Grant, 1956] which states that the time to failure times the steady state creep rate is a constant, except Equation (6.14) uses the matrix creep rate. (The number of cycles to failure times the time per cycle is the time to failure, and the matrix creep strain per cycle divided by the time per cycle is the matrix creep rate. Thus Equation (6.14) is the Monkman Grant expression in terms of the matrix creep only.)

It is logical that since only the matrix creep is considered, this approach is applicable at high strain rates where matrix rather than grain boundary deformation predominates and frequency dependence is not observed. Equation (6.15) shows that the amount of matrix creep decreases when the stress is reduced, such as when low frequency ramp cycling is used. Figure 6-48 therefore predicts that at low frequencies the fatigue life should increase (because of the decrease in the amount of matrix creep), whereas Figures 6-20 to 6-25 show that the opposite is the case. Hence, this approach is not applicable when frequency dependence is noted.

The constant in Equation (6.14) is a function of the material being studied. It is recommended [Knecht and Fox, 1991] that the parameters in Equation (6.15) (which determine the constant in Equation (6.14)) be determined for the solder being used and in this way material variations can be incorporated into the life prediction.

The fact that the calculation of the matrix creep involves integrating over the stress cycle is similar to the energy approaches described below. Both approaches also utilize the flow stress of the solder in predicting the fatigue life and may be a way of incorporating material variations into the fatigue life prediction, but more data is required to determine if this is so.

6.6 ENERGY APPROACHES FOR FATIGUE LIFE PREDICTION

The Coffin-Manson approach to predicting the fatigue life follows the view that it is the plastic strain that determines the fatigue life. This view holds that fatigue results from an irreversibility in plastic deformation during cycling. In going through mechanical hysteresis, after straining and reversing the load to restrain back to the initial condition, all of the dislocations which produce the plastic deformation do not go back to their initial position, even though macroscopically the deformation restores the strain to its initial value. Locally, generally at the surface, the irreversibility leads to the formation of intrusions and extrusions which eventually lead to the formation of a fatigue crack [Grosskreutz, 1971], which then will eventually propagate to cause failure. This view is not theoretical, but is based upon over 75 years of observations.

A second approach is to view the fatigue process in terms of the mechanical hysteresis energy, W, which is being applied in each fatigue cycle, i.e., by:

$$N_f^m W = \theta_E \tag{6.16}$$

where m and θ_E are constants. Morrow[1965] was one of the first modern fatigue researchers to show that the fatigue life could be correlated with the mechanical energy of the hysteresis loop. Morrow points out that this is not a new idea and cites a long line of workers, going back to 1911, who correlated the fatigue life with the energy of the hysteresis loop. This approach recognizes that stress is required to move the dislocations required for irreversible local plastic deformation. While most of the energy imparted goes into heat, a fraction of it is assumed to damage the material (although the nature of this damage is not explicitly defined) and to finally result in failure. This approach to fatigue life prediction finds favor with those who are used to using energy methods for stress and strain analysis.

Energy methods have already been mentioned in a previous section, where it was shown that they produced the same degree of correlation as plastic strain correlations of the influence of the stress state. Calculations of the strain energy stored in the leads of leaded packages have been successfully used to correlate fatigue behavior [Jahsman et al.; Pan, 1991; Engelmaier and Attarwala, 1989]. The calculation of strain energy also forms the basis of the Engelmaier Reliability Model, which is discussed later in this chapter. An energy approach will also be considered when crack propagation is discussed, as the energy is also used to correlate crack growth (which can be used to determine the fatigue life). The importance of energy correlations was one of the significant points of agreement reached at the Workshop on the Mechanics of Solder Alloy held in September 1993. It was felt that differences in the flow stress could normalize material effects such as grain size effects and small variations in solder chemistry. For instance, Vaynman and McKeown [Vaynman and McKeown, 1991] have shown that fatigue life can be correlated with the hysteresis loop energy, which can, in turn, be approximated by the product of the maximum stress and the plastic strain. This energy correlation is shown in Figure 6-49a. Here, the product of the maximum stress and the plastic strain, which is termed the damage per cycle, is plotted vs. the cycles to failure. The 60Sn-40Pb data is that shown for Figure 6-4, which was obtained in shear. The 62Sn-36Pb-2Ag was obtained in tension. For the shear data the maximum shear stress and shear strain were used for the damage function. For the tensile data, the tensile stress and tensile strain were used for the damage function. The data of Figure 6-49a were previously shown in Figure 6-44a where the correlation was made in terms of the von Mises effective strains. (The product of the von Mises stress and von Mises strain is the same as the product of the stress and strain because the correction terms cancel out). Figure 6-49b compares the von Mises plastic strain and the von Mises stress. With the results correlated with only the von Mises strain (Figure 6-44a) the data does not plot on a single curve. With the energy damage function both data sets lie on a single curve (Figure 6-49a). The lower stress for the 60Sn-40Pb data translates to a lower damage function and a longer fatigue life, but this is corrected for when the fatigue life correlation is made in terms of the damage function.

This energy damage function may account for factors which alter the flow stress, such as when the solder is aged, even at 25°C [Vaynman and McKeown, 1991]. Figure 6-50 shows the increase in the fatigue life with aging time at 25°C. Aging softens the solder (probably by increasing the grain size). This decreases the flow stress and increases the amount of plastic strain in the tests shown in Figure 6-50, which were run in total strain control. (The lower the flow stress the lower the elastic strain and, because the total strain is being kept constant, the plastic

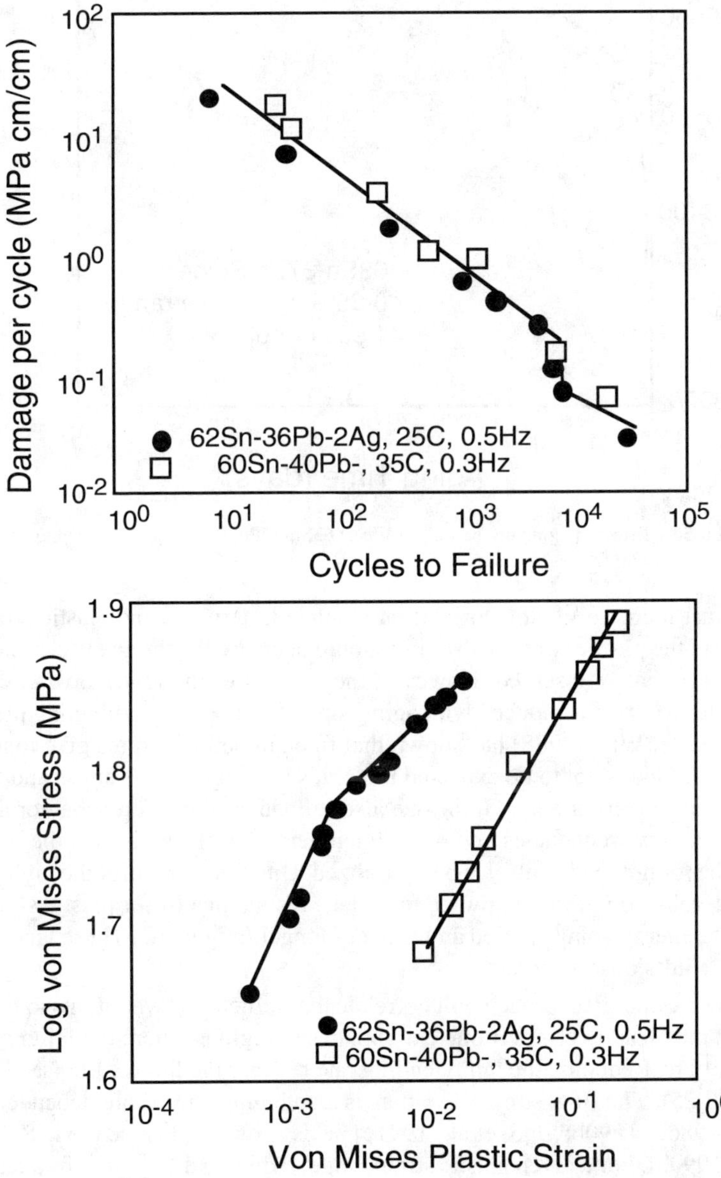

FIGURE 6-49 (a) Fatigue life of 60Sn-40Pb and 62Sn-36Pb-2Ag versus the damage per cycle and (b) von Mises tensile stress versus von Mises plastic strain. [Vaynman and McKeown, ©1991, IEEE]

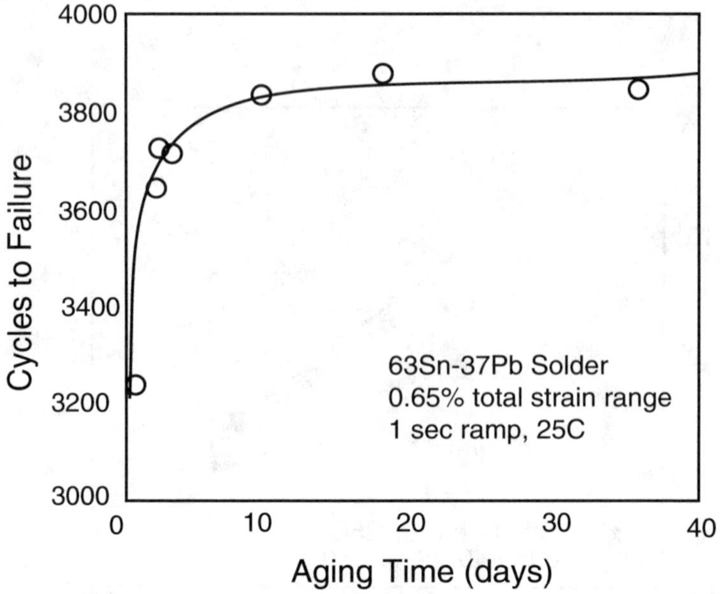

FIGURE 6-50 Effect of aging on the fatigue life of 63Sn-37Pb. [Cutiongco et al.,1990]

strain must increase.) If the correlation is made in terms of the plastic strain, an increase in the plastic strain with aging should decrease the fatigue life, rather than the increase that is observed. However, the product of the lower flow stress and higher plastic strain is reduced with aging time, thus possibly explaining the longer fatigue life. Wild [1988] has shown that fine grained structures give longer fatigue lives. This is not to be expected if it leads to a higher flow stress and thus to a greater hysteresis energy. Grain size also influences the creep behavior and the behavior at low frequencies and/or high temperatures. There is also the question of whether or not load control is being utilized. Under load control the higher flow stress, developed with fine grained materials, results in a lower plastic strain and hysteresis energy being applied and hence, a longer fatigue life. These factors may explain Wild's observation.

While the energy approach will correlate the influence of variations in the flow stress, it may lead to errors if one extrapolates to regions where the failure mechanism differs. Reducing the ramp loading rate reduces the fatigue life (see Figures 6-20 to 6-25). The flow stress of solders is also highly strain-rate dependent [Solomon, 1986b; 1990b; Guo, et al., 1991c], as is shown in Figure 6-51 [Solomon, 1986b; 1990b]. For a given plastic strain range, reducing the cycling frequency reduces the hysteresis loop energy. Thus, since the energy is reduced, reducing the cycling frequency should increase the fatigue life. Figures 6-20 to 6-25 clearly show that this is not the case. The error in this approach lies in using energy vs.

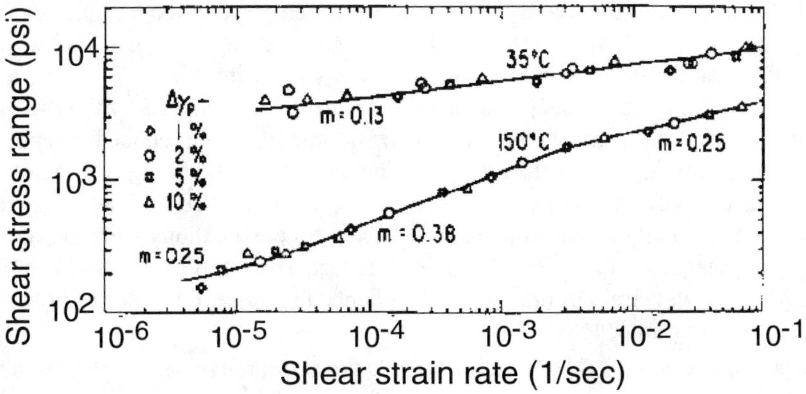

FIGURE 6-51a Shear stress versus strain rate for tests on 60Sn-40Pb tested at 35°C and 150°C. [Solomon,1986b]

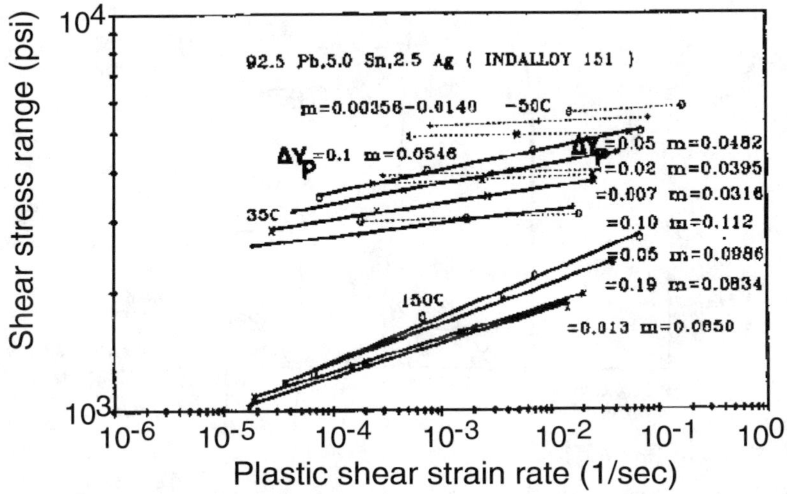

FIGURE 6-51b Shear stress versus strain rate for tests on 92.5Pb-5.0Sn-2.0Ag solder tested at –50°C, 35°C and 150°C. [Solomon,1990b]

fatigue life data generated at high strain rates and applying it to a low strain rate regime. Separate energy vs. strain rate curves are required for different strain rates, in the same way as different plastic strain vs. fatigue life curves are required. The strain energy alone cannot be used to predict the fatigue life when different failure mechanisms are operational, such as the transition from transgranular to intergranular failure that accompanies reductions in the ramp rate.

Differences in the strain rate at which strain energy is applied may also explain why, in Figure 6-22, the hold time did not produce a greater frequency effect than that of the ramp rate. While the hysteresis energy is greater for the hold time tests, than for the ramp rate tests (because of the high stress produced by the high rate straining prior to starting the hold time), the fatigue life is not reduced. Because of the large strain imposed in the initial loading, prior to the hold, Figure 6-52 shows that most of the strain energy in the hold time test was applied at a high strain rate, hence the relatively small influence of the hold time even though a larger hysteresis loop energy was applied. The hold time tests shown in Figures 6-29 to 6-31 employed a total strain of only about 1%, hence the larger influence of hold time relative to ramp rate.

Dasgupta et al. [1992] have reformulated the strain range partitioning model in terms of the hysteresis energy. This is a generalized creep-fatigue approach. Unfortunately, no low frequency or hold time data was provided, so one cannot use the strain energy approach to predict the fatigue life when low frequencies or long

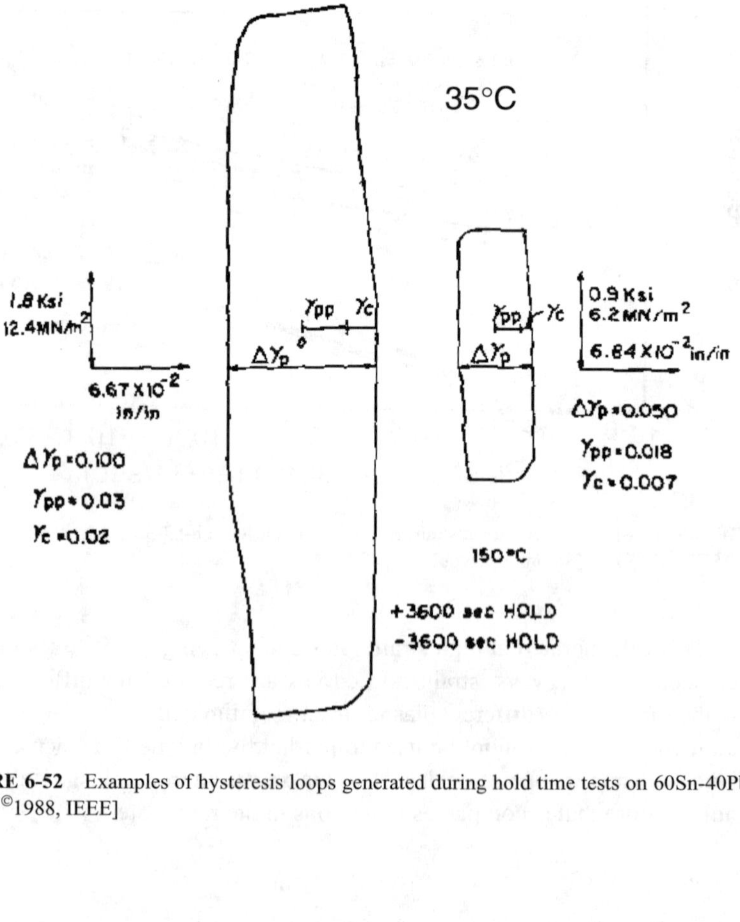

FIGURE 6-52 Examples of hysteresis loops generated during hold time tests on 60Sn-40Pb. [Solomon, ©1988, IEEE]

hold times are used. Hence this approach is currently not as useful as those which use the plastic strain range.

There is a philosophical difference between the use of the plastic strain and the use of the strain energy to predict the fatigue life. Both approaches give similar results. Strain is the controlling variable for crack initiation. What is important is the strain that is produced, not the stress required to do this. In fact, increasing the flow stress would increase the fatigue life if it meant that less plastic strain was developed. This correlates with the fact that in the high cycle fatigue regime, for a given applied stress, increasing the strength of a material increases the fatigue life. (The Basquin curve is shifted upward according to the ultimate tensile strength. This also shifts upward the high cycle region of the total strain vs. fatigue life curve.) Unfortunately, if the strength of the solder is increased too much, then the stresses developed during a thermal cycle may be large enough to crack the component to which the solder is attached.

Crack propagation is generally given as a function of the stress intensity or in the high strain regime as a function of the J integral. As such, the strain energy would be the more appropriate parameter to correlate the fatigue life. However, in the low cycle fatigue regime the crack growth rate can also be expressed as a function of the plastic strain range, rather than the stress [Manson, 1975; Coffin, 1975; Solomon, 1972a,b; Tomkins, 1968; Kaisand and Mowbray, 1979], making the plastic strain again the most appropriate parameter.

6.7 ENGELMAIER RELIABILITY MODEL

Engelmaier [1983; 1985b; 1991; 1992a] has developed a reliability model originally targeted for applications in the telecommunications industry. The approach, however, has generally applicability [Engelmaier, 1991; 1992a,b]. This model is based on Wild's fatigue data, shown in Figure 6-21 [Wild, 1971]. Combined with a lead compliance analyses developed by Kotlowitz [Kotlowitz and Engelmaier, 1986; Kotlowitz, 1989a,b; 1991a,b], it forms the basis of an approach to designing joints by evaluating "Figures of Merit" which can be used to compare one design with another and insure that the joint has the required reliability [Clech et al, 1989a,b; Clech and Augis, 1991; Engelmaier, 1992b].

Presenting all the features of this model would involve a recapitulation of much of what has already been covered. Rather than do this, only the most salient points will be discussed.

The basis of this model is the energy approach to determine the fatigue life (Equation (6.16)), with some consideration of the relative behavior of leaded vs. leadless joints. The hysteresis loops (showing total strain vs. applied stress) for the solder of each type of interconnection are compared schematically in Figure 6-53. Figure 6-53a shows the behavior for thermally cycled leadless joints. Leadless

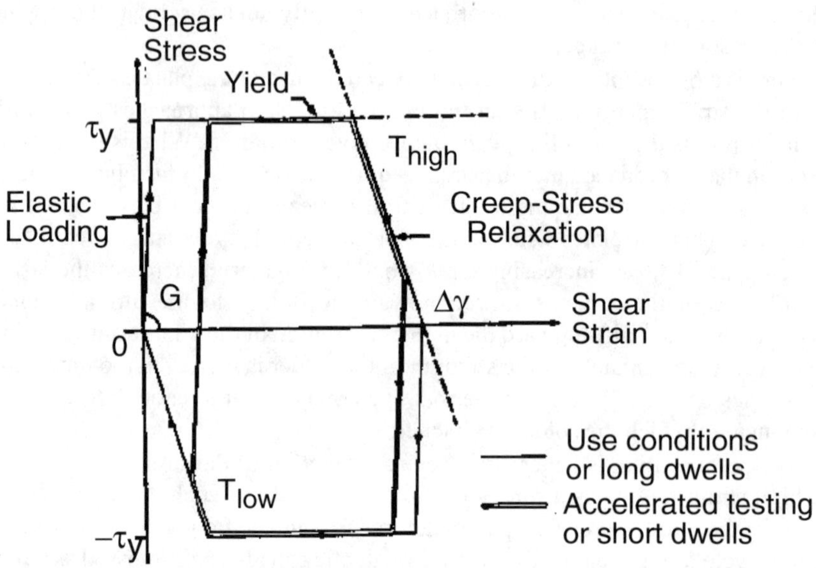

FIGURE 6-53a Hysteresis loops characterizing the solder joint response is the shear stress-strain during the thermal cycling of surface mounted leadless components. [Clech et al., ©1989, IEEE; Engelmaier, ©1989b, IEEE]

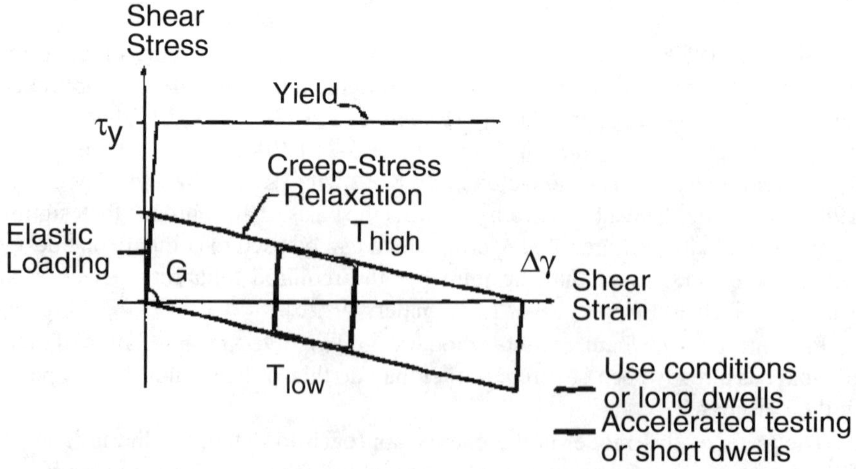

FIGURE 6-53b Hysteresis loops characterizing the solder joint response is the shear stress-strain during the thermal cycling of surface mounted leaded components, with compliant leads. [Clech et al., ©1989, IEEE; Engelmaier, ©1989b, IEEE]

joints are relatively stiff and the maximum stress that is reached during a temperature change is limited by the yield strength of the solder. During the hold time the stress relaxes. Figure 6-53a shows two types of loops. For long hold times the stress is assumed to relax to zero. For shorter hold times (such as in accelerated testing when short hold times are applied) the stress does not relax to zero before the stress is reversed by the change of the temperature back to its initial value. The hysteresis loop is completed by a hold at the initial temperature.

Figure 6-53b shows the behavior of leaded assemblies, where the lead stiffness determines the maximum stress, S. Once again two loops are drawn: one where the hold time allowed complete relaxation, and another where shorter hold times are considered and there is incomplete relaxation, and a much smaller hysteresis loop is established.

The loops of both Figures 6-53a and 6-53b are symmetric, but symmetric loops are actually never achieved in thermal cycling and are only approximated if the temperature excursion is small, making the creep rates at the temperature extremes similar. This model was developed for the thermal cycles of telecommunications equipment, which are relatively small (ΔT of 20° to 40°C, with a mean temperature of about 20°C). Such modest temperature excursions are also valid for computers kept in temperature-controlled buildings and properly cooled. For larger thermal cycles, where the peak temperature may be above 100°C and the minimum temperature below 0°C, this assumption of symmetry is definitely not valid (see Figure 6-2). However, the assumption is made that by basing the model on the average cycle temperature (T_{ave} in Equation (6.17)) the errors developed by underestimating the stress relaxation at the highest temperature and overestimating it at the lowest temperature cancel, allowing the result to be reasonably accurate. Such large thermal cycles are observed in applications where the joint is exposed to severe heating and to severe extremes in the ambient temperature. Particularly severe are under-hood automobile applications, exterior aircraft applications and military applications in general (where a –55° to +125°C thermal cycle has been required as a qualification test). Figure 6-53a shows that for leadless joints the yield stress is assumed to control the stress of the hysteresis loop. In this case the hysteresis loop can be assumed to be rectangular so the energy is just twice the yield stress times the sum of strain developed during the temperature change and that developed during the hold. Since the yield stress is highly strain-rate dependent (see Figure 6-51a for the behavior of near-eutectic solder), the yield stress will depend upon the ramp rate. The strain range will depend upon the magnitude of the temperature change and the hold time. For a given ramp rate and assuming complete relaxation, the hysteresis energy is just a constant times the strain. The fatigue life vs. energy relation (Equation (6.16)) reduces to the total strain vs. fatigue life expression. With the plastic strain larger than the elastic strain and the elastic strain converted to plastic strain by stress relaxation during the hold, the total strain can be treated as a plastic strain and the fatigue life predicted by the Coffin-Manson relation (Equation (6.3)).

Figure 6-53b shows that for leaded assemblies the lead flexure prevents the solder from reaching the yield stress. However, during the hold time the solder can creep, reducing the stress. The hysteresis loop is clearly a parallelogram, with an energy (loop area) equal to S times the strain range and with the fatigue life given by Equation (6.16).

The data of Figure 6-21 (due to Wild [1971; 1972; 1975]) was used to correct for temperature and time effects. The data was curve-fit to give a Coffin-Manson exponent of:

$$c = -0.42 - 6 \times 10^{-4} T_{ave} + 1.74 \times 10^{-2} (\ln(1+360/t_D) \qquad (6.17)$$

where T_{ave} is the average temperature and t_D is the half cycle dwell time, in minutes. The fatigue ductility coefficient was assumed to be a constant equal to 0.65 (all the curves of Figure 6-21 appear to merge). The Coffin-Manson expression can be written in different ways. In this formulation of the Coffin-Manson expression, $c = -\alpha$ of Equation (6.3).

Equation (6.17) and the observed consistency of the fatigue ductility coefficient are not in agreement with the data of Figures 6-14 and 6-15, which were obtained on a similar solder as was used by Wild (see Figure 6-21) and on a high Pb solder. There are several possible reasons for this discrepancy. Wild used displacement-limited lap shear whereas the data of Figure 6-14 were obtained with plastic strain-limited shear and that of Figures 6-15a and 6-15b in total strain-limited tension. In both cases there were strain measurements and limits based on these measurements. Wild used a simple rotating fixture with an elliptical cam to produce a cycling deflection, with total strains being calculated rather than measured as was the case with the data of Figures 6-14 and 6-15. This is stated not to criticize Wild's data, which was virtually all that was available for many years, but to point to why a different temperature dependence was noted. Wild produced total strain-fatigue life plots, which explains why the exponent is generally lower than that of Figure 6-14, which (properly for the Coffin-Manson expression) uses plastic strain as the correlating parameter. Increasing the test temperature should result in more plastic strain, which increases the slope of the total strain life plot (see Figure 6-6). In contrast, the data of Figure 6-14 were obtained with a constant plastic strain.

Equation (6.17) predicts that the total strain-fatigue life exponent increases a small amount with increases in t_D. This is not a frequency effect, *per se.* Rather it is due to the fact that increasing the dwell time allows for more stress relaxation and an increase in the plastic strain at the expense of the elastic strain. This reduces the fatigue life. This effect is also more pronounced the smaller the total strain because initially the plastic strain is smaller. Hence, increases in the plastic strain have a larger effect. This increases the slope. When the dwell time reaches about 360 minutes (Equation (6.17)) shows the beginning of a saturation in the dwell time effect. With a dwell time of 3600 minutes, the effect of t_D in Equation (6.17) is just about eliminated. This is illustrated in Figure 6-54a [Engelmaier, 1984] which plots the results of Figure 6-21 in terms of the half-cycle dwell time and compares the model results to the results or isothermal power cycling tests and thermal cycling oven tests. Figure 6-54b [Engelmaier, 1984] plots the same data as Figure 6-54a, but here in terms of the cyclic frequency instead of the dwell time. The data of Figure 6-22a shows that the frequency effect does not begin to get significant at 35C until a dwell time of about 2000 seconds (33 minutes) (corresponding to a frequency of 2.5×10^{-4} Hz when the inverse of twice this dwell time is considered, the factor of 2 is required because t_D is defined as the half cycle dwell time). This fatigue life reduction, and those shown in Figures 6-21 and 6-53b, are larger than that shown in Figure 6-54..

Equation (6.17) is the exponent for a total strain vs. fatigue life expression, with the total strain approximating the plastic strain so that the Coffin-Manson expression could be used. This approximation is strictly valid only to the extent that the elastic strains are small and can be neglected. Furthermore, Equation (6.17) is valid only in a limited temperature range (20° to 100°C) for half-cycle dwell times only up to about 360 minutes. It does not describe the data beyond this regime, where frequency and temperature effects are more pronounced.

For the leadless joints where the approximation of a rectangular loop is made, the number of cycles to failure, N_f, is given by:

$$N_f = 1/2\ [F/\theta\ (L_D\ \Delta\alpha\Delta T_e/h]^{1/c} \qquad (6.18)$$

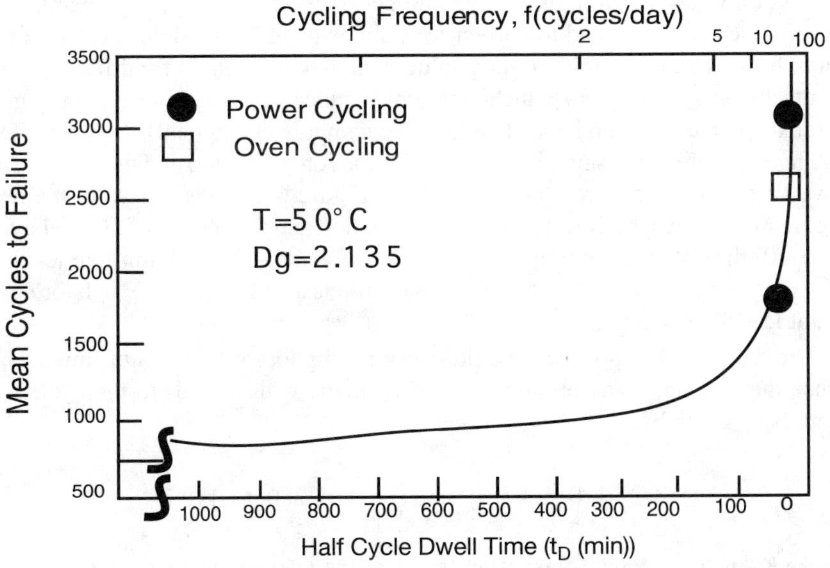

FIGURE 6-54a Dwell time correlation for predicted and experimental cyclic lives. [Engelmaier, 1984]

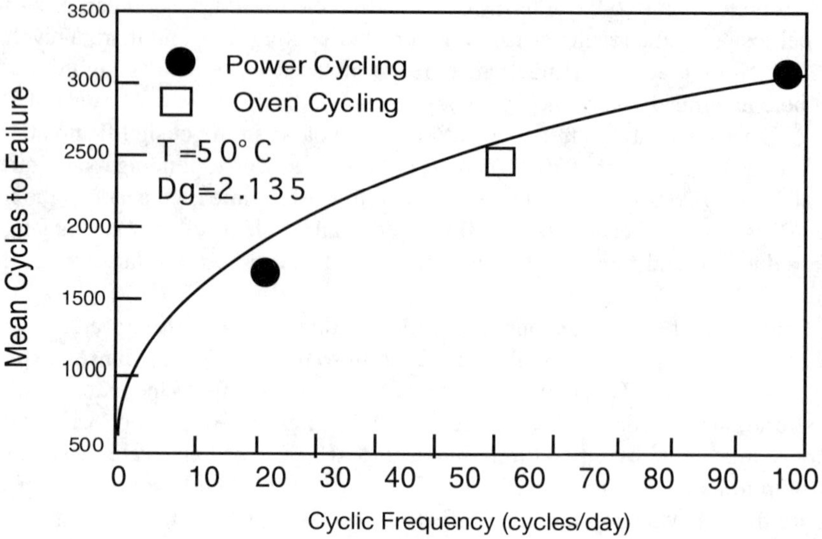

FIGURE 6-54b Frequency correlation of predicted and experimental cyclic lives. [Engelmaier, 1984]

where the 1/2 term is introduced because the formulation of the Coffin-Manson expression was given in terms of the number of reversals, $2N_f$; F is an empirical factor which is introduced to correct for unknowns in the model (in effect it is a curve fitting factor, but with limited values which have been set for different types of attachments) and L_D is 1/2 the maximum distance between joints (L/2 in Figure 6-1, except that L should be the length of the component diagonal), T_e is the effective temperature range (including power dissipation gradients) and the other terms have been defined before with reference to Figure 6-1 or elsewhere in the text. Figure 6-55 shows the correlation of this equation and the results of thermal cycle tests. All the data was fit with $1.12<F<1.27$. In general, for column-like leadless joints $1.5>F>1.0$, for castellated joints with fillets $1.2>F>0.7$ and $F = 1$ for compliant leaded attachments.

The fatigue life expression is a little more complex for leaded attachments because the rectangular assumption cannot be made with regards to the hysteresis loop. N_f is given by:

$$N_f = 1/2\ [(F/\theta)\ K_D\ (L_D\ \Delta\alpha\Delta T_e)^2/\ (\Delta h\ 200\ \text{psi})]^{\ 1/c} \tag{6.19}$$

where K_D is the diagonal flexural stiffness of the lead and is the solder joint load bearing area. The exponent 2 comes into the expression because expression for the shear strain has a ($L_D\ \Delta\alpha\Delta T_e$) term and so does the expression for the stress

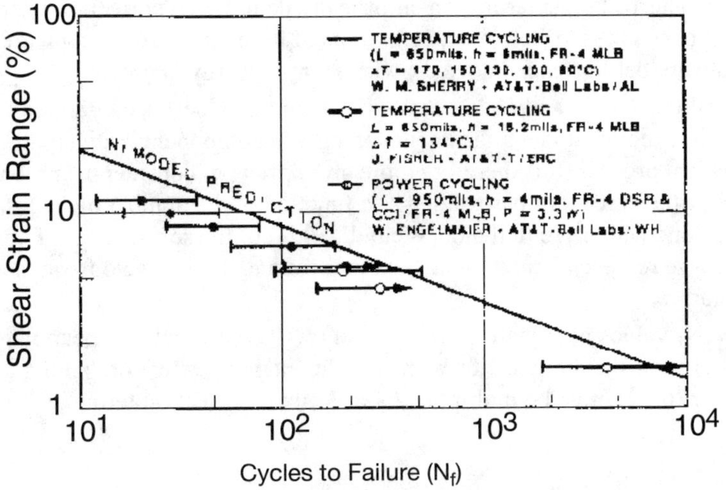

FIGURE 6-55a Coffin-Manson fatigue life plot showing the mean fatigue life and range of fatigue lives for three cycling experiments. The results have been normalized the same mean temperature (45°C) and dwell time (6.2 in.). [Engelmaier, 1984]

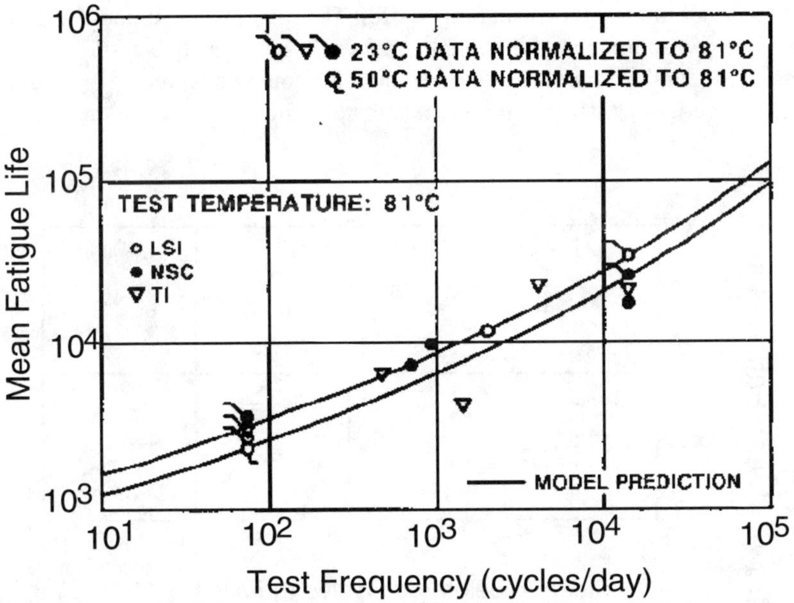

FIGURE 6-55b Comparison of the Engelmaier reliability model with test data taken at different cycle frequencies for 68 I/O leaded chip carriers subjected to mechanical bending at 81°C. [Engelmaier, 1984]

applied. The 1/200 psi term is an empirically determined correction factor. It has units of psi so the stiffness must be in units of lb/in. Figure 6-56 compares results with this model for leaded devices, where strain energy is used.

Equations (6.18) and (6.19) each give a single value for the fatigue life of all the surface mount solder attachments of a given component. Fatigue is, however, a statistical process. It is therefore more appropriate to consider the probability of failure or the percent of parts failed as a function of the number of cycles applied. This is done with a Weibull plot [Weibull, 1961] of the sort shown in Figure 6-57. Each curve represents the fatigue behavior for a different cycle frequency and/or temperature.

The N_f values of Equations (6.18) and (6.19) represent the mean fatigue life. To determine the fatigue life with a different probability of failure, Equation (6.18) or (6.19) must be multiplied by a Weibull correction term, i.e., multiplied by:

$$[\ln(1 - F(N))/\ln(0.5]^{1/\beta} \tag{6.20}$$

where β is the Weibull shape parameter, and F(N) is the cumulative failure probability being designed for (say the fatigue life when 5% of the joints would have failed). The 0.5 term occurs because the data being corrected represent a 50% probability of failure (the mean fatigue life). The Engelmaier Reliability Model is one of the few approaches to life prediction which specifically incorporates failure statistics. If another approach is utilized the life should be corrected to account for statistical variations. Unfortunately, this is not generally done, so the number of

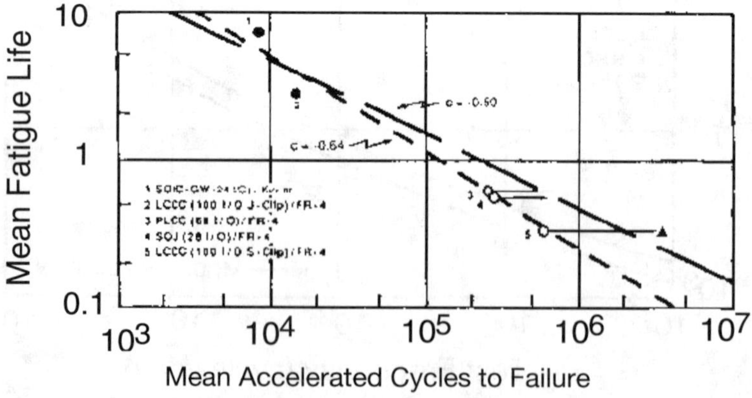

FIGURE 6-56 Coffin-Manson fatigue life plot showing the results of five tests of different leaded components. [Engelmaier, 1991]

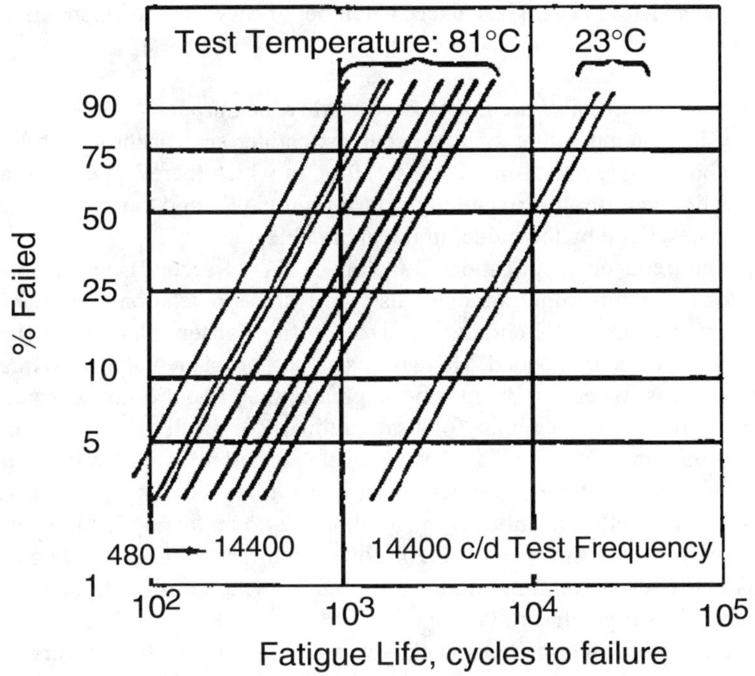

FIGURE 6-57 Weibull statistics of test results from accelerated isothermal mechanical cycling at two temperatures. Dwell times ranging from 3 to 90 seconds were also employed. [Engelmaier, 1991]

cycles to the first joint failure is not calculated. With data curves that reflect best fit lines, the mean failure is calculated. Unfortunately, in critical designs this is not sufficient.

The shape parameter, b, is characteristic of the joint type and type of solder. A large value means a tight distribution in the fatigue life with relatively little difference between the number of cycles to achieve a given % failed. A low value means that there is a relatively large distribution in the fatigue lives. Engelmaier [1991; 1992a,b] has found that for eutectic solder the $b \approx 2$ for attachments with compliant leads and ≈ 4 for leadless or when the leads are non-compliant.

This model has been utilized to establish Figures of Merit which enable an engineer to determine the relative effect of changing variables on the resulting fatigue life [Clech et al., 1989a,b; Engelmaier, 1992b, Clech and Augis, 1991]. These Figures of Merit are determined by considering the important variables in the Weibull corrected versions of Equations (6.18) and (6.19). For example, these Figures of Merit can be used to determine the influence of a dimension such as the solder height, the magnitude of a temperature range, etc. The higher the Figure of Merit, the greater the reliability.

Engelmaier [1989; 1992a] has placed a number of caveats as to when this model is not applicable:

1. when the solder joints are not of good quality, i.e., properly wetted;
2. when large temperature swings occur, extending significantly both below and above the region from –20°C to 20°C, in which there is a change from stress driven to strain-driven fatigue (see Figure 6-2), and the damage mechanism described by the model in not applicable;
3. for high frequency applications (>0.5 Hz or $T_D < 1$ second), e.g., vibration, and/or low temperature applications where stress relaxation and creep are not a significant factor and the standard Coffin-Manson relationship (Equation (6.3)) is recommended. There may also be a problem with low frequency applications, when the dwell time is greater than about 6 hours, since the model predicts the beginning of a saturation of the dwell time effect for this time, and other data predicts that the cycles to failure continue to decrease with longer hold times. The data upon which this model is based utilized a half-cycle dwell time only as long as about 12 hours (at 65°C) so extrapolations beyond this time should not be considered. The 25°C data only goes up to a half-cycle dwell time of about 4 minutes, which further limits the hold time region of greatest reliability;
4. when the global thermal mismatch is very small, e.g., ceramic on ceramic or silicon on silicon and the local mismatch between the solder and the soldered material becomes the primary cause of fatigue failure;
5. when the leads are very stiff (K_D> 500 lb/in) where the damage estimates, developed for more compliant leads, are not applicable;
6. when there are very large thermal expansion mismatches resulting in the generation of strains of >10%;
7. when other than eutectic or near eutectic solders are used, and the coefficients used in the model may not be applicable, although even here reasonable results may be obtained if the coefficients do not differ by too large an extent;
8. when the geometry differs significantly from that of the joints utilized for model's development. The model incorporates geometric variables such as fillet or lead shapes into the constants, and new constants should be developed for significantly different geometries.

6.8 CRACK PROPAGATION APPROACHES

The Coffin-Manson relation or the energy vs. the fatigue life directly determine the fatigue life from the applied strain or energy. Another widely used approach involves determining the fatigue life from the crack growth r ate. Fatigue is a process of crack initiation and propagation. At high strains (such as is the case in the fatigue of solder in some leadless joints) crack propagation is generally dominant in determining the fatigue life. It is therefore logical that crack propagation should be used to determine the life of a solder joint. (As shall be discussed, there are situations where microstructural changes control the formation of fatigue cracks and the time to achieve these changes may determine the fatigue life. In this case the crack propagation approach would only give the time to grow a crack to failure, not the total time (or cycles) to failure. In this case even though the strain is high crack initiation may dominate).

The crack growth approach is also important when a crack is found, and it is necessary to determine the number of cycles required to propagate it to failure. If the crack growth rate is known as a function of the crack length, applied stress or strain, and other relevant parameters than the fatigue life can be determined by integrating the crack growth expression between the initial and final crack lengths. The final crack length is the length associated with failure. It might be the total length of the solder joint or some lesser dimension. The crack growth approach does not handle crack initiation, so it is necessary to assume that a crack of some initial size exists. In critical applications, where a non-destructive examination is utilized, this initial crack length might be the largest crack which could escape detection. This is a very conservative approach. A less conservative approach would use a smaller initial crack size, say the size of an inclusion or the grain size. Since most formulations for the crack growth rate predict that the crack growth rate is proportional to the crack size, most of the fatigue life is spent when the crack is small; and the smaller the assumed size of the initial crack, the longer the predicted fatigue life.

The crack growth rate can be determined by direct and indirect methods. An example of an indirect approach has already been discussed in reference to Equations (6.6) and (6.7). The greater the load drop definition for failure the greater the crack growth. The number of cycles to get to this load drop can be correlated to an average crack growth and this could be then used to predict the fatigue life [Solomon, et al., 1990, 1991c]. In effect, an average crack growth rate was determined and used to determine the fatigue life of a castellated solder joint between a leadless chip carrier and a PWB [Solomon et al., 1990]. This approach allowed different load drops to be considered and accounted for the joint geometry. Guo and co-workers [1991b; 1992a] have also used load drop data to estimate the crack growth rate.

Figure 6-11 shows the load drop data used by Guo et al. [1991a,b; 1992a]. The square root of the cracked area, determined from the load drop, was used for the crack length, a, and the crack growth rate determined from the rate of the load drop (in cycles). This crack growth rate was then related to the Paris law [Paris and Edrogan, 1963] expression:

$$da/dn = C\Delta K_{II}{}^{n} = C\,(\tau\sqrt{a})^{n} \tag{6.21}$$

where K_{II} is the linear elastic mode II stress intensity range [Liu, 1985], which is related to the shear stress, and the square root of the crack length. The number of cycles to failure can be determined by integrating this expression, i.e., as:

$$N_f = \int_{a_o}^{a_f} \frac{da}{C(\tau\sqrt{a})^n} \tag{6.22}$$

where a_o is the initial crack length and a_f is the final crack length defining failure. If the stress range is constant throughout the cycling, and the exponent remains constant, the integration is straightforward.

Figure 6-58 shows Guo's correlation of the crack growth rate and the stress intensity range, for different load drops. The steepness of the slope should be kept in mind for a future discussion.

Li et al. [Subrahamayan, et al., 1989a,b; Li, et al. 1990; Wilcox, et al., 1990] also used load drop data to estimate the crack growth rate and correlated it with the stress intensity. They used the data of Figure 6-10a (and load drop data gathered at other temperatures [Solomon, 1989a]) to estimate the crack length and cyclic crack growth rate. Figure 6-59 shows their correlation with the stress intensity. Once again the slope of the curve becomes very large at higher stresses. Furthermore, the exponent is not constant; it increases as the stress intensity increases.

Instead of working with the cyclic growth rate, Li et al. used the time based form of Equation (6.21). The cycle number and cycle time are simply related by the cycle frequency, υ, i.e., $t = N/\upsilon$, so da/dN can be easily converted to da/dt. They integrated to get:

$$\int_{a_o}^{a_f} \frac{da}{(\sqrt{a})^n} = \int_0^{\frac{1}{\upsilon}} \upsilon_o e^{-\frac{Q}{RT}} \tau(t)^n dt \tag{6.23}$$

where υ_o is a constant, Q_c is an activation energy, R the gas constant (the constant and exponential expression are the constant, C, in Equation (6.21), and define its temperature change by a thermally activated process). The shear stress is given as a function of time through the cycle. Equation (6.23) uses the integral over one cycle where the time for the cycle is $1/\upsilon$. This integral is termed the "Damage Integral" and defines the damage (crack growth) over one cycle. This single cycle damage times the number of cycles to failure defines the total damage. The value of Equation (6.22) lies in the fact that N_f can now be determined by evaluating two

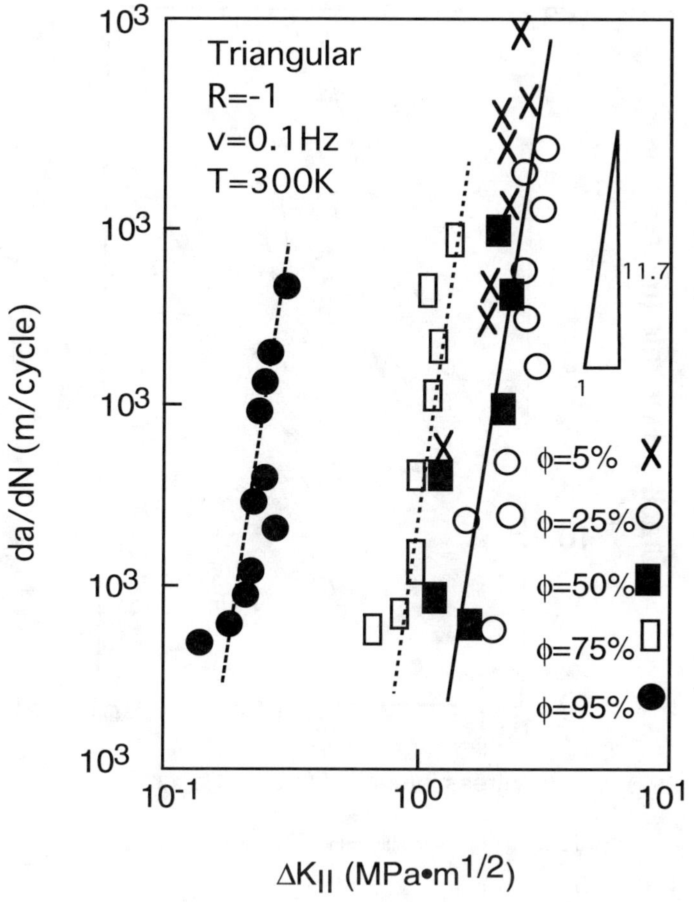

FIGURE 6-58 Crack growth rate as function of the mode II stress intensity, for 60Sn-40Pb at different load drops. [Guo et al.,1992a]

integrals. The value of one integral depends only upon the initial crack size and the final crack size, which are geometric variables. The value of the second integral depends upon the stress and temperature, which are cycle variables. This second integral can thus be determined from a knowledge of how the stress and temperature vary with time, for any loading cycle. Equation (6.23) differs from Equation (6.22) in that with Equation (6.22) the stress range is used, which is usually assumed to be a constant, whereas in Equation (6.23) the stress is assumed to vary with time. The damage (crack growth) is produced as the stress rises throughout each cycle. Li et al. were successful in utilizing this approach even though the separation of variables required that the exponent, n, be a constant, i.e., not a func-

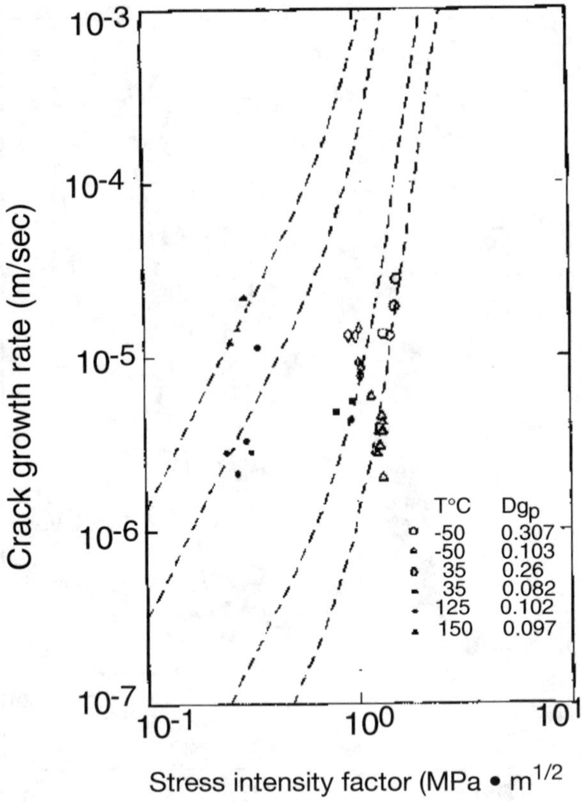

FIGURE 6-59 Calculated crack growth rate from the damage integral model. [Li et al.,1991]

tion of the stress and crack length. Unfortunately, their growth rate curve (Figure 6-59) showed a non-constant n, so this is true only if the stress intensity does not change much.

Both Figures 6-58 and 6-59 showed n to be considerably larger than the typical values of 2-4. The reason for this discrepancy may lie in the fact that the stress intensity is a linear elastic quantity which should generally not be applied when gross plastic strains are present. This problem was acknowledged by others [Guo, et al., 1991b, 1992a; Subrahamayan, et al., 1989a,b; Li et al., 1991; Wilcox et al., 1990]. The stress intensity was used because it appeared to correlate their data, even though theoretically it should not necessarily do so. This apparent correlation may, however, be illusory; this is illustrated in Figure 6-60 [Solomon, 1972c]. These curves were generated on 1018 steel which was cycled under total strain control, with plastic strain limits, and at stress levels which produced gross yielding. The plastic strain ranges that were used are listed in the figures. Figure 6-60a

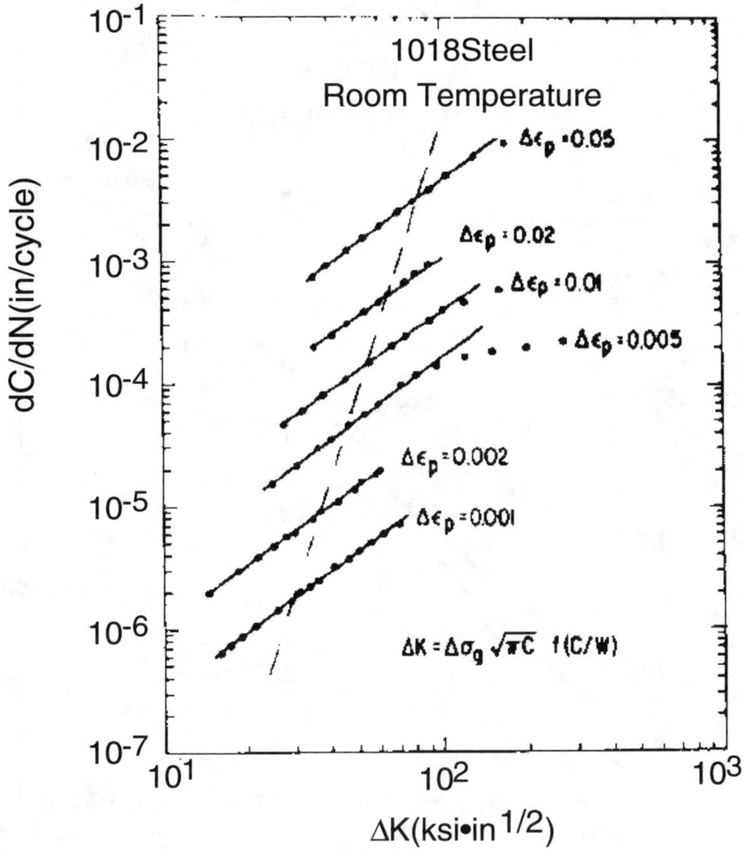

FIGURE 6-60a Crack growth rate as a function of the linear elastic stress intensity range, for 1010 steel tested at room temperature. [Solomon,1972c]

plots the measured crack growth rate vs. the stress intensity which was calculated from the measured stresses and crack lengths (which were measured optically throughout the test). For any experiment, where the plastic strain limits were kept constant, there is a good correlation with the stress intensity. However, when a different plastic strain range is used, a different curve is generated. The crack growth rate is NOT a single valued function of the stress intensity; it also depends upon the plastic strain range. The dotted line of Figure 6-58a represents the stress intensity correlation for all the data, which has a large slope. The correlation, depicted by the dotted line, is not valid.

This data was gathered over a wide enough range of plastic strains and crack growth rates at each plastic strain range to clearly show that this is not the correct

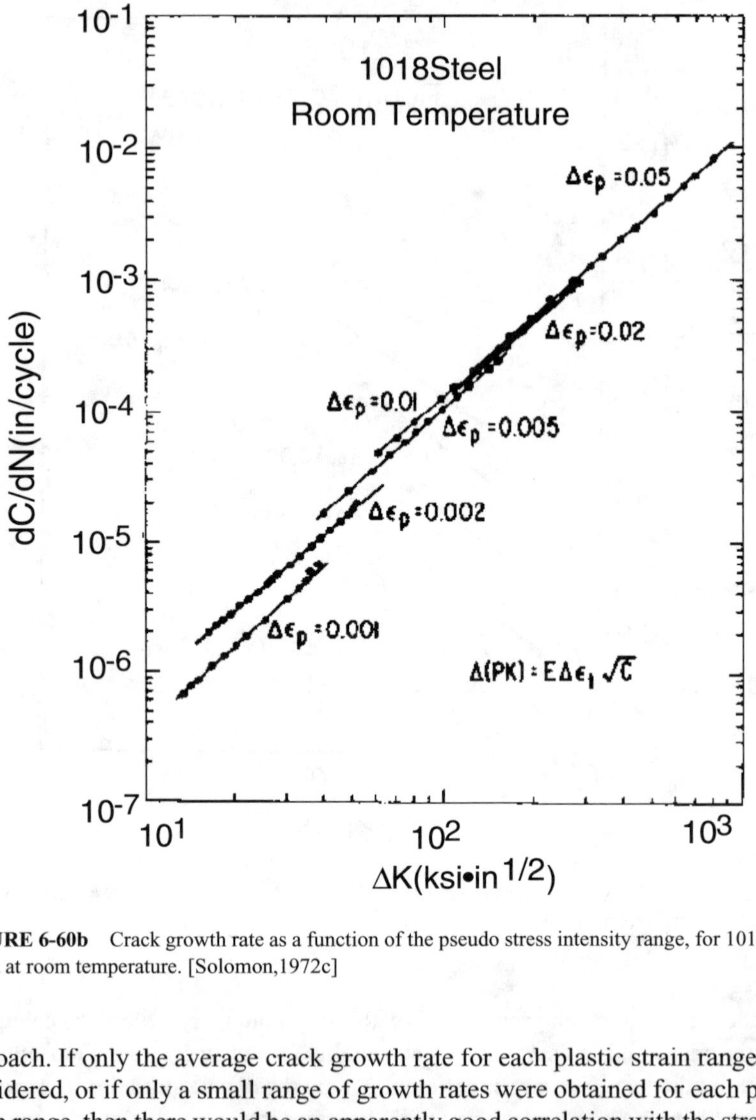

FIGURE 6-60b Crack growth rate as a function of the pseudo stress intensity range, for 1010 steel tested at room temperature. [Solomon,1972c]

approach. If only the average crack growth rate for each plastic strain range were considered, or if only a small range of growth rates were obtained for each plastic strain range, then there would be an apparently good correlation with the stress intensity. Furthermore, this erroneous correlation would yield an erroneously large exponent, n, in Equation (6.22).

Figure 6-60b shows that the proper correlating parameter is a pseudo stress intensity factor, i.e., the pseudo stress instead of the real stress is used. The pseudostress is just the total strain times the elastic modulus. The pseudo stress is often used in problems where a strain such as the thermal strain, rather than stress, is calculated. Since most of the strain is plastic, in effect this becomes a strain intensity approach. If the loading were entirely elastic, the total strain would be the elastic strain and the pseudo stress would be the real stress, and the pseudo-stress intensity factor would reduce to the real stress intensity factor. Figure 6-60b shows that the crack growth rate is a single valued function of the pseudo stress intensity factor.

Figure 6-60 is included to show that elastic fracture mechanics should not be used in situations where plastic strains predominate. Currently the most often used approach to fracture mechanics when plastic strain predominates is not to use the pseudo stress, but the J integral [Dowling and Begley, 1976; Dowling, 1977; 1979].

The J integral is a path-independent integral which describes the energy in the region ahead of the crack [Rice, 1968]. It is really not strictly applicable for fatigue because it is not defined for unloading. Nonetheless, it is often used to describe fatigue data [Dowling and Begley, 1976; Dowling, 1977; 1979a; Liaw et al., 1989; Legson et al. 1990; Guo and Conrad; 1992; Yamada, 1987; Pao et al., 1992; Pao and Pan, 1990; Raman and Reily, 1982; Sklenicka, 1990]. Guo et al. [1992a] have used this approach for the previously cited data obtained on 60Sn-40Pb solder (see Figure 6-58 for the stress intensity correlation using this data). The correlation using the J integral is shown in Figure 6-61a. Figure 6-61b shows the correlation using a slightly different energy calculation. Both results are similar, with similar low slopes. This is to be expected because a more realistic parameter that takes into account non-elastic loading. Guo et al.[1992a] obtained similar results using 63Sn-37Pb solder.

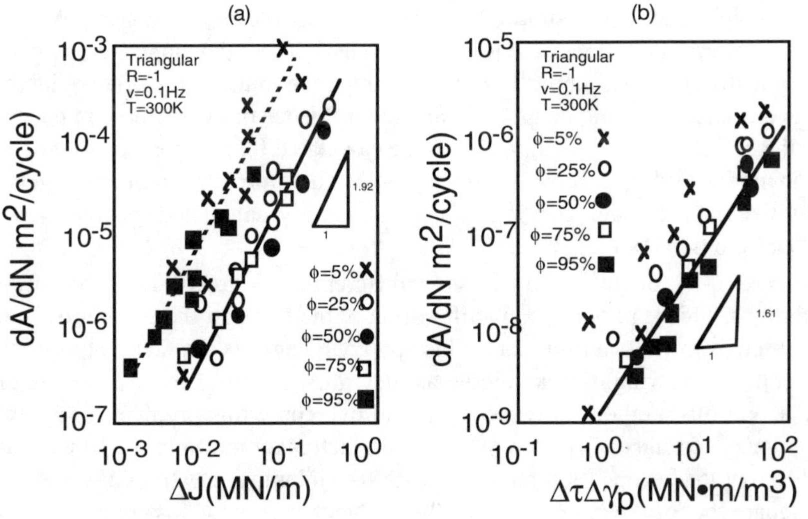

FIGURE 6-61 (a) Crack growth rate as a function of the J integral range (for shear loading), for 60Sn-40Pb solder defined by different load drops and (b) crack growth rate as a function of the hysteresis energy, for 60Sn-40Pb solder defined by different load drops. [Guo et al., 1992a].

Both the stress intensity and J integral are functions of the crack length. Thus as the crack grows longer it grows faster (providing that the stress or strain is constant). If the stress or strain field into which the crack grows is not constant then the integration of the crack growth expression can be very complex. This is the case for most real solder joints, especially castellated ones. Sometimes it is necessary to solve the problem incrementally. The number of cycles to grow a crack a short distance is determined. The stress or strain field is re-determined for the new crack, the growth rate determined for the new crack size and stress or strain; the number of cycles to grow the crack another increment is then determined. This process is continued until the final crack size is reached.

A special simplifying situation existed with regard to the approach discussed with Equations (6.6) and (6.7). The data used with these equations were obtained with tests run in simple shear, under boundary conditions where the J integral is just a function of the applied strain and not also of the crack length [Solomon, 1986a]. In this case, when the plastic strain is kept constant the crack growth rate does not increase as the crack length increases. This simplifies the application of crack growth data. This simplified approach was successfully applied to estimating the fatigue life in a castellated leadless solder joint [Solomon, et al., 1990a; Solomon, 1991c]. While the load drop data, from which the crack growth rate was inferred, were obtained in simple shear under boundary conditions where J did not depend upon the crack length, it was successfully applied to a situation where the stress state was much more complex. The von Mises effective strain approach was used to account for the stress state variations. Initially crack growth occurred under the chip carrier where simple shear was a reasonable approximation. When the crack grew into the fillet of the joint, more tensile stress developed. However, by this time the crack was about 1/3 to 1/2 through the joint, so the relative increase in the crack length could be ignored, and the influence of the increase in the crack length neglected. The growth rate could be calculated for any position in the strain field without having to know the crack length. Therefore, the number of cycles to grow the crack through this region could be easily calculated [Solomon et al., 1990a; Solomon, 1991c].

Figures 6-58, 6-61a, and 6-61b show different crack growth rates for different load drops. This is probably a manifestation of the fact that the crack growth rate was inferred from load drop data. This approach neglects the fact that more than one crack is growing. Cracks merge as they grow and the cracks have irregular fronts. Various methods have been used to overcome this problem. In applying Equations (6.6) and (6.7) different correction factors for multiple cracks were used at different load drops [Solomon et al., 1990a; Solomon, 1991c]. Guo et al. used the square root of the cracked area [1991b]. Neither approach is entirely satisfactory. What is needed is crack growth data taken on specimens with a single growing crack with a well defined crack front. Such tests have recently been performed [Liaw et al., 1989; Logson et al., 1990].

Figure 6-62 shows the fatigue crack growth data of Logsdon et al. [1990] They used conventional 1T compact tension specimens made of 63Sn-37Pb. This figure shows that with limited plasticity the crack growth behavior followed the conventional Paris law behavior with an exponent of about 4. The crack growth was conducted at low enough stress intensities to develop threshold behavior (the rapid reduction in the crack growth rate at very low stress intensities).

Figure 6-63 compares the data of Figures 6-62 and 6-61a. Guo and Conrad [1992] re-plotted Logsdon et al.'s data in terms of J. This data could be easily converted, from the standard correlation of a linear elastic K and J, i.e. $K^2 = J/E$ where E is the elastic modulus. This could be done because Logsdon et.al.'s specimen was reasonably linear elastic. The specimens used by Guo and Conrad were fully plastic and K is not applicable in this case. It is not surprising that when Guo and Conrad compared their data with that of Logsdon et al., on the basis of K, there was a discrepancy of about three orders of magnitude. On the basis of J there is a much more modest difference between the Mode I (tension) data of Logsdon et al. and the Mode II (shear) data of Guo and Conrad. All of this difference cannot be ascribed to the difference in the deformation mode. Guo and Conrad inferred the crack growth rate from load drop measurements in specimens with irregular multiple cracks, whereas Logsdon et al. had a single, relatively straight, crack.

Figure 6-62 shows that reducing the temperature reduced the crack growth rate and thus increases the fatigue life, in agreement with the data of Figure 6-14. Decreasing the cycling frequency from 10 Hz to 2 Hz had no influence on the crack

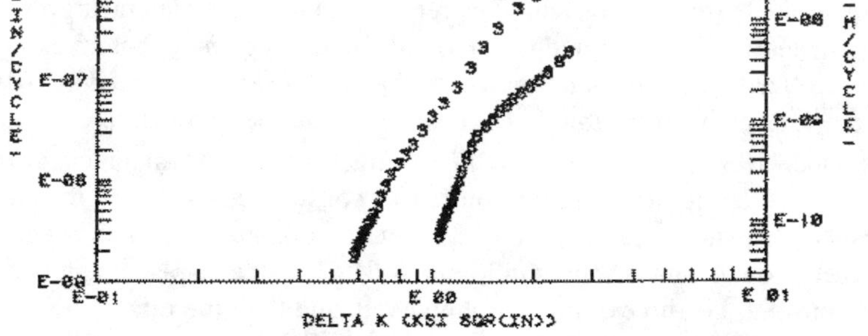

FIGURE 6-62 Fatigue crack growth rate as a function of the linear elastic stress intensity, measured on compact tension specimens of eutectic (63Sn-37Pb) solder. Reprinted from Logsdon et al., *Engineering Fracture Mechanics,* ©1990, Vol. 36, p. 183, with kind permission from Pergamon Press Ltd., Headington Hill Hall, Oxford OX3 OBW, U.K.

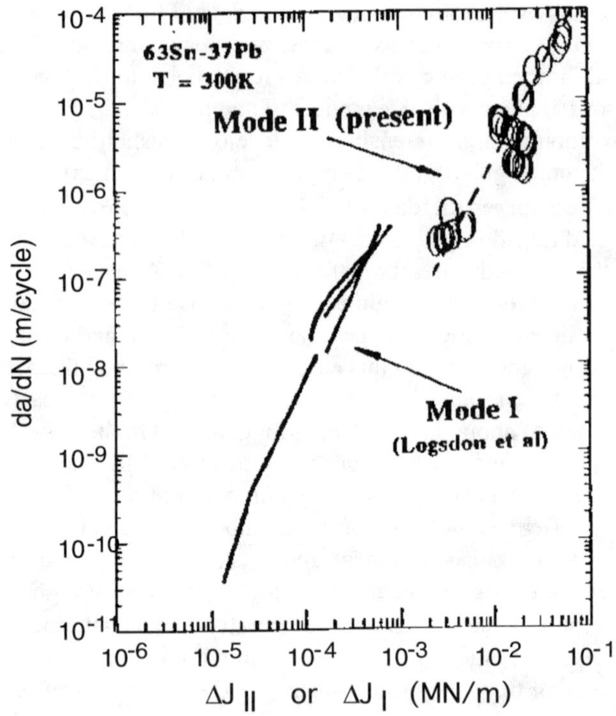

FIGURE 6-63 Fatigue crack growth for mode I and mode II loading as correlated by the J integral range. [Guo and Conrad.,1992)]; the mode I data reprinted from Logsdon et al., *Engineering Fracture Mechanics,* ©1990, Vol. 36, p. 183, with kind permission from Pergamon Press Ltd., Headington Hill Hall, Oxford OX3 OBW, U.K.

growth rate. This is in general agreement with the results of Figures 6-22 and 6-23. Lower cycle frequencies would be required to increase the crack growth rate. Hold time tests were attempted to try to reduce the frequency, but crack blunting and generalized creep deformation was observed. More work is needed to develop good crack growth data at low frequencies and at higher temperatures.

Yamada [1987] was successful in getting time-dependent crack growth in 60Sn-40Pb solder joints. Here the solder was constrained as a joint, in contrast to the tests of Logsdon, described above, where a bulk specimen was used. Yamada's crack growth data is shown in Figure 6-64. Here the displacement was maintained constant, i.e., no cycling was employed, and the time rate of crack growth was determined as a function of the strain energy release rate.

Pao [1990; 1992; Pao, et al., 1992] has recently approached the problem of applying the crack growth rate approach to predicting the fatigue life of solder joints by breaking the crack growth rate into time-dependent and time-independent

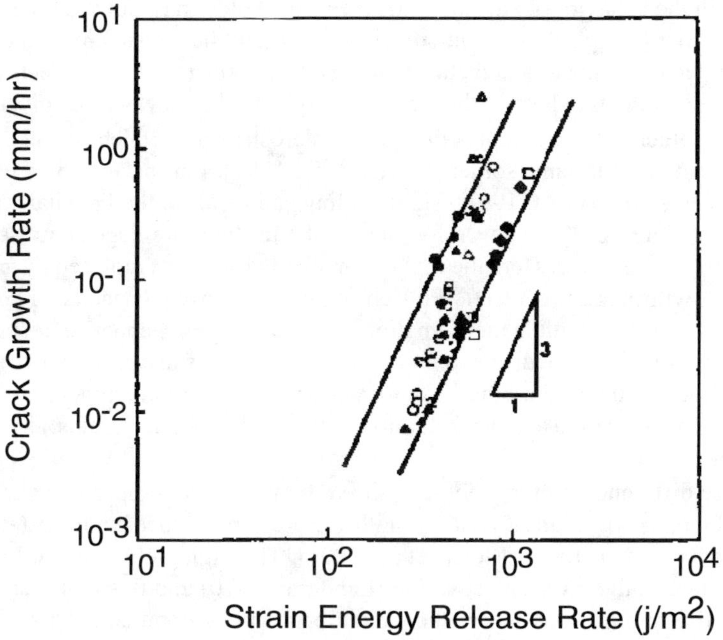

FIGURE 6-64 Correlation between the strain energy release rate and the time rate of crack growth, for solder joints tested under constant displacement. [Yamada, 1987]

terms. The time-independent term was just the J integral crack growth rate expression. The time-dependent crack growth was associated with creep crack growth [Raman and Reiley, 1987; Sklenicka, 1990; Barker and Weertman, 1990]. The C* integral was used to define the creep crack growth term. The C* integral is the viscous analog of the J integral [Reidel and Rice, 1990; Reidel, 1981; 1987; 1989]. This approach was used to determine the life of joints subjected to thermal cycles. The shear strain rate during the thermal cycle and hold time was calculated and used to determine the shear stress which, in turn, was used to determine the crack growth rate. Failure was defined when a critical value of the J integral was reached or when the crack was calculated to grow to half the joint thickness. As will be shown in the section on thermal cycling, a good correlation was achieved with this approach and literature thermal fatigue data.

6.9 INFLUENCE OF COMPOSITION AND MICROSTRUCTURE

All of the above discussion has treated solder as a continuum with little or no regard for the solder microstructure or for small variations in the composition (although the behavior of eutectic or near eutectic solder has, however, been compared to that of high Pb content solders). There are, however, important microstructural and compositional factors that must be considered. These effects are discussed at length elsewhere in this book so only a brief review will be given here.

Most solders have a complex two phase microstructure, which can take on different forms for the same solder [Morris et al., 1991]. For instance, in a eutectic, 63Sn-37Pb solder [Frost, 1991], slow cooling can result in the lamellar structure shown in Figure 6-65a. Slower cooling results in the more degenerate structure shown in Figure 6-65b. Here the regular lamellar structure is replaced by an irregular one, with large proeutectic Pb rich regions, and irregular eutectic colonies. Rapid cooling can result in the formation of a fine "salt and pepper" eutectic structure, consisting of a fine dispersion of Pb rich precipitates in a Sn rich matrix. This is shown in Figure 6-65c. Finally, deformation or just time at temperature can result in recrystallization and the formation of Pb rich and Sn rich grains, as is shown in Figure 6-65d.

These different structures possess different mechanical properties. The lamellar structure has the highest yield strength and creep resistance but lowest ductility and strain rate sensitivity [Morris et al., 1991]. The high yield strength increases the constant in the Basquine equation [Landgraf, 1970] and this will increase the fatigue life in situations where the elastic behavior is dominant. This effect is, however, modified by the strain rate sensitivity. The lower strain rate sensitivity

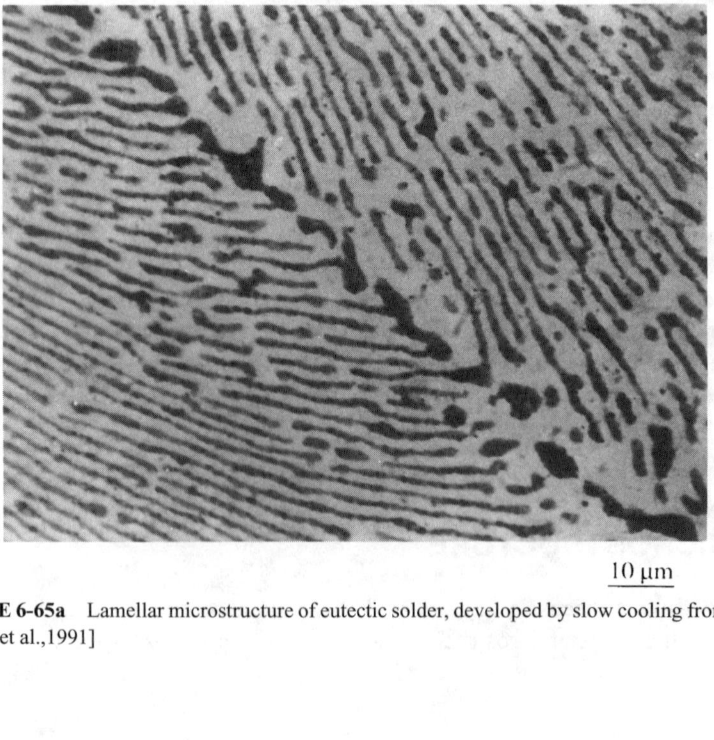

FIGURE 6-65a Lamellar microstructure of eutectic solder, developed by slow cooling from 250°C. [Morris et al.,1991]

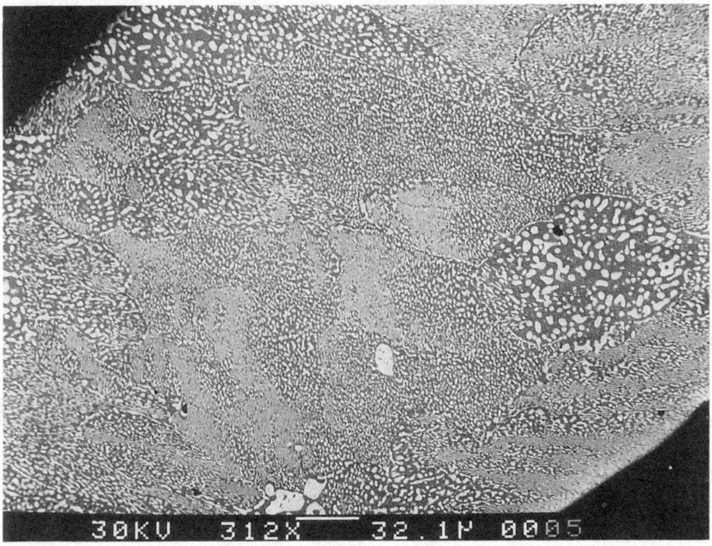

FIGURE 6-65b Degenerate lamellar microstructure in a "real" solder joint. [Morris et al.,1991]

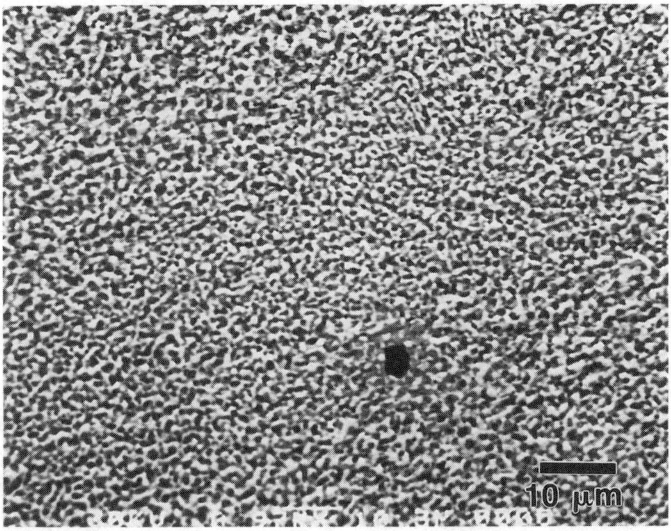

FIGURE 6-65c Fine eutectic microstructure developed by water quenching form 250°C. [Morris et al.,1991]

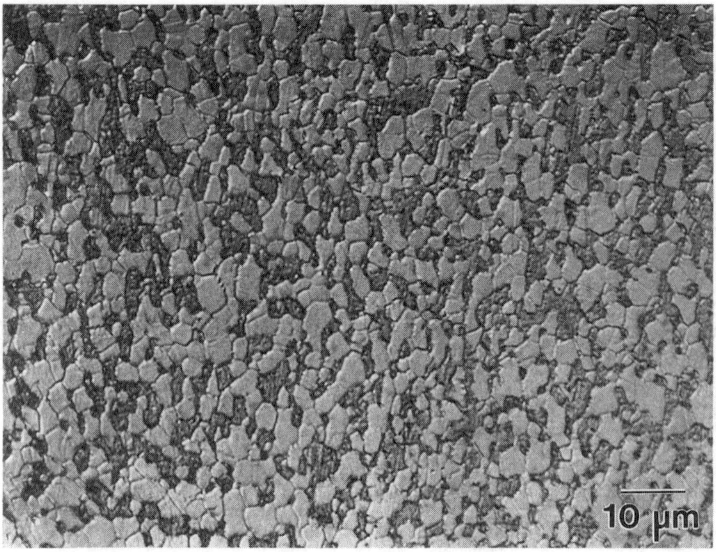

FIGURE 6-65d Worked and annealed 60Sn-40Pb solder showing a recrystallized microstructure. [Morris et al.,1991]

does not make the strain rate increase as much when the strain rate is increased, and this may actually decrease the strength of this structure relative to that of one with a greater strain rate sensitivity. The lower ductility reduces the fatigue ductility coefficient [Landgraf, 1970] which reduces the fatigue life when the plastic strain behavior is dominant. The superior creep resistance should improve the fatigue life when long hold times are applied (the amount of plastic strain developed will be less).

The degenerate structure is softer because of the relatively weak intercolony material. The recrystallized structure can be superplastic, allowing for elongations of hundreds of percent, with a possible attendant improvement in the low cycle fatigue behavior. Mei and Morris [1992] and Wild [1972] have shown that increasing the cooling rate to develop the structure of the sort shown in Figure 6-65c, as expected, improved the isothermal fatigue life of type 60Sn-40Pb solder specimens. With relatively high strains of 8.8 and 16.3% (where plastic strain dominates) the fatigue life of quenched specimens, compared to that of furnace cooled specimens, was increased by over a factor of 3 at 65°C. The effect was somewhat less for tests run at 20°C: only a factor of 2 was observed for the 8.8% strain tests with no difference for the 16.3% strain tests. Air cooled specimens generally had intermediate lives. It is not yet known if this behavior will also be seen under other isothermal conditions or in thermal fatigue (where there is a tendency to coarsen the structure).

The difficulty in applying all of the observations regarding the influence of the solder microstructure lies in the fact that the structure of most solders is unstable. The high homologous temperature means that grain growth and phase coarsening can and does take place at room temperature. Changes are observed even hours after solidification [Chilton et al., 1989]. Figure 6-50c shows that room temperature aging can alter the fatigue life—in the case of the data of Figure 6-50c, by reducing the flow stress, presumably by coarsening the microstructure.

The question arises as to what structure correlates with the fatigue behavior, the as-solidified structure, the structure when a test specimen or component is tested or put into service, that which exists during the test or in service, that which exists at failure, or that which is present when the metallography is performed. Unfortunately there is no systematic, quantitative study which correlates the fatigue behavior with all of these structures.

In some cases the change of the structure with thermal cycling is intimately related to the fatigue behavior. Morris and his co-workers [Morris et al., 1991; Frear et al., 1987; 1988; Tribula et al., 1989] and others [Bangs and Beal, 1978; Wolverton, 1987] have shown that in thermal fatigue of eutectic solders and some high Pb solders, a coarsened band develops and that the fatigue failure occurs in this coarsened region. This is illustrated in Figure 6-66. Shear deformation occurs in the region of the joint that is adjacent to the intermetallic layer that exists between the solder and substrate (perhaps because this region is Pb rich owing to the formation of Sn rich intermetallic). This deformation causes coarsening. This coarsened band is softer than the rest of the solder, causing more of the deformation to be concentrated in this region. This localized deformation leads to crack initiation and ultimately to failure. It has been found that the addition of Cd or In to the solder tended to cause relatively uniform deformation and the prevention of the formation of this coarsened band, thereby improving the thermal fatigue resistance [Morris et al., 1991]. In high Pb solders it is possible to heat the solder into a single phase region during a thermal cycle and if the frequency is low enough all of the Sn rich second phase can dissolve. Upon cooling, this phase can reprecipitate [Frost, 1991]. This should have a profound effect on the fatigue life, which could be studied.

The literature contains numerous observations about the relative fatigue resistance of different solders [Wild, 1972; 1988; Solomon, 1991b; Stone, 1990; Jarboe, 1980; Samash, 1968; Rathore et al., 1973; Denlinger et al., 1977; Inoue et al., 1986; Dunn 1978; Rack and Maurin, 1974; Pearson, 1988; Miyazaki, 1980; Lau and Rice, 1985; Stephens and Bourcier, 1986; Coombs, 1972; Olsen and Berg, 1975; Korb and Ross, 1983; Pankratz and Plumlee, 1970; Kong et al., 1977; Zommer et al., 1976; Levine and Ordonez, 1977; Desaulnier, 1986; Goldman et al., 1977; Devore, 1980; Devore and Tillet, 1979; 1981; Howard, 1982; Marshall and Wolter, 1987; Marshall, 1991; Tanak, et al., 1987; Summers and Morris, Guo et al., 1992c; Kinser et al., 1976]. However, care must be taken when assessing some of this data. Tests run at high frequencies, in load or displacement control, will

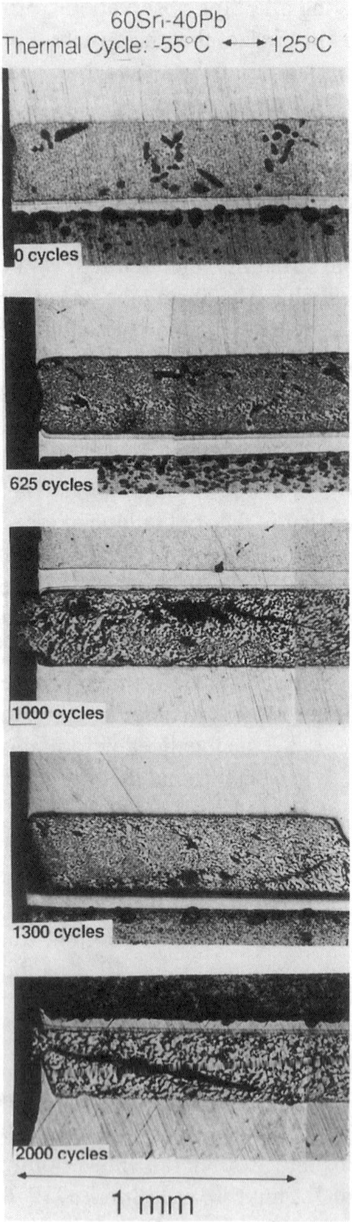

FIGURE 6-66 A series of micrographs showing the evolution of the microstructure developed in a eutectic solder joint cycled between –55°C and 125°C. The top micrograph shows the initial structure. A coarsened band develops with cycling. The structure developed with successive cycling is shown, with the cycling increasing top to bottom. [Morris et al.,1991]

tend to show that the higher yield strength solders have the longest fatigue lives, whereas tests run with plastic strain limits and large plastic strains generally show the opposite (because the ductility is the controlling material parameter). The size of the specimen, whether it is a bulk specimen or a solder joint and the definition used for failure all influence the results. The use of a large load drop definition for failure or of a resistivity increase all lengthen the apparent fatigue life [Solomon, et al. 1990a]. Those who model the behavior of solder joints should fully understand the nature and limitations of the data that they use, lest they create erroneous models. This is especially true when comparing data from different sources. The best comparisons utilize the same test specimen design and test protocols. For instance, Figures 6-67a and b compare the fatigue resistance of 60Sn-40Pb, In151 (92.5Pb-5Sn-2.Ag) and Sn96 (96.5Sn-3.5Ag) for tests in simple shear, at 35°C and 150°C, with a 50% load drop definition of failure [Solomon, 1991b]. As can be seen, both the fatigue ductility exponent and fatigue ductility coefficient differ for different solders. Thus, a solder may be superior at one strain level and not at another, or at one temperature and not another.

Figure 6-68a [Tanaka et al., 1987] compares the results of torsional fatigue tests run at room temperature, for a wide range of solders. The fatigue life is shown as a function of the total strain. There is less than an order of magnitude spread in the data. If the pure Sn and Pb curves are disregarded, the spread is even less. This observation is in general agreement with the law of universal slopes

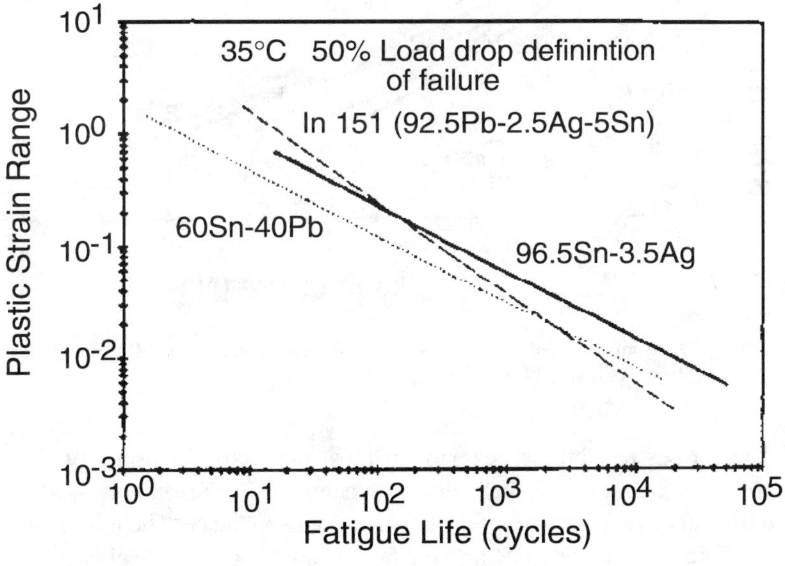

FIGURE 6-67a Comparison of the 35°C fatigue life behavior of 96Sn-4Ag, 60Sn-40Pb, and 92.5Pb-5.0Sn-2.5Ag. [Solomon,1991b]

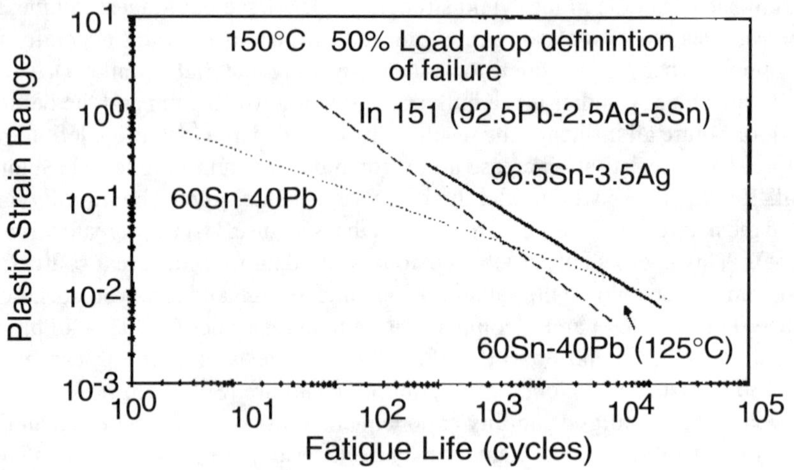

FIGURE 6-67b Comparison of the 150°C fatigue life behavior of 96Sn-4Ag, 60Sn-40Pb, and 92.5Pb-5.0Sn-2.5Ag. [Solomon,1991b]

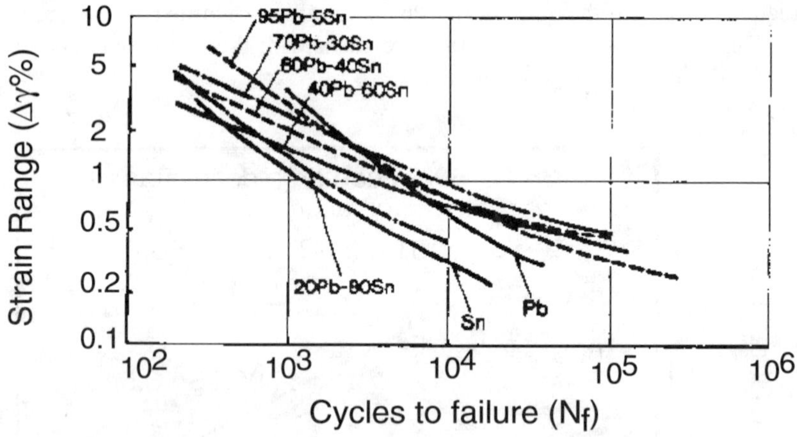

FIGURE 6-68a Comparison of the fatigue behavior of several solders tested at room temperature and a cycle frequency of 0.1Hz. [Tanaka et al.,1987]

which describes the fatigue behavior of most materials [Manson, 1975; 1985]. When the plastic strain life behavior dominates (as it does in solders), the total strain life curve is approximately the Coffin-Manson curve. The fatigue ductility exponent is approximately 0.6, and the fatigue ductility coefficient is equal to the ductility to the 0.6 power. When the plastic strain is smaller than the elastic strain, the Basquine equation defines the total strain vs. fatigue life curve. The exponent of the Basquine is approximately 0.12 and the coefficient is proportional to 3.5 times the ultimate tensile strength divided by the elastic modulus. Thus, the plastic part of the curve scales with the ductility and the elastic part with the ultimate tensile strength. These parameters are close enough for the solders tested (especially the ductility) that the fatigue curves do not vary by more than an order of magnitude. The lack of a variation in the parameters of the Coffin-Manson expression is verified by the data of Guo et al. [1992c], although they found the exponent to be only about 0.3 to 0.4.

The law of universal slopes describes time-independent behavior, so for solders it is true only when rapid cycling is employed. For all the data of Figure 6-68a [Tanaka et al., 1987] the cycle frequency was 1.0Hz. The data of Guo et al. [1992c], cited above, was generated at 0.1HZ. At low cycle frequencies (10^{-3} to 10^{-4} Hz at room temperature) it is expected that more of a divergence in the data would be observed. Such data does exist for eutectic and near eutectic solders, and some high Pb solders, but such low frequency data is sorely needed for the other solders shown in Figure 6-68a. The data of Figure 6-68a can be replotted vs. Pb content, as shown in Figure 6-68b [Tanaka et al., 1987]. For this isothermal data, there is a peak in the fatigue life for solders with about 75–100 Pb.

Guo et al. [1992a] have correlated their load drop determined crack growth rates with the volume fraction of Sn rich phase that was present. With a volume fraction of 0.75 (which was generated in eutectic 63Sn-37Pb solder) the crack growth exponent and pre-exponential coefficient were both a minimum, leading to a reduced crack growth rate and thus an increased fatigue life. At higher and lower volume fractions of Sn phase the crack growth rates were higher.

This result is similar to that of Summers and Morris who also found a maximum in the fatigue life as a function of the Sn content. However, they found the maximum in the fatigue life to occur at 50Sn-50Pb and found a much smaller effect (there was no significant difference between solders with 5%, 20% or 40% Sn (bal. Pb) and only about a factor of 2 increase with the 50% Sn, compared to these. The eutectic 63Sn/37Pb had the shortest life (about a factor of 2–10 less than the 50Sn-50Pb. Inoue, et al. [1986] also found 50Sn-50Pb to have the best fatigue life, this time for thermal cycling tests. Summers and Morris ascribed the superiority of the 50Sn-50Pb to the development of more uniform deformation. The 40Sn-60Pb solder had, however, an even more uniform deformation, but a shorter fatigue life. The differences in the behavior of these two solders was ascribed to different influences (due to differences in the Sn content in the pro-eutectic) of the pro-eutectic Pb phase which was present in both. Guo et al. also ascribed their results to the effect of proeutectic Pb phase or Sn phase (depending upon which side of the eutectic the solder is on). However, it is believed that the presence of the pro-eutectic increased the crack growth rate, which decreases the fatigue life. Summers and Morris felt that pro-eutectic was beneficial, in that it appeared to retard the development of the coarsened band. Clearly further work in this area is required.

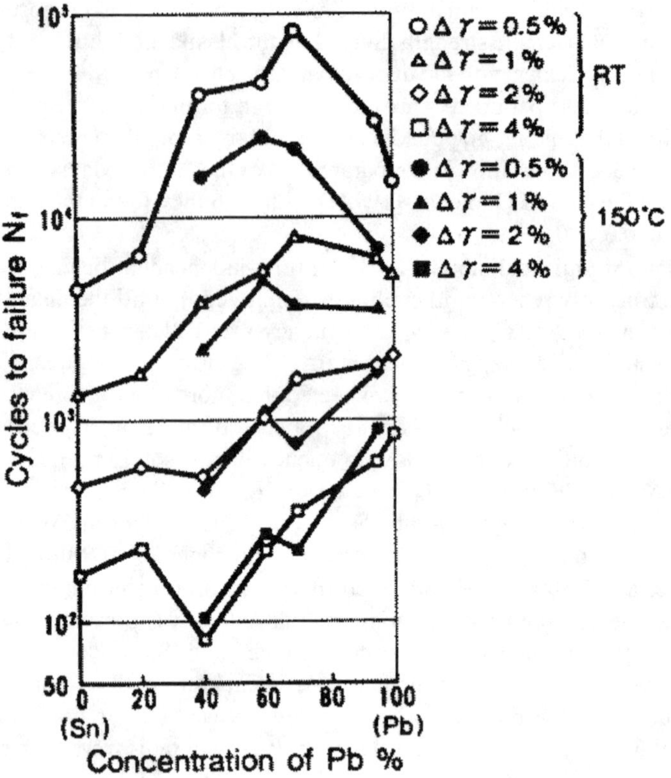

FIGURE 6-68b Compositional variation of the fatigue life (repeat of the data from the previous figure) [Tanaka et al.,1987]

The work of Summers and Morris and that of Guo et al. are at some variance with that of Tanaka et al. [1987], shown in Figure 6-68b. The data of Summers and Morris, and Tanaka et.al., and also Inoue et al. [1986] all agree, however, that the difference between the fatigue lives is at most about an order of magnitude. Given the differences in experimental procedures, it is not surprising that there would be differences between the behavior of individual solders. Guo et al. made the correlation with load drop calculated crack growth data, so even a larger difference in the specific results is not surprising.

6.10 ISOTHERMAL, THERMAL, AND THERMOMECHANICAL FATIGUE

There was a general consensus at the Workshop on the Mechanics of Solder Alloy Interconnects that isothermal, mechanical, fatigue tests could be used to predict the fatigue life due to thermal cycling, providing there is not a significant change in the deformation mechanism over the temperature range of the thermal cycling and that of the isothermal temperature used for the prediction. Given the critical importance of correlating these isothermal models with actual thermal fatigue data, it is surprising that there are relatively few studies which support the use of isothermal models for thermal fatigue, although there are virtually none which do not support it. Figures 6-54 to 6-56 show agreement to within about a factor of 2–3 between the Engelmaier Reliability Model and thermal cycle data. An even better correlation could have been achieved with a different value for the F parameter. Figure 6-69 shows data for which the support is not quite as good. The Engelmaier Reliability Model version of the Coffin-Manson relationship overestimated the thermal cycle results by about a factor of 5. The relatively poor agreement with the Engelmaier Reliability Model approach can probably be explained by the fact that the temperature cycle range and cycle frequency variations exceeded the values used in developing the model. As such, the model overestimates the fatigue life. An excellent correlation was obtained using the time-independent and time dependent crack growth expression, with the time-dependent crack growth rate being described by the C* integral [Reidel and Rice, 1980; Reidel, 1987; 1989].

Real thermal cycles often involve different types of loading. For instance, there may be a cycle due to daily temperature changes and those due to on-off cycles of

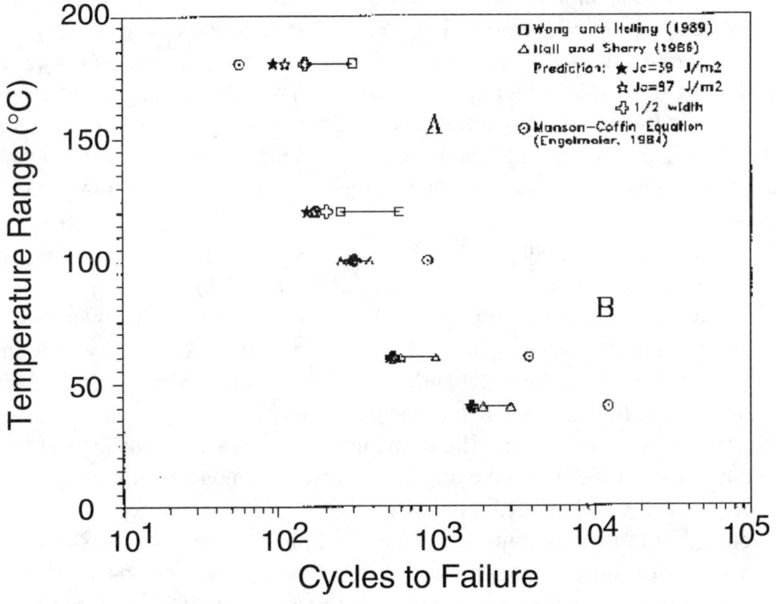

FIGURE 6-69 Comparison between experimental and predicted (from the Engelmaier reliability model). [Pao,1992]

the device. The daily temperature changes may vary with the season of the year. These factors are of particular importance in automotive devices or for military applications. One commonly used isothermal approach is to use Palmgren-Miner's rule [Palmer, 1924; Miner, 1945] which is simple linear damage summation given by:

$$\sum_{i} \frac{n_i}{N_{fi}} = K \tag{6.24}$$

where n_i is the number of cycles applied for the ith condition and N_{fi} is the number of cycles to failure for this condition, if it were applied separately and K is a constant. Failure occurs when the sum of the damage ratios equals K. The simplest case, $K = 1$, denotes no interaction. In general, $K<1$, denotes some interaction and a reduction in the life, compared to what it would have been had there been no interaction. Overloads can, however, produce compressive stresses which actually lengthen the fatigue life, i.e., make $K>1$. This approach can also be made more complex by considering nonlinear summations, where the damage sum is raised to a power [Richart and Newmark, 1948; Marko and Starkey, 1954], or other more complex approaches [Landgraf, 1973]. The time- and cycle-dependent behavior can be handled separately, as is done in Equation (6.11). As can be surmised complex cycling leads to a very complex analysis.

Multiple cycles can also be handled by crack growth approaches. Here, the influence of each cycle on the crack length must be considered sequentially and the crack followed to failure. Care must be taken, however, to properly account for the effect of overloads on the crack growth rate. Thermal cycling can also be used to simulate multiple cycles, but this is generally not done because of the complexity of the cycling. This would require one test to incorporate variations in the temperature range, maximum or minimum temperature, temperature ramp rate, hold time, etc.

While the literature generally supports the use of isothermal data to describe thermal fatigue, several reservations were raised at the Workshop on the Mechanics of Solder Alloy Interconnects. High frequency isothermal data can be used to describe time-independent cracking due to the strain developed by thermal cycling. High temperature time-dependent models can describe cracking during a high temperature hold time. Difficulties may arise, however, because the models neglect to account for any specific damage that might occur because of the low temperature part of the thermal cycle. Some of this can be accounted for by using asymmetric data. Crack growth models may be able to account for the high stresses developed at low temperatures and the resulting large stress intensity of J integral that is developed. These models will not, however, account for the development of a grain coarsened zone in thermal fatigue if it has not occurred in the isothermal tests being employed. A larger coarsening effect develops in thermal fatigue than in high temperature isothermal tests. This may be due to the influence of different deformation mechanisms operating at different temperatures [Frear and Jones, 1990; Frear et al, 1991a].This difference may also be due to local stress which develops because of the difference in the coefficient of thermal expansion between the Pb-rich and Sn-rich phases. This possible mechanism is supported by the observation that the development of a coarsened band is primarily a phenomenon that develops in solders with a two phase microstructure. The surface tension also helps to drive the coarsening process in two phase solders. To minimize the surface energy each phase would like to agglomerate into a single region, and to do this each phase tries to coarsen. Deformation at low temperatures provides dislocation tangles which increases the internal energy. This helps to promote high temperature coarsening, thereby accentuating the coarsening observed in thermal fatigue tests. The high stresses that develop in the low temperature part of the thermal cycle may also help to promote this coarsening, as they can provide elastic energy for the high temperature coarsening. When the fatigue life is dominated by the time to develop this coarsened zone and initiate a crack, the isothermal data may be inaccurate. As a general rule, if possible, some thermal fatigue tests should be run to validate any isothermal modeling which is used for the design-critical joints.

In addition to isothermal and thermal fatigue tests there is a third type of testing which should be considered, the thermomechanical test [Frear et al., 1987; 1988a,b; 1989; 1991a,b; Frear, 1989, 1990a,b; Frear and Jones, 1990; Morris, et al., 1989; Tribula et al., 1989; Hacke et al., 1991a,b; Lawson et al., 1991; Guo et al., 1992b]. Here, thermal cycling is employed with an independent application of mechanical loading. The stress is not developed by thermal stresses rather, they are applied by a testing machine, while the temperature is cycled independently. These tests can be run with the stress being cycled in-phase with the temperature, i.e., with the peak stress developed at the peak temperature, or out of phase, or with the peak stress being developed at any desired part of the thermal cycle. Thermomechanical tests are very complex, but they can provide information about the process of the fatigue failure that is available with no other testing procedure.

As a general rule the thermomechanical fatigue life is less than the isothermal fatigue life for tests in the same temperature range. Lawson et.al. [1991] and Guo et al. [1992b] have shown this to be the case for 97Pb-3Sn and 63Sn-37Pb respectively, when in-phase cycling was used (i.e., when the peak stress was applied at the peak temperature). Furthermore, Lawson et al. have also shown when out of phase cycling (when the peak stress was applied at the minimum temperature) was used, the fatigue life was actually longer than that of isothermal fatigue tests run at the peak temperature. This is illustrated in Figure 6-70, which shows the lowest fatigue life at a 270 degree phase angle. An in-phase test corresponds to a 0 degree phase angle, an out of phase test to a 180 degree phase angle, and a 270 degree phase angle to a test where the peak stress is applied midway through the cooling cycle. Lawson et al. also observed a strange inversion of the frequency effect with in-phase cycling. Here, the fatigue life increased as the cycle frequency decreased.

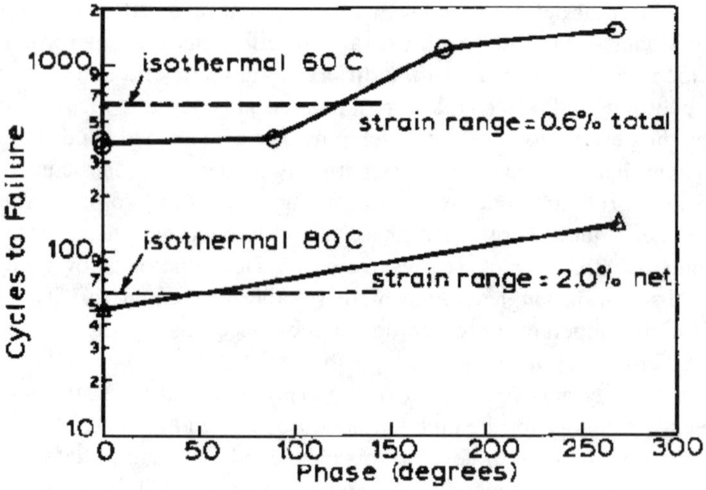

FIGURE 6-70 The effect on the phase angle between the temperature cycle and the strain cycle in a TMF test of 96.5Pb-3.5Sn solder. The upper curve is for 15°C to 60°C cycling with a 0.6% strain range. The lower curve is for 25°C to 80°C cycling with a 2% strain range and a period of 240 seconds. The dotted curves represent isothermal data. [Lawson et al.,1991]

Guo et al.[1992b], described the reduction in the fatigue life for in-phase cycling of eutectic solder in terms of a model developed by Mura and Nakasone which described crack initiation in terms of the free energy of the system. The fatigue life for in-phase cycling was determined [1992b] as a function of the fatigue life for isothermal cycling as:

$$N_{f\,TMF} = (\Delta\sigma_T - 4\,\sigma_{fT}/\sqrt{3})/(\Delta\sigma_{TMF} - 4\sigma_{fT}/\sqrt{3}) \tag{6.25}$$

where $\Delta\sigma_T$ is the stress range, 4 σ_{fT} is the flow stress for isothermal cycling at the peak temperature, and $\Delta\sigma_{TMF}$ is the stress range for the TMF cycling. Since $\Delta\sigma_{TMF}$ is greater than $\Delta\sigma_T$, the fatigue life is reduced for TMF relative to the isothermal fatigue at the peak temperature. It is interesting to speculate that if the stress range at the minimum temperature were used to describe the behavior of out of phase cycling, and if $\Delta\sigma_T$ was greater than $\Delta\sigma_{TMF}$, then Equation (6.25) would predict the observed increase in the fatigue life. Tests should be run to see if this is so.

6.10.1 Accelerated Life Testing

Let us now jump ahead in this process and assume that the displacements in the joint have been determined and that one or more of the approaches outlined above has been used to calculate the joint life. The final step in the design process is the testing of the joint to verify the life prediction calculation. Usually, the fatigue life for a properly designed joint is so long that exactly duplicating the service condition leads to an unacceptably long test time. Accelerated testing will therefore be required. In a sense, all of the previous information provided in this chapter is the required background for the consideration of accelerated testing [Engelmaier, 1985a; 1989; Lange et al., 1970; Waller et al., Howard, Goldman, 1969; Luke and Wild, 1983; Solomon, 1989c; 1991d]. As with the question of life prediction, there are two basic approaches, isothermal tests and thermal tests.

6.10.2 Isothermal Mechanical Acceleration Factors

To mechanically test a joint under isothermal conditions, it is necessary to be able to load it. One way that this can be accomplished is by using a split board technique [Solomon, 1989d; 1992]. A prototype joint between a component and printed wiring board (PWB) is made. The board is assembled in a testing machine and the board is cut apart so that the joints will carry all of the load. This is facilitated by using a special board which is cut out under the component [Solomon, 1989d; Solomon, 1992]. In this test a chip carrier is soldered to a PWB. For simplicity only two opposite rows of joints are used; however, some side joints could have also been made. After the board is mounted in the testing machine it can be cut apart by extending the slot under the chip carrier. A displacement or load can be applied to the assembly and the fatigue life determined for any given displacement or load history desired.

For an accelerated test it is required that the total testing time be less than that which would be encountered in service, with an extrapolation to the service condition. There are several ways to do this: the displacement can be increased, the cycle frequency can be altered, a different temperature could be used, a more aggressive environment used, or a different definition for failure employed.

Acceleration by Displacement Changes

Figure 6-71a and b compares the results of split board tests that were run on both leadless and leaded assemblies. These tests were run in displacement control and the results were correlated according to the total displacement applied. These displacements correspond to that developed for different ranges of temperature change. Tests can be run at high displacements where the fatigue life is short and the result extrapolated to the smaller displacement corresponding to the service temperature range. It should be noted that Figure 6-71 is similar to a strain vs. fatigue life plot, but it correlates the displacement not the strain. The strain distribution can be determined, but this is not necessary if prototype joints are used. The influence of variations in the joint geometry can be easily determined by making different joints and comparing them. Figure 6-71 shows a clear superiority of leaded joints over leadless ones. This is because at the high frequency employed

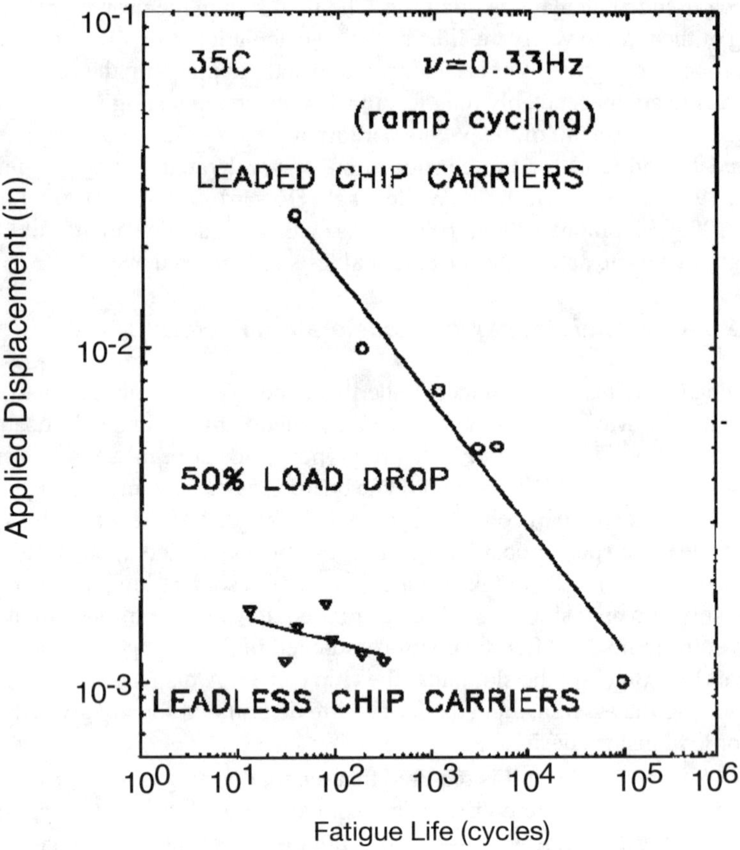

FIGURE 6-71a Comparison of the fatigue life vs. the applied displacement for leaded and leadless chip carrier joints tested at 35°C. [Solomon,1992]

much of the displacement is taken up by the lead flexure. At low frequencies, or with long hold times, the elastic strain stored in this lead flexure can be converted into plastic strain by stress relaxation. This reduces the degree to which leaded configurations are superior to leadless ones [Solomon, 1989d; 1992; 1993]. The geometry of the solder joint in a leaded configuration is different from that of a leadless one and this can also increase the fatigue life of the leaded ones relative to the leadless ones; this improvement should also be present at low frequencies.

Acceleration by Cycle Frequency Changes

Perhaps the most common method used to accelerate a fatigue failure is to perform the cycling at a higher cycle frequency. While this may lead to a greater number

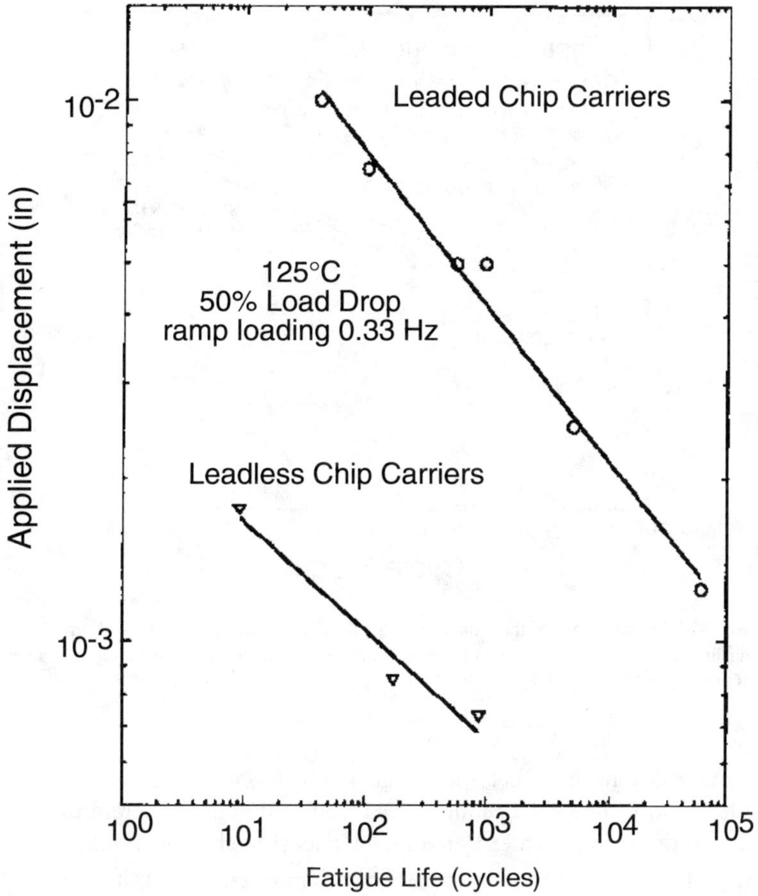

FIGURE 6-71b Comparison of the fatigue life vs. the applied displacement for leaded and leadless chip carrier joints tested at 150°C. [Solomon,1992]

of cycles required for failure, the higher cycle frequency reduces the time needed to apply these cycles. This is illustrated in Figure 6-38. If the appropriate displacement, for the design thermal cycle, is used at different cycle frequencies, then the test data can be extrapolated to the design cycle frequency. This is illustrated in Figure 6-72, which shows split board data, obtained with leaded joints and a displacement of 0.005 inches (0.13mm) [Solomon, 1993]. These data have been corrected for the additional plastic strain that develops at low frequencies, so that they show only the inherent frequency effect. In keeping with previous discussions, Figure 6-72 also shows that asymmetric loading produces a lower fatigue life, for the same plastic strain and cycle frequency, than symmetric loading, although for

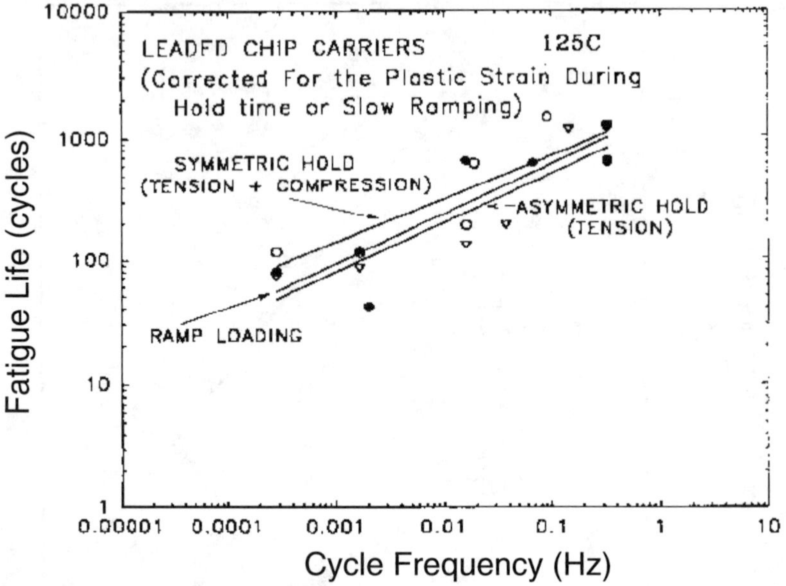

FIGURE 6-72 Comparison of the reduction in the fatigue life, corrected for variations in the plastic strain during hold times or slow ramp loading, with reductions in the cycle frequency for different types of cycling at 125°C. [Solomon, 1993]

these test conditions the effect is not large. Figure 6-55b also shows hold time data generated with leaded solder joints. Data from different temperatures were normalized using the Engelmaier Reliability Model [Engelmaier, 1992a].

The problem of accelerated testing becomes more complex if it is desired to accelerate the testing by increasing both the displacement and the cycle frequency. It is generally assumed that the fatigue ductility exponent is not affected by the cycle frequency and the frequency modified Coffin-Manson expression (Equation (6.8)) is then used [Engelmaier, 1983; 1985b; Kotlowitz and Engelmaier, 1986; Kotlowitz, 1989a,b; 1991; Kotlowitz and Taylor, 1991; Clech et al., 1989a; Clech et al., 1989b; Engelmaier, 1989, 1992a,b; Engelmaier and Attarwala, 1989; Clech and Augis, 1991; Denlinger et al.,1977]. As has been discussed, this assumption is generally not valid. An example of this, using the split board technique, is shown in Figure 6-73. Here data is shown for leadless joints. The plastic component of the displacement is used, analogous to the Coffin-Manson equation. The tests on the leadless joints were run with plastic displacement limits [Solomon, 1989d]. This could not be done for the leaded joints because of the larger compliance [Solomon, 1992]. With a displacement vs. fatigue life curve run at the proper frequency and a high strain, it is possible to extrapolate to a lower frequency of interest.

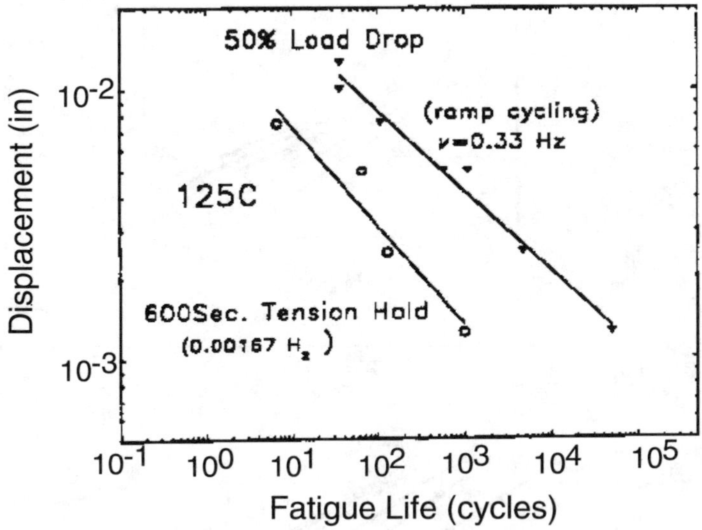

FIGURE 6-73 Comparison of the fatigue life vs. the total displacement for tests with and without a 600 second tensile hold at the maximum tensile displacement. [Solomon,1992]

Accelerations by Temperature and Environment Changes

As Figures 6-14 and 6-15 show, the fatigue life can be reduced by raising the test temperature. The maximum temperature to consider should be the maximum temperature that is expected. It is dangerous to accelerate the testing by going above this temperature because a different fatigue mechanism may be dominant and the accelerated test results would not be very meaningful. It is also dangerous to utilize too low a temperature, because of the drastic differences in solder properties and microstructures that can develop.

It has also been shown that the environment can influence fatigue life. Tests in air—or the actual environment, if it is more aggressive—would accelerate failure relative to a less aggressive environment (for instance, inert gas or nitrogen or even very low moisture-content air). Tests in an unrealistically harsh environment are not recommended because they might produce unique failure behavior.

Acceleration by Using Different Failure Criteria

Figure 6-74 shows fatigue data, obtained on leadless joints [Solomon, 1989d] for different load drop and resistance definitions of failure. A test that might take a very long time could be accelerated by using a very small load drop or resistance increase definition for failure [Kilinsky et al., 1989]. Care should be taken in doing this, however, because the load drop and resistance behavior is a function of the type of joint being studied [Solomon, 1992] and the compliance [Wilcox et al., 1990]. This problem can be eliminated by testing prototype joints, providing that

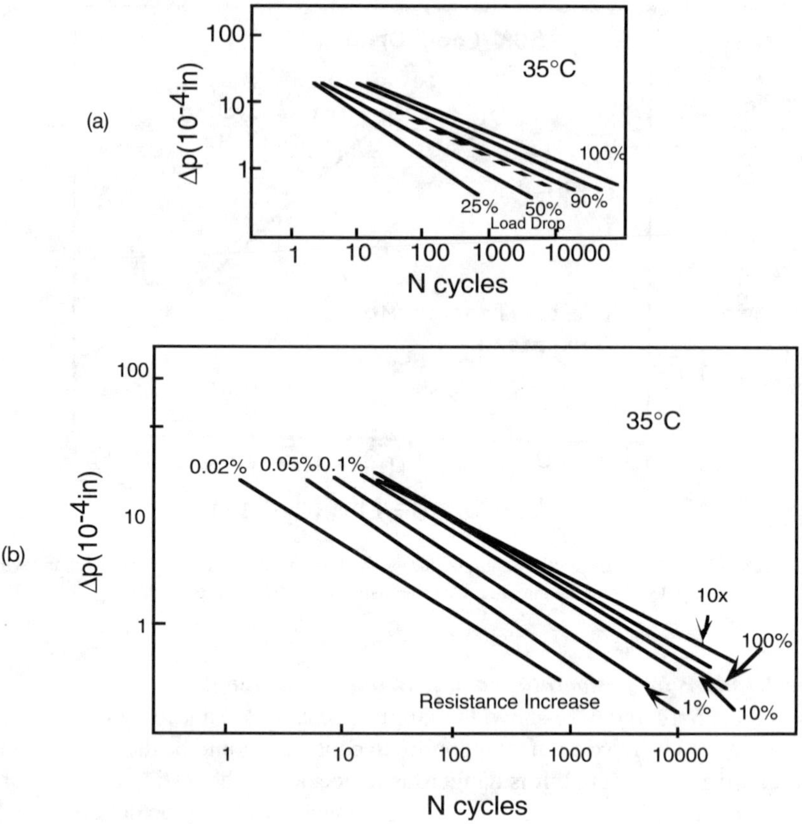

FIGURE 6-74 Fatigue life versus the plastic displacement across both rows of a leadless chip carrier joints, with the fatigue life measured at (a) different load drops and (b) different resistance increases. [Solomon, ©1989d, IEEE]

data are obtained which allow a small load drop or resistance change to be correlated with one more appropriate to the actual field condition.

6.10.3 Acceleration of Thermal and Thermomechanical Fatigue Tests

Thermal fatigue tests utilize only temperature changes to produce failure. In thermomechanical fatigue tests both the temperature and stress (or strain) can be independently varied. As such, there are a great many more ways to accelerate a thermomechanical fatigue test than a thermal fatigue test. Thermomechanical fatigue is discussed in a separate chapter, so here we shall concentrate on the acceleration of thermal fatigue tests.

The methods of accelerating thermal fatigue tests are analogous to that of the isothermal fatigue tests. There are, however, some additional complications. To increase the displacement that is applied, the temperature range of cycling must be increased. Unfortunately, this means changing at least two of the three possible variables: the maximum temperature, the minimum temperature, or the mean temperature. Different mechanisms can develop if the maximum or minimum temperatures are set too far beyond their expected service limits. One approach would be to increase the maximum temperature and decrease the minimum temperature by only modest amounts and keep the mean temperature near the expected value.

As with isothermal tests, the cycle frequency can be increased, although limitations imposed by heat transfer and thermal mass place a limit on the maximum possible frequency that can be achieved. Care should also be taken to insure that enough time has been provided to allow thermal equilibrium to be reached.

Thermal cycling tests do not allow for load drop measurements. The test is either stopped periodically and the joints visually inspected or resistance changes are used. Both approaches present problems. Stopping the test for a visual inspection greatly lengthens the test time. Furthermore, depending upon the joint configuration, cracks might be hard to locate. It is also often the case that when cracks can be easily located, they are observed very early in the fatigue life. Data will therefore be required to correlate these observations with the larger cracks which propagate through the joint to cause failure (as opposed to surface cracks which may be easily observed, but are not yet significant). The change in resistance provides a method of continuous measurement. It may, however, not be very sensitive to the extent of cracking because if the crack faces are touching they will still pass electricity. A fast-response (~1 μsecond) event detector is more likely to pick up the electrical intermittents produced by cracking than a simple DC resistance technique.

Load drop measurements can be made in thermomechanical fatigue tests, so the test can be accelerated by using a small load drop definition for failure. The temperature and temperature range can be varied as discussed above. As with isothermal fatigue and thermal fatigue, one can also accelerate a thermomechanical fatigue test by altering the ramp rate and hold time. In addition, in thermomechanical fatigue, one can increase the stress (or strain) that is being applied independently of the thermal cycle. The phase of the peak stress (or strain) can be altered with respect to the peak temperature, producing in-phase or out-of-phase cycling. Great care should be taken, however, in accelerating thermomechanical tests. The interactions between stress (or strain) and temperature are quite complex, and a test may be accelerated but not be related to what might go on under actual conditions.

6.11 CONCLUSION

This chapter has attempted to critically review the literature and summarize the available solder joint life prediction techniques, and methods for performing accelerated life testing. It should be remembered, however, that this review represents only a snapshot of a rapidly expanding field. The reader is advised to study the literature presented here and to search out the additional data that is becoming available.

An attempt has been made to be as comprehensive as possible, with the attendant problem of presenting many different approaches to life prediction. The discussions at the Workshop on the Mechanics of Solder Alloy Interconnects did not reveal an approach which was unanimously agreed upon as the best. Rather, many will probably give equivalent results, with the best approach being defined in terms of the specific problem and available data. It was suggested that a life prediction "bake off" be organized, with one group performing thermal cycle tests, and others (without access to the thermal cycle data) attempting to predict the life from the conditions of the tests and the geometry of the joints being tested.

While most of the complicating factors which control the fatigue behavior have been considered, there are still factors which have not been considered. For instance, random loads (or temperature changes) have not been discussed, nor has the effect of mean stresses (which, due to the extensive creep prevalent in solder joints, is generally not a problem). Complex load histories, over-stressing and threshold crack growth behavior have only been briefly mentioned. No discussion has been made of crack closure effects or the behavior of short cracks. It is therefore suggested that before beginning an analysis the reader should consider the general fatigue literature, as well as the literature specific to the fatigue of solder joints. The reader is also advised to remember that the literature is only a guide and should not be applied blindly. There is no substitute for a well thought out and properly carried out testing program.

• • •

The author is grateful to Darrel Frear, Werner Engelmaier and Semyon Vaynman for their reviews of this work. Their comments and suggestions greatly improved both the accuracy and readability of this chapter. The author would also like to thank George Lucey, Jr., and US Army Laboratory Command for providing the travel and living support which made his attendance at the workshop, and therefore this chapter, possible.

REFERENCES

Bae, K., Sprecher, A.F., Guo.Z., Conrad, H., and Jung, Y., 1989, *Micro Electronic Packaging Technology—Materials and Processes*, ASM p109.

Bangs, E.R., and Beal, R.E., 1978, *Weld. J. Res. Supp.* 54:377.

Barker, D., Pecht, M., Dasgupta, A., and Naqvi, S., 1991, *ASME J. of Electronic Packaging*, 113:149.

Barker, D.B., Shaarif, I., Dasgupta, A., and Pecht, M. G., 1992, *ASME J. of Electronic Packaging*, 114:177.

Basquin, O.H., 1910, *Proc. ASTM*, 10:625.

Berriche, R., Vaynman, S., Fine, M.E., and Jeannotte, D.A., 1987, *Proc. 3rd Intl. Conf. on Electronic Packaging and Corrosion in Microelectronics*, ASM, pp. 169-174.

Berriche, R., Fine, M.E., and Jeannotte, D.A., 1991, *Met. Trans.*, 22A:357.

Brown, M.W., and Miller, K.J., 1973, *Proc.Inst. of Mech. Eng.*, 187:745.

Burgess, J.F., Carson, R.O., Glascock, H.H., Neugebauer, C.A., and Webster, H.F., 1984, *IEEE Trans. CHMT*, 7:405.

Charles, H.K., and Clatterbaugh, G.V., 1990, *ASME J. of Electronic Packaging*, 112:135.

Chilton, A.C., Whitmore, M.A., and Hampshire, W, 1989, *Proceedings of the 2nd ASM Electronic Materials and Processing Congress*, Philadelphia, p. 159.

Clech, J.-P., and Augis, J.A., 1991, *Joint Reliability*, edited by J.H.Lau, Van Nostrand Reinhold, New York, p. 588.

Clech, J.-P., Engelmaier, W., Kotlowitz, R.W., and Augis, J.A., 1989a, *Proceedings of the SMART V Conference*, New Orleans, Technical Paper V-48.

Clech, J.-P., Engelmaier, W., Kotlowitz, R.W., and Augis, J.A., 1989b, *IEEE CHMT*, 12:449

Coffin, L.F.Jr., 1954, *Trans. ASME*, 76:.931.

—, 1969a, *Fracture 1969*, Chapman and Hall, London.

—, 1969b, *Proc. Air Force Conf. on Fatigue and Fracture of Aircraft Structures and Materials*, AFFDL-TR70-144, p. 301.

—, 1971, *Met Trans.*, 2:3105.

—, 1972a, *Met. Trans*. 3:1777.

—, 1972B, *Proc. Intl. Conference on Fatigue: Chemistry, Mechanics and Microstructure*, p. 590.

—, 1973, *ASTM STP-520*, p. 112.

—, 1974a, *Fatigue at High Temperature- Prediction and Interpretation*, James Clayton Memorial Lecture, Proc. Inst. Mech. Eng., 18:109.

—, Jr., 1974b, *Met. Trans.*, 5:1053.

—, 1975, *Observations and Correlations Emphasizing Frequency and Environmental Effects in Time Dependent Fatigue of Structural Alloys*, ORNL-5073, p. 37.

Coombs, V.D., 1972, ASTM-STP 515, p. 3.

Cortez, R., Fine, M.E., Daniel, I.M., and Jeannotte, D.A., 1991, *Materials Developments in Microelectronic Packaging Conference Proceedings*, ASM International, p. 147.

Cortez, R., Cutiongco, E.C., Fine, M.E., and Jeannotte, D.A., 1992, *Proc 42nd ECTC,* pp. 354-359.

Cutiongco, E.C., Vaynman, S., Fine, M.E., and Jeannotte, D.A., 1990, *ASME J. of Electronic Packaging*, 112:110.

Cutiongco, E.C., and Jeannotte, D.A., 1990, *Proceedings of the Morris E. Fine Symp.*, Ed P.K. Liaw et al., TMS, p. 435.

Dasgupta, A., Oyan, C., Barker, D., and Pecht, M., 1992, *ASME J. of Electronic Packaging,* 114:152.

Denlinger, M.C., Korb, R.W., and Lardenoit, V.F., 1977, *Development of Improved Solders for Electronic Reliability*, AFML-TR-77-93.

Desaulnier, W.E., LaFlame, T.E., Ammerman, W.B., and Binder, M.C., 1986, *Proceedings of Institute of Environmental Sciences*, p. 6.

Devore, J.A., and Tillett, S., Jr., 1979, *Fatigue Life characteristics of Various Solder Joint Designs*, GE Report R79ELS038.

Devore, J.A., 1980, *Solder and Solder Joint Properties Handbook*, GE Report # R80ELS028.

Devore, J.A., and Tillett, S., Jr., 1981, *Fatigue Characteristics of Solder and Solder Joints*, Report R81ELS018.

Dowling, N.E., 1977, *ASTM STP 637*, p. 97.

—, 1979, *Fatigue of Engineering Materials and Structures*, 2:129.

—, and Begley, J.A., 1976, *ASTM STP590*, p. 82.

Dunn, B.D., 1978, *The Resistance of Space-Quality Solder Joints To Thermal Fatigue*, ESA STM-207.

Engelmaier, W., 1983, *IEEE CHMT*, 6:232.

—, 1984, *Surface Mount Technology*, ISHM technical monograph series 6984-002, ISHM Silver Springs, MD, p. 87.

—, 1985a, *Brazing and Soldering*, 9:40.

—, 1985b, *Circuit World*, 11:61.

—, 1989, *Soldering and Surface Mount Technology*, V. 1, p. 14.

—, 1991, *Solder Joint Reliability*, edited by J.H.Lau, Van Nostrand Reinhold, New York, p. 545.

—, 1992a, *Proc. Surface Mount International Conf.*, San Jose CA, p. 433.

—, 1992b, *Soldering and Surface Mount Technology*, 10:17.

Engelmaier, W., and Attarawala, A.I., 1989, *IEEE CHMT*, 12:284.

Enke, N.F., Kilinski, T.J., Schroeder, S.A., and Lesniak, J.R., 1989, *IEEE CHMT*, 12:459.

Fine, M.E., and Jeannotte, D.A., 1989, *Micro Electronic Packaging Technology—Materials and Processes*, ASM, p. 147.

Frear, D.R., 1989, *IEEE CHMT*, 12:492.

—, 1990a, *The Metal Science of Joining*, Ed. M.Cieslak et.al., TMS Publications, p. 191.

—, 1990b, *IEEE CHMT*, 13:518.

Frear, D., Grivas, D., McCormack, M., Tribula. D., and Morris, J.W., Jr., 1987, *Effect of Load and Thermal Histories of Mechanical Behavior*, TMS, p. 113.

Frear, D., Grivas, D., and Morris, J.W., Jr., 1988a, *J. of Met.*, 40:18.

Frear, D. R., Grivas, D., McCormack, M., Tribula, D., and Morris, J.W., Jr., 1988b, *Proc 3rd. Electronic Packaging and Corrosion in Microelectronics Conf.*, ASM, p. 269.

Frear, D.R., Grivas, D., and Morris, J.W., Jr., 1989, *J. Electron Mat.*, 18:671.

Frear, D.R., and Jones, W.B., 1990, *Proc. NEPCON West '90*, Anahiem CA, p. 1324.

Frear, D.R., Jones, W.B., and Sorensen, N.R., 1991a, *Creep: Characterization, Damage and Life Assessment*, ASM, Ed. D.A.Woodford, et al., ASM International, p. 341.

Frear, D.R., Jones, W.B., Morris, J.W., Jr., and Mei, Z., 1991b, *Manufacturing Processes and Materials Challenges in Microelectronic Packaging*, ASME publications, Ed.-W.T.Chen et.al., AMD-Vol131/EEP-Vol1, p. 1.

Frost, H.J., 1991, *Solder Joint Reliability,* edited by J.H.Lau, Van Nostrand Reinhold,New York, p. 266.

Goldmann, L.S., 1969, *IBM J. Res. Dev.*, 13:251.

Goldmann, L.S., Herdzik, R.D., Koopman, N.G., and Marcotte, V.C., 1977, *IEEE Trans PHP*, 13:194.

Grivas, D., Murty, K.L., and Morris, J.W., 1979, *Acta Met.*, 27:731.

Grosskreutz, J.C., 1971, *ASTM STP495*, p. 5.

Guo, Z., Sprecher, A.F., Jung, D.Y., and Conrad, H., 1991A, *IEEE CHMT*, 14:833.

Guo, Z., Sprecher, A.F, and Conrad, H., 1991b, *41st IEEE ECTC Conference,* Atlanta GA, p. 658.

Guo, Z., Sprecher, A.F., and Conrad, H., 1991c, *ASME J. of Electronic Packaging*, 114:112.

Guo, Z., Sprecher, A.F, and Conrad, H., 1992a, *5th ASM Electronic Materials & Processing Congress*, ASM International, Cambridge MA.

Guo, Q., Cutiongco, E.C., Kear, L.M., and Fine, M.E., 1992b, *ASME J. of Electronic Packaging*, 114:145.

Guo, Z., Sprecher, A.F., and Conrad.H., 1992c, *J. of Phase Equlibria*, 13:227.

Guo, Z., and Conrad, H., 1992, *Fatigue Crack Growth in 63Sn37Pb Solder Joints*, presented at the 1992 Winter Annual ASME meeting, Symposium on Surface Mount Assemblies.

Hacke, P.L., Sprecher, A.F., and Conrad, H., 1991a, *Modeling of the Thermomechanical Fatigue Behavior of 63Sn-#7Pb Alloy*, ASTM Symp. on Thermo-Mechanical Fatigue Behavior of Materials, San Diego Ca., ASTM.

Hacke, P.L., Sprecher, A.F., and Conrad, H., 1991b, *Thermal Stress and Strain in Microelectronic Packaging*, ed J..Lau, Van Nostrand Reinhold.

Halford, G.R., and Manson, S.S, 1976, *ASTM STP 612*, p. 239.

Halford, G., 1986, *NASA-TM-87225*, NASA, Washington, DC.

Hall, P.M., 1987, *IEEE CHMT*, 12:556.

Hall, P.M., Dudderar, T.D., and Argyle, 1983, *IEEE CHMT*, 6:544.

Hall, P.M., 1984, *IEEE CHMT*, 7:314.

Hall, P.M., 1991, *Solder Joint Reliability*, edited by J.H.Lau, Van Nostrand Reinhold, New York, pp. 306-332.

Hendrix, B.C., Bretz, P.L., Tien, J.K., and Kang, S.K., 1989, *Micro Electronic Packaging Technology—Materials and Processes*, ASM p129.

Howard, R.T., 1982, *IBM J. Res. Develop.*, 26:372.

Howard, R.T., *IEEE Paper# CH1781-4/82/0000-0376.*

Inoue, H., Kurihara, Y., and Hachino, H., 1986, *IEEE Trans CHMT*, 9:190.

IPC-SM-785, 1992, *Guidelines For Accelerated Testing For Surface Mount Attachments*, IPC.

Jahsman, W.E., Lii, M.J., and Renfro, T.A., *Dependence of Solder Joint Reliability on Lead Compliance and Displacement Amplitude and Frequency*, to be published.

Jarboe, D.M., 1980, *Thermal Fatigue of Solder Alloys*, NTIS-BDX-613-2341,

Kaisand, L.R., and Mowbray, D.F., 1979, *ASTM J. of Testing and Evaluation*, 7:270.

Kandil, F.A., Brown, M.W., and Miller, K.J., 1982, *Mechanical Behavior and Nuclear Applications of Stainless Steel at Elevated Temperatures*. Book 280, London: The Metals Society, p. 203.

Kashyap, B.P., and Murty, G.S., 1981a, *Materials Science and Engineering*, 50:205.

Kashyap, B.P., and Murty, G.S., 1981b, *Met.Trans.*, 12A:2435.

Kilinsky, T.J., Lesniak, J.R., and Sandor, B.I., 1991, *Solder Joint Reliability*, edited by J.H.Lau, Van Nostrand Reinhold, New York, p. 384.

Kinser, D.L., Vaughn, J.G., and Graff, S.M., 1976, *Electronic Packaging and Production, p.* 61.

Kitano, M., Shimizu, T., and Kumazava, T., 1987, *Current Japanese Materials Research,* 2:235–50.

Knecht, S., and Fox, L.R., 1991, *Solder Joint Reliability*, edited by J.H.Lau, Van Nostrand Reinhold, New York, p. 306.

Kong, S.K., Zommer, N.D., Feutcht, D.L., and Heckel, R.W., 1977, *IEEE Trans. PHP,* 13:318.

Korb, R.W., and Ross, D.O., 1983, *IEEE Trans CHMT*, 6:227.

Kotlowitz, R.W., 1989a, *IEEE, CHMT*, 12:p431.

—, 1989b, *Proc. IEPS*, San Diego CA, p. 1071.

—, R.W., 1991, *Proc. SMTCON*, Atlantic City NJ, p. 1.

Kotlowitz, R.W., and Engelmaier, W., 1986, *6th International Packaging Conference*, p. 841.

Kotlowitz, R.W., and Taylor, L.R., 1991, *Proc 41st ECTC*, Atlanta GA,

Lacey, T.G., and Woodford, D.A., 1988, *Scripta Met*, 22:1543.

Lake, J.K., and Wild, R.N., 1983, *28th National SAMPE*, p. 1406.

Landgraf, R.W., 1970, *ASTM STP-467*, p. 3.

Landgraf, R.W., 1973, *ASTM STP519*, p. 213.

Lang, G.A., Fehder, B.J., and Williams, 1970, *IEEE Transactions on Electron Devices,* 17:787.

Lau, J.H., 1991, *Solder Joint Reliability* edited by J.H.Lau, Van Nostrand Reinhold, New York, p. 455.

Lau, J.H., and Rice, D.W., 1985, *Solid State Technology*, 28:91.

Lau, J.H., and Jeans, A.H., 1989, *Micro Electronic Packaging Technology—Materials and Processes*, ASM, p. 177.

Lau, J., Subrahmanyan, R., Rice, D., Erasmus, S., and Li, C., 1991, *ASME J.of Electronic Packaging*, 113:138.

Lawson, L., Fine, M.E., and Jeannotte, D., 1991, *Met.Trans.*, 22A:1059.

Lawson, L.R., Fine, M.E., and Jeannotte, D.A. 1987, *Fatigue 87*.

Levine, E., and Ordonez, 1977, *J. IEEE Trans. PHP*, 13:318.

Li, C.-Y., Subrahmanyan, R., Wilcox, J.R., and Stone, D., 1991, *Solder Joint Reliability*, edited by J.H.Lau, Van Nostrand Reinhold, New York, p. 361.

Liaw, P.K., Logsdon, W.A., and Burke, M.A., 1989, *Scripta Met*, 23:747.

Likhman, V.I., Rebinder, P.A., and Karpenko, G.V., 1958, *Effects of Surface Active Medium on the Deformation of Metals*, HMSO, London.

Liu, H.W., 1985, *Fatigue Fract. Eng. Mater. Struct.*, 8:295.

Logsdon, W.A., Liaw, P.K., and Burke, M.A., 1990, *Engineering Fracture Mechanics,* 36:183.

Lohr, R.D., and Ellison, E.G., 1980, *Fatigue of Engineering Materials and Structures*, 3:1.

Lord, D., and Coffin, L.F., 1974, *Met. Trans*.4:1647.

Manson, S.S., 1953, *Behavior of Materials Under Conditions of Thermal Stress*, Heat Transfer Symp., Univ. of Michigan, June 27-28, 1952, Univ. ofMich.Press;also NACA TN2933.

Manson, 1960, S.S., *Mech. Des*., 32:139.

—, 1965, *Exp. Mech*., 5:193.

—, 1966, ASTM-STP 520:744.

—, 1975, *Time Dependent Fatigue of Structural Alloy*, ORNL-5073, p. 155.

Manson, S.S., Halford, G.R., and Hirshberg, H.M., 1971, *Design for Elevated Temperature Environment*, ASME, p. 25.

Marco, S.M., and Starkey, W.L., 1954, *Trans ASME*, 76:627.

Marshall, J.L., and Walter, S.R., 1987, *International Society for Hybrid Microelectronics,* 10:11.

Marshall, J.L., 1991, *Solder Joint Reliability*, edited by J.H.Lau, Van Nostrand Reinhold, New York, p. 173.

Mei, Z., and Morris, J.W., Jr., 1992, *ASME J. of Electronic Packaging*, 114:1048.

Miner, M.A., 1945, *Trans. ASME*, 67:A159.

Miyazaki, M., Yoshioka, S., and Hijikata, A., *Zairyo*, 1980, 30:8.

Monkman, F.C., and Grant, N.J., 1956, *Proc ASTM*, 56:593.

Morris, J.W., Jr., Grivas, D., Tribula, D., Summers, T., and Frear, D. R., 1989, *Soldering and Surface Mount Tech.*, 3:4.

Morris, J.W., Jr., Tribula, D., Summers, T.S.E., and Grivas, D., 1991, *Solder Joint Reliability*, edited by J.H.Lau, Van Nostrand Reinhold, New York, p. 225.

Morrow, J., 1965, *ASTM STP-378*, p. 45.

Mura, T., Nakasone, Y., *ASME J. of Appl. Mech.*, 57:1.

Murty, G.S., and Kashyap, B.P., Trans., 1984, *Ind. Inst.Met.*, 37:509.

Neugebauer, C.A., Webster, H.F., Solomon, H.D., and Carlson, R.O., 1984, *Soldering Technology Seminar*, Naval Weapons Center, China Lake, CA.

Nir, N., Dudderar, T.D., Wong, C.C., and Storm, A.R., 1991, *ASME J. of Electronic Packaging*, 113:92.

Oberin, F.W., 1989, *Micro Electronic Packaging Technology—Materials and Processes,* ASM, p. 189.

Olsen, D.R., and Berg, H.M., 1979, *IEEE Trans CHMT*, 2: 257.

Palmgren, A., 1924, *ZVDI*, 68:339.

Pan, T-Y, 1991a, *J. of Electronic Packaging*, 113:8.

—, 1991, *Winter Annual ASME meeting Paper #91-Wa-EEP-10.*

Pankratz, J.M., and Plumlee, H.R., 1970, *IEEE Trans ED*, 17:793.

Pao, Y-H., "A Fracture Mechanics Approach to Thermal Fatigue Prediction of Solder Joints," IEEE CHMT V15, August 1992, pp. 559–570.

Pao, Y-H, and Jih, E., 1992, *Submitted to IEEE CHMT.*

Pao, Y-H., and Pan, T-Y., ASME, 1990, *J. of Electronic Packaging*, 112:154.

Pao, Y-H., Govila, R., Badgley, S., and Jih, E., 1992, *Ford Research Report SR-92-111*, to be published in J. Electronic Packaging, Dec.1992.

Paris, P.C., and Erdogan, F., 1963, *Trans ASME J. Basic Eng.*, p. 528.

Pearson, R.C., 1988, *Westinghouse Final Report # AFWAL-TR-88-4215.*

Rack, H.J, and Maurin, J.K., 1974, *J. Testing and Eval.*, 2:351.

Raman, V., and Reiley, T.C., 1987, *J. of Materials Science Letters*, 6:549.

Rathore, H.S., Yih, R.C., and Edenfeld, A.R., 1973, *J.of Testing and Evaluation*, 1:170.

Read, D.T., and Lucey, G.K., Jr., 1991, *ASME J. of Electronic Packaging*, 113:178.

Reidel, H., 1981, *J. of the Mechanics and Physics of Solids*, 29:35.

—, 1987, *Fracture at High Temperatures*, Springer-Verlag, Berlin.

—, 1989, *7th International Conference on Fracture*, Houston.

Reidel, H., and Rice, J.R., 1980, *ASTM STP 700*, p. 112.

Rice, J.R., 1968, *J. App. Mech.*, 35:379.

Richart, F.E., and Newmark, N.M., 1948, *ASTM Proc.*, 48:767.

Shamash, M.B., 1968, *NTIS N69-25697*, Westinghouse Defense and Space Center, Baltimore, 8.

Sharif, I., Barker, D.B., Dasgupta, A., and Pecht, M.G., 1991, *ASME J. of Electronic Packaging*, 113:194.
Sherry, W.M., Erich, J.S., Bartschat, M.K., and Prinz, F.B., 1985, *IEEE CHMT*, 8:417.
Shine, M.C., and Fox, L.R., and Sofia, J.W., 1984, *Proc. of the International Electronic Packaging Society*, 4:346.
Shine, M.C., and Fox, L.R., 1987, *ASTM STP942*, p. 588.
Sklenicka, V., Lukas, P., and Kenz, L., 1990, *Scripta Met.*, 24:1795.
Smith, K.N., Watson, P., and Topper, T.H., 1970, *J. of Materials*, 5:767.
Snowden, K.U., 1964a, *Acta Met.*, 12:295.
Snowden, K.U., 1964b, *Phil.Mag.*10:435.
Sochie, D.F.and Bannantine, J.A., 1989, *Materials and Engineering Design: The Next Decade*, ed. E.F.Dyson and D.R.Hayhurst, London: The London Institute of Metals, p. 301.
Solomon, H.D., 1972a, *J. of Materials*, 7:299.
—, 1972b, *Met.Trans*, 4:341.
—, 1972c, *J. of Materials*, 7:229.
—, 1986a, *Electronic Packaging Materials and Processes*, ed. J.A.Sartell, ASM, p. 29.
—, 1986b, *Brazing and Soldering*, 11:29.
—, 1986c, *IEEE CHMT-9*, p. 423.
—, 1988, *38th IEEE ECC*, p. 7.
—, 1989, *Mat. Res. Soc. Symp. Proc.*, 154:441.
—, 1989a, *ASME J. of Electronic Packaging*, 111:75.
—, 1988b, *ASTM-STP942*, p. 342.
—, 1989c, *IEEE CHMT*, 12:473.
—, 1990, *ASME J. of Electronic Packaging*, 112:123.
—, 1990b, *J. of Electronic Materials*, 19:929.
—, 1991a, *New Technology in Electronic Packaging*, Ed. B.R. Livesay and M.D. Nagarkar, ASM, p. 187.
—, 1991b, *ASME J. of Electronic Packaging*, 113:102-108.
—, 1991c, *Solder Joint Reliability*, edited by J.H.Lau, Van Nostrand Reinhold, New York, p. 406.
—, 1991d, *ASME J. of Electronic Packaging*, 113:186.
—, 1992, *ASME J. Electronic Packaging*, 114:161.
—, 1993, To Be Published, J. *of Electronic Packaging.*
Solomon, H.D., and Coffin, L.F., 1973, *ASTM STP 520*, p. 112.
Solomon, H.D., Brzozowski, V., and Thompson, D.G., 1990, *IEEE 40th ECTC*, 40:351.
Stephens, J.J., and Bourcier, R.J., 1986, *Physical Metallurgy Issues in Solder Joint Performance*, SAND-86-2080C, Sandia National Laboratories.
Stone, D., Wilson, H., Subrahmenyan, R., and Li, C.-Y., 1986, *36th IEEE ECC*, 36:630
Stone, D.S., 1990, *ASME J. of Electronic Packaging*, 112:100.
Subrahmanyaan, R., Wilcox, J.R., and Li, C.-Y, 1989a, *IEEE CHMT*, 12:480.
—, *Proc 2nd ASM Int. Electronic Materials and Processes Congress*, ASM, p. 213.
—, 1989c, *Micro Electronic Packaging Technology—Materials and Processes*, ASM, p. 213.
—, 1989d, *39th IEEE-ECC*, p. 240; also, 1989, IEEE CHMT, 12:480.
Summers, T.S.E., and Morris, J.W., Jr., *ASME J. of Electronic Packaging*, 112:94.
Taira, S., 1973, *ASTM STP-520*, p. 80.

Tanaka, T., Nishijima, S., and Ichikawa, M., 1987, *Current Japanese Materials Research*, 2:235–50.

Tomkins, B., 1968, *Phil.Mag.*, 18:1056.

Tomkins, B., 1971, *Phil.Mag.*, 23:687.

Tribula, D., Grivas, D., Frear, D. R., and Morris, J.W., Jr., 1989a, *Weld J.Res. Supp.*, 68:404s.

—, *ASME J. of Electronic Packaging*, 111:83.

Vaynman, S., and Fine, M.E., 1991, *Mat.Res.Soc. Proc.*, 226:3.

—., 1989, *Microelectronics Packing Technology: Materials and Processes*, ASM International, p. 255.

—., 1991, *Solder Joint Reliability,* edited by J.H. Lau, Van Nostrand Reinhold, New York, p. 333.

Vaynman, S., and McKeown, 1991, *41st IEEE ECTC*, p. 671.

Vaynman, S., Fine, M.E., and Jeannotte, D.A., 1990, *Scripta Met.*, 26:999.

—, 1987a, *TMS-AIME Symposium on Effect of Load and Thermal Histories on Mechanical Behavior of Materials*, and *Met.Trans,* 1988, 19A:1051.

—, 1987b, *37th Electronic Components Conference*, p. 598.

—, 1987c, *Fatigue 87.*

Vaynman, S., Fire, M.E., Jeanotte, D.A., 1988, *Metall. Trans*. 19A:1051.

Waller, D., Fox, L., and Hannemann, R., *IEEE Paper#0569-5503/83/0000-0534.*

Weibull, W., 1961, *Fatigue Testing and Analysis of Results*, Pergamon Press, NY.

Weinbel, R.C., Tien, J.K., Pollak, R.A., and Kang, S.K., 1987a, *J. of Materials Science*, 22:3901.

Weinbel, R.C., Schwartzkopf, E.A., and Tien, J.K., 1987b, *Scripta Metallurgicia*, 21:1165

White, D.J., 1969, *Proc. Inst. Mech. Eng., Appl. Mech. Group*, 184:1296.

Wilcox, J.R., Subrahmanyan, R., and Li, Che-Yu, 1989, *Micro Electronic Packaging Technology—Materials and Processes*, ASM, p. 203.

—, 1990, *ASME J.of Electronic Packaging*, 112:115.

Wild, R.N., 1971, *IBM Tech Report #71Z0004081.*

—, 1972, *Welding J.*, 51:521s.

—, 1975, *Internepcon*, (IBM# 74Z000481).

—, 1988, *Proc. 12th Solder Seminar*, China Lake CA., p. 271.

Wolverton, W.M., 1987, *Brazing and Soldering*, 13:33.

Yamada, S.E., 1987, *Fracture Mechanics Engineering*, 28:85.

Zommer, N.D., Feutch, D.L., and Heckel, R.W., 1976, *IEEE Trans.ED*, 23:843.

Zubelewicz, A., Berriche, R., Keer, L.M., and Fine, M.E., 1988, *Winter ASME meeting, paper# 88-WA/EEP-11.*

Zubelewicz, A., Guo, Q., Cutiongco, E.C., Fine, M.E., and Keer, L.M., 1990, *ASME J. of Electronic Packaging*, 112:179.

7

Thermomechanical Modeling of Solder Joints—Numerical Considerations

H. S. Morgan

7.1 INTRODUCTION

Thermomechanical modeling of solder joints means different things to different investigators. To some material scientists, it means developing materials tests and using the resulting data to formulate models that describe the constitutive behavior of solder, the evolution of microstructure, or the evolution of damage. To other experimentalists, it means defining and running tests that can be used to make life predictions. In some cases, joints are cycled until failure occurs. In others, accelerated test methods are developed that reduce testing time and still capture the failure mechanisms observed in actual service. From these tests, empirical relations are derived for predicting life. These expressions relate cycles-to-failure to quantities such as plastic shear strain and various fracture mechanics parameters. To applied mechanicians, thermomechanical modeling means reducing the physical problem of a solder joint under thermomechanical loading to a mathematical boundary value problem which can be solved to quantify stresses, strains, and other pertinent parameters. These mathematical descriptions include: (1) idealizations, and often significant simplifications, of the solder joint geometry; (2) boundary conditions; (3) initial conditions; and (4) constitutive relations that describe the deformation behavior of solder and the materials it connects. Occasionally, the geometry, loading conditions, and constitutive theories are simple enough that analytic or closed form solutions exist for the parameters of interest. However, most of the time the boundary value problems require numerical solutions for which the finite element method is usually most appropriate.

All of the above perspectives on thermomechanical modeling are important and interrelated. The obvious and already stated connections include: (1) the use of constitutive theories in boundary value models to define material behavior; and (2) the subsequent use of parameters predicted with the numerical models as input to life prediction models derived from life testing. In addition, comparison of numerical predictions and experimental data has long been accepted as the best method, if not the only method, of validating the numerical simulations. A less widely recognized but equally valuable connection is the use of numerical predictions to provide a better understanding of the experiments themselves and to verify, or at least evaluate, the accuracy of assumptions used in defining experimental geometries and test procedures. Most tests rely on assumptions about the uniformity and in some cases the uniaxiality of stress or strain fields in order to have well-controlled experiments. Thermomechanical modeling can be used to evaluate the validity of these assumptions or to determine how an experimentally measured parameter is related to the actual stress and strain fields in a test sample. Additional integration of the numerical and experimental aspects of thermomechanical modeling of solder joints will occur as the paradigm for designing solder joint interconnections changes from one of "build and test" to a more agile one in which numerical modeling is used to evaluate multiple design options before limited testing is used to verify the best one or two candidate designs.

In earlier chapters of this book, the constitutive (Chapter 4) and accelerated testing (Chapter 6) aspects of thermomechanical modeling have been discussed. Chapter 5 is devoted to the important issue of predicting solder joint geometry from the melt. In Chapters 8 and 9, modeling of through-hole and surface mount interconnects are discussed. The purpose of this chapter is to address numerical modeling issues, particularly those related to finite element modeling, that are common to all solder joint applications. It should also serve as a link between the previous chapters of the book and the applications that follow. It is not intended to be an exhaustive treatment of finite element theory or a critical evaluation of analyses already documented in the literature but rather an identification of important numerical modeling issues that should be considered in creating models for specific solder joint applications.

The chapter begins with a brief review of numerical models currently used for predicting solder joint response to thermomechanical loading. Characteristics common to the majority of the models are identified. The bulk of the chapter is devoted to the continuum mechanics and finite element aspects of reducing the physical problem of a solder joint under thermomechanical loading to a boundary value problem in which the phenomena of interest can be quantified and predicted.

This is followed by a few concluding remarks on future directions for numerical modeling of solder joint behavior.

7.2 PURPOSE OF NUMERICAL MODELING

The electronic packaging literature is replete with examples of finite element models for predicting the behavior of solder joints under thermomechanical loads [Akay, Tong, and Paydar, 1992; Burchett, 1991; Clatterbaugh and Charles, 1985; Dasgupta, Verma, and Barker, 1992; Engel and Wu, 1992; Felske and Moss, 1986; Harada and Satoh, 1990; Jahsman and Jain, 1990; Kasem and Feinstein, 1987; Lau, 1991; Lau, 1992a; Lau et al., 1987; Lau and Harkins, 1988; Lau and Moresco, 1988; Lau and Rice, 1990; Lau, Rice, and Harkins, 1989; Lau and Keely, 1989; Morgan, 1991; Ozmat, 1990; Pan, 1991; Pao, Chen, and Kuo, 1991; Pao et al., 1992; Prakash and Yang, 1992; Ross, Wen, and Mon, 1992; Sarihan, 1992; Sherry et al., 1985; Smeby, 1985; Tervalon, 1990; Tong et al., 1990; Waller et al., 1983]. The models range from two-dimensional (2-D) representations of the joint geometry with linear elastic constitutive relations to three-dimensional (3-D) models that account for both plasticity and time-dependent deformation (creep). The models have been used to qualitatively evaluate various design configurations, to investigate the effects of lead misalignment, to predict joint behavior observed in experiments, to optimize designs of leads and joints, to provide shear strains for use in predicting fatigue life, to investigate crack growth in joints, to understand the actual stress and strain fields developed in laboratory tests, and to actually predict fatigue life numerically. Sandor [1991] provides a review of the types of analyses that have been performed and points out areas where improvement is needed.

The investigations referenced above generally recognize that solder should be modeled with complex nonlinear material models that account for both plasticity and creep as well as temperature-dependent material parameters. In addition, solder joint geometries are usually complex enough to require 3-D finite element models. However, most analysts have usually used, and will continue to use, simpler geometries and constitutive theories because they are constrained by time, budget, computing systems, and/or available computer codes. Nevertheless, important information can be gained from even the simplest solder joint models as long as analysts are aware of the limitations imposed by their modeling assumptions. Regardless of complexity, the models appear to address one central issue, namely the stresses, strains, deformations, and damage induced in solder joints and the materials they connect due to mismatches in coefficients of thermal expansion during thermal cycling. The remainder of this chapter is devoted to describing the implications of many of the assumptions used in finite element modeling of solder joints. Attention is focused on models in which the solder is treated as a nonlinear material exhibiting plastic, creep, or some combination of plastic and creep behavior.

7.3 BOUNDARY VALUE REPRESENTATIONS OF PHYSICAL PROBLEMS

All models used to predict physical response, regardless of whether the body is a solder joint or some other structure, are the result of reducing the physical phenomena of interest to mathematical relations that quantify parameters that represent the physical phenomena. Conservation or equilibrium laws are used to describe the phenomena of interest over the domain or volume of interest. The conservation laws are usually expressed in the form of differential equations that have to be solved over a geometric idealization of the domain. The behavior of the material or materials in the domain must be described by mathematical expressions, and appropriate initial and boundary conditions must be applied. At this point, the boundary value problem is completely defined. Occasionally, closed form solutions to the problem exist and should be used. However, a numerical solution is usually required. The finite element method [e.g. Zienkiewicz, 1977; Hughes, 1987] has been used extensively to solve boundary value problems in solid mechanics.

In solder joint applications, the pertinent equations of motion are the stress divergence equations for deformable bodies [Malvern, 1969]:

$$T_{ij,j} + \rho b_i - \rho \ddot{u}_i = 0 \qquad \text{in } V \tag{7.1}$$

where T_{ij} represents the Cauchy stress tensor, u_i is the displacement vector, $\ddot{u}_i$ is the acceleration vector, ρ is the density, b_i is the body force vector, and V is the volume of the body. Indicial notation is used in this and subsequent equations with the subscripts i, j, which range from 1 to 3, representing one of the three orthogonal directions of the global Cartesian coordinate system. The commas appearing in subscripts denote partial differentiation with respect to the coordinate represented by the subsequent index. The accompanying boundary and initial conditions are:

$$\begin{aligned} u_i &= U_i && \text{on } S_U \\ T_{ij} n_j &= t_i && \text{on } S_T \\ (T_{ij} &= T_{ij}^0) && \text{in } V \text{ at } t = 0 \\ (\dot{u}_i &= \dot{U}_i^0) && \text{in } V \text{ at } t = 0 \end{aligned} \tag{7.2}$$

where U_i is the specified displacement vector as a function of time t, S_U is the portion of the surface over which the displacement is specified, n_i is the outward normal to the surface of the body, t_i is the specified traction vector, S_T is the portion of the surface over which t_i is specified, T_{ij}^0 is the initial stress tensor, and $\dot{U}_i^0$ is the initial velocity vector. For thermal loading, the acceleration terms in Equation (7-1) are essentially zero. However, in some iterative solution methods, such as dynamic relaxation, the acceleration terms are retained, and the problem is damped to obtain a static solution [Stone, Krieg, and Beisinger, 1985].

Constitutive equations, which relate the Cauchy stress T_{ij} to the deformations of the body, are needed to complete the boundary value problem. For large strains and rotations, several other second-order tensors must be defined in order to define the constitutive relations. First, the rate-of-deformation tensor D_{ij} is defined by

$$D_{ij} = \frac{1}{2}(\dot{u}_{i,j} + \dot{u}_{j,i}) \tag{7.3}$$

For bodies undergoing large deformations, the stresses must be computed in some reference configuration such as the unrotated configuration [Flanagan and Taylor, 1987]. The rate-of-deformation tensor in the unrotated configuration d_{ij} is defined by:

$$d_{ij} = R_{ki} D_{kl} R_{lj} \tag{7.4}$$

where R_{ij} is the second-order orthogonal rotation tensor which appears in the polar decomposition theorem [Malvern, 1969]. The Cauchy stress in the unrotated configuration σ_{ij} is defined by:

$$\sigma_{ij} = R_{ki} T_{kl} R_{lj} \tag{7.5}$$

In the reference configuration, the constitutive relations are formulated such that the Cauchy stress rate is a function of the rate of deformation tensor, the stresses, and various state variables that describe the material's behavior. That is:

$$\dot{\sigma}_{ij} = \dot{\sigma}_{ij}(d_{kl}, \sigma_{kl}, \kappa, \alpha_{kl}) \tag{7.6}$$

where κ and α_{kl} represent scalar and tensor state variables, respectively. Because the constitutive relations are expressed in the unrotated coordinate system, objectivity (material frame indifference) requirements are automatically satisfied [Flanagan and Taylor, 1987]. The material models used to describe the constitutive behavior of solder are typically the same as those used to describe the inelastic behavior of other metals. That is, the stress σ_{ij} and rate of deformation tensors are decomposed into independent volumetric and deviatoric parts as follows

$$\sigma_{ij} = s_{ij} - p\delta_{ij} \tag{7.7}$$

and

$$d_{ij} = d'_{ij} + \frac{1}{3} d_v \delta_{ij} \tag{7.8}$$

where s_{ij} is the deviatoric stress tensor, d'_{ij} is the deviatoric part of the rate of deformation tensor, δ_{ij} is the Kronecker delta, p is the pressure $(-\sigma_{ii}/3)$, and d_v is the volumetric part (d_{ii}) of the stretching tensor. All stresses and deformations are defined to be positive in extension, but positive pressure corresponds to compression.

The volumetric response is assumed to be elastic with

$$\dot{p} = -Kd_v - 3\alpha\dot{\theta} \tag{7.9}$$

where K is the elastic bulk modulus, α is the coefficient of thermal expansion, and $\dot{\theta}$ is the rate of change of the temperature. The total deviatoric strain rate d'_{ij} is decomposed into elastic $(d'_{ij})^{el}$ and inelastic $(d'_{ij})^{inel}$ parts. That is,

$$d'_{ij} = (d'_{ij})^{el} + (d'_{ij})^{inel} \tag{7.10}$$

The deviatoric stress rate $\dot{s}_{ij}$ is related to d'_{ij} by:

$$\dot{s}_{ij} = 2G(d'_{ij})^{el} = 2G[d'_{ij} - (d'_{ij})^{inel}] \tag{7.11}$$

where G is the elastic shear modulus. The elastic constants and coefficient of thermal expansion can all be temperature-dependent. In some cases, the term $(d'_{ij})^{inel}$ is further decomposed as:

$$(d'_{ij})^{inel} = d_{ij}^{pl} + d_{ij}^{cr} \tag{7.12}$$

where d_{ij}^{pl} represents the plastic strain rate tensor and d_{ij}^{cr} represents the creep strain rate tensor. In other cases, no distinction is made between plasticity and creep, and the inelastic strain rate depends on the evolution of state variables representing the microstructure of the material. The specific form of $(d'_{ij})^{inel}$ in Equation (7-12) depends on the type of model used to describe material behavior (e.g. plasticity, creep, viscoplasticity, etc.). This discussion and treatment of elastic and inelastic strain rates is more traditional but less rigorous than that presented in Chapter 4, yet the two approaches usually lead to similar results for solder joint applications.

The general boundary value problem defined above requires assumptions about material behavior, boundary and initial conditions, geometry, and finite element approximations in order to produce solutions for a particular application. For solder joint applications, assumptions about one aspect of the problem, for example the geometry or material models, can restrict assumptions that can be made about other aspects of the problem. The the role of constitutive relations is described first. Boundary conditions and initial conditions are discussed next. Then the effects of geometrical assumptions are addressed. Finite element approximations are discussed last.

7.3.1 Constitutive Relations

The constitutive equations described by Equations (7-6) to (7-12) are expressed in tensor form and are applicable to any multiaxial stress state. However, constitutive theories are usually developed from materials tests where the stress states are well defined and controllable. In these tests, samples of solder are usually loaded in uniaxial tension or compression. The resulting constitutive relations are by default scalar in form and have to be extrapolated or generalized to multiaxial stress conditions through the use of equivalent stress σ_{eq} and equivalent strain ε_{eq} measures where σ_{eq} and ε_{eq} are functions of the stress and strain invariants, respectively. The number of equivalent inelastic strain measures is limited only by the deformation response of the material, and once these are identified, the corresponding work conjugate equivalent stress measures are automatically defined [Fossum et al., 1988]. After the equivalent stress and strain measures are chosen, procedures such as the one prescribed by Fossum et al. [1988] can be used to extrapolate the scalar equations to tensor equations. That is:

$$(d'_{ij})^{inel} = \frac{\partial \sigma_{eq_1}}{\partial \sigma_{ij}} \dot{\varepsilon}_{eq_1} + \frac{\partial \sigma_{eq_2}}{\partial \sigma_{ij}} \dot{\varepsilon}_{eq_2} + \ldots + \frac{\partial \sigma_{eq_n}}{\partial \sigma_{ij}} \dot{\varepsilon}_{eq_n} \tag{7.13}$$

where n denotes the number of equivalent strain measures needed to describe the material's behavior observed in the constitutive tests. For solder, inelastic behavior is observed predominantly in shear so only one equivalent strain, namely an equivalent inelastic shear strain and its corresponding stress measure, is needed for the generalization. However, several choices for the combinations of equivalent inelastic shear strain and equivalent stress are available and are equally valid based solely on uniaxial test results. Two of the most common are: (1) the equivalent deviatoric shear strain, which is associated with the von Mises stress; and (2) the maximum shear strain, which is associated with the maximum shear stress. The equivalent deviatoric shear strain rate $\dot{\bar{\varepsilon}}$ is defined by:

$$\dot{\varepsilon}_{eq} = \dot{\bar{\varepsilon}} = \sqrt{\frac{2}{3} d'_{ij} d'_{ij}} \tag{7.14}$$

and the von Mises stress $\bar{\sigma}$ is defined by:

$$\bar{\sigma} = \sqrt{3J_2} = \sqrt{\frac{3}{2} s_{ij} s_{ij}} \tag{7.15}$$

where J_2 is the second invariant of the deviatoric stress tensor s_{ij}. The maximum shear strain rate $\dot{\gamma}$ is defined as the difference of the maximum and minimum principal strain rates. The corresponding maximum shear stress τ is defined as the difference of the maximum and minimum principal stresses and is expressed as:

$$\tau = \sigma_1 - \sigma_3 = 2\sqrt{J_2} \cos\psi \tag{7.16}$$

where σ_1 and σ_3 are the maximum and minimum principal stresses, respectively, and ψ is the Lode angle defined by:

$$\psi = \frac{1}{3}\sin\left[\frac{-3\sqrt{3}J_3}{2J_2^{3/2}}\right], \qquad \left(-\frac{\pi}{6} \le \psi \le \frac{\pi}{6}\right) \tag{7.17}$$

where $J_3 = \frac{1}{3}s_{ij}s_{jk}s_{ki}$ is the third invariant of the deviatoric stress tensor.

The von Mises measures and the maximum shear or Tresca measures are identical for uniaxial (triaxial extension or compression) loading, but they are significantly different for other stress states. The von Mises measures are associated with circular cylindrical yield or flow surfaces in deviatoric stress and strain space, whereas the flow surface for the Tresca measure is a hexahedron in deviatoric stress space with vertices on axes of uniaxial tension and compression. The intersections of the two flow surfaces with the pi-plane (the plane whose normal is the hydrostat in deviatoric stress space) are illustrated in Figure 7-1, where S_1, S_2 and, S_3 represent principal deviatoric stress directions.

The appropriate equivalent stress and strain measures can sometimes be measured experimentally by subjecting thin-walled tubes to various combinations of internal pressure and torsion until the material flows. Plots of the principal stresses at yield for the various loading combinations then provide the shape of the flow surface from which the equivalent stress measure can be determined. However, experimentally determined flow surfaces can be difficult to obtain, especially for solder, and are often ambiguous, falling between the von Mises circle and Tresca hexagon.

Few data exist for determining the appropriate equivalent stress and strain measures for solder, but the von Mises measures are assumed because of their applicability for ductile metals. The von Mises measures are also easier to implement in finite element approximations to the boundary value problem. Differences between the von Mises and Tresca measures are small for plasticity (a maximum of about 17% in pure shear), but they can be significant for creep (up to a factor of two or more in creep strain rate) when the creep strain rate is dependent on the stress raised to a power larger than one. For rock salt, a creeping geologic material with microstructure similar to metals, von Mises and Tresca measures have produced significantly different results [Munson et al., 1989; Morgan and Krieg, 1988; Morgan and Wawersik, 1991]. This example does not imply that Tresca measures should be considered for solder, but it does point out that the measures should be chosen carefully and can have subtle and significant consequences. Only the von Mises measures will be considered for the remainder of this chapter.

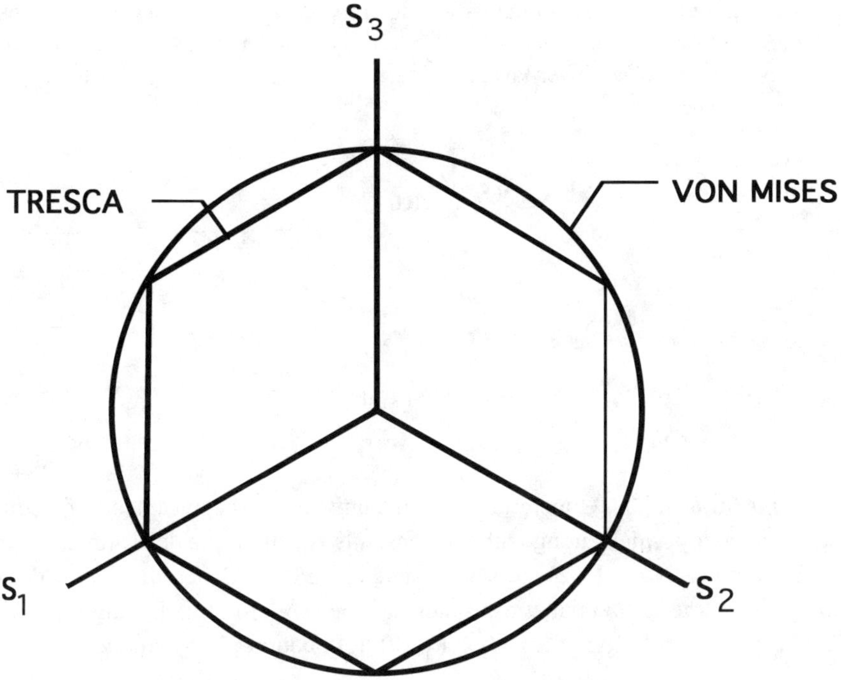

FIGURE 7-1 Von Mises and Tresca flow surfaces in deviatoric stress space.

To date, most analyses of solder joint behavior have used inelastic strain rate decompositions of the form presented in Equation (7.12). Furthermore, the inelastic behavior has usually been considered to be either entirely plasticity or entirely creep. When only plasticity is considered, the inelastic strain rate tensor in Equation (7-12) takes the form:

$$(d'_{ij})^{inel} = d_{ij}^{pl} = \frac{3}{2}\frac{s_{ij}}{\bar{\sigma}}\dot{\bar{e}} \tag{7.18}$$

after Equation (7-13) is invoked. The associated flow potential f has the form:

$$f = \bar{\sigma}^2 - k^2 \tag{7.19}$$

where k is the yield stress.

For creep, only secondary or steady state creep is typically considered in most solder joint analyses. The scalar creep strain rate-equivalent stress expression usually takes the form of a power law such as:

$$\dot{\bar{\varepsilon}} = A_0 \exp\left(-\frac{Q}{R\theta}\right)\bar{\sigma}^n \tag{7.20}$$

or

$$\dot{\bar{\varepsilon}} = A_0 \exp\left(-\frac{Q}{R\theta}\right)(\sinh\bar{\sigma})^n \tag{7.21}$$

where θ is the absolute temperature, R is the universal gas constant, and Q is the activation energy which, along with the constants A_0 and n are determined by fitting Equation (7-20) or (7-21) to steady state creep data. Generalization of these scalar relations to multiaxial stress states has the same form as Equation (7.18). For example, substitution of Equation (7-20) in Equation (7-18) produces:

$$(d'_{ij})^{inel} = d_{ij}^{cr} = \frac{3}{2}\frac{S_{ij}}{\bar{\sigma}}\dot{\bar{e}} = \frac{3}{2}A_0 \exp\left(-\frac{Q}{R\theta}\right)\bar{\sigma}^{n-1}S_{ij} \tag{7.22}$$

An important feature of the tensor relations is the multiplier 3/2 in front of the constant A_0 from the scalar strain rate expression. Conversion of almost all scalar viscoplastic strain rate relations to tensor form requires multipliers of 3/2 and 2/3 in the appropriate places. The appropriate multipliers result naturally from use of conversion relations like Equation (7-13). However, problems can arise in finite element analyses of solder joint behavior if parameters for the constitutive models are not input carefully. Analysts must know the form (either scalar or tensor) of the parameters required as input for the finite element code selected for use. This is especially important if a complicated constitutive relation requiring user subroutines, a feature common in most general purpose nonlinear finite element codes, has been chosen. The correct constitutive parameter input can be verified by running small, one-element, uniaxial loading test cases and comparing the results to those arising from closed form or numerical integration of the scalar constitutive relations.

Another important feature of the constitutive relations is the presence of the deviatoric stress invariants. Each component of the strain rate tensor depends on all of the stress components. As will be discussed later, this requirement must be considered carefully when making 2-D modeling assumptions.

Under most thermomechanical loading conditions, the behavior of a solder joint cannot be characterized as either pure plasticity or pure creep. As should be expected, for a given loading condition, behavior predicted with plasticity models is significantly different from behavior predicted with secondary creep models [Stephens et al., 1991], and neither is probably correct by itself. It is hard to justify modeling time-dependent behavior with time independent plasticity and transient behavior with secondary creep models. Thus, some analysts have used the combination of both plasticity and creep represented in Equation (7-12) [Pao, Chen, and Kuo, 1991]. The decomposition of the inelastic strain rate into plastic and creep components in Equation (7-12) is computationally appealing because it allows the use of reasonably simple constitutive models. Implementations of conventional metal plasticity and secondary creep models are usually found in most finite element codes. Furthermore, model parameters are usually well known quantities, e.g., the yield stress, that can be easily determined from conventional materials tests. However, from a theoretical and experimental perspective, plastic and creep strains are indistinguishable, and the decomposition in Equation (7-12) is hard to justify as more than a mere mathematical convenience.

A theoretically more appealing approach is to use viscoplastic models in which the inelastic strain rate is a function of state variables that can be related to deformations on a microstructural level [Miller, 1987; Krieg, 1977; Morgan and Krieg, 1988]. The constitutive relations describe how these state variables evolve with time. These models make no distinction between plasticity and creep but are capable of reproducing constant strain rate tests, creep tests (including transients), or relaxation tests depending on how the state variables evolve. Coarsening of the microstructure of solder joints with time can be included in models of this type (see Chapter 4) by defining the appropriate state variables. However, models of this type also have their drawbacks when used for making engineering predictions. One is the large number of material parameters required by the models. These parameters are hard to extract from simple materials tests, and determination of the right type and number of tests needed to define each of the parameters uniquely is extremely difficult. Another drawback is the mathematical stiffness of the constitutive equations. Either extremely small time steps, that greatly increase the computational cost of solder joint simulations, are required to integrate the constitutive equations, or alternate finite element solution schemes designed to integrate the equations as efficiently as possible are required. Most engineering investigations are usually constrained by computing resources and limited access to specialized finite element codes. As a result, the use of internal state variable models is currently focused on research where the models are used to provide a fundamental understanding of solder joint behavior. Once this understanding is obtained, they can be condensed into simpler models or used to develop design guides for engineering applications.

7.3.2 Boundary Conditions

Boundary conditions describe the loads applied to the solder joint interconnect and essentially define the type of model that is used to predict its response. If the electronic package is subjected to shock loading, a transient dynamics analysis would probably be required. Similarly, if vibrations of gullwing or J leads are important, eigenanalyses would be in order. For thermomechanical loading, the problem of interest here, the temperature field and its history must be known and idealized. Temperature field uniformity, the range of temperatures over which the interconnect is cycled, ramp times between temperature extremes, and hold times at a specified temperature are all important in defining the type of analysis that is performed. For example, if the hold times are short, creep is often ignored in the constitutive relations. Conversely, if hold times are long, only creep is usually considered.

In the most rigorous case, initial temperatures are specified for all parts of the interconnect, and a thermal analysis is performed to determine the temperature gradients throughout the interconnect for the specified temperature cycles. The temperature field history is then used as input to the thermal stress model. However, in most solder joint applications, the temperature fields are assumed to be uniform throughout the interconnect as the temperature is cycled between its specified extremes. This assumption eliminates the need for the thermal analysis.

The specified temperature ranges are also important in determining the finite element model that is used in simulating solder joint response. Military and automotive industry specifications usually require that solder joints survive cycling between extremes of –55°C and 125°C with long hold times at these extremes. Under this type of loading, observed joint failures are associated with bands of coarsened microstructure that form during cycling (see Chapters 2 and 4). Predicting failure for this type of loading would probably require a continuum mechanics model with constitutive relations such as those in Chapter 4 that account for the evolution of the microstructure. Many commercial electronic specifications only require that solder joints survive cycling between 20°C and 60°C. Under these conditions, surface cracks have been observed to develop early in the life of the joint. These cracks coalesce and grow until the joint fails, as discussed in Chapter 4. Simulating this type of response might require fracture mechanics models quite different from the continuum models used to simulate the response to military specifications. Simulating the response of solder joints to military, automotive, or commercial specifications is useful for ensuring that a joint will meet design criteria, but it might not provide good life predictions because temperature fluctuations when the device is in actual service are random. Simulating the behavior of solder joints under actual service conditions would obviously require statistical solution methods. Other sources of randomness, including the solder material itself, reinforce the need for statistical methods and further experimental results.

7.3.3 Initial Conditions

In addition to the temperature at the start of the analysis, the initial stress field in the joint and possibly the initial microstructure of the joint must be defined. Analyses are usually initiated at points in the manufacturing process where the joint is stress free, for example at the melt temperature. The stress field at any other temperature can be computed from this reference condition. However, the joint may also be considered stress free at any other starting temperature if the joint is held at that temperature long enough for the stresses to relax due to creep. This assumption is usually valid as long as the geometry of the joint is not constrained so much that relaxation cannot occur [Stephens, Burchett, and Hosking, 1991].

7.3.4 Geometrical Assumptions

In most electronic packages with solder joint interconnections, multiple solder joints are used and spaced at various intervals around the package. The finite element model in Figure 7-2 illustrates a typical solder joint arrangement. This geometry is obviously three-dimensional, and the behavior of a specific joint is affected by the number and spacing of neighboring joints. Thus, a three-dimensional ide-

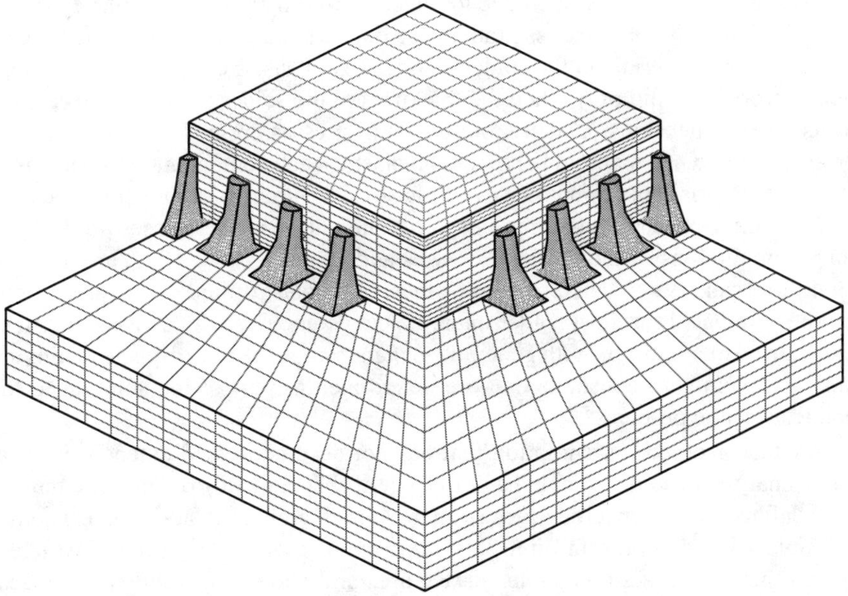

FIGURE 7-2 3-D finite element geometry of finite element simulations.

alization of the entire package would be most appropriate. The resulting finite element model would be enormous even if geometrical symmetries could be invoked to reduce the model to its smallest possible size. Some analyses have been performed with 3-D models of this type [Burchett, 1991], but they are currently beyond the resources of most investigators, especially if the solder is treated with inelastic constitutive models. Further idealizations of the actual geometry are usually required. The next level of reduction is to consider the 3-D geometry of a single interconnection with appropriate boundary conditions to represent its connection to the rest of the package. Even with these size reductions, the resulting 3-D finite element models needed to provide accurate predictions of joint behavior are quite large. Consequently, most solder joint analyses are based on 2-D idealizations of complicated 3-D package geometries.

Use of 2-D assumptions can provide valuable information about joint behavior if they are used properly. Typically, the 2-D model represents a cross-section taken at the center of the solder joint. These models are capable of capturing stress gradients in this plane but not those normal to the plane. Furthermore, the out-of-plane stresses vary with the 2D modeling assumption chosen. If plane strain is assumed, predictions of the out-of-plane normal stress are probably incorrect for thermomechanical loading. For example, an unrestrained layered assembly consisting of a thin layer of solder connecting a chip to a carrier plate, as shown in Figure 7-3, will have some layers in tension and others in compression in both the in-plane and out-of-plane directions. With plane strain models, the out-of-plane stresses in all layers are tensile due to the constraint that the out-of-plane normal strain is zero. With plane stress models, the out-of-plane stresses are zero, by definition, for all materials in the soldered assembly. Thus, the plane stress assumption is probably valid only when the thickness of the solder joint is small relative to its other dimensions. Usually, out-of-plane stresses are not of interest in 2-D problems, but when material behavior is inelastic, each component of the inelastic strain rate tensor depends on the deviatoric stress invariants, as shown in Equation (7-22). Thus, inaccurate assumptions about the out-of-plane stresses produce errors in predictions of the in-plane stresses and strains [Akay, Tong, and Paydar, 1992; Morgan, 1991]. The problem of inaccurate out-of-plane stress predictions can be eliminated for most solder interconnect geometries with an assumption of generalized plane strain. With generalized plane strain, the out-of-plane strains are unknown, non-zero constants that are determined as part of the solution to the boundary value problem.

Another geometric assumption that has not been addressed extensively, if at all, in analysis of solder joint behavior is that of shape. Most structures are built to well defined design specifications so there is little ambiguity about their geometry. Solder joints do not fall into this category. They begin as a liquid which is spread over a specified wettable surface area and allowed to solidify. The final shape of the solder fillet is determined by a balance of forces acting on the system. These forces, which can include interfacial tension, gravity, electromagnetic forc-

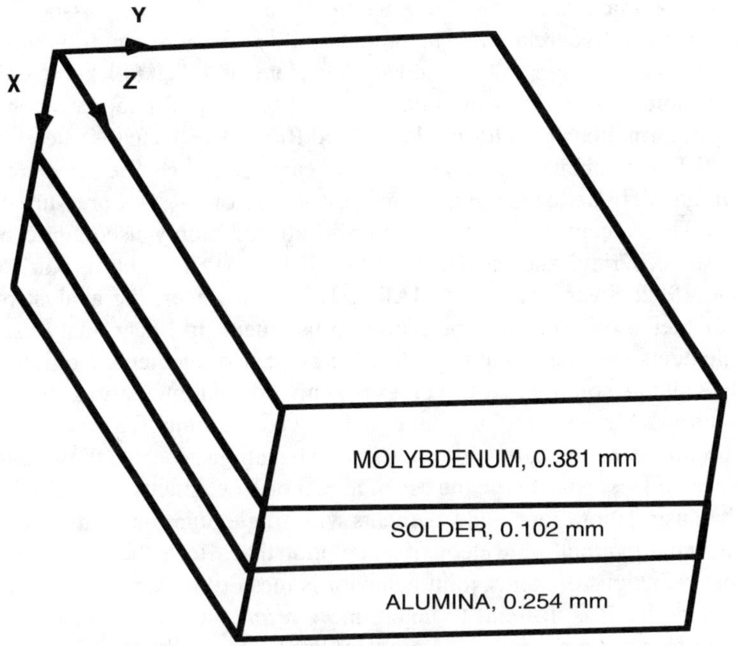

FIGURE 7-3 Layered assembly with solder connection.

es, and forces due to fluid motion, are discussed extensively in Chapter 5 along with methods of predicting actual solder joint shapes from the melt. In finite element models, solder joint shapes are typically assumed to be very smooth with a well defined heights, widths, and radii of curvature, as shown in Figure 7-2. However, actual joint shapes can be highly irregular. The consequences of the smooth shape assumptions on interconnect response are currently unknown although methods of evaluating them are becoming available. The ability to predict the joint shape resulting from the melt already exists as discussed in Chapter 5. In addition, robust mesh generation tools for automatically generating good meshes for arbitrary 2-D shapes exist and are being extended to 3-D geometries [Blacker and Stephenson, 1990; Stephenson, Canaan, and Blacker, 1992]. Thus, only the integration of the shape predictors with the finite element mesh generators needs to be completed.

7.3.5 Finite Element Assumptions

As illustrated above, many of the assumptions about boundary and initial conditions, constitutive theories, and geometry affect each other. The same is true of finite element assumptions such as element choice and discretization. Element choice is important because of the incompressibility constraint associated with inelastic constitutive relations. Incompressibility imposed on fully integrated finite elements results in an over-constrained element which blocks several shear modes of deformation that would otherwise be available for capturing shear and bending deformations [Nagtegaal, Parks, and Rice, 1974]. That is, the element is overly stiff. This phenomenon, known as shear locking, can be prevented, or at least mitigated, by reduced integration or by the use of constant pressure over the element. These techniques are available in all commonly used finite element codes [Stone, Krieg, and Beisinger, 1985; Biffle, 1984; Hibbitt, Karlson, and Sorenen, 1992; Swanson, 1992; MARC, 1992]. However, the analyst must be aware of their existence and know how to use them. In linear analyses, higher order elements are often used to reduce the number of elements required to produce an accurate solution. However, even with reduced integration, the constitutive relations must still be integrated at four Gauss points for eight-noded elements. With mathematically stiff constitutive equations, such as those associated with creep and viscoplasticity, the use of higher order elements is usually less efficient than using more four-node elements with single point integration.

A second important finite element assumption that affects the accuracy of finite element predictions of solder joint behavior is mesh refinement once an element type is selected. It is difficult to determine *a priori* where the highest levels of mesh refinement are needed. This problem has traditionally been solved by running multiple analyses with each successive analysis having increased mesh refinement until changes in the solutions produced by the changes in meshing become acceptably small. Adaptive analysis techniques, which use error measures to automatically refine or coarsen regions of the mesh, are currently becoming available. Another problem related to meshing is the existence of singularities or large gradients at the free edges of solder joint interconnections. These regions are prime locations for joint failures and thus need to be modeled accurately. Extremely fine meshes in these areas are usually needed to determine stress intensity factors associated with these singularities. A possible remedy to this problem is to use boundary element instead of finite element techniques to predict solder joint behavior [Lau, 1992b]. Boundary element methods require only the boundary to be discretized, and they handle singularities very well. However, most successes with boundary element methods have been associated with linear problems. Although they have been applied to problems involving viscoplastic material response [Mukherjee, 1982], they appear to be cumbersome and only marginally more efficient than the finite element method, if at all.

7.4 CONCLUSIONS

In this chapter, a brief overview has been presented that describes the level of complexity in finite element modeling that is required to accurately capture the behavior of solder joints under thermomechanical loading. The boundary value problem, which is the mathematical representation of the physical problem, requires idealizations of the material behavior, boundary and initial conditions, and the geometry. Accurate assumptions about all components of the problem are required to capture joint behavior quantitatively. Furthermore, assumptions about one component of the problem can affect or limit assumptions that can be made about the other components. The uses of the models are limited only by the complexity of the assumptions. For example, a three-dimensional model, with a constitutive model that treats both plasticity and creep as the evolution of microstructural parameters, can probably capture the behavior of joints observed in thermomechanical loading tests. Simpler models, that use 2-D idealizations of the geometry and elastic material models, can also provide valuable information as long as their limitations are well understood. These models are particularly useful if quick, qualitative information about various design options are needed.

The modeling techniques discussed above are based on the overlying assumption that solder joints can be modeled as a continuum. This may be valid for most packages, but as they get smaller and smaller, solder joint dimensions begin to approach the size of individual grains within the joint. Consequently, continuum models may cease to be valid, and other models in which the grains are treated individually will need to be developed.

TABLE 7-1 Nomenclature

Quantity	*Description*
A_0	Secondary creep parameter determined by fitting creep model to data
b_i	Body force vector
d_v	Volumetric part of the unrotated rate of deformation tensor
d_{ij}	Rate of deformation tensor in the unrotated configuration
D_{ij}	Rate of deformation tensor in the current configuration
d_{ij}^{cr}	Creep strain rate tensor
$(d'_{ij})^{el}$	Elastic part of the deviatoric strain rate tensor
$(d'_{ij})^{inel}$	Inelastic part of the deviatoric strain rate tensor
d_{ij}^{pl}	Plastic strain rate tensor
d'_{ij}	Deviatoric strain rate tensor
f	Plastic flow potential
G	Elastic shear modulus
J_2	Second invariant of the deviatoric stress tensor

TABLE 7-1 Nomenclature (continued)

Quantity	*Description*
J_3	Third invariant of the deviatoric stress tensor
k	Radius of the yield surface
K	Elastic bulk modulus
n	Secondary creep parameter determined by fitting creep model to data
n_j	Outward normal vector
p	Pressure
$\dot{p}$	Rate of change of pressure
Q	Activation energy
R	Universal gas constant
R_{ij}	Orthogonal rotation tensor from the polar decomposition theorem
S_i	Principal deviatoric stress
s_{ij}	Deviatoric stress tensor
S_T	Portion of a surface over which tractions are specified
S_U	Portion of a surface over which displacements are specified
t	Time
t_i	Traction vector
T_{ij}	Cauchy stress tensor
T_{ij}^0	Initial Cauchy stress tensor
u_i	Displacement vector
U_i	Specified displacement vector
$\dot{u}_i$	Velocity vector
$\dot{U}_i^0$	Initial velocity vector
$\dot{u}_{i,j}$	Partial derivative of the velocity vector with respect to the j coordinate
$\ddot{u}_i$	Acceleration vector
V	Volume
α	Coefficient of thermal expansion
α_{ij}	State variables tensor
δ_{ij}	Kronecker delta
ε_{eq}	Equivalent strain

TABLE 7-1 Nomenclature (continued)

Quantity	*Description*
$\dot{\varepsilon}_{eq}$	Equivalent strain rate
$\dot{\bar{\varepsilon}}$	Equivalent deviatoric shear strain rate
θ	Temperature
$\dot{\theta}$	Rate of change of temperature
κ	Scalar state variable
ρ	Density
σ_i	Principal Cauchy stress in the unrotated configuration
σ_{ij}	Cauchy stress tensor in the unrotated configuration
$\dot{\sigma}_{ij}$	Cauchy stress rate tensor in the unrotated configuration
σ_{eq}	Equivalent stress
$\bar{\sigma}$	Von Mises stress
τ	Maximum shear stress
ψ	Lode angle

REFERENCES

Akay, H. U., Y. Tong, and N. Paydar, 1992, *Proc. of the International Electronics Packaging Conference* (Sept. 27-30). Austin, TX.

Biffle, J. H., 1984. *SAND81-0998*, Albuquerque, NM: Sandia National Laboratories.

Blacker, T. D. and M. B. Stephenson, 1990, SAND90-0249. Albuquerque, NM: Sandia National Laboratories.

Burchett, S. N., 1991, Paper No. 91-WA-EEP-1. New York, NY: The American Society of Mechanical Engineers.

Clatterbaugh, G. V. and H. K. Charles, Jr., 1985. *Proceedings of IEEE 35th Electronic Components & Technology Conference*. pp. 60-71.Washington, D. C.

Dasgupta, A., S. Verma, and D. Barker, 1992, Paper No. 92-WA/EEP-28. New York, NY: The American Society of Mechanical Engineers.

Engel, P. A. and K. R. Wu, 1992, *Transactions of the ASME, Journal of Electronic Packaging*. Vol. 114. pp. 314-321.

Felske, K. E., and J., Moss, 1986, *Proc. of IEEE 36th Electronic Components & Technology Conf.* pp. 641-648. Seattle, WA.

Flanagan, D.P. and L. M. Taylor, 1987,*Computer Methods in Applied Mechanics and Engineering*. 62, pp. 305-320.

Fossum, A. F., G. D. Callahan, L. L. Van Sambeek, and P. E. Senseny, 1988, *Key Questions in Rock Mechanics, Proceedings of the 29th U.S. Symposium on Rock Mechanics*. Rotterdam: Balkema.

Harada, M. and R. Satoh, 1990, *Proc. of IEEE 40th Electronic Components & Technology Conference*. pp. 510-517. Las Vegas, NV.

Hibbitt, Karlsson, & Sorensen, Inc., 1992, *ABAQUS User's Manual*. Pawtucket, RI.

Hughes,T. R. J., 1987. *The Finite Element Method, Linear Static and Dynamic Finite Element Analysis*. Englewood Cliffs, NJ: Prentice-Hall.

Jahsman, W. E. and P. Jain, 1990, *Proceedings of IEEE 40th Electronic Components & Technology Conference*. pp. 926-932. Las Vegas, NV.

Kasem, Y. M. and L. G. Feinstein, 1987, *Proc.of IEEE 37th Electronic Components & Technology Conference*. pp. 96-104. Boston, MA.

Krieg, R. D., 1977, Paper M6/4. *Proceedings of the 4th SMIRT Conference*. San Francisco, CA.

Lau, J. H., ed., 1991. *Solder Joint Reliability - Theory and Applications*. New York: Van Nostrand Reinhold.

Lau, J. H.,1992a. Private Communication. Third International Workshop on Materials and Mechanics Issues of Solder Alloy Applications (September 9-11). Santa Fe, NM.

Lau, J. H., 1992b, Paper No. 92-WA/EEP-34. New York, NY: American Society of Mechanical Engineers.

Lau, J.H. and G. Harkins, 1987, *Proc. of IEEE 38th Electronic Components & Technology Conference*. pp. 23-31. Los Angeles, CA.

Lau, J., G. Harkins, D. Rice, J. Kral, and B. Wells, 1987, *Proc. of IEEE 37th Electronic Components & Technology Conference*. pp. 589-597. Boston, MA.

Lau, J. H. and C. A. Keely, 1989, *Proc. of IEEE 39th Electronic Components & Technology Conference*. pp. 603-615. Houston, TX.

Lau, J. H. and L. L. Moresco, 1988, *Proc. of IEEE 38th Electronic Components & Technology Conference*. pp.168-175. Los Angeles, CA.

Lau, J. H. and D. W. Rice, 1990, *Proc. of IEEE 40th Electronic Components & Technology Conference*. pp. 824-834. Las Vegas, NV.

Lau, J. H., D. W. Rice, and C. G. Harkins, 1989, *Proc. of IEEE 39th Electronic Components & Technology Conference*. pp.456-463. Houston, TX.

Malvern, L. E., 1969, *Introduction to the Mechanics of a Continuous Medium*, Englewood Cliffs, NJ: Prentice-Hall.

MARC Analysis Research Corp., 1992. MARC. Palo Alto, CA.

Miller, A. K., ed., 1987, *Unified Constitutive Equations for Creep and Plasticity*. Barking Essex, England: Elsevier Applied Mechanics Publications.

Morgan, H. S., 1991, *Transactions of the ASME, Journal of Electronic Packaging*. Vol. 113. pp. 350-354.

Morgan, H. S. and R. D. Krieg, 1988, *A Comparison of Unified Creep-Plasticity and Conventional Creep Models for Rock Salt Based on Predictions of Creep Behavior Measured in Several In Situ and Bench-Scale Experiments*. SAND87-1867. Albuquerque, NM: Sandia National Laboratories.

Morgan, H. S. and W. R. Wawersik, 1991, *Proc. of the 7th International Congress on Rock Mechanics*. Aachen, Germany.

Mukherjee, S., 1982, *Boundary Element Methods in Creep and Fracture*. London & New York: Applied Science Publishers.

Munson, D. E., A. F. Fossum, and P. E. Senseny, 1989, *Advances in Resolution of Discrepancies Between Model Prediction and Measured In Situ WIPP Room Closures*. SAND88-2948. Albuquerque, NM: Sandia National Laboratories.

Nagtegaal, J. C., D. M. Parks, and J. R. Rice, 1974,*Computer Methods in Applied Mechanics and Engineering*. 4. pp. 153-157.

Ozmat, B., 1990, *Proc. of IEEE 40th Electronic Components & Technology Conference*. pp. 959-972. Las Vegas, NV.

Pan, T.-Y., 1991, *Transactions of the ASME, Journal of Electronic Packaging*. Vol.113. pp. 8-15.

Pao, Y. H., K. L. Chen, and A. Y. Kuo, 1991, *Materials Research Society Proc.* Vol. 226. The Materials Research Society.

Pao, Y. H., R. Govila, S. Badgley, E. Jih, 1992, *Transaction of the ASME, Journal of Electronic Packaging*, Vol. 115.

Prakash, V. and Y. P. Yang, 1992, Paper No. 92-WA/EEP-20. New York, NY: American Society of Mechanical Engineers.

Ross, R. G., L. C. Wen, and G. R. Mon, 1992, Paper No. 92-WA/EEP-6. New York, NY: American Society of Mechanical Engineers.

Sandor, B.I., 1991, *Solder Mechanics - A State of the Art Assessment*. D. B. Frear, W. B. Jones, and K. R. Kinsman, ed. Warrendale, PA: The Minerals, Metals, and Materials Society.

Sarihan, V., 1992, Paper No. 92-WA/EEP-38. New York, NY: American Society of Mechanical Engineers.

Sherry, W. M., J. S. Erich, M. K. Bartschat, and F. B. Prinz., *Proc. of IEEE 35th Electronic Components & Technology Conference*. pp. 81-90. Washington, D.C.

Smeby, J. M., 1985, *IEEE Transactions on Components, Hybrids, and Manufacturing Technology*. CHMT-8, No. 3. pp. 391-396.

Stephens, J. J., S. N. Burchett, and F. M. Hosking, 1991, *Proc. of the Metals, Minerals, and Materials Society: Metal/Ceramic Joining*.

Stephenson, M. B., S. A. Canann, and T. D. Blacker, 1992, *Plastering: A New Approach to Automated, 3D Hexahedral Mesh Generation, Progress Report I*. SAND89-2192. Albuquerque, NM: Sandia National Laboratories.

Stone, C. M., R. D. Krieg, and Z. E. Beisinger, 1985, *SANCHO - A Finite Element Computer Program for the Quasistatic, Large Deformation, Inelastic Response of Two-Dimensional Solids*. SAND84-2618. Albuquerque, NM: Sandia National Laboratories.

Swanson Analysis Systems, 1992, ANSYS. Houston, TX.

Tervalon, M. J., 1990, *Proc. of IEEE 40th Electronic Components & Technology Conference*. pp. 954-957. Las Vegas, NV.

Tong, H. M., L. Mok, K. R. Grebe, and H. L. Yeh, 1990, *Proc. of IEEE 40th Electronic Components & Technology Conference*. pp. 345-350. Las Vegas, NV.

Waller, D. L., L. R. Fox, R. J. Hannemann, 1983, *IEEE Transactions on Components, Hybrids, and Manufacturing Technology*. CHMT-6, No. 3. pp. 257-266.

Zienkiewicz, O. C., 1977, *The Finite Element Method*. New York: McGraw-Hill.

8

Applications—Through-Hole

S. N. Burchett

8.1 INTRODUCTION

Through-hole solder joint interconnection technology has been the mainstay of the electronics industry. Although the current trend has been to implement surface mount technology (SMT) for increased density leading to improved performance in smaller and smaller packages, through-hole solder joint interconnections continue to be essential. It has been estimated [Pecht,1991] that in 1992 only 30 to 50% of new and revised circuit designs will incorporate SMT; the rest of the circuits will utilize through-hole technology. Even in SMT technology, through-hole vias are increasingly required.

The printed wiring board (PWB) is the basic platform for an electronics assembly and must be manufactured to accept the components, either discrete through-hole or SMT. The printed wiring board is a complex laminated structure consisting of one or more layers of fabric-reinforced polymers. Each layer initially has a copper coating on each side which is subsequently processed (coated with photoresist, imaged, and etched) to form the circuits. The layers are then stacked together with B-staged fabric-reinforced polymers (called prepregs) in between each layer, laminated, and cured using heat and pressure. After lamination, holes are drilled through the PWB and subsequently plated with copper to interconnect between appropriate circuit layers. After pretinning, discrete component leads are inserted into the plated PWB holes and soldered into place (Figure 8-1). The solder mechanically and electrically joins the discrete component to the appropriate electrical circuit.

Since through-hole solder interconnection has been used for many years, it is considered by many to be a "mature technology" with established high reliability. The reliability was primarily established through the use of extensive experimental testing programs in the 1960's and 1970's [Berkebile, 1967; Van Orden, 1968; Shamash, 1968; Mohler, 1971; Wild, 1972; Munford, 1975; Kobayashi, 1975; Bangs and Beal, 1975; Baker, 1975]. Literature references to mechanical modeling of through-hole solder interconnections are, however, very limited. The reason for this apparent gap is understandable since the technology was developed at a time when only "strength of materials" type approaches were available and advanced continuum analysis techniques were in their infancy. The technology to computationally evaluate solder interconnections in a thorough, realistic, and verifiable manner without significant simplifying assumptions is just coming into place primarily through the use of the finite element method. The application of finite element analysis (FEA) to solder joints is increasing rapidly. However, the majority of this work is in SMT [Sandor, 1991]. The history of through-hole solder joint interconnection technology and the drive in the electronics industry toward SMT suggest that it might not be necessary to pursue mechanical modeling of the through-hole solder joint. Evolutionary technology advances, such as increased thickness multilayer PWBs, new materials for PWBs, increased component density, and pin-grid arrays (PGA), coupled with the projection that through-hole solder joint technology will be utilized for many years suggests that mechanical modeling of these type of solder joints will be essential, however.

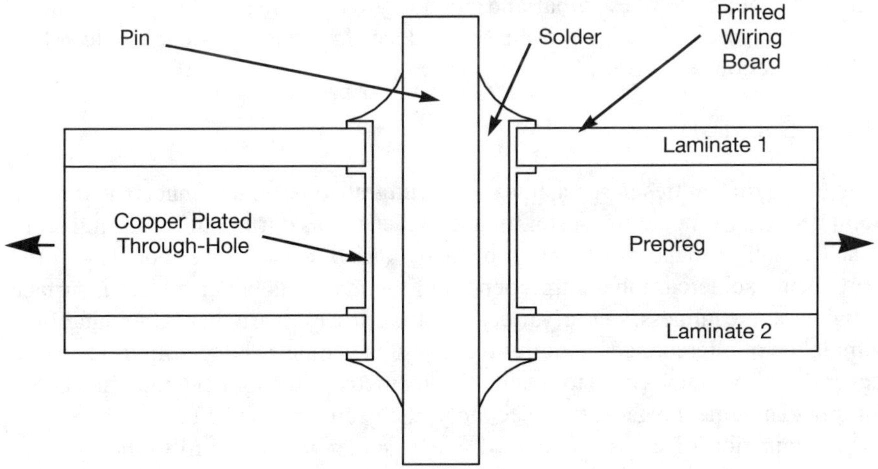

FIGURE 8-1 Idealized through-hole solder joint.

In earlier chapters of this book the three essential aspects of modeling were discussed: the interconnection material, the geometry of the joint, and thermomechanical modeling considerations. Three chapters dealt with the interconnection material in detail: the basic material properties of solder (Chapter 2), the interface materials formed during the soldering process (Chapter 3), and the constitutive modeling (mathematical representation) of solder (Chapter 4). One chapter was devoted to the issue of predicting solder joint shapes from the melt (Chapter 5) and one chapter was devoted to mechanical modeling considerations (Chapter 7). In the next two chapters, the mechanics of specific solder joint applications including through-hole solder joints (Chapter 8) and surface mount solder joints (Chapter 9) will be discussed in detail. The purpose of these chapters is to apply the aspects of mechanical modeling specifically to the through-hole solder joint. The chapter begins with a discussion of modeling considerations and will characterize the problems associated with through-hole solder joints. Next, specific modeling examples will be discussed and will include the analysis of the stresses and strains in a plated through-hole without solder, an analysis of local stresses and strains in a through-hole solder interconnection, analysis of an encapsulated (or conformal coated) component, and an analysis of a PGA structure. The chapter concludes with a discussion of lessons learned and questions for the future.

8.2 MODELING CONSIDERATIONS

As stated previously (Chapter 7), three key ingredients must be specified in order to perform any computational analysis: the geometry, the material response, and the loading that includes initial and boundary conditions. In this section, these three key ingredients will be discussed in detail for a plated through-hole solder interconnection.

8.2.1 Geometry

One of the difficulties in mechanical modeling of solder interconnects is that the joint geometry is not known prior to (and in fact, forms during) the soldering process. The solder shape is a function of many variables including geometry of the parts being soldered, inherent properties of the materials being soldered, surface roughness, cleanliness, and a variety of other process variables. Computational simulation of interconnects requires that the joint geometry be specified. There are several methodologies used to specify the geometry of the joint with the first being simply to assume the geometry. Secondly, the geometry can be defined by physical examination of cross-sections of actual solder joints. Thirdly, the external geometry can be obtained from advanced inspection techniques [Read and Bayba, 1991; Paydar et al.,1991]. Finally, methodologies are being developed to a priori predict the geometry of the joint (Chapter 5). Of these methodologies, physical cross sections and simple assumed geometries are the most widely used; however, the coupling of predictive joint geometry capability with thermomechanical modeling offers the greatest potential for the future.

In an idealized through-hole solder joint, shown schematically in Figure 8-1, the lead of a discrete component is inserted into the copper plated through-hole (PTH) of the printed wiring board which is interconnected to the circuit layers. Solder fills the annulus between the component lead and the PTH, thus, electrically and mechanically joining the component to the PWB. Assuming that the component lead is circular in cross section, the lead axis is on the centerline, and solder completely fills the annulus, the geometry of the solder joint structure is axisymmetric (two-dimensional). Chu [1975] and Rooks et al. [1991] present analytical models for predicting axisymmetric solder joint shape profiles for through-hole applications (Chapter 4, Figures 4-23 and 4-24). Without additional specific information, these figures can be used to obtain "idealized" joint geometries for analysis.

Actual through-hole geometries rarely match the idealized geometry. Geometry variations include the leads not being on the centerline, the leads being tilted, and solder defects such as voids in the solder annulus (Figure 8-2). All of these geometrical variations can significantly affect the reliability of the through-hole solder joint. From a modeling standpoint, the variations have the effect of making the geometry three-dimensional. If the effects of these variations are to be considered, a three-dimensional analysis is required.

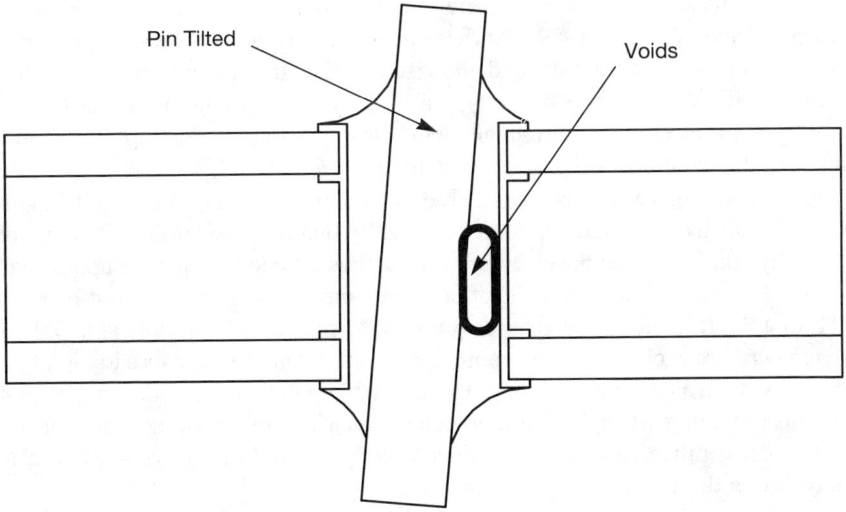

FIGURE 8-2 Schematic of "typical" through-hole solder joint.

Figure 8-1 also idealizes the thickness of the through-hole plating to be constant and uniform which, again, is rarely the case. Variations in the thickness of the plating can significantly affect the mechanical or thermomechanical response of a through-hole joint.

8.2.2 Materials and Material Response

While this book is intended to primarily examine solder interconnects, the response of the solder in an interconnection cannot be divorced from the other materials in the assembly, especially in a through-hole joint. In fact, the reliability of through-hole joints is more closely associated with failures in the copper plated through-hole or between the copper plating and the circuit inner layers. Returning to the sketch in Figure 8-1, the printed wiring board is a complicated structure formed by impregnating a low thermal expansion woven fabric with a polymeric material. The purpose of the woven fabric is to strengthen the laminate and to lower the coefficient of thermal expansion (CTE) characteristics primarily in the plane of the laminate. The PWB is therefore neither homogeneous nor isotropic. For a multilayer PWB, several copper circuit layers and non-functional pads may also be included. The CTE in the plane of the laminate can be significantly less than the CTE out of plane. Furthermore, the in-plane modulus is typically much higher than the out-of-plane modulus. Another complicating factor is that mechanical properties of PWBs are highly temperature-dependent and possibly process history dependent. The temperature dependence is caused by the polymeric material which, during normal thermal processing such as soldering and normal in-service environmental temperatures, may be operating both above and below the glass transition temperature of the polymer (especially for epoxy boards) [Klein Wassink, 1989]. At these temperatures, the mechanical properties of a polymer are also time-dependent (viscoelastic). In most mechanical or thermomechanical models the PWB is typically assumed to be a homogeneous, transversely isotropic, temperature-dependent, elastic material. For highly accurate analyses, the additional effects must be included.

The second material to be considered is the copper used to plate the through-hole. The electrodeposited copper mechanically bonds to the laminate and metallurgically bonds to the exposed circuit layers, thus interconnecting the appropriate electrical circuits. Since the CTE of copper is much lower than the out-of-plane CTE of a PWB laminate, stresses develop in the plated through-hole assembly as the temperature is changed from room temperature. During exposure to high temperatures, such as during soldering, tensile yielding can occur in the copper plating (this problem will be discussed in detail later). Therefore, an inelastic (elastic-plastic) constitutive model for the ductile copper should be incorporated in an analysis of a through-hole solder joint.

The pin is typically a ductile metal such as Kovar, Alloy-42, or copper; it can be modeled using a temperature dependent linear elastic constitutive model.

The basic material properties of solder, the constitutive models for solder, and the implementation of the solder constitutive model into FEA programs were discussed previously (Chapters 2, 4, 7). These discussions amplify the difficulty in predicting the response of solder interconnections since the behavior of solder is time- (creep), temperature-, and microstructure-dependent.

8.2.3 Loading and Initial Conditions

Thermal Loading

Three sources of thermal loading for through-hole solder joints have been previously identified by Munford [1975] and an additional source by Jeannotte et al. [1989]. Thermal loadings for component assemblies which are not encapsulated or conformal coated are:

Type I: Thermomechanical loading due to the thermal expansion mismatch between the conductor pin, the solder, the copper plated through-hole and the PWB, Figure 8-3. This type of loading is of primary importance in solder-filled through-hole via.

Type II: Thermomechanical bending loads in the conductor pins due to the horizontal thermal expansion mismatch between a component which has two or more leads and the PWB, Figure 8-4. This thermal loading is particularly of interest for pin grid arrays.

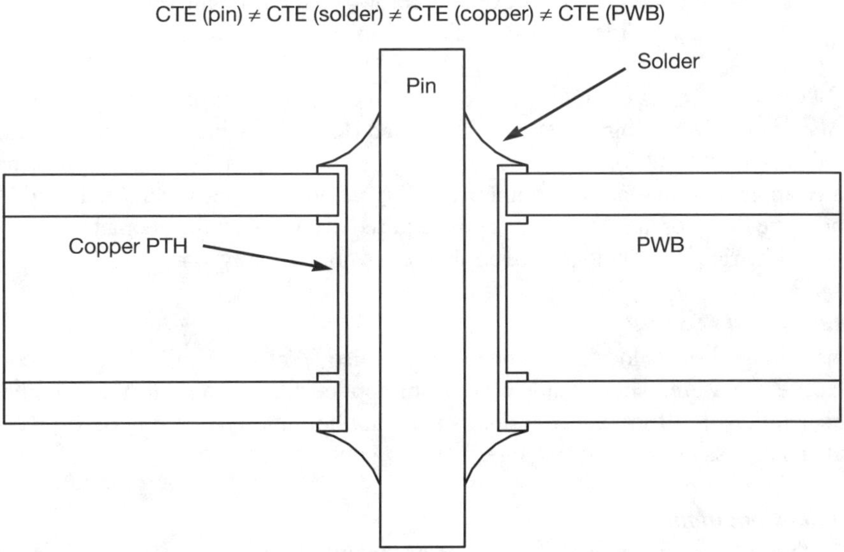

FIGURE 8-3 Thermal loading Type I.

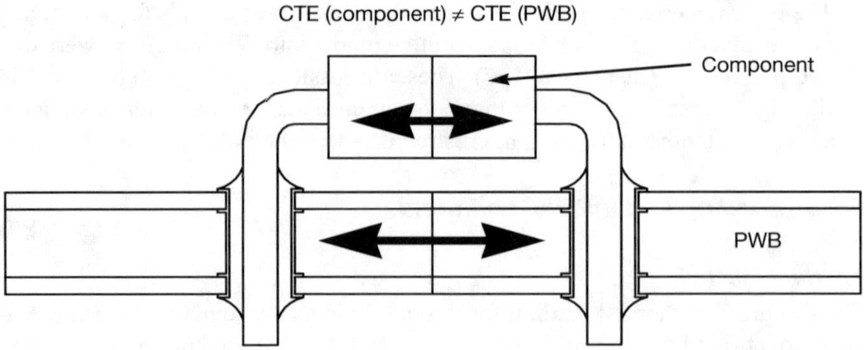

FIGURE 8-4 Thermal loading Type II.

For component assemblies which are encapsulated or conformal coated, thermal loads arise due to:

Type III: Thermomechanical vertical loads (perpendicular to the PWB) in the conductor pins due to the thermal expansion mismatch between the component leads and the encapsulant or conformal coating, Figure 8-5.

Type IV: Thermomechanical bending loads (parallel to the PWB) in the conductor pins due to the thermal expansion mismatch between the encapsulant and the PWB (in-plane), Figure 8-6. If the thickness of the encapsulant is not equal above and below the PWB (symmetric), substantial PWB bending can also occur.

Type I loading is obviously inherent to all of the other types of loading but is often ignored when joint fracture is evaluated [Klein Wassink, 1989]. If an idealized axisymmetric geometry is assumed, Type I and Type III loads are symmetrically applied so that an axisymmetric (2-D) analysis is appropriate. For Type II and Type IV loads, a 3-D evaluation is required even though the geometry may be axisymmetric since the loads are applied nonsymmetrically.

Mechanical Loading

The through-hole solder interconnection can also be mechanically loaded at constant temperature. These loads arise from applied loads (handling) and applied accelerations (uniform acceleration, shock, and vibration). Bending of the PWB can also stress the solder joint significantly, Figure 8-7.

Initial Conditions

In a thermomechanical analysis, choice of the initial conditions can be important, especially for a through-hole joint. The plated through-holes are manufactured at

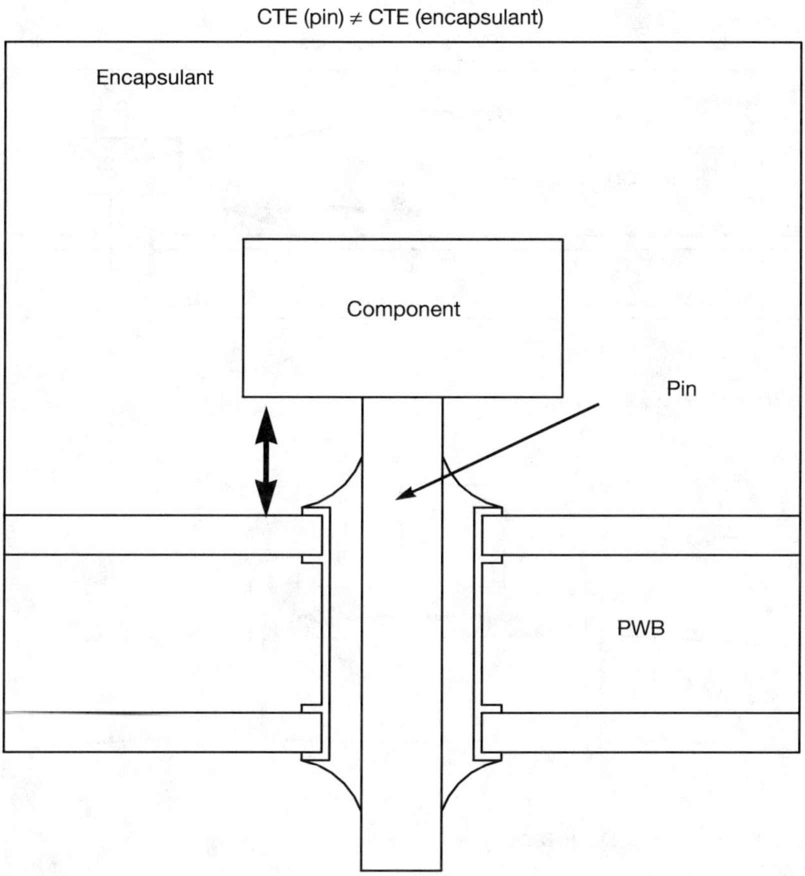

FIGURE 8-5 Thermal loading Type III.

room temperature and should be relatively stress-free at room temperature. If the PWB is heated to soldering temperatures, significant stresses are developed in the copper plated through-hole, as will be discussed later. The stresses can actually exceed the yield strength of copper. The solder joint is formed and the solder is in a relatively stress-free state. As the assembly is cooled to room temperature, stresses develop in the assembly. Many investigators choose to assume a stress-free temperature at room temperature or some other arbitrary temperature. This assumption is based upon the belief that any stresses that develop will relax due to the creep characteristics of solder. This assumption may be appropriate for many SMT applications; however, for a plated through-hole joint, the displacements are highly constrained so that creep relaxation may not occur. The choice of an initial condition must be chosen with great care.

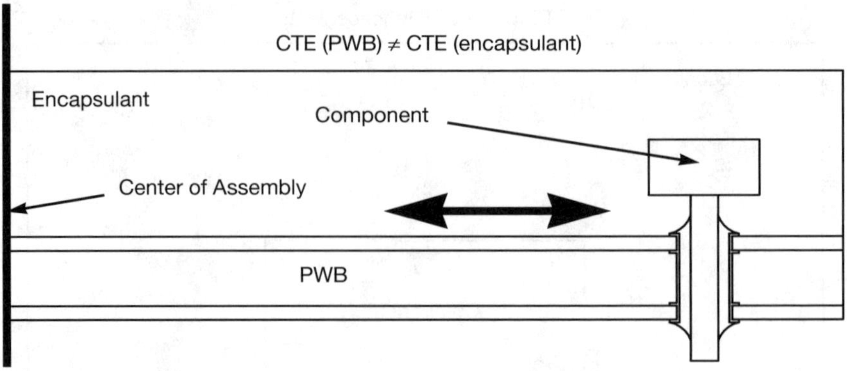

FIGURE 8-6 Thermal loading Type VI.

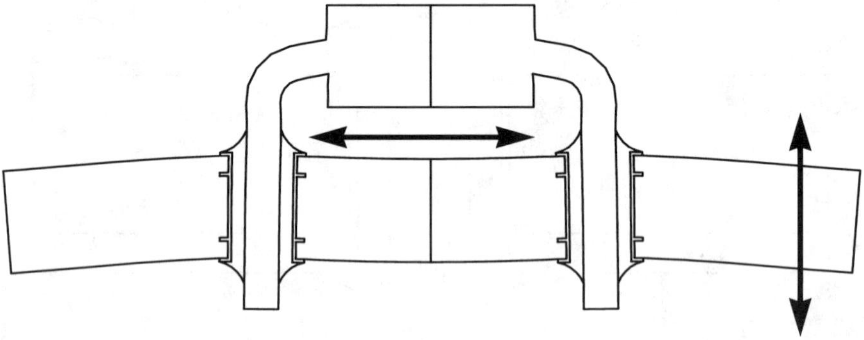

FIGURE 8-7 Bending Due to Mechanical Loading.

8.2.4 Miscellaneous Assumptions

In addition to the modeling considerations above, there are miscellaneous assumptions that may be important. For instance, it is generally assumed in mechanical modeling that material interfaces are perfectly bonded and infinitely strong. This assumption can be invalid. If interface separations do occur, the geometry changes thus affecting the response of the assembly.

8.2.5 Interpretation of Results—Failure Prediction

The desired result of a mechanical or thermomechanical model is the ability to predict the response of a structure when it is subjected to manufacturing and/or operational environments. Ultimately, the goal is to predict failure (reliability). The results of most mechanical modeling of structures is the prediction of loads, displacements, stresses, and strains. The analyst must apply the results of analysis to predict failure. For thermomechanical analysis of solder joints, this is typically accomplished by computing the maximum total strain range in the solder for a few thermal cycles and using this value as input to an empirically based fatigue model (Chapter 6). Empirical fatigue models are based upon test results in which actual solder joints are thermally cycled to failure, "shear-type" test specimens are mechanically cycled to failure at various temperatures, or "shear-type" test specimens are thermomechanically cycled to failure and an empirical relationship is developed which relates a strain range to cycles to failure. The problem with this approach is that the empirically derived strain range is actually an average shear displacement per characteristic length, not strain; whereas in the finite element analysis, actual multidimensional strain distributions in the solder are predicted. These two quantities can be vastly different except in very specific geometries. Additionally, this procedure does not take into account the fact that if cracks develop in the solder (which they do), the strain distributions change drastically. Currently, progressive failure is rarely included in any analysis.

8.3 EXAMPLES

8.3.1 Plated Through-Hole—Without Solder

The increased demands on package density have led to very complex PWB assemblies with numerous circuit layers. Multilayer, rigid/flex, and polymer on metal (POM) PWBs are increasing in complexity and are finding greater use. Interconnection between circuit layers is accomplished with copper plated through-holes. With increased use of SMT, the trend for PTHs is decreasing hole diameters that may or may not contain solder [Bhandarkar, et al.1992] and for POM that may contain polymer resin [Ozmat et al., 1990]. The reliability of unfilled PTHs in PWBs when subjected to thermomechanical loading is of substantial concern. Failure mechanisms include (see Figure 8-8):

1. barrel cracking,
2. lifted lands and corner cracking,
3. inner layer foil cracks, and
4. separation of the plating from the inside of the hole.

Mechanisms 1 and 3 can lead to electrical open or intermittent circuits that are often not detected until the assembly process has proceeded to a very high value added state. This problem may limit increased PWB complexity and/or force development of new interconnection strategies or improved PWB laminate materials. While this interconnection is not solder, the problem is intimately related to, and is a subset of, the through-hole solder interconnection.

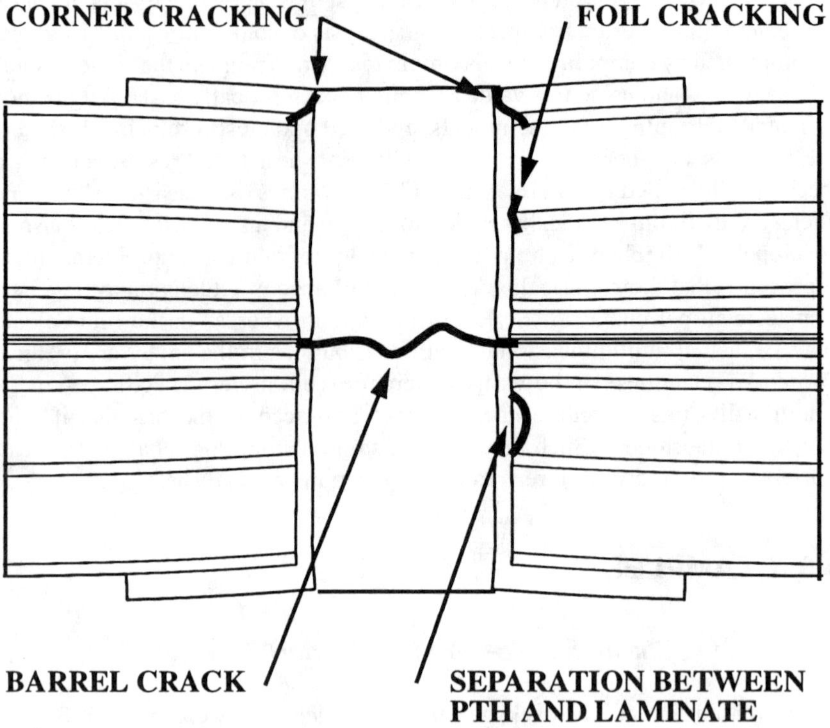

FIGURE 8-8 Possible failure modes in plated-through-hole.

A majority of PTH research has been in the experimental arena [Ammann and Jocher, 1976; Hagge, 1980; Hodges, 1987; Nicewarner, 1987; IPC-TR-579, 1988; Yager, 1988; Tossell and Ashbee, 1989; Munikoti and Dhar, 1990; Davidson, 1991; Engelmaier, 1991]. The research has been directed toward: measurement of displacements and strains in PTHs; and thermal cycling PWBs to failure to determine fatigue reliability (cycles to failure). Fatigue reliability data are used to develop empirical relationships to predict cycles to failure.

Models developed to simulate thermomechanical behavior of PTHs have followed two separate approaches [Bhandarkar, et al., 1992]. The first is simplified closed form expressions to estimate critical values of stress and strain [Oien, 1976; Hagge, 1980; Mirman, 1988; Yager, 1988; Engelmaier, 1991], and the second is numerical approaches using finite elements [Lee et al., 1984; Vecchio and Hertzberg, 1986; Ozmat et al., 1990; Iannuzzelli, 1991; Barker et al., 1991; Burchett, 1992; Bhandarkar et al., 1992)]. Results from the simplified closed form expressions are generally used as input for empirical fatigue reliability equations. The simplified expressions are extremely useful in initial designs. However, they do not and cannot capture the nonlinear thermomechanical response of a PTH interconnection. The accurate determination of the stresses and strains in the PTH-via barrel requires a complex finite element analysis (FEM) [Engelmaier, 1991].

Multilayer Printed Wiring Boards

Multilayer boards (MLB) are commonly used PWB technology in the electronics industry. Studies by Iannuzzelli [1991], Bhandarkar et al. [1992], and Barker et al. [1991] are examples of finite element analysis (FEA) of PTHs in MLBs. Iannuzzelli uses the FEM to analyze the inelastic response of the PTH/laminate to predict the strains in the PTH barrel. Results from the analysis, called the strain model, are used in experimentally determined empirical relationships to predict low cycle fatigue reliability (LCF). Specific details about the finite element approximation are somewhat sketchy but it appears that a 2-D axisymmetric model with mid-plane symmetry (symmetric laminate) was used. The copper was characterized as temperature-dependent, time-independent nonlinear material with properties of copper foil and copper plating being equivalent. The laminate was characterized as a temperature-dependent, time-independent elastic material. Although not specifically clear, it is assumed that the laminate was treated as a homogeneous isotropic material. The model was loaded with a uniform temperature cycle up to a maximum temperature of 232°C. In Part I of this study, Iannuzzelli [1991] modified the geometry of the model to specifically agree with test coupons and predicted LCF life for these specimens. He reports an uncertainty tolerance of a factor of 2 in predicting the lifetime for these specimens. In Part II, Iannuzzelli presents a model sensitivity study. The variables examined include laminate glass transition temperature, laminate CTE, plating thickness, plating uniformity, hole diameter, and pad diameter. He reported increased PTH reliability with increasing copper ductility, increasing plating thickness, increasing hole diameter, decreasing plating non-uniformity, decreasing pad diameter, decreasing temperature excursions, and decreasing laminate CTE.

In practice, PTHs are closely spaced and the trend is toward even closer spacing. At some critical hole spacing, the thermomechanical response of a PTH begins to be influenced by the neighboring PTH. To address this problem, Bhandarkar et al. [1992] examined modeling considerations for MLBs by comparing a 2-D axisymmetric geometry and a 3-D quarter symmetry geometry, Figure 8-9. In this study, the copper was characterized as a temperature-independent, linear elastic material. The authors recognized the limitations of this assumption and deferred quantitative predictions of stresses and failure to future work. The laminate was characterized as a temperature-independent, orthotropic, linear elastic material. The model was thermally loaded. However, specifics were not given, and it is assumed that a uniform temperature increase was applied. This study concluded that for barrel stresses, a 2-D axisymmetric approximation is appropriate for PTH separation distances, L/ro, greater than 10, where 2L is center-

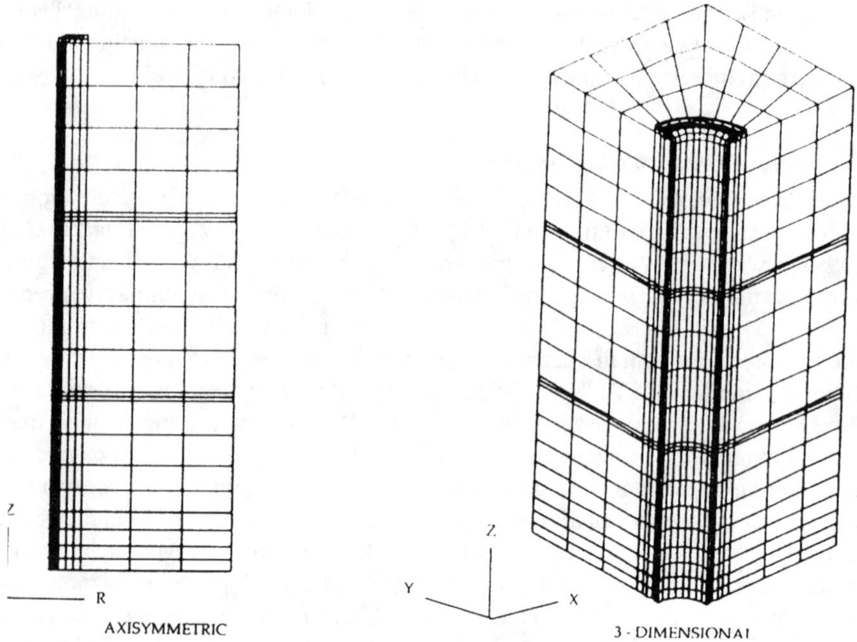

FIGURE 8-9 Plated-through-hole finite element models [Bhandarkar et al., 1992].

line to centerline hole spacing and ro is PTH diameter. In this study, the laminate was assumed to be transversely isotropic (isotropic in-plane). The authors report that the laminate orthotrophy is important in modeling interfacial de-adhesion failure between the plating and the surrounding laminate, and a 3-D model should be used. The authors also present results of parameter sensitivity studies including the effects of plating thickness, hole aspect ratio, presence and location of copper innerplanes, and presence of non-functional pads for both FR-4 MLBs and Kevlar/polyimide MLBs. The authors report that barrel stresses decrease with increased plating thickness, increased PTH spacing, and decreased aspect ratio.

The most severe thermal environment that a PWB must survive is solder processing. Wave soldering is a common manufacturing technique described earlier (Chapter 4). Mechanical modeling studies examining the response of PTHs due to exposure to soldering environments typically assume that temperature is applied uniformly to the structure and neglect any heat transfer effects. Barker [Barker et al., 1991] examined the transient thermal stresses developed in a PTH during the heating phase of wave soldering. It was assumed that the PTH was masked so that solder did not fill the hole. The copper was characterized as a temperature-independent, nonlinear material with linear work hardening. The laminate was characterized as a temperature-independent, transversely isotropic, linear elastic material. A step increase in temperature was applied to the bottom surface of the model and the transient temperature of the assembly was computed as the laminate heated to the solder temperature. The methodology used to predict the transient thermal history was, however, somewhat vague. The predicted thermal history was then applied to structural finite element models to predict the transient thermomechanical response of the assembly. Time-dependent stresses and strains were computed as the assembly heated to the solder temperature. The response was computed for three separate structural finite element models: a 3D model with innerplanes (Figure 8-10), 2-D axisymmetric model with innerplanes which corresponded to the 3-D model, and a 2-D axisymmetric model without innerplanes. Barker reported that maximum equivalent plastic strains developed in the barrel using the 2-D axisymmetric models were approximately 8% higher than those predicted for the 3-D model. The results also indicated that while the temperature and the peak strains for both 2-D axisymmetric models reached the same value, the rate for the model with innerplanes was much slower, implying that the copper innerplanes diffuse the heat more efficiently into the laminate. Thermal cycling of the laminate was left for future studies.

Polymer on Metal Printed Wiring Boards

High reliability PWBs that are designed primarily for SMT applications that will experience severe environments are often bonded to a constraining low CTE metal

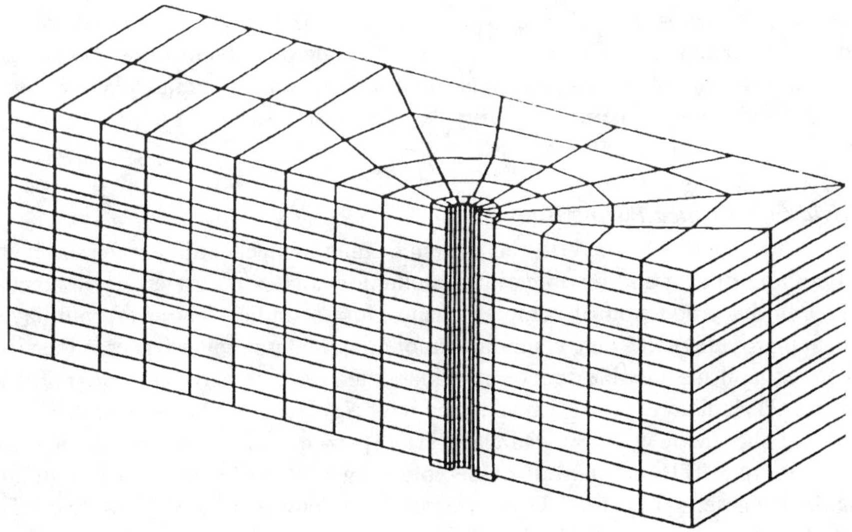

FIGURE 8-10 3-D plated-through-hole finite element model [Barker et al., 1991].

core. In this technology, standard PWBs are bonded to both sides of the metal core. The (POM) technology has plated-through-holes which are often filled with resin material which comes from the core laminating process. Ozmat and associates [Ozmat et al., 1990] examined the nonlinear thermomechancal response of PTHs for a POM assembly subjected to a thermal cycle. In this study, a quarter-symmetry 2-D axisymmetric model neglected edge effects and interactions with neighboring PTHs. The constraining metal core was laminated copper/Invar/copper and the PWB consisted of eight layers with internal pads. The Invar in the constraining core was characterized as a temperature independent, linear elastic material. The copper in the constraining core, PTHs, and PWB pads was characterized as temperature-independent, elastic plastic material with kinematic, linear work hardening. The yield stress of the copper in the constraining core and in the MLB were different. Two laminate materials were evaluated in this study: epoxy/glass and polymide/glass. The laminates were characterized as temperature-dependent, transversely isotropic, linear elastic materials. The properties for the laminates were computed from the constituent materials but the procedure was not discussed. The model was loaded with a single, uniform temperature cycle starting at 25°C, heating to 125°C, cooling to –55°C, and then heating back to 25°C. The authors used this model to examine the effects of PTH thickness, PTH copper yield stress, and hole fill materials including no fill. The authors conclude that for epoxy/glass POM assemblies: 1) the PTH copper yield stress has the maximum influence on low cycle fatigue of the PTH, with higher yield stress increasing life however, the copper must have sufficiently high ductility; 2) decreasing the PTH thickness below 0.0508 mm significantly reduced the LCF life; and 3) filling the hole with epoxy resin, polymide resin, and DAP slightly decreased, had no effect, and slightly increased LCF life, respectively. For polyimide/glass POM assemblies, no yielding in the PTH copper plating was predicted unless the PTH thickness was decreased and the hole was filled with epoxy. Polyimide/glass systems were found to have superior reliability for these systems, but it is implied that they are much more difficult to process.

Rigid Flex Printed Wiring Boards

Rigid flex PWBs are a hybridized system of interconnects utilizing the concepts of both multilayer and flexible circuit technology. They offer significant advantages in the production of complex electronic assemblies. The rigid multilayer section provides electronic component mounting and innerplane innerconnection features, and the flexible section provides interconnects between various MLB layers. PTH reliability is of critical concern for rigid flex PWBs due to high expansion of the flexible layers. Burchett [1992] examined the thermomechanical response of a PTH in a rigid flex assembly subjected to a solder float test. In this study, a quarter-symmetry 2-D axisymmetric model was used where edge effects and interactions with neighboring PTHs were neglected, Figure 8-11. The three types of copper were characterized as temperature-independent, elastic plastic

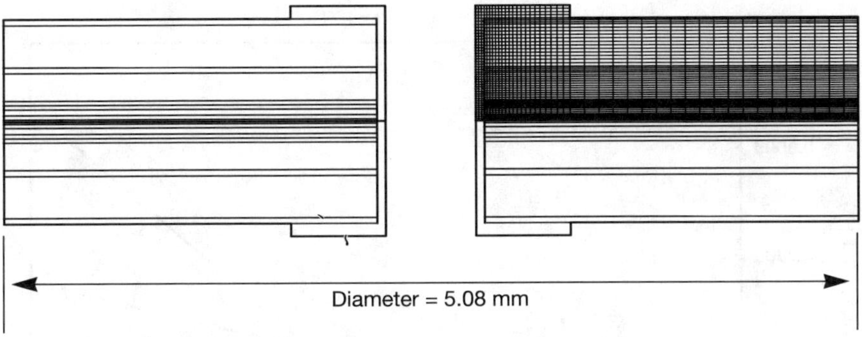

FIGURE 8-11 Rigid/flex plated-through-hole without solder, finite element model.

material with isotropic, linear work hardening with different values for the yield stress. The two laminate materials in this layup, modified epoxy/glass and polyimide/glass, were characterized as temperature-independent, transversely isotropic, linear elastic materials. The properties for the laminates were computed from the constituent materials and not obtained from experimental testing. The Kapton and acrylic adhesive were characterized as temperature-independent, linear elastic materials. For all materials, the CTE was temperature-dependent. The assembly was uniformly heated from room temperature to 287°C and then cooled to room temperature. In Figure 8-12, the computed through thickness displacement response of two points on the top surface of the assembly, Point A located at the PTH and Point B located away from the PTH, are plotted. This plot characterizes the nonlinear response of a PTH in a PWB when subjected to severe temperature excursions. As the temperature increases, the z-displacement of the PTH lags the displacement of the laminate. However, at approximately 150°C, the rate of displacement of the PTH increases significantly. At this point the stresses in the copper PTH exceed the yield strength, and the copper plastically deforms. When the assembly is cooled from 287°C to room temperature, the PTH has permanent plastic deformation while the laminate returns elastically to room temperature. At 287°C, large plastic strains are developed in the PTH, Figure 8-13. An interesting computed effect is the development of very high interfacial stresses between the PTH and laminate, Figure 8-14. The author concludes that barrel cracking is the most likely failure mode, but interfacial failure between the PTH and the laminate is also possible. This study also examined the effects of several parameters that indicated that barrel cracking could be minimized by increasing the ductility of the plating, insuring that the plating thickness is uniform, increasing the plating thickness, and using larger holes. Increasing the plating thickness increases both the laminate radial stress and the innerplane bending stresses. Variations in prepreg thickness, rigid laminate thickness, and plating yield stress have minimal effects on critical failure parameters.

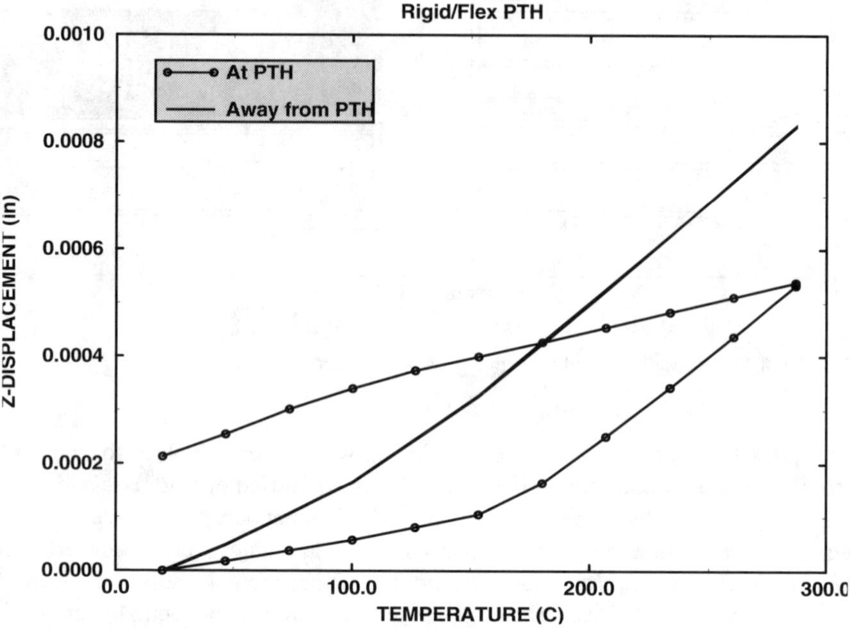

FIGURE 8-12 Computed displacement response of top surface of laminate.

8.3.2 Through-Hole Solder Interconnection

Through-Hole Solder Interconnection—No External Loads

Stresses and strains develop in a through-hole solder interconnection subjected to thermal cycling due to the local mismatch in materials (Type I loading) even if no external loading is applied. In evaluating solder joint fracture and/or reliability, the local stresses and strains are often ignored [Klien Wassink, 1989]. However, studies by Lau et al. [1991] address this issue. In this study, the authors used the finite element method to evaluate the thermomechanical response of a single through-hole solder joint interconnection in an MLB subjected to thermal cycling. The geometry consists of a 0.43 mm diameter alloy-42 pin soldered into a 0.635 mm hole of a 2.36 mm thick two layer FR-4 epoxy/glass multilayer board with eutectic Sn/Pb solder. Three solder conditions were examined: a joint with a typical fillet, a joint with no fillet, and a half-filled joint. The thickness of the copper plated through-hole barrel was 0.03 mm. This geometry was modeled assuming quarter symmetry using 2-D axisymmetric elements. The Alloy 42 pins were characterized as a temperature-independent, linear elastic material. The copper (PTH and pads) was characterized as temperature-independent, elastoplastic material

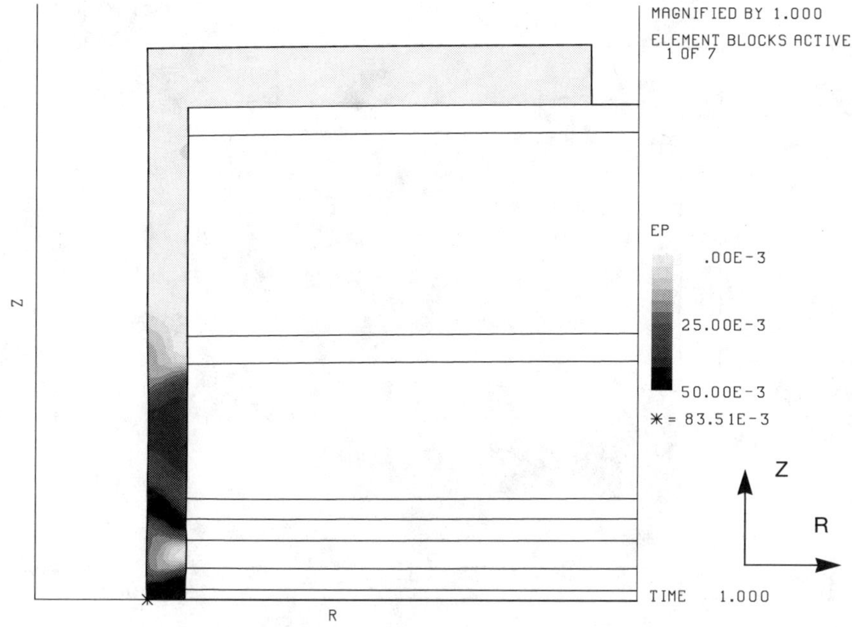

FIGURE 8-13 Computed equivalent plastic strain in rigid/flex PWB barrel @ 287 °C.

with linear strain hardening. The FR-4 epoxy/glass laminate (MLB) was characterized as a temperature-independent, transversely isotropic, linear elastic material. The solder was characterized as a temperature-independent, elastoplastic material with linear strain hardening. The value chosen for the yield strength of copper corresponded to a yield strength at 120°C. The thermomechanical response was computed for a uniform temperature increase of 100°C. The authors then used the computed peak strain values in the solder to predict cycles to failure. Initial cracking in the typical solder joint with fillet was reported to be 520 cycles. By assuming that the strain distribution does not change as a crack propagates, the authors predict total solder separation at 4500 cycles. This assumption neglects any stress and strain redistribution as damage is developed in the solder. Similarly, for the joint without fillet, failure was predicted at 3400 cycles; for the half-filled solder joint, failure was predicted to be 1200 cycles. The authors conclude that no failures develop in the FR-4 epoxy/glass.

Through-Hole Solder Interconnection—Encapsulated Component

Electronic assemblies are often encapsulated or conformal coated with a polymeric material to enhance shock, vibration, and high voltage survival and to pro-

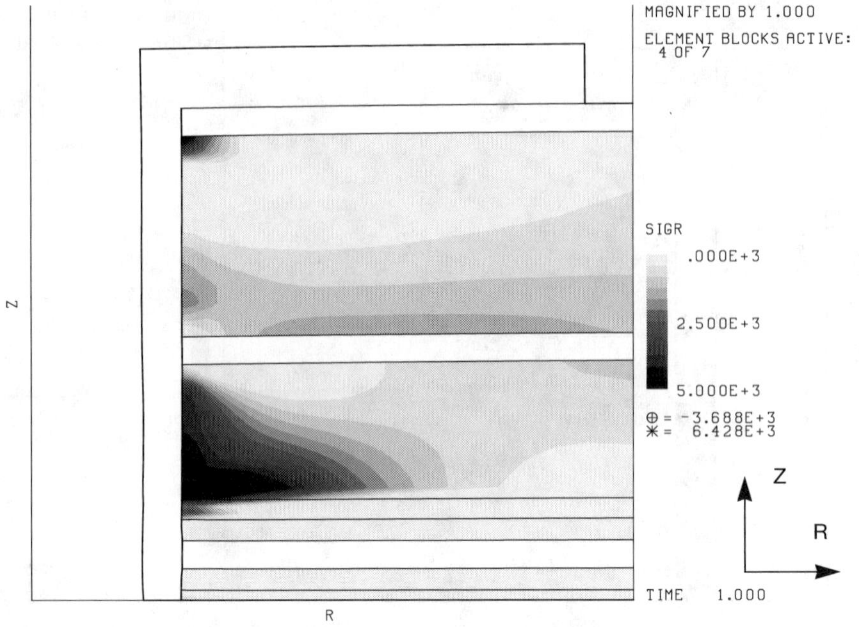

FIGURE 8-14 Computed radial stress in rigid/flex PWB barrel @ 287 °C.

tect against environmental contaminants. The polymeric encapsulant can, however, generate substantial Type III and Type IV thermal loads on through-hole solder interconnections due to the high CTE of the encapsulant in addition to Type I loads. Studies by Pan [1991] and Dasgupda et al.[1992] examined the through-hole solder response of encapsulated discrete components. Pan employed the finite element method to compute the deformation and strain distribution in the solder interconnection of a three lead encapsulated discrete component, Figure 8-15, subjected to thermal cycling. Pan modeled the basic geometry with a 2-D planar model with a plane of symmetry through the center pin. The author did not specify which 2-D planar assumption was used in the analysis (plane stress, plane strain, or generalized plane strain). Two specific 2-D geometric models were created for the solder interconnect; one with perfect homogeneous solder and one with solder voiding. The material for the component body and the copper component leads was characterized with a temperature-independent linear elastic model; and the epoxy encapsulant was characterized with a temperature-dependent, linear elastic model. The FR-4 epoxy/glass PWB laminate was characterized as a temperature-dependent, transversely isotropic, linear elastic material. The eutectic Sn/Pb solder was characterized as a temperature-dependent, elastic plastic material with an addition of a steady state creep law. Although not specified by Pan, it appears that the assembly was assumed stress

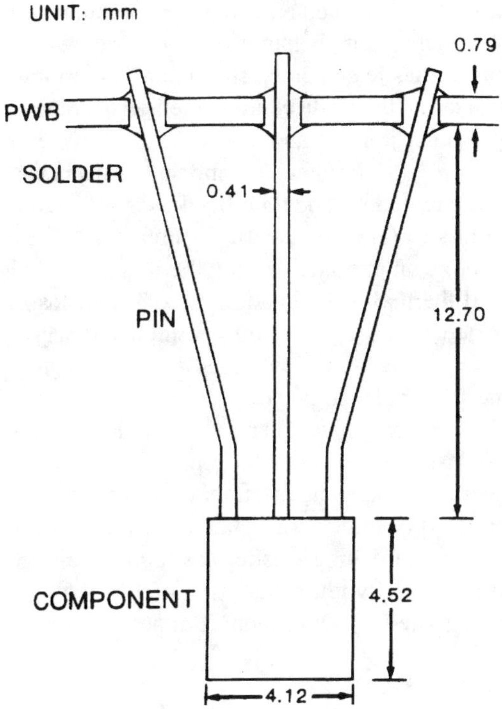

FIGURE 8-15 Encapsulated solder interconnect evaluated by Pan, ©1991, IEEE.

free at –40°C. A temperature cycle was then applied to the assembly ranging up to a peak temperature of 140°C and then cooling back to –40°C (DT = 180°C) in approximately 45 min. followed by a hold for 15 min. at –40°C. Computations were completed for a total of 20 thermal cycles. Computationally, the smooth thermal cycle was approximated with a series of instantaneous temperature steps at constant time followed by a hold in temperature for a period of time. During the instantaneous temperature step, the solder is characterized with the elastic plastic model; during the temperature hold, the solder was characterized with the steady state creep law (Chapter 7). The author monitored the displacement response in the assembly and reported that if the accumulated permanent displacements per cycle are extrapolated to 1000 cycles, there was agreement with experimental observations. Good experimental/numerical correlation of the deformation response of solder voids was observed. Strain distributions in the solder were presented and found to be non-uniform and constantly changing. The author concludes that the deformation response is accurately predicted. The author also suggests that thermal cycling fatigue life predictions which are based upon uniform shear are not applicable for this solder interconnect.

Dasgupda et al. [1992] proposed an energy-partitioning technique to examine the dependence of solder fatigue behavior on temperature-dependent changes in the relative amounts of plastic and creep strain in a solder joint. In evaluating this technique, the authors used the finite element method to compute the strain history in the through-hole solder joint of an encapsulated simulated component shown sketched in Figure 8-16. The simulated component was a brass pin with a circular flange, resembling a thumbtack. The thumbtack was soldered into a through-hole of an FR-4 epoxy/glass PWB with eutectic Sn/Pb solder. The assembly was then encapsulated with epoxy on one side. During thermal cycling, the solder joint was subjected to Type III thermal loads in addition to Type I loads. In this study, the assembly was modeled with 2-D axisymmetric finite elements. The brass component and the copper PTHs were characterized as isotropic, temperature-independent, power-law hardening elastic-plastic materials with a Ramberg-Osgood type constitutive model. The epoxy encapsulant was characterized with a temperature-independent, linear elastic model. The FR-4 epoxy/glass PWB laminate was characterized as a temperature-independent, transversely isotropic, linear elastic material. The eutectic Sn/Pb solder was characterized as isotropic, temperature independent, power-law hardening elastic-plastic materials with a Ramberg-Osgood type constitutive model with the addition of a Weertman steady state creep law. Although not specified by Dasgupda, it appears that the assembly was as-

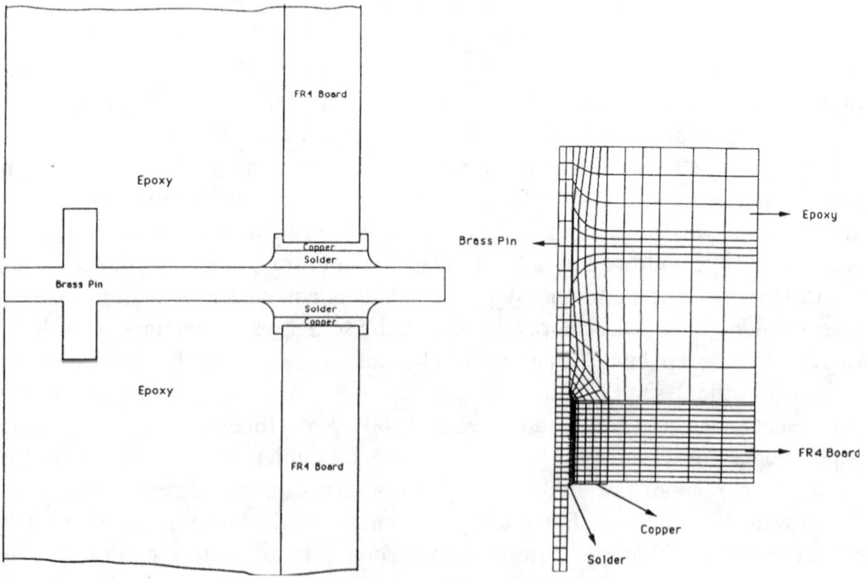

FIGURE 8-16 geometry and Finite Element Idealization Employed by Dasgupda et al., 1992.

sumed stress free at the cold temperature extreme. Two temperature cycles were applied to the assembly, each consisting of a 6°C/min ramp to hot temperature, a 20 min. hold at hot temperature, a 6°C/min. ramp to cold temperature, and a 20 min. hold at cold temperature. Several temperature cycle ranges were examined from $\Delta T = 31°C$ to $\Delta T = 113°C$; however, the mean temperature of the thermal cycle was maintained at 40°C. Computationally, the smooth thermal cycle was approximated with a series of instantaneous temperature steps at constant time followed by a hold in temperature for a period of time. During the instantaneous temperature step the solder was characterized with the elasto-plastic model. During the temperature hold, the solder was characterized with a steady state creep law (Chapter 7). The authors computed the thermomechanical response on the assembly and present contour plots of the maximum shear strain in the solder at various stages in the thermal cycle. These results show that the shear strain in the solder is not uniform or constant with time. The authors also present plots of effective (von Mises) stress as a function of equivalent strain. From these curves the plastic and creep energy dissipated per thermal cycle can be calculated. This information is then used to predict fatigue life. Dasgupda and associates conclude that the state of strain in a through-hole solder joint is multidimensional rather than pure shear, implying that additional laboratory-generated multidimensional fatigue damage constants are needed. The authors also conclude that the finite element method is extremely useful to obtain necessary information for solder fatigue analyses.

Through-Hole Solder Interconnection—PGA

The reliability of through-hole solder joints is crucial in pin grid array (PGA) interconnection technology. However, literature references to mechanical modeling of PGA through-hole solder interconnects are minimal. Experimental studies to determine fatigue life of a PGA through-hole joint subjected to lateral displacements (Type II loading) are presented by Lau et al.[1991a]. Lau [Lau et al., 1991b] also experimentally examined the effects of rework on the reliability of PGA interconnects. Engel and associates [Engel et al., 1984, Engel et al., 1991a, Engel et al., 1991b] present a series of analytical and finite element studies to examine the interaction between a PGA module and the PWB through the pins. These studies examine the effect of pin diameter, pin length, pin material, and module stiffness on transverse and axial force, and bending moments transmitted into the solder interconnection. The through-hole solder interconnection is treated as a beam on an elastic foundation. Although some load distribution is computed in the solder elastic foundation, detailed stress and strain distributions in the solder are not computed. Results from this study are very useful in sizing pins, selecting pin materials, and defining the standoff height of the module to minimize loads transmitted to the solder joints. Results from this study could also be used as input to a detailed finite element model of a single pin in a PGA.

8.4 CONCLUSION AND RECOMMENDATIONS

In this chapter, printed wiring board through-hole interconnections were examined. Specific through-hole modeling considerations were first discussed. This was followed by discussions of current applicable research found in the literature. While the thrust of this book is mechanical modeling of solder interconnections, modeling studies of PTHs without solder are included due to the fact that they are a subset of the through-hole solder interconnect. Finally, applicable mechanical modeling studies of through-hole solder interconnects are discussed. As was stated in the introduction, the technology to computationally evaluate solder interconnects in a thorough, realistic, and verifiable manner without significant simplifying assumptions is just now coming into place. The current literature verifies this statement. Mechanical modeling of a plated through-hole with or without solder is a difficult nonlinear thermomechanical problem. Addition of time-, temperature-, and microstructure-dependent solder, three-dimensional geometries and three-dimensional loading makes the thermomechanical loading of through-hole solder interconnects very challenging but essential for high quality, high reliability applications.

Based upon this review, it is clear that there are shortcomings in mechanical modeling of through-hole interconnects. First, additional research is needed to characterize the response of the PWB laminates. Currently, the laminates are modeled as homogeneous, temperature-dependent, time-independent, orthotropic (transversely isotropic) linear elastic materials. Literature values given for the properties of the same laminate vary significantly. These differences may be due to differences in fiber fill. At the very least, some methodology is needed to standardize material properties and material tests. Since a significant percentage of the laminates are polymeric materials and the manufacturing and operational temperatures are near or in excess of the glass transition, the laminate response is most likely time dependent. None of the current research employ time-dependent laminate response.

Second, solder constitutive models need to be improved. Research (Chapter 4) to develop microstructurally based constitutive models is currently in progress. However, these models need to be verified and incorporated into analysis procedures. These models also must be capable of predicting failure of solder.

Third, an improved methodology to generate realistic geometries is required. Coupling predictive solder joint geometry techniques to mechanical analysis techniques appears to offer the most promise.

Fourth, failure in materials other than the solder need to be examined. There seems to be a fixation that failures occur only in solder. This fixation has the tendency to direct attention away from other possible failure mechanisms.

Finally, computational techniques need improvement. There is great need to extend analysis to 3-D. This extension should not exclude using 3-D analysis to verify the applicability of 2-D approximations.

REFERENCES

Ammann, H. H., and Jocher, R. W., 1976, *14th Annual IEEE International Reliability Physics Symposium*, Las Vegas, NV.

Barker, D., Pecht, M., Dasgupta, A., and Naqvi, S., 1991, *Transactions of the ASME, Journal of Electronic Packaging*, 113.

Burchett, S. N., 1992, *The Institute for Interconnecting and Packaging Electronic Circuits 35th Annual Meeting*, Bal Harbor, Fl.

Bhandarkar, S.M., Dasgupta, A., Barker, D., Pecht, M., Engelmeier, W., *Transactions of the ASME, Journal of Electronic Packaging*, 114:8-13.

Berkebile, M. J., 1967, *NASA TMX 53653*.

Baker, E., 1975, *Insulation/Circuits*, 21:49-56.

Bangs, E. R.; Beal, R. E., 1975, *Welding J. Res. Suppl.* 54:377S-383S.

Chu, T. Y., 1975, *Western Electric Engineer*, 19:31-42.

Davidson, D. L., 1991, *Materials Developments in Microelectronic Packaging: Performance and Reliability, Proceedings of the Fourth Electronic Materials and Processing Congress, Electronic Materials and Processing Division of ASM International*, edited by Prabjit Singh. ASM, Materials Park, OH.

Dasgupda, A., Oyan, C., Barkar, D., Pecht, M., 1992, *Transactions of the ASME, Journal of Electronic Packaging*, 114.

Engel, P. A., Lim, C. K., Toda, M. D., Gjone, R., 1984, *Computers in Mechanical Engineering*, 2(6):59-69.

Engel, P. A., Wu, K. R., 1991, *ASME 91-WA-EEP-39, ASME Winter Annual Meeting*, Atlanta, GA.

Engel, P. A., Webb, J. R., 1991, *ASME 91-WA-EEP-47, ASME Winter Annual Meeting*, Atlanta, GA.

Engelmaier, W., 1991, *NEPCON West 91, National Electronic Packaging and Production Conference*, Anaheim, CA.

Hagge, J. K., 1980, *Proc. Printed Circuit World Expo '80* .

Hodges, R. M., I*PC-TC-623, The Institute for Interconnecting and Packaging Electronic Circuits.*

Iannuzzelli, R., 1991, *The Institute for Interconnecting and Packaging Electronic Circuits*, Boston, MA.

IPC-TR-579, 1988, *The Institute for Interconnecting and Packaging Electronic Circuits,* Lincolnwood, IL.

Jeannotte, D.A., Goldmann, L.S., and Howard, R.T., 1989, *Microelectronics Packaging Handbook,* edited by R. R. Tummala and E. J. Rymaszewski, Published by Van Norstrand Reinhold.

Klein Wassink, R. J., 1989, *Soldering in Electronics*, 2nd edition Electrochemical Publications Limited.

Kobayashi, M., 1975, *Welding J. Res. Suppl.*, 54(10):363s-369s.

Lau, J., Subrahmanyan, R., Rice, D., Erasmus, S., Li, C., 1991, *Transactions of the ASME, Journal of Electronic Packaging*, 113.

Lau, J., Leung, S., Subrahmanyan, R., Rice, D., Erasmus, S., Li, C. Y., 1991, *Circuit World*, 17.

Lee, L. C., Darekar, V. S., Lim, C. K., 1984, *IBM Journal of Research and Development*, 28:711-728.

Mirman, B. A., 1991, *IEEE Transactions on Components, Hybrids, and Manufacturing Technology*, 11(4):506-511.

Mohler, J. B., 1971, *Machine Design*, p. 84.

Munford, J. W., 1975, *IEEE Trans.*, PHP-11(4):296-304.

Munikoti, R. and Dhar, P., 1990, *Proc. of the 40th Electronic Components & Technology Conference*, Las Vegas, NV.

Nicewarner, E., 1987, *Circuit World*, 13(4).

Oien, M. A., 1976, *14th Annual IEEE International Reliability Physics Symposium*, Las Vegas, NV.

Ozmat, B., Walker, H., and Elkins, M., 1990, *Proc. Electronics Packaging Conference*, (IEPC), Marlborough, MA., pp. 359-381.

Pan, T., 1991, *IEEE Trans. on Components, Hybrids, and Manufacturing Technology*, 14(4).

Paydar, N., Hasan, A. U., Tong, Y., 1991, *Electrecon '91 Proceedings*, Indianapolis, IN, pp 113-125.

Pecht, Michael, 1991, *Handbook of Electronic Package Design*, Published by Marcel Dekker, Inc..

Read, D. T., Bayba, A., 1991, *91-WA-EEP-17, ASME Winter Annual Meeting*, Atlanta, GA.

Rooks, S., Racz, L.M., Szekely, J., Benhabib, B., and Neumann, A.W., *LANGMUIR*, 7:3222-3228.

Sandor, B.I., 1991, *SolderMechanics - A State of the Art Assessment*, D. R. Frear, W. B. Jones, and K. R. Kinsman, eds, TMS, Warrendale, PA.

Shamash, M.B., 1968, *Report No. 8402A, Westinghouse Defense and Space Center*, Baltimore, MD.

Tossell, D. A., Ashbee, K. G. H., 1989, *Journal of Electronic Materials*, 18(2).

Van Orden, E. Ray, 1968, NASA TMX 53713.

Vecchio, K. S. and Hertzberg, R. W., 1986, *Microelectronics Reliability,* 26:715-732.

Wild, R. N., 1972, *Welding Research Supplement to Welding Journal.*

Yager, T. A., 1988, *Printed Circuits Handbook*, edited by Clyde F. Coombs, Jr., McGraw-Hill Book Company, pp 30.1-30.14.

9

Surface Mount Solder Joints Under Thermal, Mechanical, and Vibration Conditions

J. H. Lau

9.1 INTRODUCTION

Surface mount technology (SMT) is a mature technology. For a detailed definition, please refer to Lau [1991, 1992, 1993]. SMT allows placement of more surface mount components (SMC) into smaller and tighter printed wiring board (PWB) areas. This increased density means increased performance and power in smaller packaging systems, and it allows manufacturing of smaller and higher performance products at lower cost.

The advent of integrated circuit (IC) technology and the requirements of high density for high-speed circuitry is driving the design of SMC to higher pin count and smaller package size. In general, the higher pin count and smaller package size are accomplished by reducing the bond pad size and spacing (pitch) on the chip level, and the lead dimensions and pitch on the chip carrier (module) level. When the lead spacing of an SMC is equal to or less than 0.65 mm, then it is called a fine pitch SMC. Any SMT with fine pitch SMCs is called fine pitch technology (FPT), see Lau [1989]. Figure 9-1 shows a 0.2 mm pitch, 32 mm body size, 584-pin plastic quad flat pack (QFP).

In the past few years, many companies have been designing and assembling fine pitch SMCs into their products. The small amount of solder volume of the fine pitch solder joints and the concerns associated with the fine pitch assembly processes have brought solder joint reliability of fine pitch SMCs under scrutiny. In

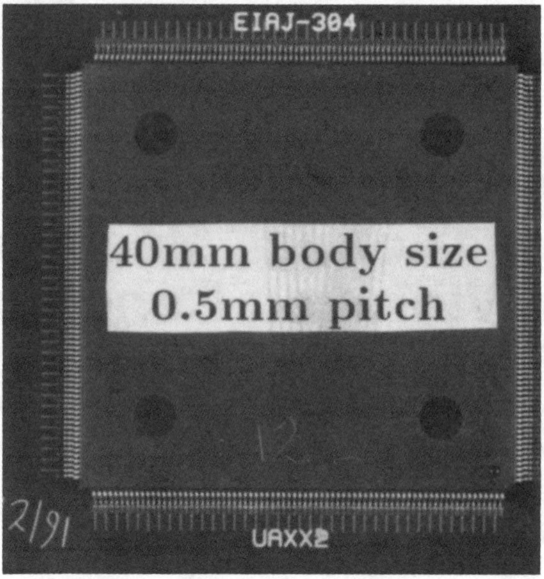

FIGURE 9-1 Hitachi's 0.2 mm pitch, 584-Pin QFP.

this chapter, some experimental results of the 0.5 mm fine pitch solder joints are presented.

In order to achieve 100% SMT/FPT, suppliers are developing many types of surface mountable components to fill the diverse needs of SMT. Connector technology is one area where these new components are evolving. Surface mount connectors are becoming available in different configurations to meet the specific needs of surface mount products.

Some of the advantages of the surface mount connectors are: high density, low profile, light weight, fine pitch, smaller PWB area, circuit operation at higher speeds with less crosstalk and interference, better signal transmission due to shorter trace routes, exposed leads for visual inspection, vacuum pick-up, and compatibility with SMT processes. Some of the disadvantages of surface mount connectors are: high cost, inadequate availability, and less reliable solder joints. However, for high volume applications, the availability and cost of surface mount connectors is expected to become similar to the through-hole connectors they replace. Solder joint reliability of five surface mount connectors is another focus of the present chapter.

A variety of environmental stress factors such as temperature and humidity, thermal and functional cycles, corrosion, shock and vibration, handling, shipping, reworking, mating and unmating, etc. may lead to solder joint reliability problems (failures). The most commonly observed failure modes are overload and fatigue.

Overload failure occurs whenever the stress or strain in the solder joint, brought about by the imposed environmental stress factors, is greater than the strength or fracture toughness of the solder materials. Examples of overload failure conditions include excessive bending and twisting of a PWB with large surface mount components or connectors soldered to its surfaces, and excessive forces for mating/unmating a surface mount connector. On the other hand, fatigue failure of a solder joint takes place via the initiation and slow propagation of a crack until it becomes unstable. The environmental stress factors that typically cause failure by fatigue are much less severe than the overload failure levels. Examples presented of fatigue failures include: the vibration of a PWB with large and heavy surface mount components or connectors soldered to its surfaces, and the temperature cycling of surface mount components and connectors which are soldered to the surfaces of a PWB.

This chapter presents, by means of testing and statistical analysis, the life distribution, reliability function, and failure rate of the fine pitch thin small outline package (TSOP) solder joints under thermal cycling conditions; the mechanical responses of the 304-pin fine pitch QFP solder joints under bending and twisting conditions; and the life distribution, reliability function, and failure rate of the solder joints of five different surface mount connectors under vibration conditions. Results are summarized in the format of tables and figures and can be used as a reference data base for the solder joints of other surface mount components and connectors.

9.2 FINE PITCH TSOP SOLDER JOINTS UNDER THERMAL CONDITIONS

9.2.1 Test Board Assembly

Figure 9-2 shows a schematic of a TSOP surface mount assembly [Lau et al. 1992a]. It consists of four major parts: the package body (Figure 9-3), the alloy-42 leads (Figure 9-4), the 63Sn-37Pb solder joints, and the FR-4 epoxy/glass PWB.

Fine Pitch Component

The fine pitch component used in the test was a 28 lead, 0.5 mm pitch Type-I TSOP (with live memory devices), Figures 9-2 through 9-4. The dimensions of this TSOP were 11.8 mm × 108 mm × 101.025 mm. The ratio of the volume of silicon to plastic was very large. The thermal coefficient of expansion of the silicon (2.9×10^{-6}/°C) and the plastic (22×10^{-6}/°C) differed greatly (Table 9-1). Consequently, the effective thermal coefficient of expansion of the whole package body was of interest and was measured.

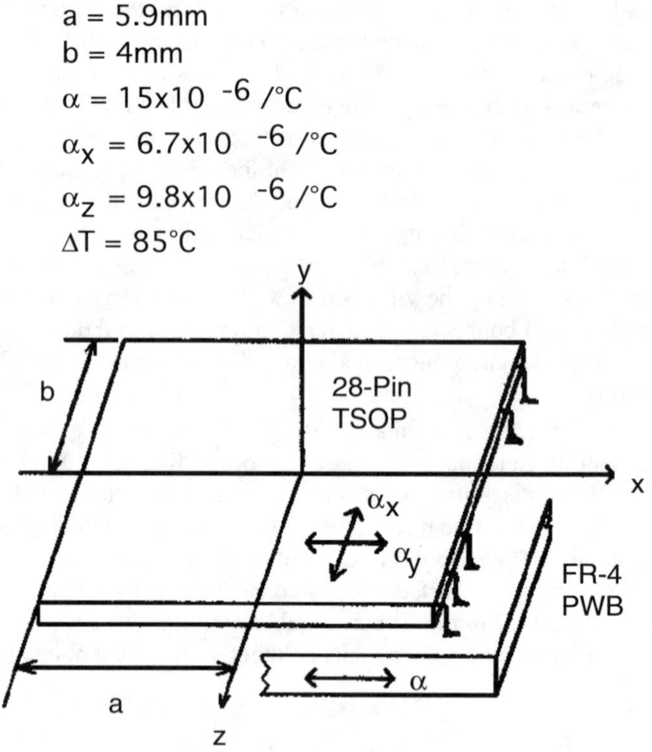

FIGURE 9-2 Schematic of a 28-Pin Type-I TSOP Assembly that Shows the Three Major Sources to Produce Thermal Stress in the 28-pin TSOP Solder Joint: (1) Global TCE mismatch between the TSOP and PWB, (2) local TCE mismatch between the lead and solder, and (3) lead stiffness.

Figure 9-5 shows the effective thermal coefficient of expansion of the 28-pin TSOP body in the long or x-direction (α_x) and the short or z-direction (α_z). For temperatures less than 150°C, $\alpha_x = 6.7 \times 10^{-6}$/°C and $\alpha_z = 9.8 \times 10^{-6}$/°C. The reasons for $\alpha_z > \alpha_x$ are: (1) the ratio of silicon to plastic (Figure 9-3) in the x-direction (68%) is larger than that in the z-direction (59%); (2) the thermal coefficient of expansion of silicon is smaller than that of plastic; and (3) the leads are oriented in the x-direction. Comparing the thermal coefficient of expansion of the TSOP body α_x and α_z) and the PWB ($\alpha = 15 \times 10^{-6}$/°C), the difference (global thermal expansion mismatch) was very large.

Furthermore, the lead frame material of the TSOP was made of alloy-42 with a 300 microinches (7.62 microns) thick 90Sn-10Pb plated surface. The thermal coefficient of linear expansion of alloy-42 was 5×10^{-6}/°C. Thus, the local thermal expansion mismatch between the gull-wing lead and the Sn/Pb solder joint (21×10^{-6}/°C) was large.

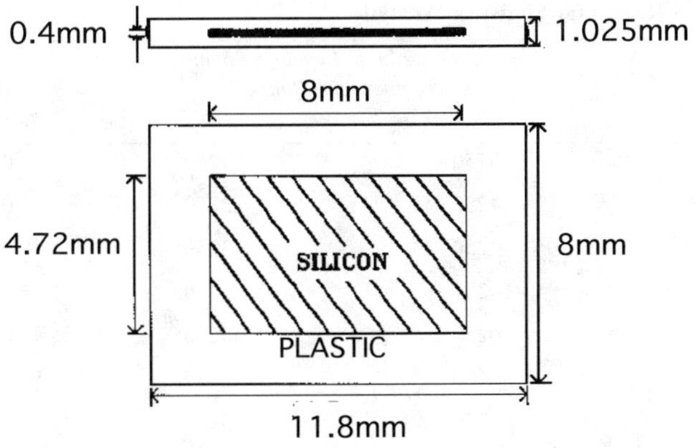

FIGURE 9-3 Side and top views of the TSOP (leads not shown).

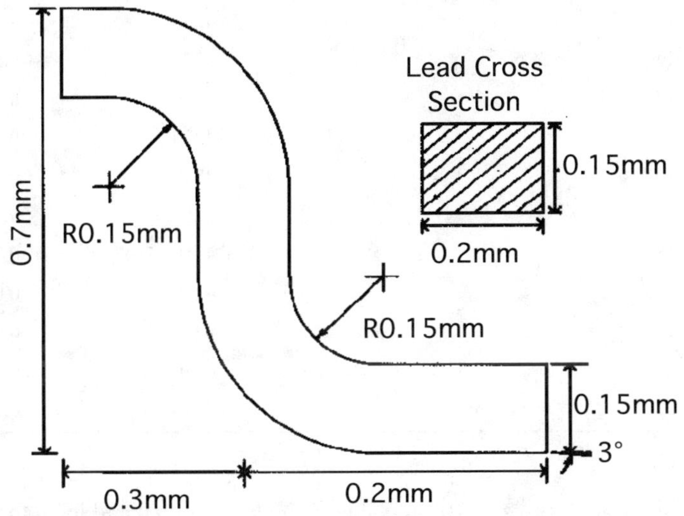

FIGURE 9-4 Lead geometry of the TSOP.

Finally, the critical dimensions of the gull-wing shaped leads: the foot length, width, and thickness were 0.28 mm, 0.2 mm, and 0.15 mm, respectively, Figure 9-4. The standoff of the gull-wing lead was short compared with the other surface mountable fine pitch components; for examples, see Lau [1989]. Thus, the TSOP lead was stiffer (i.e., has less compliance). Consequently, when the TSOP assembly was subjected to a temperature change, the leads and solder joints were

TABLE 9-1 Test Results for Statistical Analysis

	Cycles to failure (1 cycle = 26 min.)		
	Temperature Range		*Median*
Failure Number	*180°C*	*215°C*	*(50%) Rank*
1	3750	1250	0.0315
2	4250	2750	0.0761
3	4250	3250	0.1207
4	4250	3250	0.1653
5	4250	3250	0.2099
6	4750	3250	0.1653
7	4750	3250	0.2992
8	5750	4250	0.3438
9	5750	4250	0.3884
10	5750	4250	0.4330
11		4250	0.4776
12		4250	0.5223
13		4250	0.5669
14		4250	0.6115
15		4250	0.6561
16		4250	0.7007
17		4250	0.7454
18		4250	0.7900
19		4250	0.8346
20		4250	0.8792
21		4250	0.9238
22		4250	0.9684

subjected to large stresses and strains due to the large global and local thermal expansion mismatches and the low profile of the TSOP.

Test Board

Because of the small size of the test board and cost saving, multiple board manufacturing method was adopted. The dimensions of the large panel were 406.4 mm × 10304.8 mm × 101.57 mm. It consisted of 40 small test boards, each with dimensions of 50.8 mm × 1025.4 mm × 101.57 mm. These test boards were connected by tiny connecting tabs. The test board material was made of single-layer FR-4 epoxy/glass.

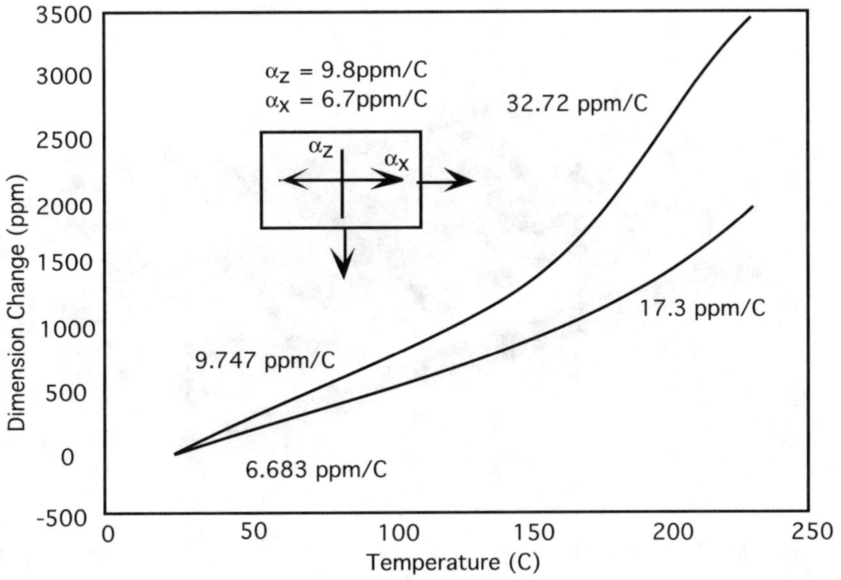

FIGURE 9-5 TCE of the TSOP.

Paste, Printing, Pick & Place, and Solder Reflow

The solder paste contained 50 percent by volume (88 percent by weight) solder balls in an RMA (rosin, mildly activated) flux. The solder ball composition was the eutectic 63Sn-37Pb with particle size (–325, +500), i.e., the diameter was no greater than 0.0432 mm or less than 0.0254 mm. The viscosity was 1000±102 kcps. This solder paste was printed through a metal stencil with a squeegee. The stencil had an opening of 0.61 mm × 100.25 mm and the wet paste thickness was 0.15 mm. The TSOPs were then placed in position on the panel using a pick and place machine.

The solder joints for all the TSOPs were mass reflowed by infrared soldering with a peak temperature of 240°C. Immediately after the solder was reflowed, the assembled panel was solvent cleaned and inspected. Finally, the panel was depanelized into 40 small test boards, each with a 28-pin TSOP (Figure 9-6). For electrical measurement purposes, the test board was soldered to a 28-pin header for socketing.

9.2.2 Test Set-Up and Boundary Conditions

The test boards were tested for full electrical functionality including an opens/shorts test routine using a tester for memory components before and during intermittent readout of the temperature cycling. The units that showed an electrical

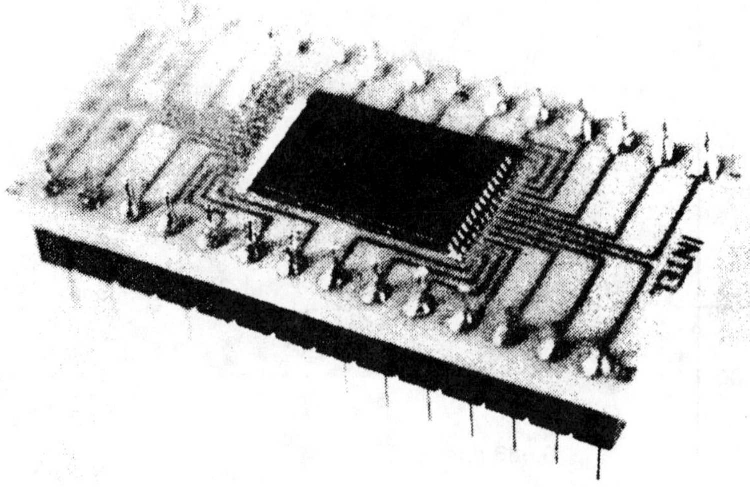

FIGURE 9-6 TSOP test board.

open were retested on the curve tracer to verify the opens. Solder joint opens were confirmed by pressing slightly on the open pin, which resulted in full continuity and full functionality of all the failed units. Units showing an electrical open were counted as failures and were removed from the population during further testing. The number of cycles to failure (N) was defined as the first solder joint failure in a 28-pin TSOP component. The solder joints were inspected under SEM for the quality of the joints as well as for documenting the failure modes.

The interposer test boards were tested in an air-to-air thermal cycling chamber. They were subjected to two Temperature Cycling (T/C) profiles shown in Figure 9-7 and below:

T/C "B": –55° to 125°C
T/C "C": –65° to 150°C

For both conditions, the ramp time, dwell time, and period (time to complete 1 cycle) were 1 minute, 12 minutes, and 26 minutes, respectively. At each condition, 22 interposer test boards were tested. Before testing, all solder joints were inspected and any that did not appear "good" were reworked.

9.2.3 Statistical Analysis of Test Results

Table 9-2 shows the test results under T/C "B" (ΔT = 180°C) and "C" (ΔT = 215°C) conditions. In this chapter, reliability of the solder joints was modeled by

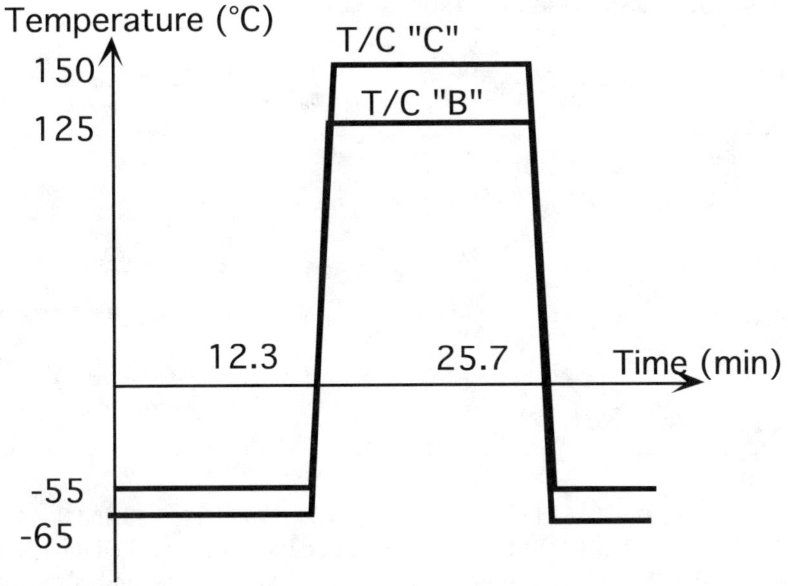

FIGURE 9-7 Temperature cycling conditions.

Weibull's probability density function [1952, 1961] and analyzed by Johnson's statistical method with ranking [1951, 1964]. Furthermore, an acceleration model was developed to bridge the temperature and frequency gaps between the testing and operating conditions.

Solder Joint Reliability Under Testing Conditions

The two-parameter Weibull probability density function has the following form:

$$f(x) = \left(\frac{\beta}{x}\right)\left(\frac{x}{\theta}\right)^{\beta} e^{-\left(\frac{x}{\theta}\right)^{\beta}} \tag{9.1}$$

In Equation (9.1), x is the value of the random variable (or cycles to failure (N) in the present study), β is the shape parameter (Weibull slope), and θ is the scale parameter (characteristic value). The Weibull cumulative distribution function has the following form:

$$F(x) = 1 - e^{-\left(\frac{x}{\theta}\right)^{\beta}} \tag{9.2}$$

Applying the principles of least squares and ranking to the temperature cycling test results (Table 9-2), the "best fit" Weibull parameters (for the median rank)

TABLE 9-2 Material Properties of the TSOP Assembly

	Young's Modulus (GPa)	*Poisson's Ratio*	*Thermal Coefficient of Linear Expansion 10^{-6}(m/m°C)*
Sn-Pb Solder	10	0.40	21
Alloy 42	148	0.30	5
Silicon	131	0.30	2.9
Plastic	14	0.30	22
Gold	78	0.30	15
FR-4	11	0.28	15

are: $\beta = 5.14$ and $\theta = 6,363$ for T/C "C" ($\Delta T = 180$°C) and $\beta = 3.48$ and $\theta = 4,340$ for T/C "C" ($\Delta T = 215$°C). With these values, Equation (9.2) is called the life distribution of the 28-pin TSOP solder joints under test conditions, and is shown in Figure 9-8.

Mathematically, reliability is the probability of a product (or system) performing its intended purpose without failure for a specified period of time under a given operating condition. Numerically, reliability is the percent survivors, i.e.:

$$R(x) = 1 - F(x) \tag{9.3}$$

Figure 9-9 shows the reliability of the 28-pin TSOP solder joints under test conditions.

The Weibull failure rate is defined as follows:

$$h(x) = \frac{f(x)}{R(x)} = \left(\frac{\beta}{\theta}\right)\left(\frac{x}{\theta}\right)^{\beta - 1} \tag{9.4}$$

One of the units of failure rate is parts per million/per thousand hours (PPM/K or FIT). One PPM/K means one failure is expected out of one million components operating for 1,000 hours.

Acceleration Factor

The temperature cycling test frequency (f_t) was at 1 cycle/26min (f_t = 1 cycle/ 26min = 55.38 cycle/day) and the temperature ranges were $\Delta T = 180$°C (T/C "B") and $\Delta T = 215$°C (T/C "C"). The operating conditions were assumed to be: $f_0 = 1$

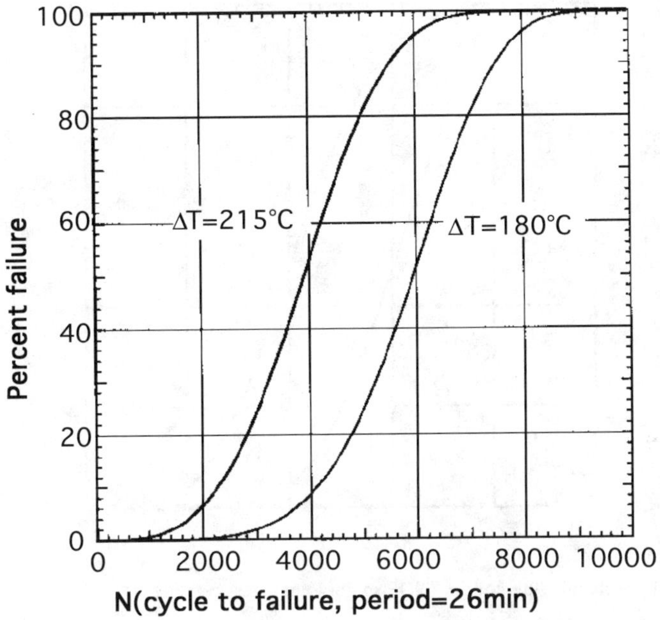

FIGURE 9-8 Weibull plot of the TSOP solder joint at test conditions.

cycle/day and $\Delta T_0 = 85°C f_0$ is the cycling frequency at operating condition, ΔT_0 is the temperature range at operating condition, and the maximum junction temperature at the solder joint was assumed to be 85°C). Consequently, an acceleration factor is needed to transfer the life distribution, reliability function, and failure rate at the higher testing conditions to the lower operating condition.

A linear acceleration may be assumed, i.e.:

$$x_0 = \eta x_t = \eta x \tag{9.5}$$

The Weibull life distribution at operating condition ($F_0(X_0)$) is given by:

$$F_0(x_0) = 1 - e^{-\left(\frac{x_0}{\eta\theta}\right)^{\beta}} \tag{9.6}$$

The reliability function at operating condition ($R_0(x_0)$) is given by:

$$R_0(x_0) = e^{-\left(\frac{x_0}{\eta\theta}\right)^{\beta}} \tag{9.7}$$

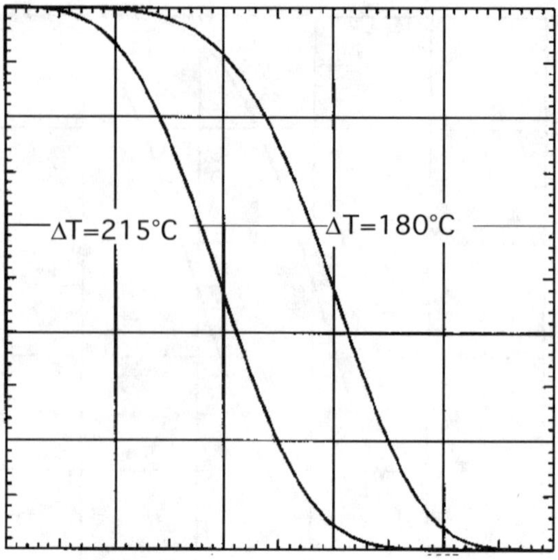

FIGURE 9-9 Reliability plot of the TSOP solder joint at test conditions.

The failure rate at operating condition ($h_0(x_0)$) is given by:

$$h_0(x_0) = \eta^{-\beta}\left(\frac{\beta}{\theta}\right)\left(\frac{x_0}{\theta}\right) \tag{9.8}$$

In Equation (9.5) to (9.8), x_0 is the cycles to failure at operating condition, $x_t =$ x is the cycles to failure at testing condition, and η is the acceleration constant (factor) relating cycles to failure at the test and operating conditions.

In the present study, we proposed the follow acceleration factor:

$$\eta = \left(\frac{f_0}{f_t}\right)^{\frac{1}{3}}\left(\frac{\Delta T}{215^{\circ}C}\right)^{-2.38} \tag{9.9}$$

For $f_0 = 1$ cycle/day, $f_t = 55.38$ cycle/day, $\Delta T = \Delta T_0 = 85^{\circ}C$, Equation (9.9) becomes:

$$\eta = \left(\frac{1}{55.38}\right)^{\frac{1}{3}}\left(\frac{85}{215}\right)^{-2.38} \tag{9.10}$$

In Equation (9.9), the acceleration factor is a function of the product of the cycling frequency effect and the cycling temperature effect. The cycling frequency term $(f_0/F_t)^{1/3}$ is well documented in the literature and requires no further explanation (for examples, see Lau et.al., [1992]; Norris and Landzberg, [1969]). The cycling temperature term $(\Delta T/215^\circ C)^{-2.38}$ was generated as follows:

1. In Figure 9-8, the number of cycles to failure at 50% failure was recorded from both test conditions, i.e., $N_{50\%}(\Delta T = 180^\circ C) = 5{,}800$ and $N_{50\%}\ (\Delta T = 215^\circ C) = 3{,}800$.
2. In Figure 9-10, the points (180, 5800) and (215, 3800) were plotted; and a straight line was constructed through these two points. Then $N_{50\% at \Delta T} = (\Delta T/215^\circ C)^{-2.38}(3800)$.

Solder Joint Reliability Under Operating Conditions

Once the acceleration factor, $\eta = 2.39$, was determined, the life distribution, reliability function, and failure rate under operation condition can be determined by Equations (9.6) to (9.8), with $\beta = 3.48$ and $\theta = 4{,}340$.

$$F_0(x_0) = 1 - e^{-\left(\frac{x_0}{10373}\right)^{3.48}} \tag{9.11}$$

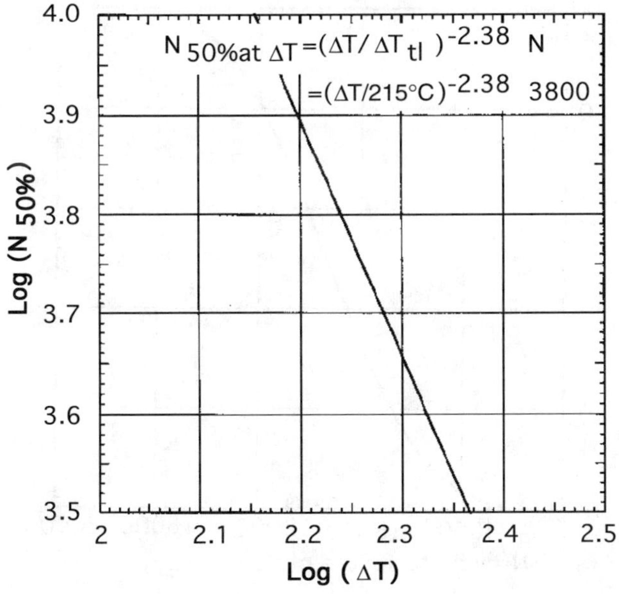

FIGURE 9-10 Temperature range vs. cycles to failure at 50%.

$$R_0(x_0) = e^{-\left(\frac{x_0}{10373}\right)^{3.48}} \qquad (9.12)$$

$$h_0(x_0) = 0.0000387\left(\frac{x_0}{4340}\right)^{2.48} \qquad (9.13)$$

Figure 9-11 shows the life distribution $F_0(x_0)$ of the 28-pin TSOP solder joints under operating conditions. The reliability $R_0(x_0)$ and failure rate $h_0(x_0)$ of the TSOP solder joints operating at 1 cycle/day and with a temperature range of 85°C are shown in Figures 9-12 and 9-13, respectively. The 1% cumulative failure point occurred at approximately 66,500 hours or 7.5 years, and the 50% cumulative failure point occurred at about 224,000 hours or 25.5 years. These values are beyond a typical design lifetime.

9.2.4 Failure Analysis of TSOP Solder Joints

In view of Figure 9-8, the solder joint performed better at T/C "B" ($\Delta T = 180$°C) than at T/C "C" ($\Delta T = 215$°C). Also, beyond 1,500 cycles the solder joint failure

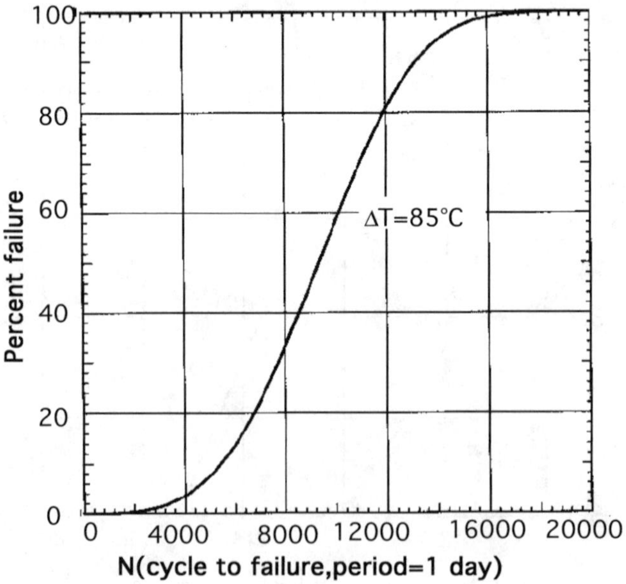

FIGURE 9-11 Weibull plot of the TSOP solder joint under operating conditions.

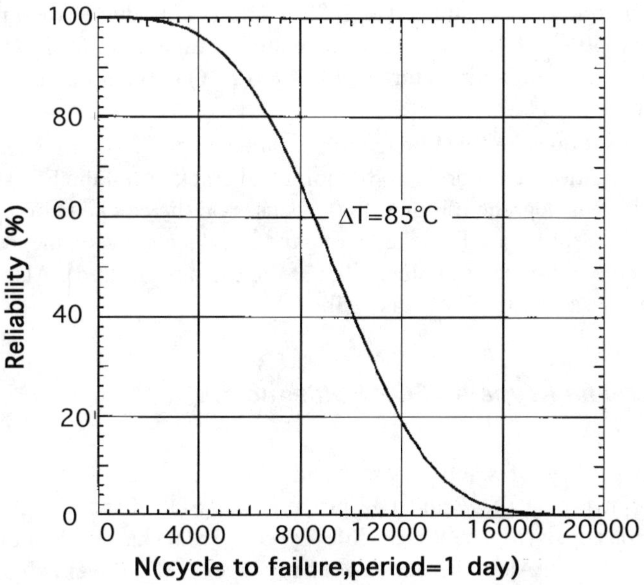

FIGURE 9-12 Reliability plot of the TSOP solder joint under operating conditions.

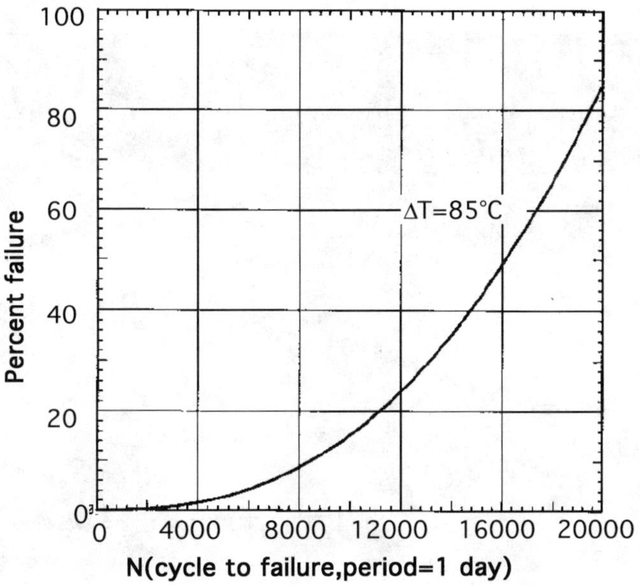

FIGURE 9-13 Failure rate plot of the TSOP solder joint under operating conditions.

rate was higher at T/C "C" than at T/C "B". These were due to the higher peak temperature (150°C) at T/C "C", the low melting temperature (183°C) of solder, and the glass transition temperature of FR-4 (~125°C in the horizontal direction) [Lau, 1992].

Figure 9-14 shows the SEM image of the failed solder joints after 1,500 cycles at T/C "C" condition. The corner solder joint had cracked and the FR-4 test board (solder mask) was degraded. Figure 9-15 focuses on the cracked corner lead and solder joint. The solder joint cracked along the interface between the lead and the solder joint (the tiny piece of solder fillet at the toe disappeared). Also, cracking of the solder mask on the PWB was visible.

9.2.5 Type-I and Type-II TSOP Solder Joints

There are two types of TSOP, Type I and Type II. Figure 9-16 schematically shows a 32-pin Type-I TSOP and a 32-pin Type-II TSOP. Type-I TSOP has leads on the shorter edges (Figure 9-6), while Type II TSOP has leads on the longer edges of the TSOPs. Also, the dimensions of Type-II TSOP were slightly larger than that of Type-I TSOP. However, the lead width of Type-II TSOP (0.4 mm) was twice that of Type-I TSOP (0.2 mm); therefore, the lead stiffness of Type-II TSOP was much larger than that of Type-I TSOP. Which type of TSOP's solder

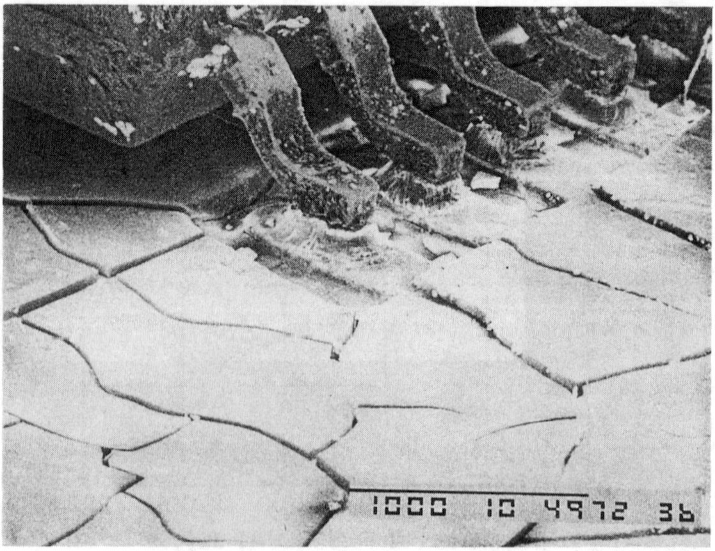

FIGURE 9-14 SEM of the TSOP assembly (after 1,500 T/C "C" cycles).

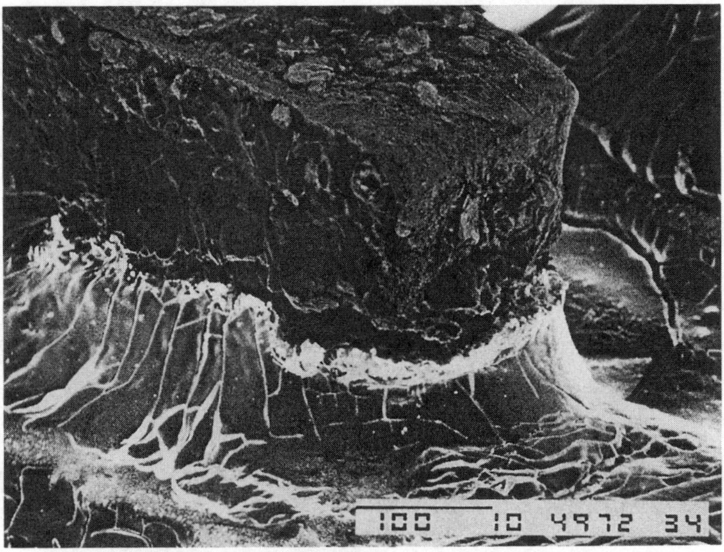

FIGURE 9-15 Exaggerated corner lead and solder joint (after 1,500 T/C "C" cycles).

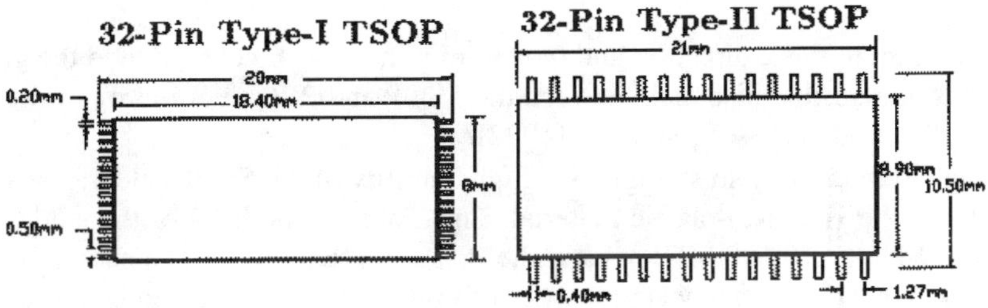

FIGURE 9-16 Type-I and Type-II TSOP solder joints.

joint has a better thermal performance? A quantitative comparison is given by the following nonlinear finite element analyses.

Figure 9-17 shows the finite element model (dashed lines) for the analysis of the corner lead and solder joint of the 32-pin Type-I TSOP assembly. It consisted of 1,024 3-D solid elements (544 for the alloy-42 lead and 480 for the Sn/Pb solder joint). Each element had 20 nodes, and each node had 3 degrees of freedom. The boundary conditions were: (1) the TSOP assembly was subjected to a temperature change from 0°C to 85°C (DT = 85°C); (2) the upper end of the lead does not move in the x- and z-directions; and (3) at the bottom of the solder joint all points moved

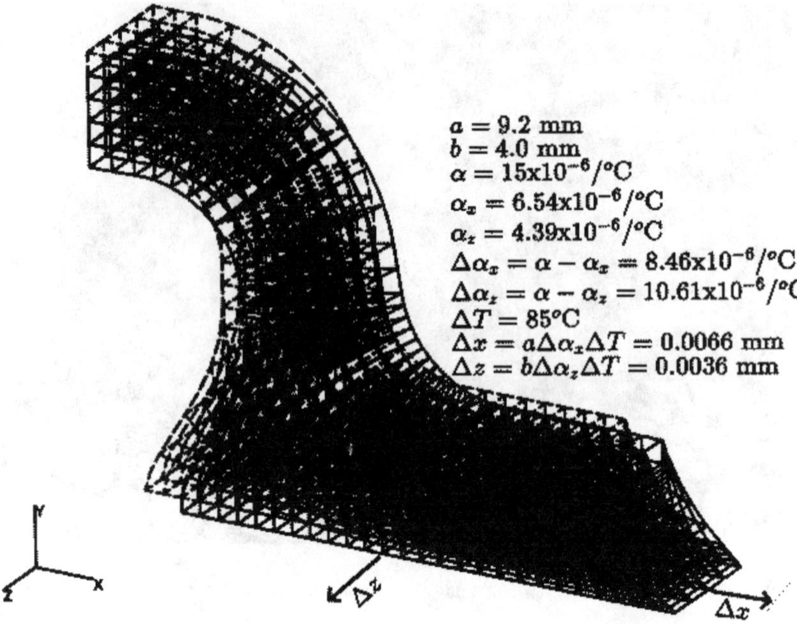

FIGURE 9-17 Model and deformation of the Type-I TSOP lead and solder joint.

0.0066 mm in the x-direction and 0.0036 mm in the z-direction due to the global thermal expansion mismatch between the TSOP and PWB. For more information about the modeling, see Lau et al. [1992].

Due to the low yield strength and high ductility of the Sn/Pb solder, it was assumed to be an elasto-plastic material. The Young's modulus of the solder was 1.5×10^6 psi (10,000 MN/m^2) and the Poisson's ratio was 0.4. It should be emphasized that the mechanical properties of solders were strongly temperature-, frequency-, rate-, and time-dependent as discussed in this volumes previous chapters. These variables have to be included in the analysis to have an accurate estimation of the strains, and consequently the number of cycles to failure, of the solder joint. However, for the sake of simplicity and due to the objective of this study, all the material properties were assumed to be temperature-, frequency-, rate-, and time-independent. Furthermore, the stress-strain curve at 120°C was used (yield stress = 1,200 psi or 8.3 MN/m^2; yield strain = 0.0008; strain-hardening parameter = 0.1) [Lau et.al., 1992] (Figure 9-18).

In this chapter, the yield surface (i.e., yield criterion) of the solder was assumed to obey the distortion energy theory (i.e., von Mises yield criterion) and the plastic flow rule (i.e., plastic stress-strain relation) was assumed to obey the incremental deformation theory (i.e., Prandtl-Reuss theory). The fundamental equations governing the present boundary-value problem were as follows.

Equilibrium equations

$$\frac{\delta\sigma_{ij}}{\delta x_j} = 0 \tag{9.14}$$

Strain-displacement equations

$$\varepsilon_{ij} = \frac{1}{2}\left(\frac{\delta\mu_i}{\delta x_j} + \frac{\delta\mu_j}{\delta x_i}\right) \tag{9.15}$$

Constitutive equations

$$d\varepsilon_{ij} = d\varepsilon^e_{ij} + d\varepsilon^p_{ij} \tag{9.16}$$

where

$$d\varepsilon_{ij} = \frac{dS_{ij}}{2G} + (1-2\upsilon)\,\delta_{ij}\frac{d\sigma_{ij}}{E} \tag{9.17}$$

$$d\varepsilon^p_{ij} = d\varepsilon^e_{ij} + d\varepsilon^p_{ij} \tag{9.18}$$

and

$$S_{ij} = \sigma_{ij} - \frac{1}{3}\sigma_{\beta\beta}\delta_{ij} \tag{9.19}$$

If

$$f = \frac{1}{2}S_{ij}S_{ij} - \frac{\sigma_y^2}{3} \tag{9.20}$$

$$\bar{\sigma} = \sqrt{\frac{3}{2}S_{ij}S_{ij}} \tag{9.21}$$

and

$$\overline{d\varepsilon^p} = \sqrt{\frac{2}{3}d\varepsilon^p_{ij}\,(d\varepsilon^p_{ij})} \tag{9.22}$$

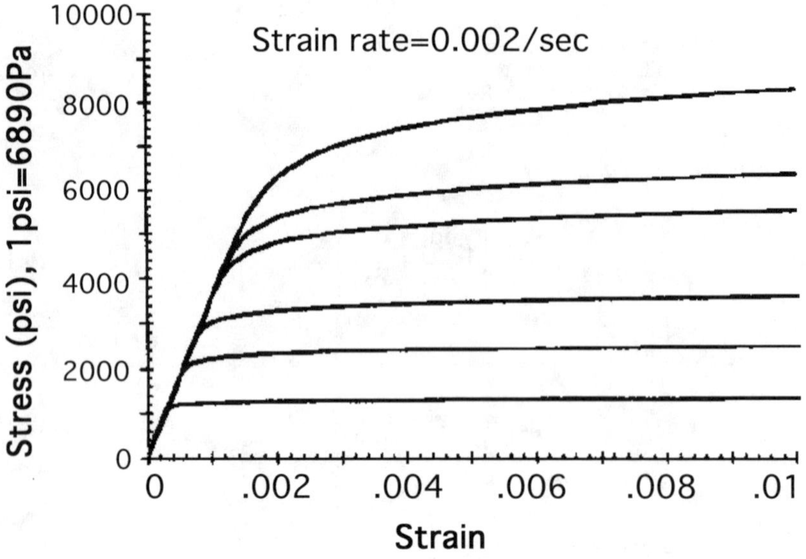

FIGURE 9-18 Stress-strain curves of 63Sn-37Pb solders.

then

$$d\lambda = \frac{3}{2}\frac{\overline{d\varepsilon^p}}{\bar{\sigma}} \tag{9.23}$$

and $\bar{\sigma}$ and $\overline{d\varepsilon^p}$ coincided with the uniaxial stress-strain curve, Figure 9-18.

In Equations (9.14) to (9.23), σ_{ij} was the stress tensor, x_i were the coordinates of each particle in the original configuration, ε_{ij} was the strain tensor, u_i are the displacement components, $d\varepsilon_{ij}$ was the incremental strain tensor, $d\varepsilon^e_{ij}$ was the incremental elastic strain tensor, $d\varepsilon^p_{ij}$ was the incremental plastic strain tensor, S_{ij} was the stress deviation tensor, G was the shear modulus, ν was the Poisson's ratio for elastic deformation, δ_{ij} was the Kronecker delta, $d\sigma_{ij}$ was the incremental stress tensor, E was the modulus of elasticity, $d\lambda$ was a nonnegative function which depends on stress, strain, and strain history, $\bar{\sigma}$ was the effective stress, $\overline{d\varepsilon^p}$ was the incremental effective plastic strain, and σ_y was the uniaxial yield stress. All the indices ranged over 1, 2, 3 and the summation convention for repeated indices was used.

The boundary-value problem was to solve Equation (9.14) to (9.23) for the elasto-plastic strains of the corner lead and solder joint shown in Figure 9-17 when it was subjected to the boundary conditions described earlier. A total of 11 increments were applied with about 10 iterations per increment.

Figure 9-17 shows the whole-field displacements (deflections) of the TSOP assembly during heating. The solid lines are for the displaced mesh. It can be seen that the gull-wing lead and solder joint not only expanded in the x-direction but also expanded in the z-direction and resulted in a very complex state of stresses and strains. This was due to: (1) the horizontal thermal expansion mismatch between the TSOP body and the PWB; (2) the local thermal expansion mismatch between the alloy-42 lead and the Sn/Pb solder joint; and (3) the especially low standoff of the lead.

The von Mises (effective) stress ($\bar{\sigma}$) distribution, Equation (9.21), in the Type-I TSOP solder joint is shown in Figure 9-19. The von Mises stress distribution in the solder joint was not uniform. The maximum von Mises stress occurred at the edge (in the positive z-direction) near the middle portion of the solder joint. This maximum stress was concentrated in a very small area and declined rapidly away from this location. Thus, any solder joint damage (cracks) should initiate at this location and propagate in the solder joint along the interface between the gull-wing lead and the solder joint and through the width of the solder joint.

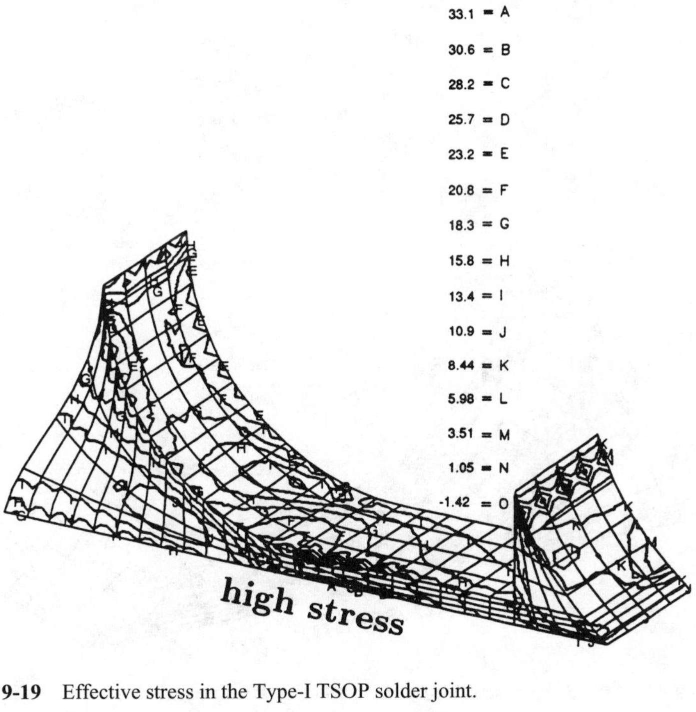

FIGURE 9-19 Effective stress in the Type-I TSOP solder joint.

The accumulated plastic strain components in all directions (ε^p_x, ε^p_y, ε^p_z, γ^p_{xy}, γ^p_{xz}, γ^p_{yz}) acting in the solder joint are not presented here. However, the accumulated effective plastic strain:

$$(\overline{\varepsilon_p} = \int \overline{d\varepsilon_p})$$

is shown in Figure 9-20. The strain contours of $(\overline{\varepsilon_p})$ in the solder joint were not uniform. The maximum $\overline{\varepsilon_p}$ (0.0227) was concentrated at a very small area at the edge in the positive z-direction near the center portion of the solder joint. This maximum value decreased rapidly to a much smaller value, a short distance away from the center portion.

Figure 9-21 shows the finite element model (dashed lines) for the analysis of the corner lead and solder joint of the 32-pin Type-II TSOP assembly (Figure 9-17). It consisted of 1,446 3-D solid elements (770 for the alloy-42 lead and 676 for the Sn/Pb solder joint). The element type was exactly the same as those for the Type-I TSOP lead and solder joint. The boundary conditions were also the same, except the displacement boundary condition at the bottom of the solder joint

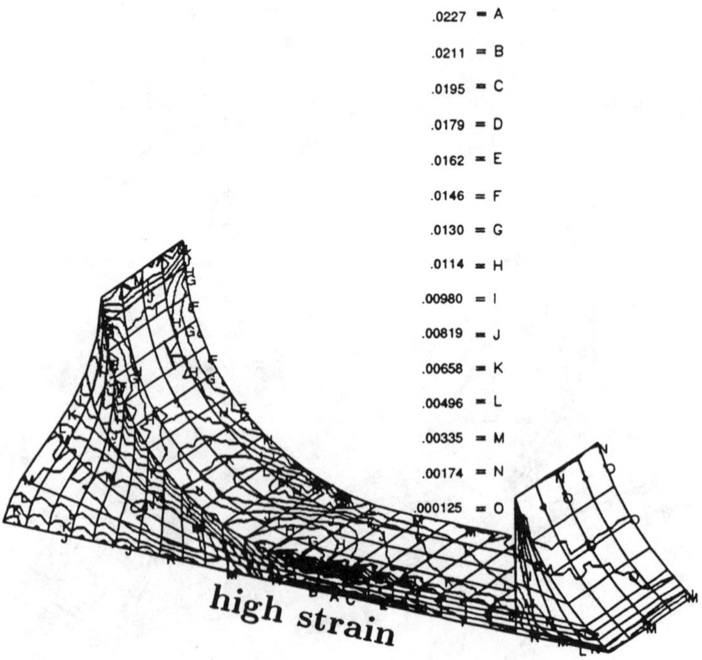

FIGURE 9-20 Effective plastic strain in the Type-I TSOP solder joint.

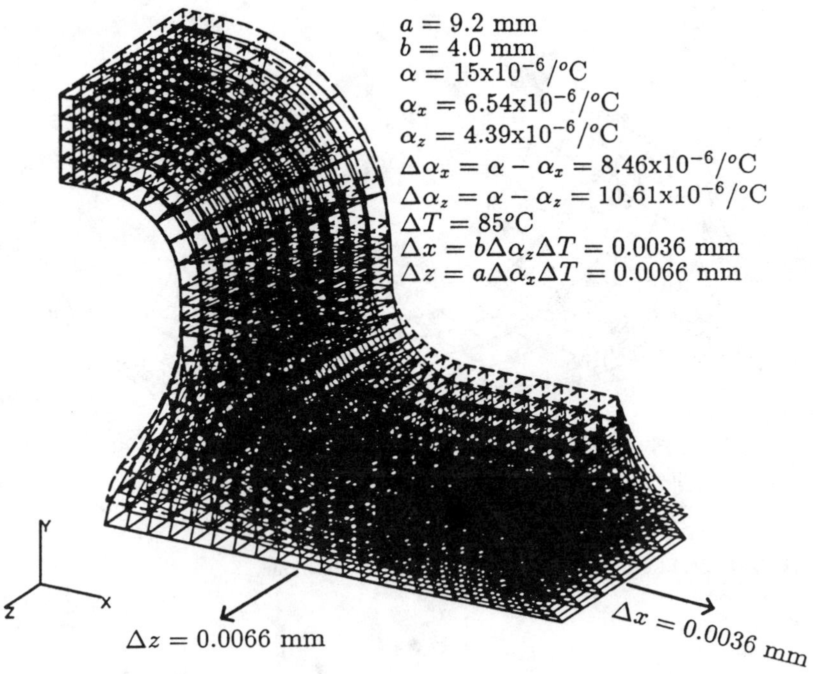

FIGURE 9-21 Model and deformation of the Type-II TSOP lead and solder joint.

where all points moved 0.0036 mm in the x-direction and 0.0066 mm in the z-direction.

The whole-field displacements (solid lines) of the Type-II TSOP corner lead and joint under heating were very similar to that of the Type-I TSOP corner leads and joint. Figures 9-22 and 9-23 show, respectively, that the effective stress ($\bar{\sigma}$) and accumulated effective plastic strain ($\bar{\varepsilon}^p$) in the Type-II TSOP solder joint follow the functions:

$$\bar{\sigma} = \sqrt{\frac{3}{2} S_{ij} S_{ij}}.$$

$$\overline{\varepsilon^p} = \sqrt{\frac{3}{2} \varepsilon_{ij}^p d\varepsilon_{ij}^p}$$

These values were much larger than those of the Type-I TSOP solder joint. This was due to the larger stiffness of the Type-II TSOP lead than that of the Type-I

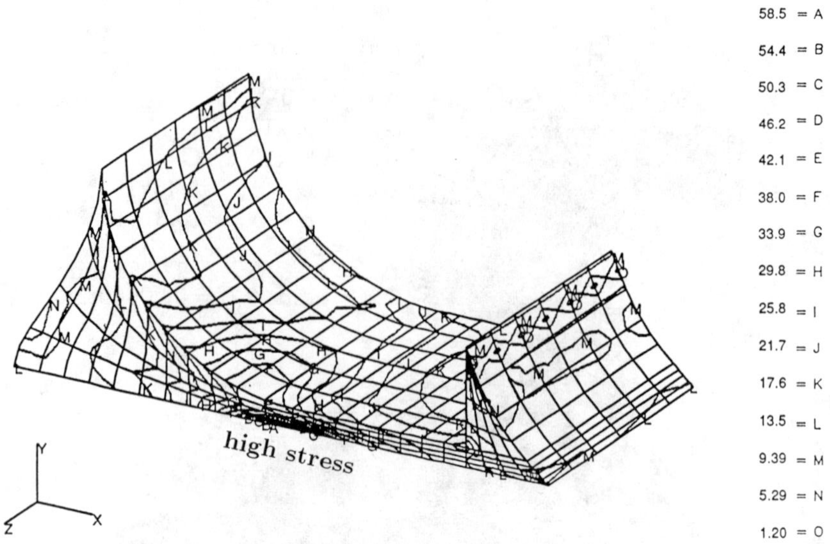

FIGURE 9-22 Effective stress in the Type-II TSOP solder joint.

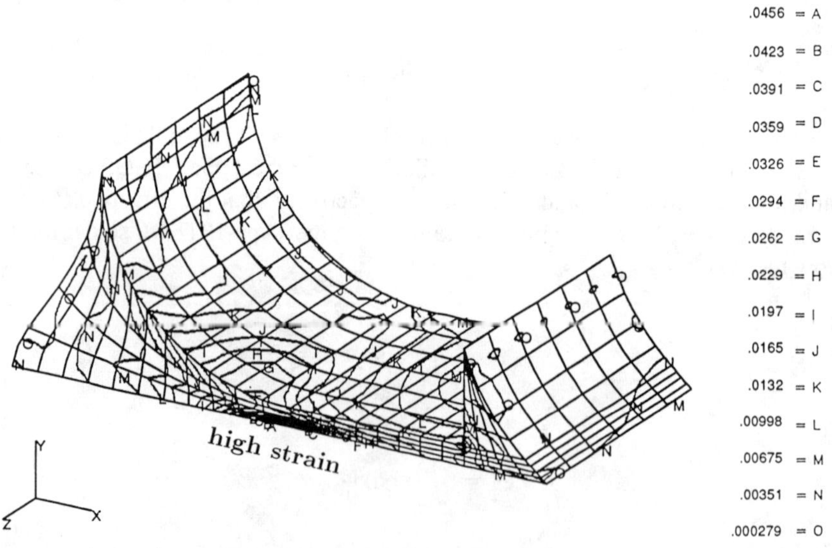

FIGURE 9-23 Effective plastic strain in the Type-II TSOP solder joint.

TSOP leads. In order for Type-II TSOP solder joint to have the same level of thermal reliability as Type-I TSOP joints, one way is to reduce the Type-II TSOP lead width by a factor of 2, i.e., 0.2 mm.

9.3 FINE PITCH LARGE QFP SOLDER JOINTS UNDER BENDING AND TWISTING CONDITIONS

9.3.1 Test Board Assembly

Figure 9-24 schematically illustrates a 304-pin QFP assembly. It consists of four major parts, namely the FR-4 epoxy/glass PWB, the 63Sn-37Pb solder joints, the copper lead, and the plastic QFP. Because of the QFP's large body size, mechanical responses of its solder joints under bending and twisting conditions (e.g., due to handling) were of interest. The solder joint thermal responses of this QFP can be found in Lau and Erasmus [1992].

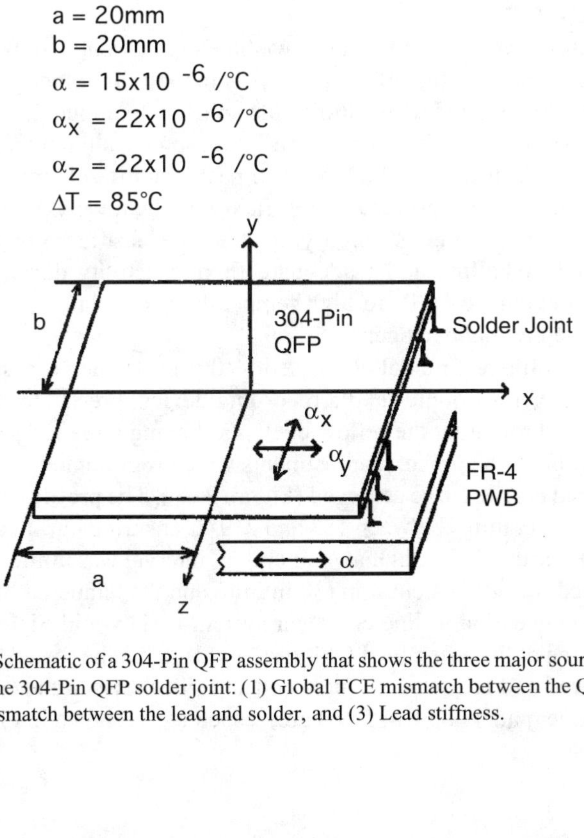

FIGURE 9-24 Schematic of a 304-Pin QFP assembly that shows the three major sources to produce thermal stress in the 304-Pin QFP solder joint: (1) Global TCE mismatch between the QFP and PWB, (2) Local TCE mismatch between the lead and solder, and (3) Lead stiffness.

Fine Pitch Component

The surface mount component used in the test was a 304-pin, 0.5 mm pitch plastic QFP shown in Figure 9-25. The linear coefficient of thermal expansion of the plastic is 22×10^{-6}/°C. The Young's modulus of the plastic is 2×10^{-6}psi (14,000 MN/m^2) and the Poisson's ratio is 0.3. It has gull-wing shaped leads which are spaced on a 0.5 mm pitch with a 0.5 mm foot length. These leads provide both electrical and mechanical interconnections of the package to the outside world and are soldered to the surface of a PWB. The lead frame material is made of copper. The linear coefficient of thermal expansion of the copper is 17×10^{-6}/°C. The Young's modulus of the copper is 17.5×10^6 psi (121,000 MN/m^2)and the Poisson's ratio is 0.35. The horizontal dimensions of the package are 40 mm by 40 mm, Figures 9-24 and 9-25.

Test Board

The thickness of the test PWB is 0.062" (1.58 mm); it is made of FR-4 epoxy/glass. The linear coefficient of thermal expansion of the FR-4 material in the in-plane direction is 15×10^{-6}/°C. The Young's modulus of the PWB is 22 GPa and the Poisson's ratio is 0.24. The dimensions of the bending and twisting test boards were 200×76.2 mm and 152.4×152.4 mm, respectively.

Solder Reflow

The test boards used for bending and twisting were assembled by a mass reflow process. The purpose of the reflow process is to melt the solder particles, wet the surfaces to be joined, and then solidify the solder into a metallurgical bond. The primary areas of concern during reflow are flux spread, slumping, and solder particle dimension. As heat is applied to solder paste prior to solder melting, the paste volatile solvents will evaporate and the flux will chemically activate. Excessive heat during the preheating and predrying stage or a fast ramp-up of temperature can create solder balling and inadequate fluxing activity during reflow. Also, exposure of the plastic QFPs to high temperature for relatively long periods of time can cause internal damage.

A reflow profile test model was assembled prior to the processing of the test boards. Using thermocouples and a recording device, precise temperature levels were recorded throughout the reflow cycle. Following a few passes of the profile test model through the process, adjustments were programmed into the controller and the desired profile curve achieved (Figure 9-26). The profile exposed the plastic QFPs to temperatures between 183 and 219 °C ensuring liquification of the solder paste, but the dwell time at that temperature interval was limited to 48 seconds. The belt speed was at 86.4 cm/min (38 in/min) and the temperature (°C) at the ten zone forced convection in-line conveyer system is shown in Figure 9-26. Figure 9-27 shows one of the assembled QFP test boards. It can be seen that the QFP was properly placed and reflowed. Typical end and side views of the copper lead, solder joint, copper pad, and FR-4 PWB are, respectively, shown in Figures 9-27 and

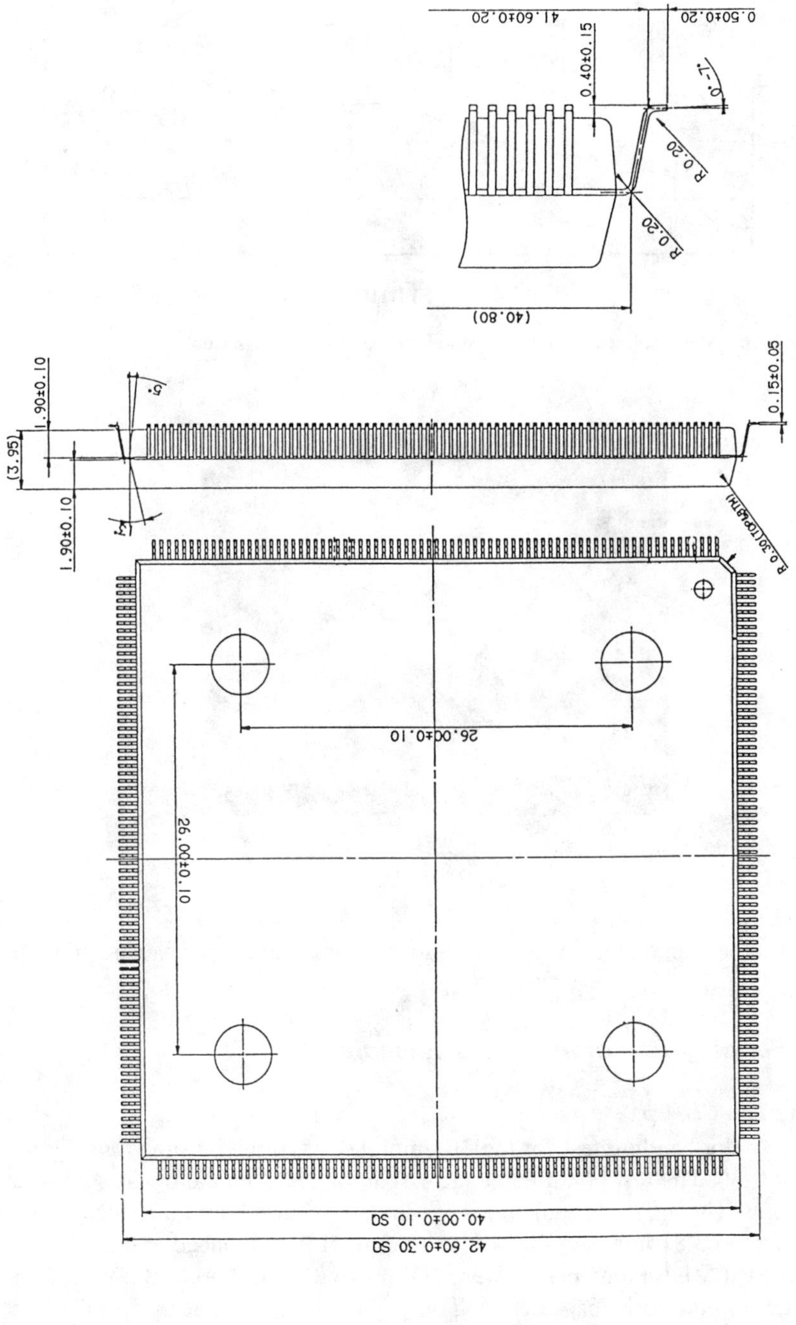

FIGURE 9-25 Dimensions of a 304-Pin QFP.

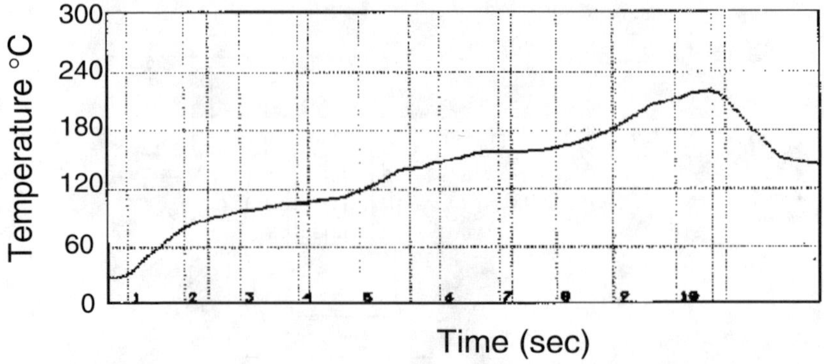

FIGURE 9-26 Mass reflow temperature profile for the 304-Pin QFP assembly.

FIGURE 9-27 304-Pin QFP assembly.

9-28. It can be seen that the copper lead, solder joint, and copper pad are all in the right position.

9.3.2 Bending Test of the QFP Assembly

Bending Test Set-Up

The three point bending test for the 304-pin QFP assembly is shown in Figures 9-29 and 9-30. The test board had a clear span of 152.4 mm and was supported at both ends. The load was applied at the center of the test board with a crosshead speed equal to 3.81 mm/sec. The 40 × 40 mm QFP was placed at the center of the test board, whose dimensions were 200 × 76.2 × 1.58 mm. Before testing, all solder joints were inspected and any that did not appear "good" were

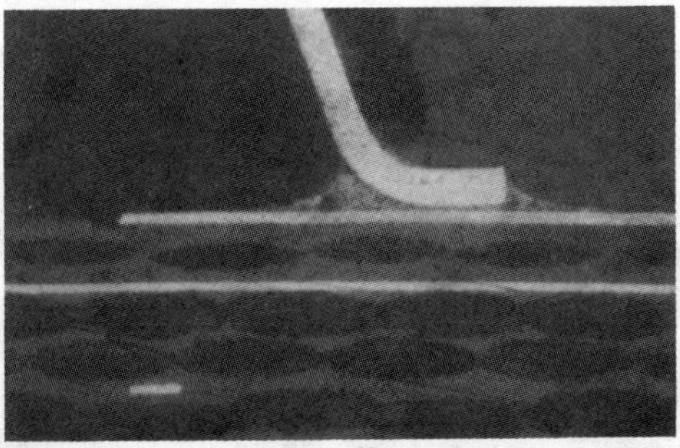

FIGURE 9-28 Typical side view of the 304-Pin QFP assembly.

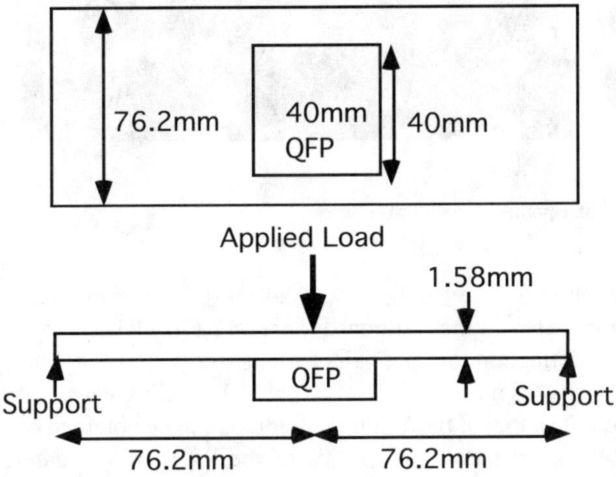

FIGURE 9-29 Three points bending test board.

reworked. The sample size was planned to be 10; however, because of the repeatability of the test results, most of the tests stopped after the fifth specimen.

Bending Test Results

Figure 9-31 shows the load-deflection curves of the test board with and without the 304-pin QFP. The QFP did increase the stiffness of the test board. The Young's modulus of the FR-4 PCB can be determined from the load-deflection curve without the QFP component using the following equation:

$$E = \frac{L^3}{48I}\left(\frac{F}{\Delta}\right) \tag{9.24}$$

FIGURE 9-30 Three points bending test set-up.

where L is the length of the test board (152.4 mm), I is the moment of inertia of the cross-sectional area of the test board ($I = (1/12)(76.2)(1.58)^3 = 25\ mm^3$), and F (force) and Δ (deflection) are the pairs (in this study, we used 5 pairs) of values from the test results (Figure 9-31). Thus, we have E = 22 GPa. It should be noted that a more accurate value of the Young's modulus can be obtained by $(1 - \nu_1\nu_2)E$, where ν_1 and ν_2 are the Poisson's ratios of the PWB in the material principal directions 1 and 2, respectively.

Figure 9-32 shows the load- and unload-deflection curves of the test board with the 304-QFP SMC. It can be seen that, after unloaded, the FR-4 PWB did not return to its original position. Furthermore, the larger the deflections the greater the offset (permanent) deflections. This is due to plastic deformation of the PWB. However, Figure 9-32 also shows that there were no solder joint nor lead failures, even with a load at 120 N and a deflection at 12.7 mm. For this size of board, this magnitude of deflection is beyond most of the manufacturing, shipping, handling, and rework conditions.

Another set of tests (to failure) showed that failures did not occur until the applied load reached 158.7 N and the deflection reached 19.5 mm (Figure 9-33). For a board of this size, this magnitude of deflection is way beyond all the manufacturing, handling, shipping, and rework conditions. Cross sections of the failure specimen showed that a few leads were broken (near the shoulder of the gull-wing, Figure 9-34) on both sides (near the middle) of the QFP in the longer direction of

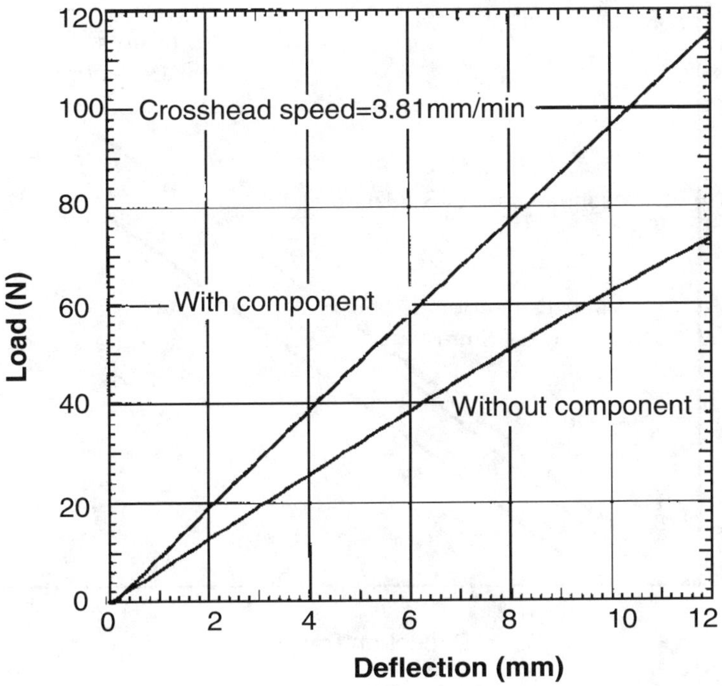

FIGURE 9-31 Three points bending load-deflection curves.

the test board. Also, a few leads (near the shoulder of the gull-wing) and solder joints (near the interface between the gull-wing lead and the solder joint) were broken on both sides (near the corners) of the QFP in the long direction of the test board.

9.3.3 Twisting Test of the QFP Assembly

Twisting Test Set-Up

The four point twisting test of the 304-pin QFP assembly is shown in Figures 9-35 and 9-36. The overall dimensions of the board were 152.4 × 152.4 × 1.58 mm and the QFP was located at its center. The board was loaded by two forces at its two opposite "corners" (the other two "corners" were supported). These "corners" were 3.2 mm from the actual corners (for the purpose of applying the forces and supports). The clear distance between the forces and supports was 146 mm. In this study, the edges of the board were not reinforced, i.e., bending and shearing deformations in the edge planes were possible. The crosshead speed was 3.81 mm/sec.

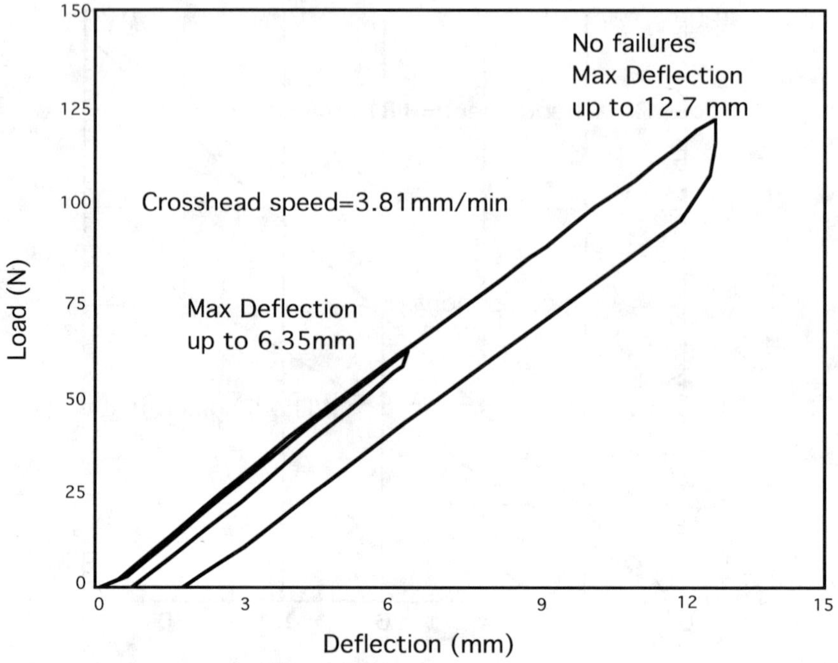

FIGURE 9-32 Three points bending load- and unload-deflection curves.

Twisting Test Results

Figure 9-37 shows the load-deflection curves of the test board with and without the 304-pin QFP. For both cases most of the curves were nonlinear except the small initial portion of the curves which were linear. This was due to the test boundary condition (by allowing the deformation of the edges of the board and large deflection and rotation of the board). The twisting (or torsional) stiffness of the FR-4 PWB can be determined by the test result curve (without QFP) in Figure 9-37. The linear twisting stiffness (T) can be determined by the following equation [Timoshenko and Woinowsky-Krieger 1959]:

$$T = \frac{c^2}{8} \left(\frac{F}{\Delta}\right) \qquad (9.25)$$

where c = 146, and F and Δ can be obtained from the initial linear portion of the curve without the QFP in Figure 9-37 ($F\Delta = 2.2$). Thus, we have T = 5,860 N/mm.

The load- and unload-deflection curves of the 304-pin QFP test board are shown in Figures 9-38 and 9-39. Similar to the bending tests, because of the plastic deformation of the FR-4 PWB, after unloading there was permanent deflection of

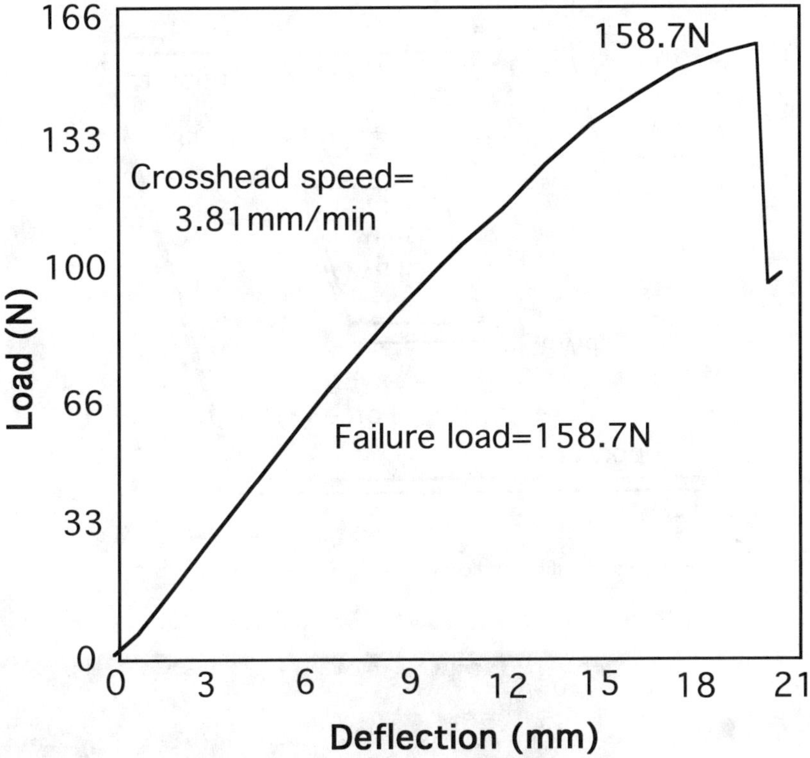

FIGURE 9-33 Three points bending load-deflection curves (to failure).

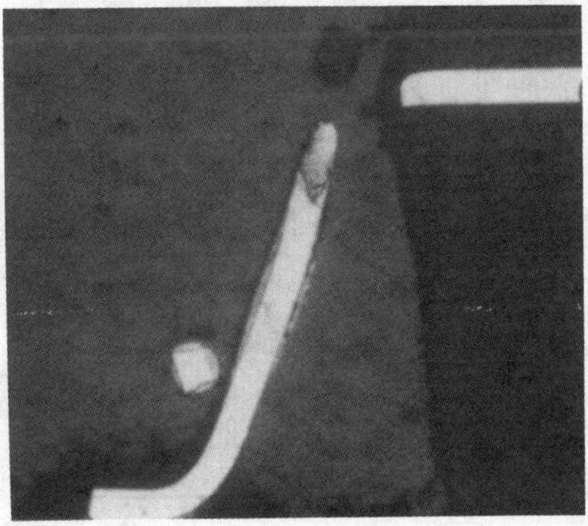

FIGURE 9-34 Three points bending failure mode (broken lead).

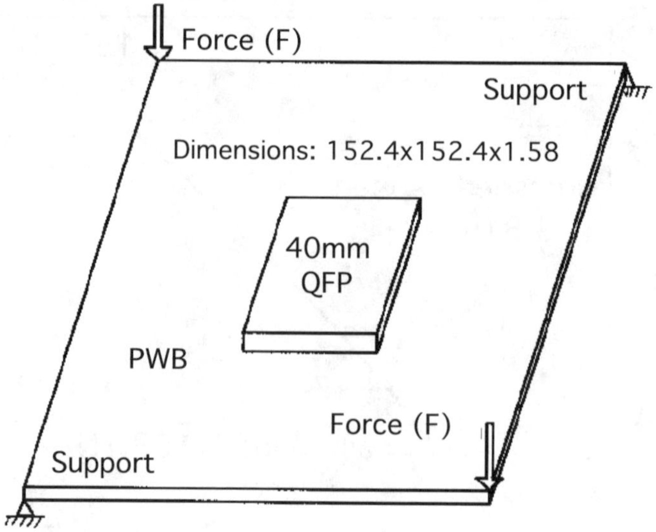

FIGURE 9-35 Four points twisting test board.

FIGURE 9-36 Four pointstwisting test set-up.

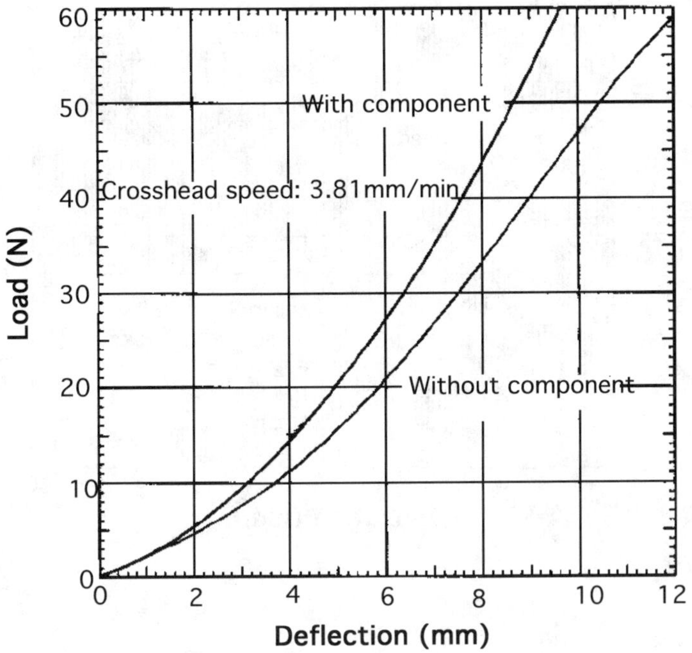

FIGURE 9-37 Four points twisting load-deflection curves.

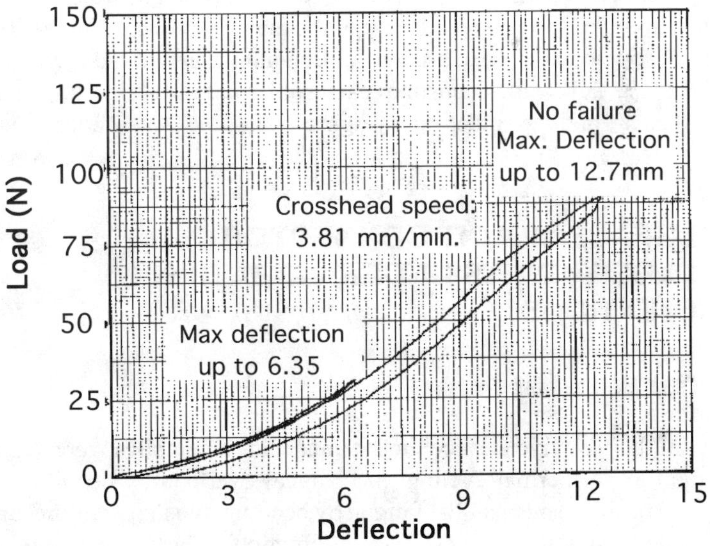

FIGURE 9-38 Four points twisting load- and unload-deflection curves.

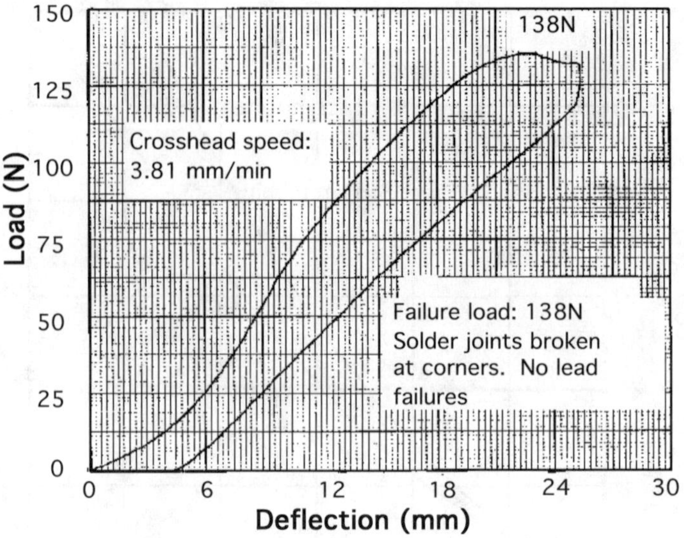

FIGURE 9-39 Four points twisting load-deflection curves (to failure).

the board. However, as far as the lead and solder joint reliability were concerned, there were no failures at a load equal to 80 N and a deflection equal to 12.7 mm (Figure 9-38). In fact, the solder joints were not cracked until the load reached 138 N and the deflection reached 22.2 mm, Figure 9-39. (For a board of this size, this magnitude of deflection is beyond most of the manufacturing, shipping, handling, and rework conditions.) These broken solder joints were around the four corners of the QFP. The failure mechanism of each solder joint was close to the interface between the lead and the solder joint (Figure 9-40). No lead failures were observed.

9.4 SURFACE MOUNT CONNECTOR SOLDER JOINTS UNDER VIBRATION CONDITIONS

9.4.1 Test Board Assembly

In Lau et al. [1992], five surface mount connector assemblies were subjected to 11 different tests (thermal cycling; mechanical overload by bending, twisting, pulling, and pushing; mechanical fatigue by bending, twisting, pulling, and pushing; in-plane vibration; and out-of-plane vibration). These connectors were the 1.27 mm pitch vertical SIMM (Single In-line Memory Module) socket, the

FIGURE 9-40 Three points bending failure mode (broken lead).

1 mm pitch right angle ZIF (Zero Insertion Force) flex circuit, the 2.54 mm pitch right angle unshrouded Header, the double row vertical Receptacle, and the right angle female DSUB (D-SUBminiature). In this chapter, only the in-plane and out-of-plane vibration tests of these five surface mount connector solder joints are presented.

Surface Mount Connectors and Mating Connectors

Figure 9-41 shows the 114.3 mm long, 72-pin SIMM socket. The connector housing was made of liquid crystal polymer (LCP) plastic; it had a plastic post at the middle and metal pins at both ends. The dimensions of the lower portion of the

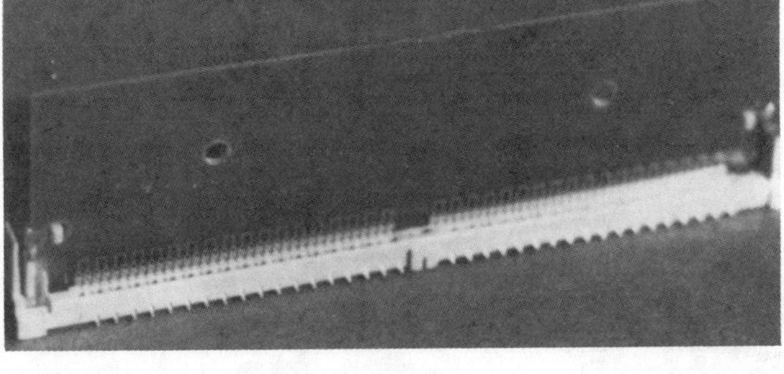

FIGURE 9-41 SIMM socket and mating connector.

leads were 0.38 mm in width and 1.42 mm in length. The weight of the SIMM socket was 7.89 grams. A 72-position printed circuit card was used for the insertion into the SIMM connector, Figure 9-41. The mating card dimensions were 108 mm × 1031 mm × 101 mm. These contacts were interconnected via traces on the card in an alternating pattern so as to provide daisy-chained connections when the SIMM socket was soldered to the PWB. Weights were added on each side of the card to make the module simulate the 35 g module commonly used. The weight of the mating connector (card and weights) was 36.3 grams.

Figure 9-42 shows the 28.58 mm long (metal foot tip to metal foot tip), 23-pin ZIF surface mount connector. The connector housing was made of 46 Nylon plastic and had metal feet soldered to the pad on a PWB at both ends. The dimensions of the lower portion of the leads were 0.3 mm in width and 1.83 mm in length. The weight of the connector was 0.84 grams. A 23-position mating connector (flex circuit) for insertion into the ZIF connector is shown in Figure 9-42. The mating connector dimensions were 24 mm × 1050 mm. The non-contact ends of the flex circuit were manually soldered in an alternating pattern so as to provide daisy-chained connections when the ZIF connector was soldered to the PWB. The weight of the mating connector was 1.06 grams.

Figure 9-43 shows the 38.1 mm long (metal foot tip to metal foot tip), 26-pin dual Header. The connector housing was made of polyphenylene sulfide (PPS)

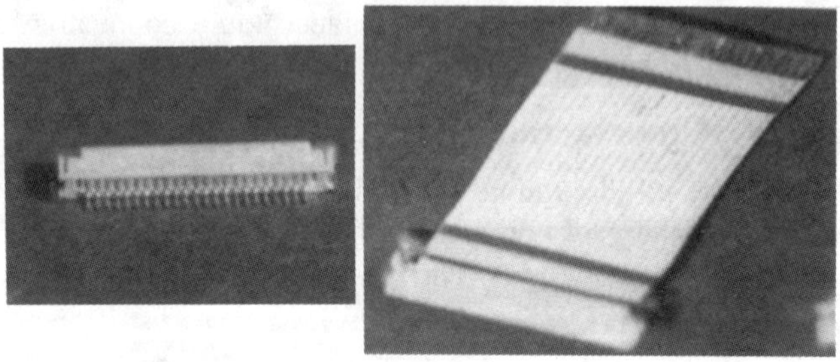

FIGURE 9-42 ZIF and flex circuit mating connector.

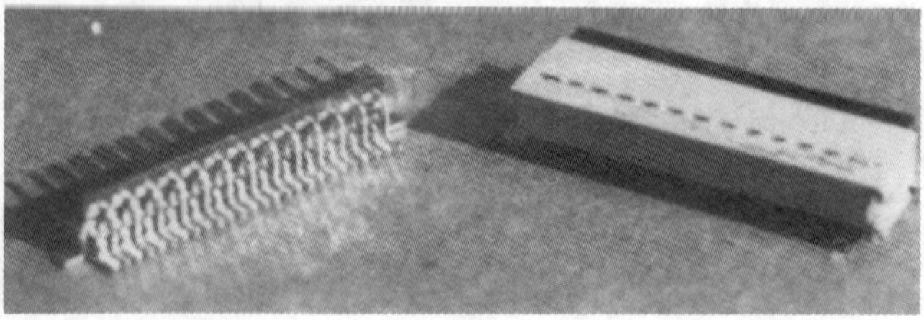

FIGURE 9-43 Dual header and mating connector.

plastic and had metal feet soldered to the pad of the PWB at both ends. The dimensions of the lower portion of the leads were 0.51 mm in width and 1.98 mm in length. The weight of the header was 2.15 grams. A 26 position double row mating connector for insertion into the header is shown in Figure 9-43. The mating connector dimensions were 37 mm × 1011 mm × 106 mm. These contacts were interconnected in an alternating pattern so as to provide daisy-chained connections when the header was soldered to the PWB. The weight of the mating connector was 3.54 grams.

Figure 9-44 shows the 38.1 mm long, 26-pin double row vertical Receptacle surface mount connector manufactured by Dupont. The connector housing was made of PPS plastic and had plastic pins at both ends. The dimensions of the lower portion of the leads were 0.61 mm in width and 1.02 mm in length. The weight of the connector was 2.23 grams. A 26-pin double interconnect strip was used for insertion through the holes of the PWB into the vertical Receptacle connector. (Figure 9-44). The non-insertion pin ends were manually soldered in an alternating pattern so as to provide daisy-chained connections when the vertical Receptacle connector was soldered to the PWB. The weight of the mating connector was 4.01 grams.

Figure 9-45 shows the 50.8 mm long, 25-pin DSUB connector. The connector housing was made of PPS plastic and had a nut and a bolt at each end. The dimensions of the lower portion of the leads were 0.51 mm in width and 1.38 mm in

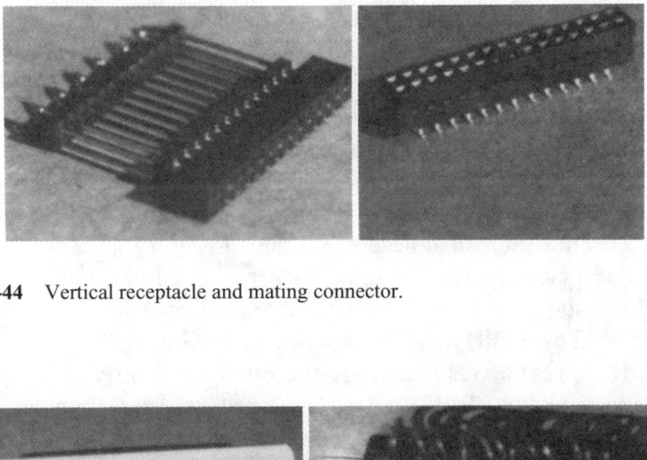

FIGURE 9-44 Vertical receptacle and mating connector.

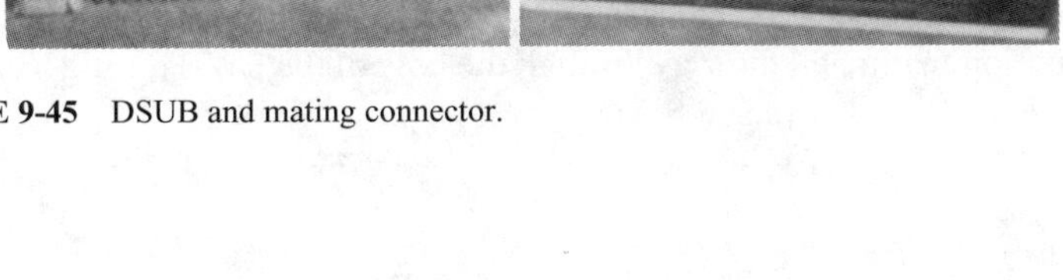

FIGURE 9-45 DSUB and mating connector.

length. The weight of the DSUB was 11.29 grams. A 25 position metal shrouded mating connector was used for the DSUB and is shown in Figure 9-45. The mating connector dimensions were 52 mm × 1019 mm × 1012 mm. Wires on the back of the mating connector were interconnected in an alternating pattern so as to provide daisy-chained connections when the DSUB was soldered to the PWB. The weight of the mating connector was 13.23 grams.

Test Board Designs

Figure 9-46 schematically shows the test board. It was designed for all five surface mount connectors and the in-plane and out-of-plane vibration tests. Consequently, only one set of layout, stencil, and board needed to be made. Figure 9-46 shows that there were 30 surface mount connectors (6 for each connector type: 3 in the x-direction and 3 in the y-direction). Traces were routed on the test board to allow for daisy chaining from the pad on the board surface through the connector lead, into the mating connector, and then over to the next pad and lead. The board was made of FR-4 epoxy/glass and its dimensions were 355.6 mm × 10292.1 mm × 101.6 mm. The pad dimensions were 3.1 mm × 101 mm for the SIMM socket, 0.7 mm × 102.7 mm for the ZIF flex circuit, 0.9 mm × 103.4 mm for the dual header, 1.3 mm × 101.7 mm for the vertical Receptacle, and 0.8 mm × 104.7 mm for the DSUB. The test board had a liquid photo imageable solder mask with hot-air-leveled Sn-Pb coating.

The entire PWB was used intact for the in-plane vibration test. However, for the out-of-plane vibration test, the board was cut into smaller test specimens (coupons) with only one connector at the center of each coupon. The specimen dimensions for out-of-plane vibration test were 228.6 mm × 1030.5 mm × 101.6 mm.

Paste, Printing, Pick & Place, and Solder Reflow

The solder paste used in this study was eutectic 63Sn-37Pb. The stencil opening dimensions were: 3.1 mm × 101 mm × 100.25 mm for the SIMM socket, 0.7 mm × 102.7 mm × 100.25 mm for the ZIF flex circuit, 0.9 mm × 103.4 mm × 100.25 mm for the dual header, 1.3 mm × 101.7 mm × 100.25 mm for the vertical Receptacle, and 0.8 mm × 104.7 mm × 100.25 mm for the DSUB. All the surface mount connectors were hand loaded, except the ZIF flex circuit, which was placed by a fully automated pick and place system.

The DSUB connector was retained to the PWB via two sets of nuts and bolts torqued to approximately 10 lbs (44.5 N). The other surface mount connectors were held in place by the retention device provided with the connectors by the vendor. In this study, some major issues with hand-loaded surface mount connectors were: (1) the smearing of solder paste, which subsequently contributed to bridges and insufficient solder; (2) the damaging of connectors which contributed to bent leads, misalignment, etc.; and (3) the failure to push the center SIMM socket retention feature all the way through the hole of the PWB during hand loading, which contributed to raised SIMMs.

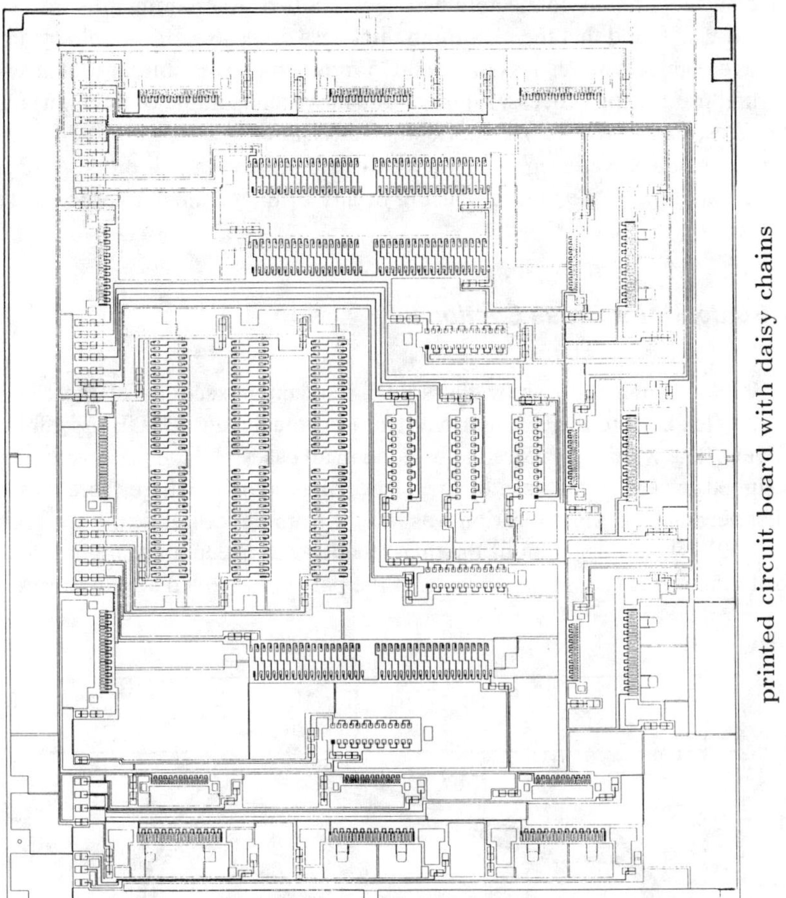

FIGURE 9-46 Test board.

Figure 9-47 shows the reflow temperature profile for assembling the populated surface mount connector PWB in a Vitronics Infrared Reflow oven. It can be seen that the conveyor speed was 305 mm/min (12 in/min), which was much slower than that of the true production process. (The reason for this slow speed was to have a complete reflow of the DSUB solder joints). However, this would not occur with forced convection ovens.

The effect of slow conveyor speed on the microstructure of solder joints of these five surface mount connectors had been studied by scanning electron microscopy. It was found that the maximum thickness of Cu_3Sn for the solder joints of these five connectors was less than 0.0025 mm, which was similar to that seen for a normal production process. Figure 9-48 shows an example of the interfacial intermetallics, Cu_3Sn (DSUB). Cu_6Sn_5 was not visible.

The flux residues were removed via a standard aqueous wash process. There was no degradation of the plastic housing of any of the connectors processed in this manner.

9.4.2 Solder Joint Cross Sections

Figures 9-49 through 9-53 show the solder joint cross sections of the SIMM socket, ZIF flex circuit, dual Header, vertical Receptacle, and DSUB assemblies, respectively. The wetting appeared to be good in all cases. All the pads were properly designed except in the case of the vertical Receptacle, where there was no heel fillet because the pad's location was too far from the center of the connector (Figure 9-52). The solder joint of the dual Header was acceptable (Figure 9-51), but a better heel fillet could be achieved if the pad were moved a little toward to

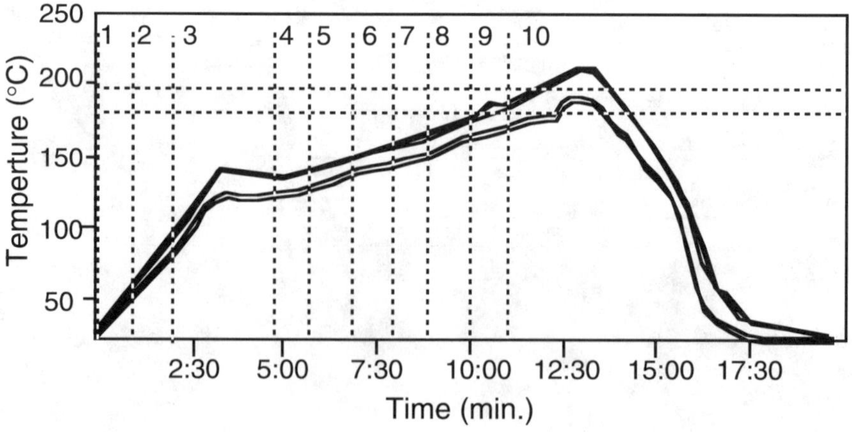

FIGURE 9-47 Mass reflow temperature profile.

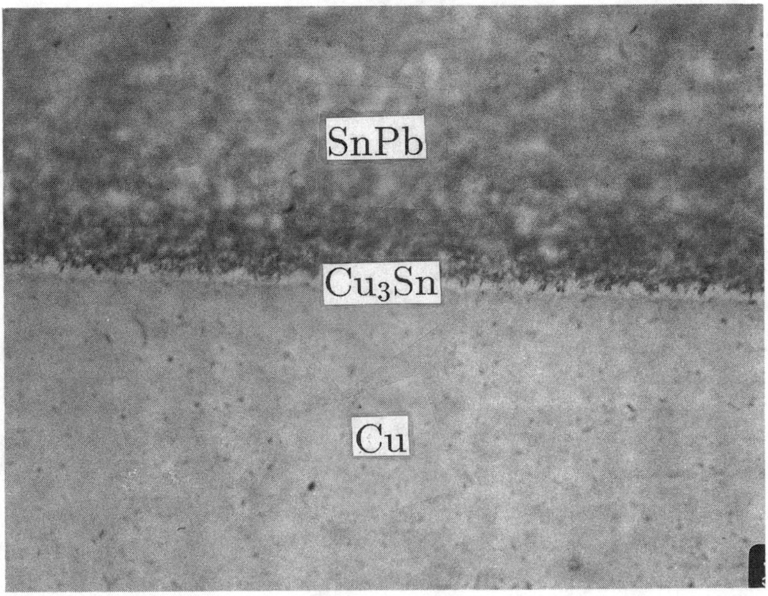

FIGURE 9-48 Intermetallic metallurgy (Cu_3Sn) in DSUB solder joint.

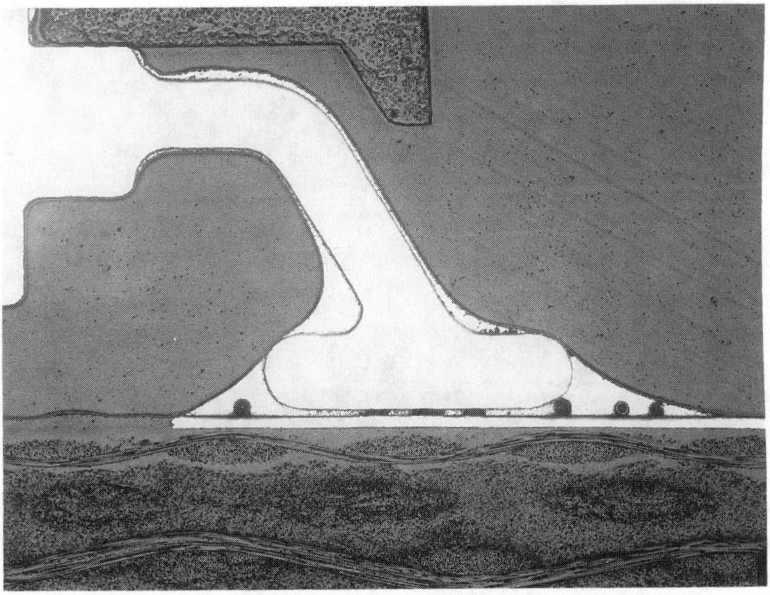

FIGURE 9-49 Cross section of SIMM socket joint.

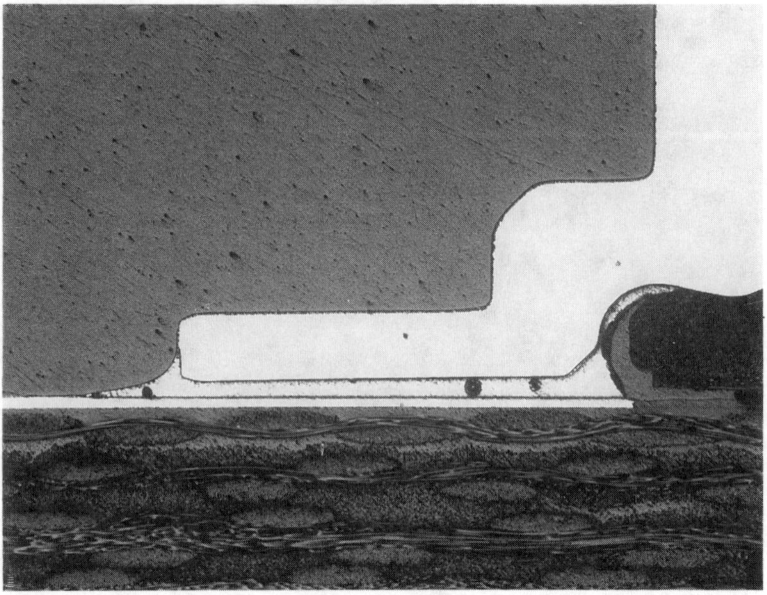

FIGURE 9-50 Cross section of ZIF joint.

FIGURE 9-51 Cross section of dual header joint.

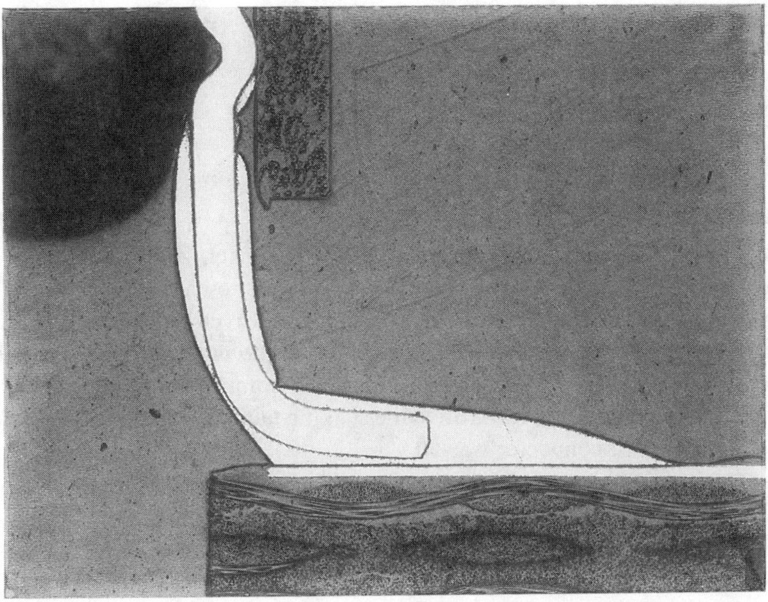

FIGURE 9-52 Cross section of vertical Receptacle joint.

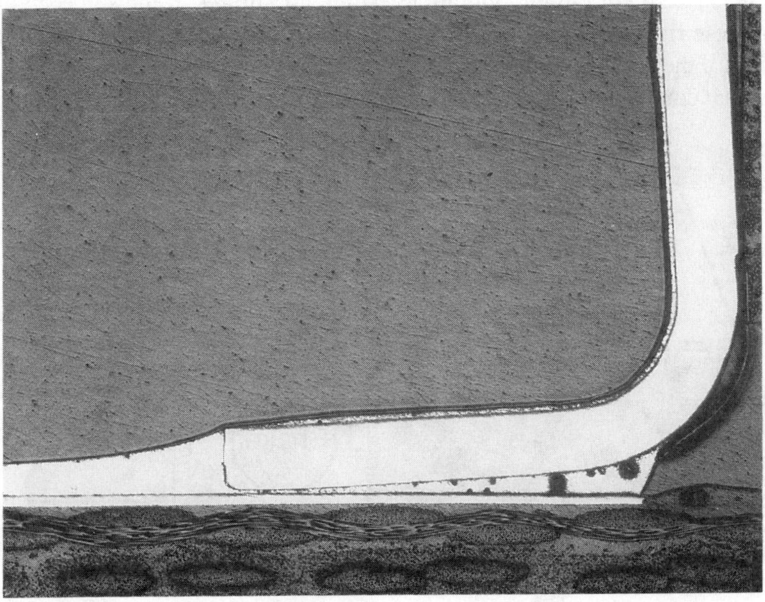

FIGURE 9-53 Cross section of DSUB joint.

the right. Before testing, all solder joints were inspected and any that did not appear "good" were reworked.

9.4.3 In-Plane Vibration Test of the Connector Assemblies

In-Plane Vibration Test Set-Up

The test for the in-plane vibration of the SIMM socket, ZIF flex circuit, dual Header, vertical Receptacle, and DSUB assemblies is shown schematically in Figure 9-54. It consisted of a vibration table which was driven with a random excitation u(t). The power spectral density corresponding to u(t) can be analyzed by a Fourier transform to G(w). Ten identical in-plane vibration test boards (Figure 9-46) were bolted down to the fixture and the shaking table, Figure 9-55. The sample size for each type of connector was 50.

The shaking table (Figures 9-54 and 9-55) performed random vibration with frequencies ranging from 5 to 2000 Hz, and with a magnitude of $0.048g^2/Hz$. The excitation spectrum is shown in Figure 9-56. A data acquisition system continually monitored the electrical resistance of the connector solder joints and recorded the failure times.

In-Plane Vibration Test Results

After seven hours of random vibration testing, no failures of the solder joints on any of these five connectors were observed. This was due to the high natural frequencies of the surface mount connectors in the in-plane direction. In general, surface mount connectors and their solder joints are not excited by the in-plane vibra-

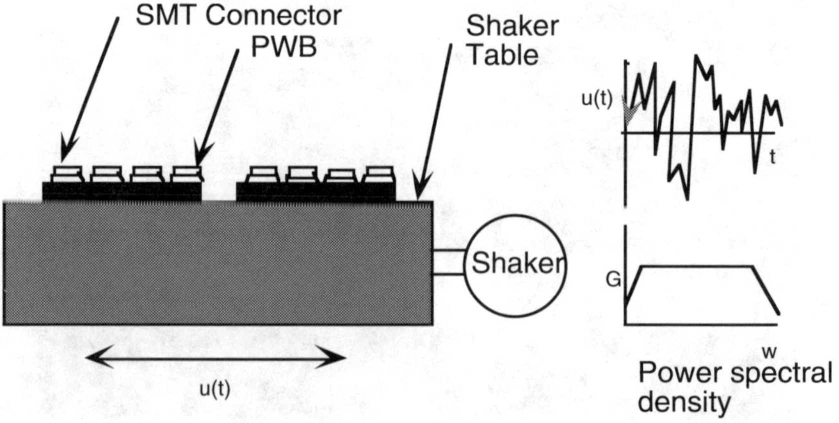

FIGURE 9-54 In-plane vibration test.

FIGURE 9-55 In-Plane Vibration Test Set-Up.

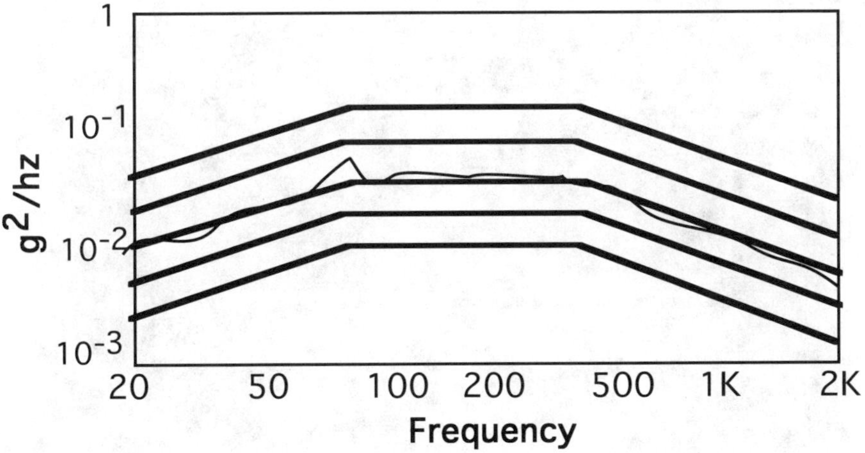

FIGURE 9-56 In-plane vibration input spectrum.

tions occurring under normal in-plane operating conditions, which were typically well below the present in-plane vibration test conditions.

9.4.4 Out-of-Plane Vibration Test of the Connector Assemblies

Out-Of-Plane Vibration Test Set-Up

The test set-up for the out-of-plane vibration of the SIMM socket, ZIF flex circuit, dual Header, vertical Receptacle, and DSUB assemblies is shown in Figure 9-57. It consisted of a table which was vertically driven by a shaker. A special test frame was designed and constructed to provide 2-point bending of the test board, as shown in Figure 9-58. The actual board deflection was caused by the movement of the test frame bolted to the surface of the vertical shaker. The shaker was controlled by a data acquisition unit and produced a sinusoidal excitation. There was only one surface mount connector on each test board, and the clear (free) span of the test boards was 177.8 mm, and the sample size of each type of connector was 50, a total of 250 boards were tested.

In order to excite the test board, the excitation frequency must be close to the natural frequency of the test board. Consequently, the natural frequencies of the test boards with these five different types of connectors were estimated by meas-

FIGURE 9-57 Out-of-plane vibration test set-up.

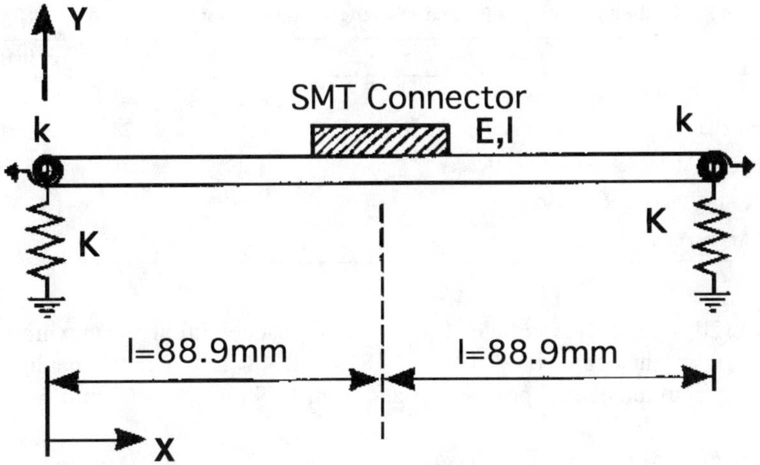

FIGURE 9-58 Out-of-plane vibration model.

uring the vibration response of the structural system with a sweep of frequencies. The natural frequency of the structural system corresponded to the maximum response. For example, the maximum response of the vertical Receptacle structural system occurred at a frequency of 95 Hz (Figure 9-59). Similarly, the natural frequencies of the SIMM socket, ZIF flex circuit, dual Header, and DSUB structural system were measured to be 100 Hz, 115 Hz, 105 Hz, and 65 Hz, respectively.

Once the natural frequencies of the test boards were estimated, the shaker was programmed to perform a sinusoidal excitation with a frequency sweep between

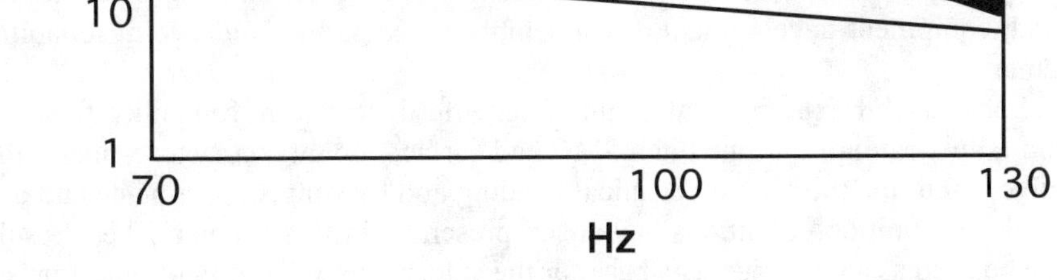

FIGURE 9-59 Vibration frequency for vertical Receptacle.

TABLE 9-3 Weibull Parameters for Out-of-Plane Vibration Test

Connector	*b*	$\theta(10^6)$
1.27 mm Vertical SIMM Socket	3.04	3.3
1.0 mm Right Angle ZIF (Flex Circuit)	3.23	6.0
2.54 mm Right Angle Unshrouded Header	3.36	4.1
Double Row Vertical Receptacle	3.40	4.8
Right Angle Female DSUB	3.12	3.8

50 and 120 Hz, and a magnitude of 6g. With this test condition, the maximum total deflections of the boards ranged from 3.18 mm to 3.81 mm, i.e., the maximum deflections from the neutral position ranged from 1.59 mm to 1.905 mm.

Out-Of-Plane Vibration Test Results

A data acquisition system continually monitored the electrical continuity of all surface mount connector solder joints and recorded the failure times. With the described severe test conditions, failure was not observed at one million cycles. The vibration testing continued to 3.24 million cycles. The two-parameter Weibull distribution, Equation (9.4), was chosen for the data analysis of the test results. The "best fit" Weibull parameters (for the median rank) are shown in Table 9-3 for all five surface mount connectors. With these values of the Weibull parameters, the life distribution, reliability function, and failure rate of the solder joints can be readily determined, respectively, by Equations (9.2) to (9.4). For example, the solder joint life distribution of the SIMM socket is shown in Figure 9-60, and the solder joint reliability of the ZIF flex circuit is shown in Figure 9-61.

9.5 CONCLUSION

For the past decade, SMT has been the most rapidly developed technology in the electronics industry. Today, more than 90% of the components in watches, cameras, memory cards, smart cards, thermal printer heads, laser-jet printers, televisions, calculators, personal, laptop, handheld, and notebook computers are surface mounted. Advances continue to be made in SMT however, challenges such as fine pitch equipment development, the assembly process, and solder joint reliability remain.

Some useful experimental results (e.g., life distribution, reliability function, and failure rate) of the fine pitch SMC and surface mount connector solder joints subjected to the thermal, mechanical bending and twisting, and in-plane and out-of-plane vibration conditions have been presented in this chapter. These results can be used as a reference data base for the solder joint reliability of other surface mount components and connectors.

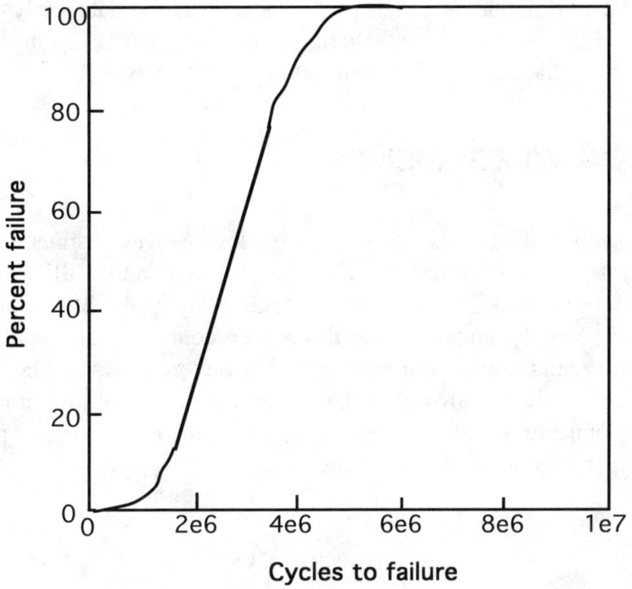

FIGURE 9-60 Life Distribution of SIMM socket solder joint.

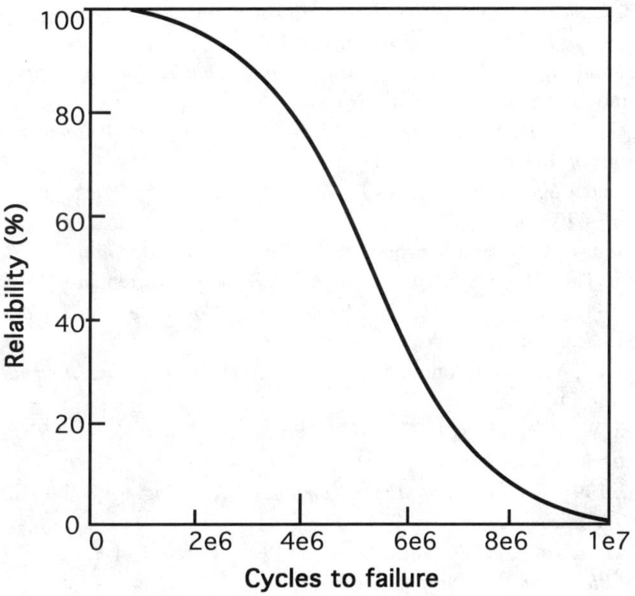

FIGURE 9-61 Reliability of ZIF flex circuit solder joint.

Detailed experimental design, set-up, procedures, and data analysis have also been given. This information can help researchers, engineers, and managers to design and execute their own solder joint reliability projects.

9.6 ACKNOWLEDGMENTS

The author would like to thank Suresh Golwalkar, Steve Erasmus, Paul Boysan, Robert Surratt, Richard Forhringer, Douglas Hoffman, and William Jahsman for their contributions to the TSOP solder joint reliability study; Steve Erasmus, Tom Chang, Kevin Moody, and Gene Smith for their contributions to the QFP solder joint reliability study; and Tom Marcotte, Jim Severine, Andy Lee, Steve Erasmus, Tom Baker, Heidi Aaron, and Gary Gruver for their contributions to the surface mount connector solder joint reliability study. It has been a great pleasure and a fruitful experience to work with them.

REFERENCES

Barker, D. B., Sharif, I., Dasgupta, A., and Pecht, M. G., 1992, *Journal of Electronic Packaging, Transactions of ASME,* 14:177.

Charles, H. K., and Clatterbaugh, G. V., 1990, *ASME Transactions, Journal of Electronic Packaging,* 112:135.

Cutiongco, E. C., Vaynman, S., Fine, M. E., and Jeannotte, D. A., 1990, *ASME Transactions, Journal of Electronic Packaging,* 112:110.

Doane, D. A., and Franzon, P. D., 1992, *Multichip Module Technologies and Alternatives,* Van Nostrand Reinhold, New York, NY.

Engel, P.A., Caletka, D. V., and Palmer, M. R., 1991, *Journal of Electronic Packaging, Transactions of ASME,* Vol. 113:129.

Engel, P. A., and Vogelmann, J. T., 1992, *Journal of Electronic Packaging, Transactions of ASME,* 114:203.

Frear, D. R., Jones, W. B., and Kinsman, K. R., 1991, *Solder Mechanics, A State of the Art Assessment,* The Minerals, Metals & Materials Society, Warrendale, PA.

Frear, D. R., 1990, *Proceedings of the 40th Electronics Components & Technology Conference,* p. 518, Las Vegas, NV.

Guo, Z., Sprecher, A. F., and Conrad, H., 1992A, *ASME Transactions, Journal of Electronic Packaging,* 114:112.

Guo, Y., Chen, W. T., and Lim, K. C., 1992B, *Proceedings of the 1st ASME/JSME Advances in Electronic Packaging Conference,* p. 779.

Harper, C. A., 1991, *Handbook of Microelectronics Packaging,* McGraw-Hill Book Company, New York, NY.

Heinrich, S. M., Elkouh, A. F., Nigro, N. J., and Lee, P. S., 1991A, *ASME Transactions, Journal of Electronic Packaging,* 112:210.

Heinrich, S. M., Nigro, N. J., Elkouh, A. F., and Lee, P. S., 1991B, *ASME Transactions, Journal of Electronic Packaging,* 112: 219.

Hinch, S. W., 1988, *Handbook of Surface Mount Technology,* Longman Scientific & Technical, New York, NY.

Hollomon, J. K., Jr., 1989, *Surface-Mount Technology,* Howard W. Sams & Company, Indianapolis, IN.

Hutchins, C., 1990, *SMT: How to Get Started,* Hutchins and Associates, Raleigh, NC.

Jahsman, W. E., and Jain, P., 1990, *Proceedings of the 40th Electronics Components & Technology Conference,* p. 926, Las Vegas, NV.

Johnson, L. G., 1951, *Industry Mathematics,* 2:1.

Johnson, L. G., 1964A, *Theory and Technique of Variation Research,* Elsevier Publishing Company, Amsterdam.

Johnson, L. G., 1964B, *The Statistical Treatment of Fatigue Experiments,* Elsevier Publishing Company, Amsterdam.

Lau, J. H., 1989, *Journal of Electronic Packaging,* 111:2.

Lau, J. H., 1991, *Solder Joint Reliability: Theory and Applications,* Van Nostrand Reinhold, New York.

Lau, J. H., and Erasmus, S., 1992, ASME paper No. 92W/EEP-33, 1992 ASME Winter Annual Meeting, Anaheim, CA.

Lau, J. H., 1992A, *Handbook of Tape Automated Bonding,* Van Nostrand Reinhold, New York, NY.

Lau, J. H., 1992B, ASME Paper No. 92W/EEP-34, 1992 ASME Winter Annual Meeting, Anaheim, CA.

Lau, J. H., 1992C, *Proceedings of the 13th IEEE International Electronics Manufacturing Technology Symposium,* p. 293.

Lau, J. H., Golwalkar, S., Rice, D., Erasmus, S., and Foehringer, R., 1992a, *Journal of Electronic Packaging, Transactions of ASME,* 114:169.

Lau, J. H., Dody, G., Chen, W., McShane, M., Rice, D., Erasmus, S., and Adamjee, W., 1992b, *Circuit World,* 18:13.

Lau, J. H., Golwalkar, S., Boysan, P., Surratt, R., Rice, D., Forhringer, R., and Erasmus, S., 1992c, *Proceedings of the 42nd Electronic Components and Technology Conference,* p. 519.

Lau, J. H., Marcotte, T., Severine, J., Lee, A., Erasmus, S., Baker, T., Moldaschel, J., Sporer, M., and Burward-Hoy, G., 1992d, ASME Paper No. 92W/EEP-39, 1992 ASME Winter Annual Meeting, Anaheim, CA.

Lau, J. H., Govila, R., Larner, C., Pao, Y., Erasmus, S., Dolot, S., and Solberg, V., 1992e, *Proceedings of the 13th IEEE International Electronics Manufacturing Technology Symposium,* p. 305.

Lau, J. H., Govila, R., Larner, C., Pao, Y., Erasmus, S., Dolot, S., Jalilian, M., and Lancaster, M., 1992F, *Circuit World,* 1.

Lau, J. H., Rice, D., and Erasmus, S., 1992g, Proceedings of the 1st ASME/JSME Electronic Packaging Conference, p. 855.

Lau, J. H., Pao, Y., Larner, C., Twerefour, S., Govila, R., Gilbert, D., Erasmus, S., and Dolot, S., 1993, to be presented at the 43rd IEEE ECTC Orlando, Florida.

Lau, J. H. and Erasmus, S., 1993, *Electronic Packaging & Production.*

Lau, J. H., 1993a, *Thermal Stresses and Strains in Microelectronics Packaging,* Van Nostrand Reinhold, New York, NY.

Lau, J. H., 1993b, *Handbook of Fine Pitch Technology,* Van Nostrand Reinhold, New York, NY, 1993.

Lau, J. H., 1993c, *An Introduction to Chip-On-Board Technology,* Van Nostrand Reinhold, New York, NY.

Lau, J. H., 1993d, *International Journal of Engineering Fracture Mechanics.*

Lau, J. H., 1993e, 43rd IEEE ECTC, Orlando, Florida.

Lea, C., 1988, *A Scientific Guide to Surface Mount Technology,* Electrochemical Publications, Scotland.

Lea, C., 1992, *After CFCs? Options for Cleaning Electronics Assemblies,* Electrochemical Publications Ltd., IOM, British Isles.

Lee, S. M., and Stone, D. S., 1992, *ASME Transactions, Journal of Electronic Packaging,* 114: 118.

Manko, H. H., 1986, *Soldering Handbook for Printed Circuits and Surface Mounting,* Van Nostrand Reinhold, New York, NY.

Marshall, J. L., Lucey, G., and Hwang, J., 1991, *Proceedings of the 41st Electronics Components & Technology Conference,* p. 647, Atlanta, GA.

Mei, Z., and Morris, J. W., 1992, *ASME Transactions, Journal of Electronic Packaging,* 114:104.

Messuer, G., Turlik, I., Balde, J., and Garrou, P., 1992, *Thin Film Multichip Modules,* International Society for Hybrid Microelectronics, Silver Spring, MD.

Morgan, H. S., 1991, *ASME Transactions, Journal of Electronic Packaging,* Vol. 113:350.

Norris, K. C., and Landzberg, A. H., 1969, *IBM Journal of Research and Development,* p. 266.

Pao, Y. H., Badgley, S., Govila, R., Baumgartner, L., Allor, R., and Cooper, R., 1992, *ASME Transactions, Journal of Electronic Packaging,* 114:135.

Pan, T. Y., 1991, *ASME Transactions, Journal of Electronic Packaging,* 113: 8.

Patra, S. K., and Lee, Y. C., 1991, *ASME Transactions, Journal of Electronic Packaging,* 113:337.

Pecht, M., 1988, *Handbook of Electronic Package Design,* Marcel Dekker, New York, NY.

Prasad, R. P., 1989, *Surface Mount Technology,* Van Nostrand Reinhold, New York, NY.

Racz, L. M., and Szekely, J., 1992, ASME Paper No. 92W/EEP-32, 1992 ASME Winter Annual Meeting, Anaheim, CA.

Read D. T., and Lucey, G. K., 1991, *ASME Transactions, Journal of Electronic Packaging,* 113:178-185.

Ross, R. G., Wen, L. C., Mon, G. R., and Jetter, E., 1992, *ASME Transactions, Journal of Electronic Packaging,* 114:185.

Sandor, B.I., *Solder Mechanics, A State of the Art Assessment,* edited by Frear, D. R., Jones, W. B., and Kinsman, K. R., p. 363, The Minerals, Metals & Materials Society, Warrendale, PA.

Seraphim, D. P., Lasky, R., and Li, C. Y.1989, *Principles of Electronic Packaging,* McGraw-Hill Book Company, New York, NY.

Solberg, V., 1990, *Design Guidelines for SMT, TAB Professional and Reference Books,* New York, NY.

Solomon, H. D., 1992, *ASME Transactions, Journal of Electronic Packaging,* Vol. 114:161.

Stephens, J. J., Burchett, S. N., and Jones, W. B., 1992, *Proceedings of the 1st ASME/JSME Advances in Electronic Packaging Conference,* p. 363-372.

Suhir, E., 1991, *ASME Transactions, Journal of Electronic Packaging,* 113: 240.

Timoshenko, S. P. and Woinowsky-Krieger, S., 1959, *Theory of Plates and Shell,* McGraw-Hill Book Company, New York, NY.

Tummala, R. R., and Rymaszewski, E., 1989, *Microelectronics Packaging Handbook,* Van Nostrand Reinhold, New York, NY.

Vardaman, J., 1992, *Surface Mount Technology, Recent Japanese Developments,* IEEE Press, New York, NY.

Wassink, R. J. K., 1989, *Soldering in Electronics, Electrochemical Publications,* Scotland,

Weibull, W., 1952, *Journal of Applied Mechanics, Transactions of ASME,* 19, No. 1.

Weibull, W., 1961, *Fatigue Testing and Analysis of Results,* The MacMillan Company, New York.

Westinghouse Defense and Electronics Center, 1985, "Hermetic Chip Carrier \ Compatible Printed Wiring Board," Air Force Wright Aeronautical Laboratories Report No. AFW-AL-TR-85-408\ 2.

Index